The German Physical Society in the Third Reich

Physicists between Autonomy and Accommodation

This is a history of one of the oldest and most important scientific societies – the German Physical Society – during the Nazi regime and immediate post-war period. The German Physical Society had Jewish scientists among its members, including the prominent Fritz Haber and Albert Einstein, when Adolf Hitler was appointed chancellor of Germany in January 1933. As Jewish scientists subsequently began to lose their jobs and emigrate, the society began to lose members. In 1938, under pressure from the Reich Ministry for Science, Education and Culture, the society forced out the last of its Jewish colleagues. This action was just the most prominent example of the tension between accommodation and autonomy that characterized the challenges facing physicists in the society. They strove to retain as much autonomy as possible but tried to achieve this by accommodating themselves to Nazi policies, which culminated in the campaign by the society's president to place physics in the service of the war effort.

Dieter Hoffmann is a Senior Fellow at the Max Planck Institute for the History of Science and a professor of history of science at Humboldt University in Berlin. Hoffmann has coauthored two books in English, *Science under Socialism: East Germany in Comparative Perspective* (1999) and *The Emergence of Modern Physics* (1996), as well as several books published in German. Hoffmann is a member of the editorial boards for the journals *Physikalische Blätter* and *Physics in Perspective*. His articles have appeared in *Physics in Perspective* and *Physics Today*. He is a member of the German Physical Society; the German Society for the History of Science, Technology, and Medicine; and the History of Science Society.

Mark Walker is the John Bigelow Professor of History at Union College. He is the author of *Nazi Science: Myth, Truth, and the German Atomic Bomb* (1995) and of *German National Socialism and the Quest for Nuclear Power, 1939–1949* (1989). He has coauthored several books, including *The Kaiser Wilhelm Society under National Socialism* (2009), *Politics and Science in Wartime: Comparative International Perspectives on the Kaiser Wilhelm Institutes* (2005), and *Science and Ideology: A Comparative History* (2003). His articles have appeared in *Journal of Contemporary History*, *Physics Today*, *Historical Studies in the Natural Sciences*, *Nature*, *Physics in Perspective*, *Minerva*, and *Metascience*.

The German Physical Society in the Third Reich

Physicists between Autonomy and Accommodation

Edited by

DIETER HOFFMANN

Max-Planck-Institut für Wissenschaftsgeschichte

MARK WALKER

Union College

Translated by

ANN M. HENTSCHEL

CAMBRIDGE
UNIVERSITY PRESS

CAMBRIDGE UNIVERSITY PRESS
Cambridge, New York, Melbourne, Madrid, Cape Town,
Singapore, São Paulo, Delhi, Tokyo, Mexico City

Cambridge University Press
32 Avenue of the Americas, New York, NY 10013-2473, USA

www.cambridge.org
Information on this title: www.cambridge.org/9781107006843

First published in the German language by Wiley-VCH Verlag GmbH & Co. KGaA, Boschenstraße 12, D-69469 Weinheim, Federal Republic of Germany under the title "Physiker zwischen Autonomie und Anpassung. Die Deutsche Physikalische Gesellschaft im Dritten Reich."

First English edition 2012

Printed in the United States of America

A catalog record for this publication is available from the British Library.

Library of Congress Cataloging in Publication data

The German Physical Society in the Third Reich : physicists between autonomy and accommodation / [edited by] Dieter Hoffmann, Mark Walker.
p. cm.
Includes bibliographical references and index.
ISBN 978-1-107-00684-3 (hardback)
1. Deutsche Physikalische Gesellschaft (1963–) 2. Science and state – Germany – History – 1933–1945. 3. National socialism and science – Germany. I. Hoffmann, Dieter, 1948– II. Walker, Mark, 1959–
Q49.G225 2011
506.43′0903–dc23 2011025075

ISBN 978-1-107-00684-3 Hardback

This book is dedicated to Paul Forman.

Contents

List of Figures

Notes on Contributors

Richard H. Beyler (r.beyler@pdx.edu) is a professor in the History Department at Portland State University, where he teaches history of science and European intellectual history and coordinates the department's graduate program. His current research focuses on how scientific institutions maneuvered through 20th-century German political history and on the history of biophysics before the advent of molecular biology. Forthcoming in the latter connection is a contribution to the volume *Towards a Physical Biology* on the 1935 "Three-Man Paper" in genetics by Timoféeff-Ressovsky, Zimmer, and Delbrück.

Ute Deichmann, Ph.D., Prof. (uted@bgu.ac.il), is the director of the Jacques Loeb Centre for the History and Philosophy of the Life Sciences at Ben-Gurion University of the Negev, Israel. She has published *Biologists under Hitler* (1996); *Flüchten, Mitmachen, Vergessen. Chemiker und Biochemiker in der NS-Zeit* (2001); and numerous articles, and she coedited, among others, *Jews and Sciences in German Contexts* (2007) and "Darwinism, Philosophy and Experimental Biology" in the *Journal for General Philosophy of Science* (2010). She is the recipient of the Ladislaus Laszt Award of Ben-Gurion University (1995) and the Gmelin Beilstein Medal of the Society of German Chemists (2005).

Michael Eckert (m.eckert@deutsches-museum.de) studied physics, receiving his Ph.D. in theoretical physics in 1979 before he turned to the history of science and technology as his true profession. He has participated in several projects on the history of physics and fluid dynamics at the Deutsches Museum. In 2000 and 2004, he edited (with Karl Märker)

Arnold Sommerfeld's scientific correspondence. At present, he is writing Sommerfeld's biography.

Klaus Hentschel (klaus.hentschel@po.hi.uni-stuttgart.de) studied at Hamburg University. After his Ph.D. thesis on the philosophical (mis)interpretations of Einstein's theory of relativity, he became an assistant at Göttingen University, where he wrote his habilitation thesis on the interplay of instrument building, experimental practice, and theory formation in astrophysics and spectroscopy in the period 1880–1960. After fellowships at MIT and Berne University, he became full professor of history of science and technology at the University of Stuttgart in 2006. For further details, see http://www.unistuttgart.de/hi/gnt/hentschel/.

Dieter Hoffmann (dh@mpiwg-berlin.mpg.de) was a research scholar at the Academy of Sciences of the German Democratic Republic. He is currently a senior research scholar at the Max Planck Institute for History of Science and teaches as an associated professor at Humboldt University in Berlin. His main areas of interest are the history of German physics and physics under National Socialism. He has been a Fellow of the Humboldt Foundation and in 2010 received the Medal of Honor of the German Physical Society. His books include (with Mark Walker) *"Fremde" Wissenschaftler im Dritten Reich. Die Debye-Affäre im Kontext* (2011).

Gerhard Rammer (gerhard.rammer@tu-berlin.de) studied physics, the history of science, and musicology at the Universities of Vienna and Göttingen. His 2004 Ph.D. dissertation is a history of the nazification and denazification of physics at the University of Göttingen. From 2005 to 2009, he was a research scholar and instructor at the University of Wuppertal, and since 2009 he has had the same position at the Technical University of Berlin. His current research project is a history of hydraulics in the 18th and 19th centuries.

Volker R. Remmert (vrr@ivs.au.dk) is associate professor of history and philosophy of mathematics at the University of Aarhus (Denmark). His main fields of interest are the history of mathematics in Germany from the late-19th century to the mid-20th century and the history of early modern science. Among his recent publications are (with Ute Schneider) *Eine Disziplin und ihre Verleger – Disziplinenkultur und Publikationswesen der Mathematik in Deutschland, 1871–1949* (2010) and *Picturing the Scientific Revolution: Title Engravings in Early Modern Scientific Publications* (2011).

Gerhard Simonsohn received his diploma and Ph.D. in physics from the Free University of Berlin. In 1970, he became professor of experimental physics at this university. His main field was modern optics (photon statistics, high-resolution photon counting spectroscopy applied to scattering processes in liquids). Because he was in charge of physics teacher education for the Physics Department, he was especially engaged in this field. He retired in 1990.

Mark Walker (walkerm@union.edu) is the John Bigelow Professor of History at Union College in Schenectady, New York, where he teaches the history of modern Europe and history of science and technology. His books include (with Dieter Hoffmann) *"Fremde" Wissenschaftler im Dritten Reich. Die Debye-Affäre im Kontext* (2011); (with Helmuth Trischler) *Physics and Politics: Research and Research Support in Twentieth Century Germany in Comparative Perspective* (2010); and (with Susanne Heim and Carola Sachse) *The Kaiser Wilhelm Society under National Socialism* (2009).

Stefan L. Wolff (s.wolff@deutsches-museum.de) has studied physics and earned a Ph.D. in the history of science at the University of Munich. He is a scholar at the Research Institute of the Deutsches Museum in Munich and lecturer in the history of physics at the University of Munich. His main area of research is the history of physics in the 19th and 20th centuries, especially thermodynamics and physics during National Socialism, including the emigration of scientists. In 2010, he coedited a book on the history of the Deutsches Museum during National Socialism.

Foreword

The German Physical Society (*Deutsche Physikalische Gesellschaft*, DPG) is the oldest and largest professional society of physics in the world and originated from the Physical Society of Berlin (*Physikalische Gesellschaft zu Berlin*). Founded in 1845, the membership of the Physical Society, along with its scientific reputation, has steadily been growing ever since. This development became particularly visible during the decades around 1900 when Germany assumed a worldwide leadership role in many areas of physical research. In that period, the society's presidents included such leading figures as Emil Warburg, Max Planck, and Albert Einstein – living symbols of excellence in physics. Indeed, Planck and Einstein counted among the most prominent scientists of the early 20th century.

The year 1933 put an abrupt end to this flowering in physics and brought on trenchant changes. The National Socialist dictatorship persecuted political opponents and nonconformists and deprived Jewish intellectuals and scientists of their livelihoods. A consequence of this racist, discriminatory, and repressive policy was the partial demise of physical research in Germany. To date, the role played by the German Physical Society in this process has not received adequate treatment. Only biographies and surveys of the overall historical processes in the physical sciences have addressed this chapter in the history of physics in Germany. The German Physical Society is aware of this deficit.

In view of this desideratum, Dieter Hoffmann, chairman of the society's history of science division, *Fachverband "Geschichte der Physik,"* proposed during the planning stage in preparing for the Year of Physics 2000 that the history of the German Physical Society in the Third Reich become the subject of a special study. The society's president at that time,

Alexander Bradshaw, welcomed this proposal immediately and without reservation. It was a matter of particular importance to the German Physical Society that it examine its own past during the National Socialist period. For this reason, the society's board established a committee to clarify the conditions necessary for a reassessment of the history of the German Physical Society. The outcome of this inquiry was the suggestion to the board in spring 2001 to call to life a research project funded by the German Physical Society and to entrust its direction to Mark Walker, the American historian of science. The expectation was that this project would also avail itself of the expertise within the society. In agreement with Mark Walker, Dieter Hoffmann was then named codirector of the project. Theo Mayer-Kuckuk, former president of the German Physical Society, was assigned the task of official contact between the society and the editors. An international group of independent authors worked on various aspects of the history of the German Physical Society in the Third Reich during the years that followed. Their results are presented here in this book.

On behalf of our society, I would like to express my hearty thanks to the two editors, Mark Walker and Dieter Hoffmann, and to all the other contributors for the completed volume. This work is more than a consistent documentation and analysis of the history of the German Physical Society and physics in Germany. It is an act against forgetting. How the future will develop depends quite crucially on our ability to continue to confront our own history and learn from it.

Eberhard Umbach
President of the *Deutsche Physikalische Gesellschaft*
Würzburg
29 October 2006

Preface

The German Physical Society (*Deutsche Physikalische Gesellschaft*, DPG), one of Germany's oldest professional associations, has a rich tradition. Its membership in the decades after its founding in 1845 kept pace with its rising renown in the scientific world. This reputation came from the fact that the research conducted in Germany in the decades straddling the turn of the 19th century to the 20th century defined global standards in many areas of physics. The year 1933 signified a grave setback for this highly developed physics culture. The racist campaigns and repressive policies of the National Socialist dictatorship barred Jewish intellectuals and scientists from making a living, consequently forcing many to emigrate. Albert Einstein is symbolic of this exodus of intelligentsia from Germany. His emigration signaled the decline of physical research in Germany.

A number of interesting and insightful studies about this phenomenon have appeared in recent decades – starting with Alan Beyerchen's pioneering work *Scientists under Hitler* (1977). They range from David Cassidy's detailed biography of Heisenberg, *Uncertainty* (1992), which has recently been updated as *Beyond Uncertainty* (2009), to Klaus and Ann Hentschel's anthology *Physics and National Socialism* (1996), a collection of important documents from this period in English translation. In these and the many other notable publications about the phenomenon of physics in the Third Reich, the DPG figures only marginally – if at all – within the general historical context of the developments in physics. Little is revealed about its specific function within the framework of active science policy and the constellations of political power in the Third Reich;

this, incidentally, applies to the role of scientific societies generally as mediators between research and politics.

This book intends to close this research lacuna. In the past few years, an international group of authors has worked on a variety of aspects of the history of the DPG during the Third Reich. The results of this research are compiled in this collection of essays. Like a mosaic, the separate sample components in the society's history are laid out and analyzed to arrive at an overall picture of its role during this period. Mark Walker provides an introduction to the general political conditions and sets the history of the society within the National Socialist context of the time. Richard H. Beyler examines general aspects of the partly successful attempt by the DPG to preserve its authority and autonomy even under the repressive conditions of the Nazi state. Stefan L. Wolff concerns himself with the emigration of physicists during the Third Reich and what this meant for the DPG; that is, what role this society played in the ostracism of Jewish colleagues in the field. Michael Eckert critically assesses the relation between the DPG and "Aryan physics" (*Deutsche Physik*) and the DPG's obdurate battle against political physics (*Parteiphysik*) that was so vehemently repudiated after the war. The Ramsauer era, which coincided with the war and bears the mark of the DPG's partial self-mobilization, is described in detail in Dieter Hoffmann's contribution. Richard H. Beyler, Michael Eckert, and Dieter Hoffmann examine the Planck medal, the highest distinction conferred by the society, because the awarding practice during the Third Reich is an exemplary indicator of the DPG's relations between autonomy and alignment. Gerhard Simonsohn offers a detailed survey of the then-current topics of physical research – mirrored in the physics conferences and other scientific activities of the DPG as well as in contemporary periodicals. The essays by Volker R. Remmert and by Ute Deichmann provide a comparative perspective on the DPG's mathematical and chemical sister societies during the Third Reich. Finally, two contributions focus on revealing continuities and discontinuities in the society's post-war history. Klaus Hentschel traces the mentality of physicists during the first post-war years by means of the technique of thick description, and Gerhard Rammer investigates the institutional recommencement of the DPG after 1945 and its policy toward, or "grappling" with, its past.

This summary shows that although this book focuses on the history of the DPG during the despotic years of National Socialism, it is discussed from a comparative perspective. This comparison relates, on one hand, to the temporal dimension, whereby the years before and after the Nazi dictatorship are taken sufficiently into account, also touching on the issue of the continuities and discontinuities in the society's history. On the other hand, the history of the DPG in the Third Reich is not treated in isolation but is placed within the general political context as well as within the history of science, and the conduct of the DPG is compared to that of other scientific societies during the Third Reich.

Three workshops helped further the necessary discussions about the topic at hand and procedural clarification among the authors. These meetings were open forums of discussion, in which others besides the participants in the research project could also take part. Other competent representatives of the field and interested members of the DPG were able to attend and make their own suggestions. The first workshop attracted particular attention and drew almost 50 members of the profession to the *Magnus-Haus* in Berlin.

In closing, we would cordially like to thank everyone involved in the production of this book. Thanks are particularly due to the German Physical Society, which not only generously funded the research project and assumed the printing costs for the original German version of this book but also gave its unwavering active support. Special thanks go to its two former presidents Alexander Bradshaw and Theo Mayer-Kuckuk for their great interest and active engagement in the furtherance of the research project. We must likewise express our gratitude to the managing directors of the DPG, Volker Häselbarth and Bernhard Nummer, and to their co-workers at the business office in Bad Honnef for many a constructive suggestion in overcoming practical bottlenecks and obstacles. Not least of all, we owe many thanks to numerous archives and libraries and, especially, to the society itself. They willingly helped make accessible to our research their – in many cases as-yet-untouched – treasures on the history of the DPG.

Uwe Hank put much effort and expertise into the editing of the majority of the contributions in the original German volume, and Ralf Hahn took care of the proofs and the production of the final manuscript; he also assisted in the research on the illustrations. We are grateful to Ann Hentschel for her very good translation of our book into English. Last but not least come the publishers, Wiley-VCH for the German version, and Cambridge University Press for the English edition.

Dieter Hoffmann and Mark Walker
Berlin and Schenectady
2011

Abbreviations

AEA	The Albert Einstein Archive of the Jewish National and University Library at the Hebrew University of Jerusalem, Israel
AEG	German General Electricity Company (*Allgemeine Elektrizitäts-Gesellschaft*)
AHQP	Archive for the History of Quantum Physics
AIP	American Institute of Physics, Niels Bohr Library, College Park, MD
BA	Federal German Archives, Berlin and Koblenz (*Bundesarchiv*)
BASF	Baden Aniline and Soda Factory (*Badische Anilin- & Soda-Fabrik*)
BLO	Bodleian Library, Oxford
BSC	Bohr Scientific Correspondence in AHQP
CITA	Archives of the California Institute of Technology, Pasadena, CA
DAL	German Academy for Aviation Research (*Deutsche Akademie für Luftfahrtforschung*)
DChG	German Chemical Society (*Deutsche Chemische Gesellschaft*)
DDR	German Democratic Republic (*Deutsche Demokratische Republik*)
DFG	German Research Foundation (*Deutsche Forschungsgemeinschaft*)
DGtP	German Society for Technical Physics (*Deutsche Gesellschaft für technische Physik*)
DMA	Archives of the Deutsches Museum, Munich (*Deutsches Museum Archiv*)

DMV German Mathematical Association (*Deutsche Mathematiker-Vereinigung*)

DPG German Physical Society (*Deutsche Physikalische Gesellschaft*)

DPGA Archives of the German Physical Society (*Archiv der Deutschen Physikalischen Gesellschaft*)

EHR Ehrenfest Scientific Correspondence in AHQP

FIAT Field Information Agency, Technical

GAMM Society for Applied Mathematics and Mechanics (*Gesellschaft für angewandte Mathematik und Mechanik*)

GDCh Society of German Chemists (*Gesellschaft Deutscher Chemiker*)

GDNÄ Society of German Scientists and Physicians (*Gesellschaft Deutscher Naturforscher und Ärzte*)

Gestapo Secret State Police (*Geheime Staatspolizei*)

HATUM Historical Archives of the Technical University of Munich (*Historisches Archiv der Technischen Universität München*)

HBA Aschaffenburg Library (*Hofbibliothek Aschaffenburg*)

KWG Kaiser Wilhelm Society for the Advancement of the Sciences (*Kaiser-Wilhelm-Gesellschaft zur Förderung der Wissenschaften*)

KWI Kaiser Wilhelm Institute (*Kaiser-Wilhelm-Institut*)

LTAMA Archives of the State Museum for Technology and Work in Mannheim (*Archiv des Landesmuseums für Technik und Arbeit in Mannheim*)

LTI *lingua tertii imperii* (Viktor Klemperer's acronym for the nazification of the German language)

MPG Max Planck Society (*Max-Planck-Gesellschaft*)

MPGA Archives of the Max Planck Society (*Archiv der Max-Planck-Gesellschaft*)

MPIfP Max Planck Institute for Physics (*Max-Planck-Institut für Physik*)

MR Reich Mathematical Federation (*Mathematische Reichsverband*)

NS National Socialist (Nazi)

NSBDT National Socialist League of German Engineers (*Nationalsozialistischer Bund Deutscher Technik*)

NSDAP National Socialist German Workers Party (*Nationalsozialistische Deutsche Arbeiterpartei*)

NSDDB National Socialist German University Lecturers League (*Nationalsozialistischer Deutscher Dozentenbund*)

OKW	Supreme Command of the Armed Forces (*Oberkommando der Wehrmacht*)
PAW	Prussian Academy of Sciences (*Preußische Akademie der Wissenschaften*)
PGzB	Physical Society of Berlin (*Physikalische Gesellschaft zu Berlin*)
PTR	Imperial Physical-Technical Institute (*Physikalisch-Technische Reichsanstalt*)
REM	Reich Ministry for Science, Education and Culture (*Reichsministerium für Wissenschaft, Erziehung und Volksbildung*)
RFR	Reich Research Council (*Reichsforschungsrat*)
RLM	Reich Aviation Ministry (*Reichsluftfahrtministerium*)
RLUC	The Josef Regenstein Library, Special Collections, University of Chicago
RTA	Reich Association for the Technical and Scientific Work (*Reichsgemeinschaft der technisch-wissenschaftlichen Arbeit*)
SA	Storm Troopers (*Sturmabteilung*)
SBPK	Prussian State Library, Berlin (*Staatsbibliothek Preußischer Kulturbesitz*)
SPSL	Society for the Protection of Science and Learning
SS	*Schutz-Staffel*
UAM	Archives of the University of Munich (*Universitätsarchiv, Ludwigs Maximilian Universität München)*
UFA	Archives of the University of Freiburg (*Archiv der Universität Freiburg i.Br.*)
UMI	Italian Mathematical Union (*Unione Matematica Italiana*)
VdCh	Association of German Chemists (*Verein deutscher Chemiker*)
VDEh	Archives of the German Iron & Steel Institute (*Verein Deutscher Eisenhüttenleute*)
VDI	Union of German Engineers (*Verein Deutscher Ingenieure*)

1

The German Physical Society under National Socialism in Context

Mark Walker

The history of the German Physical Society (*Deutsche Physikalische Gesellschaft*, DPG) is not, and cannot be, a comprehensive history of physics under National Socialism.[1] Although most physicists were members of this society, the DPG had little, if anything, to do with much of what these scientists did between 1933 and 1945. Most of these physicists had multiple affiliations – a position at a university, research institution, or private firm, perhaps membership in an academy of science, appointment to the editorial board of a journal, and so forth. Max von Laue is an example of a physicist who wore many different hats: associate professor[2] at the University of Berlin, member of the Kaiser Wilhelm Institute for Physics, member of the Prussian Academy of Sciences (*Preußische Akademie der Wissenschaften*, PAW), member of the advisory board of the Imperial Physical-Technical Institute (*Physikalisch-Technisch Reichsanstalt*, PTR), referee for the Emergency Society for German Science (*Notgemeinschaft der Deutschen Wissenschaft*; subsequently renamed the German Research Foundation[3] [*Deutsche Forschungsgemeinschaft*,

[1] The best single source for physics under National Socialism is Klaus Hentschel, *Physics and National Socialism: An Anthology of Primary Sources* (Basel: Birkhäuser, 1996), including his extensive introduction; for the history of the DPG in the Third Reich, also see Dieter Hoffmann and Mark Walker, "The German Physical Society under National Socialism," *Physics Today* (December, 2004), 52–58, Dieter Hoffmann, "Between Autonomy and Accommodation: The German Physical Society during the Third Reich," *Physics in Perspective*, 7/3 (2005), 293–329, and Hentschel, *Physics*, lxx, 407.

[2] *Außerordentliche*.

[3] For the German Research Foundation, see Karin Orth and Willi Oberkrome (eds.), *Die Deutsche Forschungsgemeinschaft 1920–1970* (Stuttgart: Franz Steiner Verlag, 2010).

DFG]), member of the editorial boards of several journals, and, of course, both a member of and an official in the DPG.

Many of the scientists who stayed in Germany during the period 1933-1945 and remained members of the DPG did not play an active role in the society. Others, including some of the most famous, such as Werner Heisenberg[4] and Carl Friedrich von Weizsäcker,[5] appear only briefly in the history of the society. Some members were not even physicists; for example, the radiochemist Otto Hahn was a member of the DPG, but his work on nuclear fission and his experiences under National Socialism as director of the Kaiser Wilhelm Institute for Chemistry are not very relevant for the history of the DPG.[6] Pascual Jordan only became a member so that he could receive the society's Max Planck Medal for Theoretical Physics. This book will focus more narrowly on the DPG, a rich subject that illuminates interesting and important aspects of the history of physics and science under National Socialism.

German history from the First World War to the post–Second World War era is an immense subject, but for the purposes of this introduction, a short list of important milestones in the history of National Socialism will be used to put the history of the DPG under Hitler into context:

1. 1933: The National Socialist "Seizure of Power" (*Machtergreifung*)
2. 1933: The purge of the civil service
3. 1934: The purge of the Storm Troopers (*Sturmabteilung*, SA), and Hitler as *Führer* ("Leader")
4. 1935: The Nuremberg Laws
5. 1936: Rearmament and the Four-Year Plan
6. 1938: "Night of Broken Glass"
7. 1939: The start of the Second World War
8. 1941: The German attack on the Soviet Union
9. 1941: The end of the Lightning War and the beginning of war with the United States
10. 1943: German defeat and surrender at Stalingrad
11. 1945: The unconditional German surrender
12. 1945: The division of Germany into zones of occupation

[4] See David Cassidy, *Beyond Uncertainty: Heisenberg, Quantum Physics, and the Bomb* (New York: Bellevue Literary Press, 2009).

[5] Konrad Lindner, *Carl Friedrich von Weizsäckers Wanderung ins Atomzeitalter. Ein dialogisches Selbstporträt* (Paderborn: Mentis, 2002).

[6] See Mark Walker, "Otto Hahn: Responsibility and Repression," *Physics in Perspective*, 8/2 (2006), 116–163.

13. 1949: The founding of the two German states
14. 1953: Full West German sovereignty

Not all of these events had a discernible effect on the history of the DPG under National Socialism, but when they did, the results were sometimes unexpected.

Adolf Hitler's appointment as German chancellor and the subsequent step-by-step consolidation of a monopoly of political power by the National Socialist movement did not significantly change the day-to-day business of the DPG until the eve of the Second World War and, with a few important exceptions, even then did not cause major changes in what the organization did or how this was carried out. The exceptions were as follows: (1) the introductory speeches made at conferences by DPG president Karl Mey, which were full of praise for Hitler and used some of the language of the National Socialist period,[7] what Victor Klemperer called the *lingua tertii imperii* (LTI)[8]; (2) the formal expulsion of Jewish members in 1938[9]; and (3) the political advocacy of the militarization of physical research during the war.[10]

The purge of the civil service caused by the National Socialist Law for the Restoration of the Professional Civil Service (*Gesetz zur Wiederherstellung des Berufsbeamtentums*) in spring 1933 had a profound effect on German physicists because most scientists outside of industry were civil servants.[11] Many physicists either lost their jobs or no longer saw any professional future in Germany and left the country.[12] However, this

[7] See Simonsohn's chapter in this volume.

[8] Victor Klemperer, *The Language of the Third Reich: LTI—Lingua Tertii Imperii. A Philologist's Notebook* (New York: Continuum, 2006).

[9] See Wolff's chapter in this volume, as well as Stefan L. Wolff, "Vertreibung und Emigration in der Physik – 1933," *Physik in unserer Zeit*, 24 (1993), 267–273 and Stefan L. Wolff, "Frederick Lindemanns Rolle bei der Emigration der aus Deutschland vertriebenen Physiker," *Yearbook of the Research Center for German and Austrian Exile Studies*, 2 (2000), 25–58, and most recently, Stefan L. Wolff, "Das Vorgehen von Debye bei dem Ausschluss der jüdischen Mitglieder aus der DPG," in Dieter Hoffmann and Mark Walker (eds.), *"Fremde" Wissenschaftler im Dritten Reich. Die Debye-Affäre im Kontext* (Göttingen: Wallstein, 2011), 106–130.

[10] See Hoffmann's chapter in this volume.

[11] See Wolff's chapter in this volume, as well as Alan Beyerchen, *Scientists under Hitler: Politics and the Physics Community in the Third Reich* (New Haven: Yale University Press, 1977), 12–50, Hentschel, *Physics*, 21–34, including the text of the Civil Service law, and Cassidy, *Beyond*, 205–217.

[12] For the emigration, see Hentschel, *Physics*, liii–lxiv, and Klaus Fischer, "Die Emigration von Wissenschaftlern nach 1933. Möglichkeiten und Grenzen einer Bilanzierung," *Vierteljahrshefte für Zeitgeschichte*, 39 (1991), 535–549.

did not necessarily have an immediate effect upon their membership in the DPG. Whereas professional organizations of chemists, engineers, and mathematicians forced their Jewish members out during the first years of the Third Reich (see later), the DPG and its officials tried hard to act as if nothing unusual was going on.[13] Indeed, very few German émigrés or foreign colleagues who were members of the DPG formally resigned; instead, those that left merely stopped paying their dues and were quietly removed from the rolls.[14]

The so-called Einstein affair was an exception to this rule. Einstein had been well known since the First World War as an outspoken pacifist and internationalist.[15] During the Weimar Republic, Einstein had become the target of anti-Semitic groups, and the physicist had publicly defended himself. For all these reasons, Einstein was a *political* threat to the National Socialist movement, and, among all German scientists (including all Jewish or politically active scientists), he was singled out for special treatment. Einstein was out of the country when the National Socialists were helped into power. He remained away and criticized National Socialist (NS) policies, including the purge of Jewish civil servants. Einstein recognized that his membership in German organizations was now a political issue, so he tried to resign voluntarily and discreetly.[16]

The nationwide boycott of Jewish businesses that began on April 1, 1933, sponsored by the National Socialist German Workers Party (*Nationalsozialistische Deutsche Arbeiterpartei*, NSDAP) and led by fanatical National Socialists and anti-Semites like Josef Goebbels and Julius Streicher, had to be cut short to a single day because of the lukewarm reception given to it by many Germans and the strong protests from outside of Germany. Immediately thereafter, many state agencies and institutions either came under pressure to make their own stances on the "Jewish question" clear or took the initiative without much prompting. The Reich Ministry for Science, Education and Culture (*Reichsministerium für Wissenschaft, Erziehung und Volksbildung*, REM), among

[13] See Deichmann's and Remmert's chapters in this volume, as well as Karl-Heinz Ludwig, *Technik und Ingenieure im Dritten Reich* (Düsseldorf: Droste, 1974), 105–160.

[14] See Wolff's chapter.

[15] For Einstein, see David Rowe and Robert Schulman (eds.), *Einstein on Politics: His Private Thoughts and Public Stands on Nationalism, Zionism, War, Peace, and the Bomb* (Princeton: Princeton University Press, 2007).

[16] For Albert Einstein and the DPG, see the documents section, "Albert Einstein, Max von Laue und Johannes Stark," in the original German version of this book, Dieter Hoffmann and Mark Walker (eds.), *Physiker zwischen Autonomie und Anpassung – Die DPG im Dritten Reich* (Weinheim: Wiley-VCH, 2007), 530–548.

other things, wanted its subsidiary organization PAW to make a public show of anti-Semitism by throwing out Einstein.

As several historians have described, unfortunately Einstein had already resigned from the PAW, so an academy official had to take the further radical step of declaring that the PAW was glad that Einstein was gone.[17] The subsequent ambivalent responses of Einstein's respected colleagues Max von Laue and Max Planck are also well known. The ever-diplomatic Planck defended Einstein's scientific reputation and legacy but agreed that Einstein, through his political conduct, had made it impossible for himself to remain in the academy. Although Planck undoubtedly did not want Einstein to leave the academy, his public statement could be interpreted as agreement with the NS insistence that he go.[18] Max von Laue, in contrast, publicly one of Einstein's staunchest supporters, privately wrote Einstein and chided him for his "political" conduct.[19]

Here the contrast between the PAW and the DPG is stark. The DPG officials quietly removed Einstein's name from the membership list, apparently without any pressure from REM to do more. The overall strategy of the DPG was to avoid conflict and confrontation with the NS government.[20] Thus Einstein, one of the few scientists to grab and hold the attention of leading National Socialists, was gone from the DPG long before the society had to deal with the issue of Jewish members.

The "Einstein affair" was not typical. Alan Beyerchen in his pathbreaking book has compared how Max Born, Richard Courant, and James Franck responded to the NS purge of the civil service.[21] In the end, all of the different responses, ranging from Born's quiet departure

[17] See Wolff's chapter in this volume, as well as Hentschel, *Physics*, 18–21, John L. Heilbron, *The Dilemmas of an Upright Man. Max Planck as Spokesman for German Science* (Berkeley: University of California Press, 1986), 155–159; Jürgen Renn, Giuseppe Castagnetti, and Peter Damerow, "Albert Einstein. Alte und neue Kontexte in Berlin," in Jürgen Kocha (ed.), *Die Königlich Preußische Akademie der Wissenschaften zu Berlin im Kaiserreich* (Berlin: Akademie-Verlag, 1999), 333–354, here 349–351; and Dieter Hoffmann, "Einsteins politische Akte," *Physik in unserer Zeit*, 35, No. 2 (2004), 64–69.

[18] For Planck, see Heilbron, Dieter Hoffmann, "Das Verhältnis der Akademie zu Republik und Diktatur. Max Planck als Sekretär," in Wolfram Fischer (ed.), *Die Preußische Akademie der Wissenschaften zu Berlin 1914–1945* (Berlin: Akademie-Verlag, 2000), 53–85, and Dieter Hoffmann, *Max Planck: Die Entstehung der modernen Physik* (Munich: Beck, 2008).

[19] See Heilbron, 70–73; also see the documents section, "Albert Einstein, Max von Laue und Johannes Stark" in Hoffmann and Walker, *Physiker*, 530–548.

[20] See Eckert's and Wolff's chapters in this volume.

[21] See Beyerchen, 15–39.

to Franck's public and defiant resignation, were ineffectual. No matter how many scientists resigned, there were competent and often quite good colleagues ready and willing to take their places. One physicist, Richard Becker, was transferred from the Berlin Technical University to the University of Göttingen against his will but nevertheless proved willing to teach once he got there.

Perhaps most disturbing is how the NS regime exploited the natural and quite justifiable efforts by the German physicists untouched by the civil service law to rebuild their discipline. Both Planck and Heisenberg, for example, sought out colleagues who were "Aryan" enough to be acceptable to the Third Reich but were good physicists. However, the unintended consequence was that Planck and Heisenberg thereby apparently accepted and justified the racist policy of firing Jews and only hiring Aryans.[22] Unfortunately, little is known about industrial research in this regard. German physicists who lost their academic jobs usually did not move to German industry, presumably because they were not welcome there. The example of Nobel laureate Gustav Hertz, who was forced out of his professorship at the Technical University of Berlin and subsequently accepted an offer to lead a research laboratory at Siemens and worked on military research during the war, was not typical.

Perhaps the one event most often mentioned as an example of scientists resisting National Socialism is The Haber Memorial Service in 1934.[23] Fritz Haber became a Nobel laureate for his work on the fixation of nitrogen from the air.[24] During the First World War, he transformed and greatly expanded his Kaiser Wilhelm Institute for Physical Chemistry into a research and development center for chemical weapons. Haber's institute had an unusually large number of Jewish chemists and physicists, including Haber himself, when the National Socialists came to power. Similar to Einstein as a person, Haber's institute became a target for the National Socialists in REM. Haber was ordered to fire almost all of his staff. He did so and then publicly resigned.[25] Haber was temporarily

[22] See Cassidy, *Beyond*, 215–217.

[23] For the Haber Memorial, see Beyerchen, 67–68, Heilbron, 162, Kristie Macrakis, *Surviving the Swastika: Scientific Research in Nazi Germany* (Cambridge, MA: Harvard University Press, 1993), 68–72, John Cornwell, *Hitler's Scientists: Science, War and the Devil's Pact* (London: Viking, 2003), 138–139; see the documents section, "Der Haber Feier," in Hoffmann and Walker, *Physiker*, 557–561.

[24] For Haber's biography, see Dietrich Stolzenberg, *Fritz Haber. Chemiker, Nobelpreisträger, Deutscher, Jude* (Weinheim: VCH, 1994), and Margit Szöllösi-Janze, *Fritz Haber 1868–1934* (Munich: Beck, 1998).

[25] For Haber's resignation, death, and subsequent reaction by colleagues, see Hentschel, *Physics*, 44–45, 63–65, 76–79.

replaced by a scientist imposed by Army Ordnance, which was very interested in using the institute for chemical weapons research. Eventually a candidate more acceptable to the Kaiser Wilhelm Society (*Kaiser-Wilhelm-Gesellschaft*, KWG), Peter Adolf Thiessen, became director and devoted a significant amount of the institute's effort to chemical weapons.

Haber died in exile in 1934. A year later the KWG, with the support of the DPG and the German Chemical Society (*Deutsche Chemische Gesellschaft*, DChG), honored his memory with a private ceremony. This was of course controversial. Officials in REM bristled at honoring a Jew who had protested their policies. REM forbade anyone under their jurisdiction from attending. Planck and DPG officials responded by insisting that no protest or criticism of governmental policies was intended. Minister Rust in turn offered to grant exemptions for scholars who wished to attend.

The university professors stayed away, although some of them sent their wives. Only one member of the DChG tried to get the exemption promised by Rust, but he was turned down. In contrast, the Union of German Chemists (*Verein deutscher Chemiker*, VdC) forbade its members from attending. Several VdC members protested against this prohibition. However, this internal protest did not translate into a public statement against NS policy.[26] Planck and Hahn (both DPG members) spoke at the private ceremony for Haber. In the end, this represented the high point of (quasi) public protest of or opposition to NS policies toward scientists. Although the DPG was listed as one of the sponsors, no DPG officials participated, but they also did not tell anyone else not to go.

Perhaps the best known and most infamous example of physics under National Socialism is the so-called Aryan physics (*Deutsche Physik*) movement founded and led by the Nobel laureates Philipp Lenard and Johannes Stark.[27] This small clique called loudly for a more "Aryan" and a less "Jewish" physics, and Stark sought to control appointments, funding, and publishing in physics – and thereby threatened the DPG. Lenard and Stark gave Hitler and his movement strong public support at a time when his fortunes appeared poor. Stark had actively campaigned for the

[26] See Deichmann's chapter in this volume.

[27] For Aryan physics, see Beyerchen, 79–167, Hentschel, *Physics*, 7–10, 100–116, 119–129, 152–161, Freddy Litten, *Mechanik und Antisemitismus. Wilhelm Müller (1880–1968)* (München: Institut für Geschichte der Naturwissenschaften, 2000), and Cornwell, 178–190; Michael Eckert, *Die Atomphysiker. Eine Geschichte der theoretischen Physik am Beispiel der Sommerfeldschule* (Braunschweig: Vieweg, 1993), 196–203; for Stark and Aryan physics, see Mark Walker, *Nazi Science: Myth, Truth, and the German Atomic Bomb* (New York, Perseus Publishing, 1995), 5–63.

National Socialists during the last hectic years of the Weimar Republic. When Hitler became chancellor, these two physicists were rewarded. Lenard, who was already retired, mostly received honors. Stark became president of both the PTR and the DFG and intended to dominate the DPG as well.[28]

Although many scientists inside and outside of Germany took Stark's influence at the start of the Third Reich and his attempts to take over physics as proof that the National Socialists wanted to dominate and transform science, it is now clear that this was not true. There was no conscious, coordinated, and deliberate attempt on the part of the NS leadership to damage, control, distort, or alter science – although contemporary observers both inside and outside of Germany can be forgiven for believing that this was so. In fact, most leading National Socialists did not consider science important enough to be a priority for their *Gleichschaltung* (coordination or synchronization) of German society. It is also true that Stark's ambitions were normal for a member of the NS elite. Throughout the German state, National Socialists fought with each other to carve out satrapies and assert a monopoly of power over a given area. For example, Josef Goebbels sought to control propaganda, and Max Amman sought to control newspaper publishing. It should have been no surprise that Stark tried to do the same in physics.

Arguably, the true business of physics takes place in its journals and other professional publications such as textbooks and handbooks. Thus, Stark was right to try to seize control of the publication of research at the beginning of the Third Reich via his attempt to dominate the DPG. Many of these journals, the *Zeitschrift für Physik*, for example, were published in the name of the DPG, and although they were really influenced more by their respective editors than by the society as a whole, this is precisely what Stark could have changed. Most journals remained remarkably free from overt political influence.[29] Research topics like the theory of relativity never disappeared. Physicists continued to cite and discuss articles by émigrés such as Einstein. The ideological debate between "Jewish" and "modern" physics rarely emerged, and when it did, it was handled in a discreet way. The few adherents of Aryan physics had their own journal, the *Zeitschrift für die gesamte Naturwissenschaft*, for ideological attacks,

[28] See the documents section, "Albert Einstein, Max von Laue und Johannes Stark," in Hoffmann and Walker, *Physiker*, 330–348.

[29] See Hentschel, *Physics*, xvi–xvii, and Simonsohn's chapter in this volume, as well as Gerhard Simonsohn, "Physiker in Deutschland 1933–1945," *Physikalische Blätter*, 48 (1992), 23–28.

but when their works did appear in the professional journals, they were limited to publishing old-fashioned physics, not politics.[30]

Along with journals, the business of physics is also expressed in funding. Here, Stark's control of the DFG was a good opportunity to steer research into particular channels, but he soon squandered his influence by fighting with other influential National Socialists in the government, the bureaucracy, and the NSDAP.[31] In 1936, he was forced to resign and was succeeded by the chemist and REM official Rudolf Mentzel. Stark's presidency of the PTR was not much better. In 1936, he lost control over his budget, and in 1939 he had to retire and was succeeded by Abraham Esau. Although Stark had stopped supporting research in modern physics, his policies were not very different from those of his successors: Stark, Mentzel, and Esau all supported some basic research while emphasizing applied, often military, research. In contrast to Stark, during the war physicists such as Esau and his successor Walther Gerlach became very influential serving as the Plenipotentiary for Physics in the Reich Research Council (*Reichsforschungsrat*, RFR), an institution founded to help coordinate scientific research for the war effort and closely linked to the DFG. Had Stark succeeded in controlling physics publications and research grants, he truly could have influenced what sort of physics was done in Germany. But he failed, mainly because he had as many enemies as friends among the NS elite and had not gained support for the reforms he proposed.

Aryan physics was a political movement of scientists within the NS movement.[32] In particular, "it was above all the local politics, that is, those of the community of physicists."[33] The very few successes of Stark,

[30] See Simonsohn's chapter and Walker, *Nazi Science*, 43–47.

[31] Walker, *Nazi Science*, 5–63; also see Sören Flachowsky, *Von der Notgemeinschaft zum Reichsforschungsrat. Wissenschaftspolitik im Kontext von Autarkie, Aufrüstung und Krieg* (Stuttgart: Franz Steiner Verlag, 2008), 163–200.

[32] See Mark Walker, *German National Socialism and the Quest for Nuclear Power, 1939–1949* (Cambridge, Cambridge University Press, 1989), 60–73.

[33] See Beyler's chapter in this volume for this quotation, as well as Richard H. Beyler, "'Reine' Wissenschaft und personelle 'Säuberungen.' Die Kaiser-Wilhelm/Max-Planck-Gesellschaft 1933 und 1945," in Carola Sachse (ed.), *Ergebnisse. Vorabdrucke aus dem Forschungsprogramm "Geschichte der Kaiser-Wilhelm-Gesellschaft im Nationalsozialismus*, No. 16 (Berlin: Forschungsprogramm, 2004), Richard H. Beyler, Alexei Kojevnikov, and Jessica Wang, "Purges in Comparative Perspective: Rules for Exclusion and Inclusion in the Scientific Community under Political Pressure," in Carola Sachse and Mark Walker (eds.), *Politics and Science in Wartime*, Volume 20 of *Osiris*, (Chicago: University of Chicago Press, 2005), 23–48, and Richard H. Beyler, "Maintaining Discipline in the Kaiser Wilhelm Society during the National Socialist Regime," *Minerva*, 44/3 (2006), 251–266.

Lenard, and their adherents were eventually revealed as Pyrrhic victories. The outcome of the struggle over Aryan physics should be seen as a successful attempt to reassert the extant patterns of authority within the boundaries of the physics community.[34] What is arguably more important is what this movement tells us about science, and particularly physics, during the Third Reich. Science policy and management under National Socialism reflected the polycratic nature of the regime, whereby many different and competing sources of authority, funding, and other forms of support swirled around immediately below Hitler's dictatorship. Thus, when Stark found a particular patron among the highest levels of the NS state, physicists threatened by Stark's ambitions had to find their own patrons, in particular individuals or groups more sympathetic to the importance of modern physics. They did this, eventually becoming very successful, far more than Stark had been.

This had a price, however, for the patrons supported these scientists because the patrons expected something in return. Stark's conflict with the DPG, just like his battles with the established physics community in general, represented a struggle for authority within the physics community in the context of the pressures exerted on the external boundaries of that community by the NS state.[35] The final fate of Aryan physics was analogous to the fate of the NS Storm Troopers, the SA. Under its leader Ernst Röhm, the SA was very useful, if not indispensable for the National Socialists in their quest to gain and consolidate political power. However, at some point the SA's calls for a "second revolution" – because in their eyes the first one had not gone far enough – became counterproductive for Hitler and the rest of the NS elite. The SA leadership was then silenced and the masses of SA men reigned in. Although no advocate of Aryan physics fared as badly as Röhm, their movement had a similar experience. At the start of the Third Reich, the calls for an "Aryan science" in physics, mathematics, and other disciplines facilitated the NS *Gleichschaltung* of these disciplines. However, within a few years, the most important and influential members of the NS elite were far more concerned about how science and engineering could be useful to them than about its ideological purity, and Stark and his followers were silenced as well.[36]

34 See Beyler's and Eckert's chapters in this volume.

35 See Beyler's chapter in this volume.

36 See Monika Renneberg and Mark Walker, "Scientists, Engineers, and National Socialism," in Monika Renneberg and Mark Walker (eds.), *Science, Technology, and National Socialism* (Cambridge: Cambridge University Press, 1993), 1–17, 339–346.

It appears that there was never a direct confrontation between the DPG and Aryan physics. When this was threatened, the DPG made sure that it had the agreement of REM before acting. The reaction of the DPG to Aryan physics was not opposition to but rather cooperation with the NS regime.[37] A good example of this is the Max Planck Medal for Theoretical Physics, the most prestigious award given out by the DPG.[38] The first winners of this award – Einstein,[39] Heisenberg,[40] and so on – were prominent representatives of precisely the type of modern physics that the advocates of Aryan physics opposed. Thus, the medal became controversial and was not awarded for several years. The DPG resumed awarding the medal when the influence of Aryan physics began to fade but then ran up against the problem of honoring émigrés or non-Aryans.[41] Eventually, this was solved by honoring German physicists who were deserving from a professional point of view but were also acceptable to the National Socialists running the REM. The problems with non-Aryan candidates for the Max Planck Medal foreshadowed the issue of Jewish scientists in the DPG.

If science was not a particular target of NS reformers, the university system certainly was. The universities were essentially purged of non-Aryans during the first 2 years of the Third Reich.[42] The exceptions granted to war veterans as a mollifying gesture to the aged President Hindenburg were in the end not honored. The Third Reich, like the Soviet Union, quickly moved to control and transform the educational system in order to train and indoctrinate the youth. This included science education. Even after the non-Aryans and few university faculty members on the left wing of the political spectrum had been purged, the remaining staff had to accommodate themselves to the new regime, but to different degrees.

Scholars who already had permanent positions and were either politically compatible with National Socialism or apolitical could often stay

37 See Eckert's chapter in this volume.

38 See Beyler, Eckert, and Hoffmann's chapter in this volume; also see the documents section, "Der Planck Medaille," in Hoffmann and Walker, *Physiker*, 579–591.

39 See Rowe and Schulman.

40 See Cassidy, *Beyond*.

41 For the 1938 award to Louis de Broglie, see Heilbron, 182.

42 See Hentschel, *Physics*, xxxiii–lii, for faculty and students at universities; see John Connelly and Michael Grüttner (eds.), *Zwischen Autonomie und Anpassung. Universitäten in den Diktaturen des 20. Jahrhunderts* (Paderborn: Schöningh, 2002), especially Michael Grüttner, "Schlussüberlegungen: Universität und Diktatur," 266–276, for universities under dictatorships in a comparative context.

where they were without major additional sacrifice. If they wanted a better position, or funding, or influence, however, they had to prove their worth and loyalty to the new politicized leadership of the universities. Young academics who wanted a permanent position, a successful habilitation,[43] or even just wanted to be able to teach had to make much greater concessions to the new regime. These included taking part in political indoctrination camps, joining the NSDAP or an ancillary NS organization, and undergoing a political evaluation by the local National Socialist University Teachers League. Students were the most politicized of all. Large portions of their time were now devoted to physical fitness exercises, political and ideological training, and public service. Often there was little time left for their studies. When the war came, many of them left for the front, never to return. After 1945, students again flooded the universities, but usually their teachers had themselves been students during the period of the Weimar Republic or before. The generation of science students trained during the Third Reich was essentially lost.

However, other scientific institutions, including some financed directly or indirectly by the German state, were not immediately forced to purge themselves of their Jewish members. Most important is the fact that there were significant differences between the various scientific–technical organizations. The Union of German Engineers (*Verein Deutscher Ingenieure*, VDI) fended off an attempt by the "old fighter" and engineer Gottfried Feder to take control of the society by quickly accommodating itself to the new NS. The VDI declared that it would help "rebuild" Germany by fighting unemployment, working to solve the problems caused by the shortage of raw materials, and increasing the military preparedness of Germany. In particular, the engineers would use science to help achieve the "goals of the national movement." The VDI also declared that it would apply the "Aryan Paragraph" of the Civil Service Law to its members, thereby eliminating any "non-Aryan members."

Feder failed in large part because the most influential figures in the NS state did not support his "socialist" brand of National Socialism and denied him the necessary political backing. This was analogous to the fates of the physicist Johannes Stark and the mathematician Ludwig Bieberbach (see later). In 1933, the VDI joined the later-named National Socialist League of German Engineers (*Nationalsozialisticher Bund Deutscher*

[43] This was analogous to a second doctoral dissertation; the traditional prerequisite for a professorship.

Technik, NSBDT), an umbrella organization that facilitated the *Gleichschaltung* of scientific and technical organizations.[44]

There were some differences between the two chemical societies DChG and VdC with regard to their accommodation to the new state. In 1933, VdC chairman Paul Duden – himself not an NSDAP member – introduced the NS leadership principle (*Führerprinzip*), incorporated political goals into the constitution, openly announced his support of the NS state, and brought his society into the NSBDT. The DChG eventually did this, but not right away. They were faced with the problem of how to get rid of the Jewish members of the governing board without provoking international protest, yet they were able to do this quietly.[45] DChG membership declined from 1932 to 1941. Departures were listed as resignations or as removal from the rolls because of unpaid dues. The "purge" of the editorial process for chemistry books and journals was so extensive and so many people were fired that the production process was severely affected. In 1936, Alfred Stock became president of the society, introduced the leadership principle, and required members in Germany to fill out an ancestry questionnaire. The DChG tried to exclude its 40 percent foreign membership – who helped pay for the expensive publications – from the racial question. In 1938, Stock led the DChG into the NSBDT as well. He was succeeded as president that same year by Richard Kuhn, who oversaw the removal of the last Jewish names from the journals, strongly embraced Hitler in public speeches, and obediently served the NS state – but did not become an NSDAP member.[46]

The various different mathematical societies also reacted differently to the NS regime. The Reich Mathematical Federation (*Mathematische Reichsverband*, MR) introduced the leadership principle right away in 1933, whereas the Society for Applied Mathematics and Mechanics (*Gesellschaft für angewandte Mathematik und Mechnik*, GAMM) was more cautious.[47] The German Mathematical Association (*Deutsche*

44 Ludwig, 113–118.

45 See Deichmann's chapter in this volume, as well as Ute Deichmann, *Flüchten, Mitmachen, Vergessen* (Weinheim: Wiley-VCH, 2001) and her earlier book, *Biologists under Hitler* (Cambridge, MA: Harvard University Press, 1996).

46 See Deichmann's chapter in this volume.

47 See Herbert Mehrtens, "Angewandte Mathematik und Anwendungen der Mathematik im nationalsozialistischen Deutschland," *Geschichte und Gesellschaft*, 12 (1986), 317–347; Herbert Mehrtens, "Die 'Gleichschaltung' der mathematischen Gesellschaften im nationalsozialistischen Deutschland," *Jahrbuch Überblicke Mathematik* (1985), 83–103.

Mathematiker-Vereinigung, DMV) had to stave off an attempt by a colleague turned NS, Ludwig Bieberbach, to control mathematics just as the DPG had rebuffed Stark.[48] Bieberbach failed to push through a strong leadership principle in the DMV. However, the DMV managed to fend off Bieberbach only by relinquishing independence vis-à-vis the REM.[49]

Many mathematicians left the society. In 1937 Wilhelm Süß, an effective political force at the University of Freiburg, NSDAP member, and former student of Bieberbach, became chair of the DMV. The next year, Süß began agitating to bar Jewish editors at the mathematical journals, including a denunciation of his colleague Issai Schur.[50] In September 1938, Süß learned that all scientific societies would be required to "implement the Aryan principle." DMV officials proposed that the Jewish members who were forced out would be listed as "resigned" or "membership expired."

The letter that was sent out read: "In the future you can no longer be a member of the German Mathematicians' Association. Therefore I advise you to declare your resignation from our association. Otherwise we will make the termination of your membership public at the next opportunity." Thus, whereas the DPG told its members to decide whether or not they had to resign (see later), the DMV identified the members affected and contacted them.[51]

During the war, the DMV developed into an effective instrument of professional politics under the leadership of Wilhelm Süß, comparable with Carl Ramsauer's role in the DPG (see later). Süß had ambitious plans for creating a Reich Institute for Mathematics. The DPG and DMV

[48] For Bieberbach, see Herbert Mehrtens, "Ludwig Bieberbach and Deutsche Mathematik," in Esther R. Phillips (ed.). *Studies in the History of Mathematics* (Washington, DC, Mathematical Association of America, 1987), 195–241, and Herbert Mehrtens, "Irresponsible Purity: The Political and Moral Structure of Mathematical Sciences in the National Socialist State," in Renneberg and Walker, 324–338, 411–413; for mathematics under National Socialism in general, see Sanford Segal, *Mathematicians under the Nazis* (Princeton: Princeton University Press, 2003) and Reinhard Siegmund-Schultze, *Mathematicians Fleeing Nazi Germany: Individual Fates and Global Impact* (Princeton: Princeton University Press, 2009).

[49] See Remmert's chapter in this volume, as well as Moritz Epple, Andreas Karachalios, and Volker R. Remmert, "Aerodynamics and Mathematics in National Socialist Germany and Fascist Italy: A Comparison of Research Institutes," in Sachse and Walker, 131–158.

[50] For mathematical publishing, also see Reinhard Siegmund-Schultze, *Mathematische Berichterstattung in Hitlerdeutschland. Der Niedergang des 'Jahrbuchs über die Fortschritte der Mathematik'* (Göttingen: Vandenhoek & Ruprecht, 1993) and Volker Remmert, "Mathematical Publishing in the Third Reich: Springer-Verlag and the Deutsche Mathematiker-Vereinigung," *The Mathematical Intelligencer*, 22/3 (2000), 22–30.

[51] See Remmert's chapter in this volume.

had common concerns about war research. Ramsauer and Süß not only cooperated, but also Süß repeated Ramsauer's comparison of Anglo-Saxon and German physics (see later) for mathematics and got similar results. When Süß sent to leading figures in the NS state a privately printed copy of the speech he gave at the 1943 Salzburger Rector Conference, he received a favorable response; for example, from Heinrich Himmler, head of the *Schutz-Staffel* (SS). Süß was able to make the DMV into an effective political instrument precisely because he had been so accommodating with regard to the "Jewish question."[52]

The DPG acted in a significantly different way during the early years of the Third Reich. It was able to retain a relatively high degree of autonomy and avoid being incorporated into the NSBDT and the nearly complete *Gleichschaltung* that would have been the result.[53] The Nuremberg Laws of 1935, which did attract international attention and condemnation, did not have a noticeable effect on the DPG's policies toward its non-Aryan members. When the DPG was forced to purge itself of its Jewish members in 1938, it did so without either public or private enthusiasm.

The academies of science in Germany retained some non-Aryan members until 1938 when, like the DPG and the other scientific societies, they were forced to get rid of these members. The Dutch physicist and DPG president Peter Debye, director of the Kaiser Wilhelm Institute for Physics, sent a letter to all DPG members in this regard shortly after the infamous "Night of Broken Glass" (*Reichskristallnacht*), the nationwide pogrom that made brutally clear that the NS government had no intention of tolerating Jews in Germany.[54] The few Jewish scientists left in the DPG by this time were generally scientists who had retired before the National Socialists came to power and had had nowhere else to go. Just because the last Jewish members had not yet been forced out, this did not mean that the DPG was a haven. The DPG policy of pretending that everything was business as usual also precluded showing any solidarity with Jews.[55]

Foreign membership in the DPG had also decreased, due in part to disturbing events like the promulgation of the Nuremberg Laws and their

[52] See Remmert's chapter in this volume.

[53] See Hoffmann's chapter in this volume.

[54] For Peter Debye and the DPG, see Horst Kant, "Peter Debye und die Deutsche Physikalische Gesellschaft," in Dieter Hoffmann, Fabio Bevilacqua, and Roger H. Stuewer (eds.), *The Emergence of Modern Physics* (Pavia: Universita degli Studi die Pavia, 1996), 505–520, and most recently, Horst Kant, "Peter Debye als Direktor des Kaiser-Wilhelm-Instituts für Physik in Berlin," in Hoffmann and Walker, *Fremde*, 76–109.

[55] See Wolff's chapter in this volume.

explicit degradation of German Jews into second-class citizens. Whereas in 1933 REM had insisted that the PAW publicly denounce Einstein, by 1938 it was an embarrassment that the PAW still had Jewish members, and the ministry wanted them out, but quietly.[56] Both the academies and the professional scientific societies were faced with the dilemma of what to do with foreign, corresponding members.[57] It was recognized that they had already lost foreign members – and the valuable foreign currency their dues represented – because of the NS persecution of Jews, and that any attempt to ask foreign members to demonstrate their Aryan status would be disastrous. In the end, the Third Reich put off purging Jews from the foreign memberships in order to help these institutions keep their international scientific relations, contacts, and resulting revenues.

It is important to recognize that the relatively few adherents of Aryan physics were in no way synonymous with the much larger group of German physicists who had embraced National Socialism, either out of conviction or as opportunists. When Debye left Germany in 1939, a small group of young NS physicists within the DPG tried to "nazify" the organization by attempting to get Esau elected DPG president.[58] This was rebuffed by a small group of senior physicists who ran the society, who instead nominated Carl Ramsauer, an industrial physicist with long-standing ties to the German military and a great deal of experience in armaments research.[59]

Ramsauer belonged to the influential technocrats who were accepted by all sides as middlemen between the interests of the scientific–technical community and the pragmatic part of the NS leadership.[60] Ramsauer instituted the NS leadership principle in the DPG, but to his own advantage and in a way that maintained some autonomy vis-à-vis the NS hierarchy. The active role the DPG came to play in armaments research was the result not of pressure from the NS leadership but of the initiative of

56 Walker, *Nazi Science*, 65–122.

57 See the documents section, "Außenpolitik," in Hoffmann and Walker, *Physiker*, 549–556.

58 See Hoffmann's chapter in this volume; also see the documents section, "Gleichschaltung," in Hoffmann and Walker, *Physiker*, 562–577.

59 For Carl Ramsauer, also see Burghard Weiss, "Rüstungsforschung am Forschungsinstitut der Allgemeinen Elektricitätsgesellschaft bis 1945," in Helmut Maier (ed.), *Rüstungsforschung im Nationalsozialismus. Organisation, Mobilisierung und Entgrenzung der Technikwissenschaften* (Göttingen: Wallstein, 2002), 109–141.

60 See Hoffmann's chapter in this volume and Helmut Maier, "Einleitung," in Maier (ed.), 7–29.

the DPG leadership and other leading physicists.[61] Thus, the DPG continued the gradual process of *Selbstgleichschaltung* (self-coordination or self-synchronization).[62]

Both the massive German rearmament that began in earnest in 1936 (in several areas this had begun as soon as the National Socialists came to power) and, in particular, the start of war in September 1939 provided German physics with its bargaining chip vis-à-vis the NS.[63] Ironically Ramsauer, who was chosen to avoid politicization of the DPG, in fact used his office to become a spokesman and advocate for the use of physicists and physics research in the service of the German war effort.

When the war began to turn sour for Germany, physicists took the initiative and argued that, if the government increased its support for physics, then physical research could help turn the tide and win the war. This was a new role for the DPG. Previously, it had not attempted to influence physics research or science policy directly, but rather had restricted itself to overseeing physics publications and organizing meetings and opportunities for scientists to present their work.[64] In 1942, Ramsauer submitted a long memorandum to REM – with copies strategically sent to other high-ranking military and political officials – on the dangerous decline of German theoretical physics and the challenge presented by Anglo-American physics. Although Ramsauer used this report to attack Aryan physics, this danger had already passed for the physics community. Ramsauer's main goal was both to mobilize German physics for the war effort and to use the war effort to increase the prestige of and material support given to physics and physicists. Aryan physics had become a straw man.

Arguably, the single greatest effect National Socialism had on physics was the mobilization of physics for war – not Aryan physics. Very many scientists were involved in various armament or armament-related projects; others had to take over the training of the scientists and engineers German industry was clamoring for to fuel and equip the German war machine. This culminated in the 1944 DPG program for the expansion of physics.[65] Here, as in Ramsauer's 1942 memorandum, the DPG argued that even a small number of researchers could make a big

[61] See Hoffmann's chapter in this volume; also see the documents section, "Selbstmobilisierung," in Hoffmann and Walker, *Physiker*, 592–635.

[62] See Beyler's and Hoffmann's chapters in this volume.

[63] See Beyler's and Hoffmann's chapters in this volume.

[64] See Simonsohn's chapter in this volume.

[65] See Hoffmann's chapter in this volume.

difference.[66] Ramsauer emphasized the central significance of physics and advocated a rational use of physicists for the war effort (i.e., as scientists not soldiers).[67] This program was also a blueprint for the rebuilding of physics after the war, although Ramsauer and the other advocates were careful to note that, of course, this would only take place after the inevitable German victory.

This was the dilemma of self-mobilization and the consequence of *Selbstgleichschaltung*: Under Ramsauer, the DPG left its quiet niche and took up an ever more active role in the social–political process of the Third Reich. Because this undoubtedly had the effect of stabilizing the system and extending the war, it can hardly be considered opposition, let alone resistance. Although Ramsauer was able to successfully push the interests of the DPG and physics in general, the NS dictatorship also in this way significantly profited from his competence as a supposedly apolitical expert – whether Ramsauer wanted this or not. "That he had thereby placed himself in an unholy alliance with the military, helped extend the war, and moreover had the effect of stabilizing the system, either did not bother scientists like Ramsauer, or they never reflected upon it."[68]

After the Allied victory and division of Germany into different occupation zones, German physics – like all parts of German society – was to be denazified and demilitarized. Despite the official rhetoric, apparently only a few German physicists greeted the Allied invasion as a liberation of Germany.[69] The DPG was dissolved. Gradually, regional physical societies were created within the various western zones. The most important one was established in Göttingen, arguably the best place to reestablish German science and physics because of the strong support provided by British officials. Eventually, these societies merged together to form a reconstituted West German DPG once the two German states had been founded.[70]

[66] See Simonsohn's chapter in this volume.

[67] See Hoffmann's chapter in this volume.

[68] See Dieter Hoffmann, "Carl Ramsauer, die Deutsche Physikalische Gesellschaft und die Selbstmobilisierungder Physikerschaft im 'Dritten Reich,'" in Maier, 273–304, here 282–283, translation by Mark Walker.

[69] See Hentschel's chapter in this volume.

[70] See Rammer's chapter, as well as Klaus Hentschel and Gerhard Rammer, "Kein Neuanfang: Physiker an der Universität Göttingen 1945–1955," *Zeitschrift für Geschichtswissenschaft*, 48 (2000), 718–741; Klaus Hentschel and Gerhard Rammer, "Physicists at the University of Göttingen, 1945–1955," *Physics in Perspective*, 3/1 (2001), 189–209; Klaus Hentschel and Gerhard Rammer, "Nachkriegsphysik an der Leine:

Norbert Frei's model of "Amnesty, Integration, and Demarcation" for the post-war era fits the physics community and the DPG well.[71] Physicists tried to overcome[72] the NS past more by denying or forgetting it than by really trying to exclude colleagues who had been part of the NS movement or punish them for what they had done under Hitler.[73] Two different things interacted: the formal denazification process and the network of colleagues lending each other support.[74] Foreign colleagues who visited Germany observed that rather than being self-critical, German scientists were full of "self-pity." "Under this sign of a flight from reality and repression, the weighing of a stranger's suffering against one's own, and an inability to grieve, a sincere and effective 'denazification' was condemned to failure."[75]

Max von Laue became the first president of the DPG in the British zone because he was politically the best choice: As a known opponent of National Socialism, he had the support of the British and the physicists outside of Germany. Ironically, he went on to betray this trust. Von Laue and the other physicists running the DPG were more concerned with scientific performance and collegiality than with what a colleague had actually done in the Third Reich, let alone justice.[76] As a result of this policy, a few uncollegial colleagues were ostracized from the university landscape and physics community while professionally highly qualified former National Socialists were welcomed back. Von Laue and others argued that exceptional scientific performance outweighed past political mistakes.

As demonstrated by the example of Otto Hahn as post-war president of the new Max Planck Society, this was typical of the German scientific community.[77] This was all about rebuilding German science, not punishment. It was correctly assumed that these scientists would not again advocate National Socialism.[78] However, when Wolfgang Finkelnburg,

Eine Göttinger Vogelperspektive," in Dieter Hoffmann (ed.), *Physik im Nachkriegs-Deutschland* (Frankfurt: Harri Deutsch, 2003), 27–56.

[71] Norbert Frei, *Vergangenheitspolitik. Die Anfänge der Bundesrepublik und die NS-Vergangenheit* (München: DTV, 1999).

[72] See Hentschel's and Rammer's chapters in this volume.

[73] See documents section, "Nachkriegszeit," in Hoffmann and Walker, *Physiker*, 636–658.

[74] See Rammer's chapter in this volume.

[75] See Hentschel's chapter in this volume, from which this quotation is taken.

[76] See Rammer's chapter in this volume.

[77] See Walker, "Otto Hahn."

[78] See Rammer's chapter in this volume.

Ramsauer's handpicked second-in-command of the DPG during the war, asked the DPG for help in getting a job, he was rebuffed. It was not the DPG that determined who got jobs but rather the informal network of influential physicists.[79]

When Finkelnburg, Ramsauer, and Ernst Brüche, editor of the informal DPG journal *Physikalische Blätter* during and after the war, faced denazification in the post-war period, they returned once again to the straw man of Aryan physics.[80] Ramsauer argued that the desperate period of National Socialism had been good for one thing: it had made German physics self-reliant and allowed the rebuilding of German physics to begin.[81] Brüche and his journal played a central role in rewriting the history of the DPG.[82] In particular, Ramsauer's memorandum was reprinted without the passages tainted by the *lingua tertii imperii* – but without telling the reader that anything had been cut out.[83] This intellectually dishonest act speaks volumes about how Ramsauer saw his own actions during the Third Reich.[84]

During the NS period, the German physics community was "able to pursue and advance its own agenda under its own self-selected leadership; not always and everywhere, but nevertheless sometimes and in some ways."[85] The DPG neither resisted nor capitulated to National Socialism, and indeed it would be misleading to assume such a black-and-white dichotomy. When Jewish colleagues lost their positions and left Germany, the DPG did not rush to throw them out, but when they did leave, the society pretended that nothing unusual had happened. When Johannes Stark tried to seize control of the society, the DPG rebuffed him both through courageous acts like von Laue's public address and through not so courageous acts like Mey's obsequious public speeches honoring the Führer. When the Max Planck Medal was politicized, the DPG took care only to take steps when its overseers in the REM approved, but also only awarded it to deserving physicists. When the DPG was forced to purge its last remaining Jewish members, it did so in an awkward way that betrayed a lack of enthusiasm. When the DPG was threatened with

[79] See Hentschel's chapter in this volume.
[80] See Eckert's chapter in this volume.
[81] See Hentschel's chapter in this volume.
[82] See Hentschel's chapter in this volume.
[83] See Klemperer.
[84] See Hoffmann's chapter in this volume.
[85] See Beyler's chapter in this volume, from which this quotation is taken.

complete nazification, it responded by mobilizing itself for the war effort. When the war was over and National Socialism was gone, the physicists who had led the DPG and who were trying to reestablish it suppressed the truth, created self-serving myths and legends, and left the telling of its history under Hitler for another day.

2

Boundaries and Authority in the Physics Community in the Third Reich

Richard H. Beyler

A commonplace in discussions of the history of science in the National Socialist era is the question whether science "retained its freedom" under National Socialism or not. This question was especially prevalent in the immediate post-war period in critiques of and apologias for conduct among the historical actors themselves, above all in the context of denazification, but the question has also spilled over into more recent historiographical discourse. There have been three predominant kinds of response: straightforward affirmation, straightforward denial, and turning the question against the applicability of the concept of "freedom" altogether. The straightforward "yes" and "no" have been conventional answers among the mainstream of historically interested scientists and the general public. Thus much of the post-war discussion of the fate of science in the Third Reich involved claims and counterclaims about whether science as a whole or specific scientific institutions constituted "islands of freedom," heroically resisting the flood tide of Nazism, or, alternatively, had become sites of capitulation to or even collaboration with Nazi crimes and stupidities.[1]

A third, "neither–nor," response has come to the fore among the professional historians of science. In this viewpoint, posing the question whether science was "free" under Nazism or not raises more problems

[1] Armin Hermann, *Werner Heisenberg 1901–1976* (Bonn: Inter Nationes, 1976), 44–53, quoting Fritz Bopp on 45; see Herbert Mehrtens, "Kollaborationsverhältnisse: Natur- und Technikwissenschaften im NS-Staat und ihre Historie," in Christoph Meinel and Peter Voswinckel (eds.), *Medizin, Naturwissenschaft, Technik und Nationalsozialismus: Kontinuitäten und Diskontinuitäten*, (Stuttgart: Verlag für Geschichte der Naturwissenschaft und der Technik, 1994), 13–31.

than it solves; inter alia, historians have good reasons to be skeptical of the notion that science was "free" either before 1933 or after 1945. From this viewpoint, even to pose the question "Did science retain its freedom?" represents an uncritical acceptance of actors' categories, which must be unpacked by historians. Although more or less applicable to the sciences as a whole, this schema of responses can be found perhaps most clearly in the historiography of German physics and the German Physical Society (*Deutsche Physikalische Gesellschaft*, DPG) as one of its institutional embodiments.

Though it is generally useful for historians to treat actors' categories with critical skepticism, it is too simple to dismiss the discourse of freedom as *mere* rhetoric. The notion had historically real effects because the historical actors believed it to be important to their professional identity. Assertions on the part of physicists that the discipline of physics maintained relative autonomy or independence during the Third Reich seem linked to a rather distinct conception of a separate community within the broader society of the Third Reich, a community that had, to some extent at least, maintained its distinctiveness vis-à-vis the ostensibly totalitarian state.[2] The question to analyze then becomes what the historical actors saw as constituting the freedom of science. If we seek to operationalize the term, to inquire what *work* the notion of freedom or independence did for scientists and their social constituency, we find some interesting results, especially at historical turning points when the functional meaning of the concept was in flux. At these historical turning points, the voluble claims

[2] I use the term "totalitarian" to refer to the self-identification of the National Socialists as the agents of a total transformation of the political and social structure of Germany. I am hereby not committing myself one way or another to the long-running debate over the general concept of "totalitarianism" derived, primarily, from the work of Hannah Arendt. It should be emphasized that the translation of the National Socialists' totalizing rhetoric into the political and social reality was, at best, problematic; a more historically accurate description of the political structure of Nazism is that it was "polycracy" and that the response of the populace to claims of total allegiance was uneven at best. For example, see Martin Broszat, *The Hitler State: The Foundation and Development of the Internal Structure of the Third Reich*, trans. John W. Hiden (London: Longman, 1981); Detlev J. K. Peukert, *Inside Nazi Germany*, trans. Richard Deveson (New Haven: Yale University Press, 1987). For discussions specific to the history of science, see, for example, Kristie Macrakis, *Surviving the Swastika: The Kaiser-Wilhelm-Society 1933–1945* (New York: Oxford University Press, 1993), 5; Monika Renneberg and Mark Walker, "Scientists, Engineers, and National Socialism," in Monika Renneberg and Mark Walker (eds.), *Science, Technology, and National Socialism* (Cambridge: Cambridge University Press, 1994), 1–29; Richard H. Beyler, "Targeting the Organism: The Scientific and Cultural Context of Pascual Jordan's Quantum Biology, 1932–1947," *Isis*, 87 (1996), 248–273.

of scientists that they were defending the freedom of science indicates that they conceived of themselves as part of a community with boundaries and patterns of authority that were distinct enough to be defended against restructuring in new and difficult political circumstances.

But what, exactly, did they conceive of themselves as defending? One well-known German physicist defined the freedom of science thus:

> [Free scientists] could choose their problems entirely according to their own criteria and carry out their work without interference from whatever side; they were not liable for an accounting thereof.... The government did not mix itself up with the research work at all. [See later for source]

Operationalized in this way, freedom in science meant the power to pursue a research agenda without external constraint of any form. As this author admitted, in a world of finite resources, this could never literally be true at least as far as matériel was concerned, but even under financial limitations the ability to determine the direction and content of work was the primary criterion of freedom. In particular, outsiders such as political figures had nothing to say to free scientists to contradict or limit their range of action within the domain of scientific inquiry. Such, at least, was the vision.

This vision of the freedom of science stems from Johannes Stark and is taken from the section of his memoirs dealing with his self-described "Struggle for Freedom of Research."[3] Of course, we should take Stark's representation with a very big grain of salt. Inter alia, what counted as governmental "interference" for Stark was not a neutral judgment: Opposition to scientific causes that he deemed correct counted as interference, but support for causes that he deemed correct did not. However, prima facie, Stark's version of the freedom of science, and of what constituted outsiders' interference with that freedom, was not so different from that of other leaders of the physics community. Both Stark and his antagonists in the physics community claimed to be defending the borders of that community against untoward outside influences. Within this shared framework of community autonomy as a goal, multiple interpretations of the same events were possible; both sides in an intramural rivalry could, and did, present their positions as defending the best interests of the freedom of science.[4]

[3] Johannes Stark, *Erinnerungen eines deutschen Naturforschers*, ed. Andreas Kleinert (Mannheim: Bionomica Verlag, 1987), 123.

[4] A thought-provoking article on the relevance of multiple perspectives in the historiography of science is Klaus Hentschel, "What History of Science Can Learn from Michael Frayn's 'Copenhagen,'" *Interdisciplinary Science Reviews*, 27 (2002), 211–216.

The statement on the freedom of science quoted earlier, if taken literally, describes a state of complete anarchy. Patently that was not the reality, however, nor the state of affairs that Stark evidently desired. There are at least three tacit moderating assumptions behind this understanding of the freedom of science. First, it depends on some consensus about what questions are worth pursuing and who are the relevant parties to this consensus.[5] If freedom of action *within* the scientific domain is ideally to be minimally constrained, demarcating the borders of the scientific community – or more precisely, deciding who has access to decision making within that domain – becomes a crucial question. A second assumption is, therefore, that not everyone within the community is intended to have this grandiose degree of freedom, but only those whose authority and responsibility is already well established. To put it bluntly, Stark is talking not about the freedom of research but about the freedom of full professors and institute directors to do research – an assumption that was widely shared among the German physics community in this period. (This is not to say the German physics community was unique in this regard.) Freedom of science meant above all freedom for leading scientists, and so defining the elite within the community was also a critical question. A third tacit assumption was that ultimately the freedom of science defined in this way was underwritten by support, material and otherwise, from members of the broader society (i.e., the various constituencies or patrons of physics). Under ideal circumstances, the terms under which this support was provided were more or less clear to all parties involved, and hence the existence and nature of the broader societal underwriting of science remained largely unspoken. In reality, however, the basis for this pact between the scientific community and the broader society might be up for renegotiation or even stronger challenges. This was particularly the case in the Nazi era: If the boundaries and internal authority of the scientific community were to be maintained, quite different kinds of quid pro quos were expected.

From this perspective, we might ask what the history of the DPG in the National Socialist era reveals about the perceived ideals, norms, and patterns of authority of the physics community. Scientists' traditional, often unspoken sense of identity as members of that community was suddenly and explicitly brought into question. The relationship between the scientific community's set of values and the formal social institutions of

[5] On this form of consensus-based demarcation" see Thomas F. Gieryn, "Boundary-Work and the Demarcation of Science from Non-Science: Strains and Interests in Professional Ideologies of Scientists," *American Sociological Review*, 48 (1983), 781–795.

science was, likewise, opened up for debate and alteration. Physicists' responses to these challenges to their community's traditional norms and patterns of authority were varied and sometimes mutually contradictory. Neither the responses of the DPG as an institution to political exigencies of the Nazi regime nor the responses of its individual members were unified phenomena. Rather, the arrival of Nazism compelled a complex set of reexaminations and renegotiations of the social role of science. These renegotiations revolved around terms of values, norms, and so forth, but they also implicitly or explicitly delineated domains of social authority. The ostensible defense of the freedom or independence of science in the Third Reich was not the defense of anarchy, but rather the attempt to affirm traditional patterns of authority within the community and traditional patterns of authority beyond the community.

As scientists and historians later sought to describe this story, they fastened upon different aspects of it, reflecting a wide range of understandings of the fundamental characteristics of the German physics community and, indeed, of scientific communities in general. Hence, rather than a single history, one can perhaps better speak of histories of the DPG under Nazism. As historians, we can examine these histories critically but also take seriously the ways they embody the social reality of professional identity. Obviously, if one examines this historical record, the notion of freedom, or the lack thereof, was central to this professional self-identity of many German physicists; what, then, did this concept mean as it was used in the social and political context of Nazism and its aftermath?

Several interrelated events in the history of the DPG in the Third Reich can illuminate this problem. This essay analyzes three in particular: the conflict over Johannes Stark's candidacy for the DPG presidency in 1933 and the opposition spearheaded by Max von Laue; the so-called Aryan physics (*Deutsche Physik*) controversy as it played out in the mid to late 1930s; and the selection of Carl Ramsauer as DPG president. (Another example, considered in a separate essay in this volume, is the background to the awarding or non-awarding of the DPG's prestigious Max Planck Medal from 1933 to 1942.[6]) These episodes suggest that the notion of freedom of science under Nazism was closely connected to the defense of community boundaries and authority patterns. But it is useful first to consider how the historiographical perspectives of skepticism, resistance, and capitulation might regard these historical episodes.

[6] See the essay by Beyler, Eckert, and Hoffmann in this volume.

HISTORIOGRAPHY OF SKEPTICISM

History of science rose as a profession roughly one-half century ago; the job of the historian of science, as it was frequently conceived of then, was as a kind of mediator between science, seen as a valuable social entity that followed its own relatively autonomous norms, and a broader public audience of nonscientists. When these autonomous norms were violated (e.g., for political reasons), science was assumed to be in trouble.[7] Later historians, however, began to challenge these notions of freedom and autonomy and to ask, skeptically, whether it is legitimate to separate science from the concerns of politics, economics, and so forth. In this approach, assertions of the "freedom" of science are treated as a rhetorical tactic, which may be useful for self-image and public relations, but which do not describe a social reality in which the choices and behavior of scientists are more or less confined by their relation to the state, industry, and other powerful constituencies.

Thus, for example, an "ideology of non-ideology" emerged in West Germany after 1945, under which science was promoted as a domain of objective, neutral inquiry and hence as such mutually incompatible with ideology per se, particularly as manifested in totalitarianism. But such assertions of political disinterestedness served patently political ends in the nascent Federal Republic: bracketing off of the Nazi past in the service of establishing a viable new political culture, and also countering the emergent Cold War enemy to the east while undergoing integration into the Western European/North Atlantic alliance. One should note that many of these expressions were, figuratively and literally, carried out under American sponsorship. For example, the Congress for Cultural Freedom, whose first meeting was demonstratively held in the divided Berlin in 1950, and whose 1952 meeting in Hamburg had the theme "Wissenschaft und Freiheit," was, it turns out, largely funded by the Central Intelligence Agency. The picture of scientific autonomy in the post-war era was painted against a background that depicted a grave and nearly fatal challenge to that autonomy under Nazism. Seeing the political charge carried by "science and freedom" in the post-war era, one

7 On the relevant history of the history of science profession, see, for example, Jessica Wang, "Merton's Shadow: Perspectives on Science and Democracy," *Historical Studies in the Physical and Biological Sciences*, 30 (1999), 279–306; Mark Walker, "Introduction: Science and Ideology," in Mark Walker (ed.), *Science and Ideology: A Comparative History* (London: Routledge, 2003), 1–16.

might have good reason to be skeptical about claims of freedom in the Nazi era.[8]

A variant of this argument suggests that "freedom" may have a certain limited reality but only within, because of, and constrained by a broader social and political context. A classic book that contributed strongly to this line of argument a generation ago was Fritz Ringer's *Decline of the German Mandarins*, which showed how the supposed "apolitical" nature of German scholars was itself the product of a kind of bargain with the state. Other scholars of German academic and professional life have elaborated this line of inquiry. Konrad Jarausch titled his study of German lawyers, teachers, and engineers *The Unfree Professions* to call attention to a disparity between the self-identification of the "free professions" (*freie Berufe*) and their close relations to the German state.[9]

Ringer and Jarausch do not look at natural scientists specifically, but others have analyzed how the notion of apolitical and therefore autonomous scholarship played out in the natural sciences, usually finding some resonances but also a need for modifications of Ringer's thesis. In his detailed study of German geneticists, Jonathan Harwood found applicability, but also limitations, of Ringer's analysis; Harwood argued that a significant minority of German geneticists did *not* espouse the apolitical idea but rather more of a socially engaged science.[10] Mitchell Ash likewise argues for a more differentiated understanding of scientific autonomy, as applicable not to science in general but to specific purposes and with specific benefits and costs. Science thus does not operate separately from politics, but rather uses politics as a resource, just as science can become, in turn, a resource for politics.[11] Even more pointedly, historian Herbert

[8] Richard H. Beyler and Morris F. Low, "Science Policy in Post-War West Germany and Japan between Ideology and Economics," in Walker, *Science and Ideology*, 97–123.

[9] Konrad H. Jarausch, *The Unfree Professions: German Lawyers, Teachers, and Engineers, 1900–1950* (New York: Oxford University Press, 1990); see also Geoffrey Cocks and Konrad H. Jarausch (eds.), *The German Professions 1800–1950* (New York: Oxford University Press, 1990); Charles E. McClelland, *The German Experience of Professionalization: Modern Learned Professions and Their Organizations from the Early Nineteenth Century to the Hitler Era* (Cambridge: Cambridge University Press, 1991).

[10] Jonathan Harwood, *Styles of Scientific Thought: The German Genetics Community 1900–1933* (Chicago: University of Chicago Press, 1993); Jonathan Harwood, "The Rise of the Party-Political Professor? Changing Self-Understandings among German Academics, 1890–1933," in Doris Kaufmann (ed.), *Geschichte der Kaiser-Wilhelm-Gesellschaft im Nationalsozialismus, Volume I* (Göttingen: Wallstein, 2000), 21–45.

[11] Among various essays by Mitchell G. Ash, exemplary here is "Wissenschaft und Politik als Ressourcen für einander," in Rüdiger vom Bruch and Brigitte Kaderas (eds.), *Wissenschaften und Wissenschaftspolitik: Bestandaufnahmen zu Formationen, Brüchen,*

Mehrtens has termed the arrangements between science and the Nazi state "relationships of collaboration" (*Kollaborationsverhältnisse*), his point being that these relations aimed at and resulted in mutual benefit to both sides.[12] In this vein, assertions of the freedom of science are simply red herrings. Empirical confirmations of a reciprocal relationship between science and the political sphere, also or especially during the Third Reich, can be found in the voluminous research produced by the Presidential Commission of the Max Planck Society.[13]

Focusing more specifically on physics, Alan Beyerchen's classic account describes that discipline in the Third Reich as pushing back against ideologically grounded interference, but at a steep price, leaving an ambivalent judgment about the viability of professional freedom. Any illusions the scientists – or Beyerchen's readers – might have about the supposed political innocence of science were quickly dispelled.[14]

In *Dilemmas of an Upright Man*, John Heilbron's portrait of Max Planck is that of someone highly committed to a strong, austere set of values centered on a sense of duty – a loyalty to state authority that transcended politics. Heilbron interprets Planck's actions in the Nazi years as an attempt to salvage, as from a shipwreck, what remained of this value system of loyalty as it came crashing down around him. The story is a tragic one because it is the very qualities that made Planck "upright" in more normal times that prevented him from opposing more directly the National Socialists; nevertheless, Heilbron's tenor is not one of condemnation. Planck appears as the guardian or "spokesman" for a community that valued objectivity and the disinterested pursuit of truth but that was crushed under a Nazi regime that decidedly did not share those values. The terms of the pact between scholarship and the state had changed

und Kontinuitäten im Deutschland des 20. Jahrhunderts (Stuttgart: Franz Steiner Verlag, 2002), 32–51.

12 Mehrtens, "Kollaborationsverhältnisse"; see his earlier path-breaking essay, "Das 'Dritte Reich' in der Naturwissenschaftsgeschichte: Literaturbericht und Problemskizze," in Herbert Mehrtens and Steffen Richter (eds.), *Naturwissenschaft, Technik, und NS-Ideologie: Beiträge zur Wissenschaftsgeschichte des Dritten Reich* (Frankfurt am Main: Suhrkamp, 1980), 15–87.

13 The results of this research have appeared in the book series Geschichte der Kaiser-Wilhelm-Gesellschaft im Nationalsozialismus (Göttingen: Wallstein, 2000–2007) and in the preprint series Ergebnisse (available on-line at www.mpiwg-berlin.mpg.de/KWG/publications.htm#Ergebnisse); for a representative summary of the findings, see Susanne Heim, Carola Sachse, and Mark Walker (eds.), *The Kaiser Wilhelm Society under National Socialism* (Cambridge: Cambridge University Press, 2009).

14 Alan D. Beyerchen, *Scientists under Hitler: Politics and the Physics Community in the Third Reich* (New Haven: Yale University Press, 1977).

too drastically for traditionalists such as Planck to be able to respond effectively.[15] In this sense, Heilbron is indebted to a historiography of heroic resistance as an abstract ideal but finds Planck unable, because of his biographical–historical circumstances, to offer this resistance. Planck, in any event, did not simply capitulate to the Nazi worldview, which is more (in Heilbron's analysis) than can be said for some of Planck's colleagues.[16]

A more critical tenor is heard in David Cassidy's reading of the career of Werner Heisenberg. As the title *Uncertainty* suggests, for Cassidy's Heisenberg, unlike Heilbron's Planck, the choices were by no means clear. Heisenberg's initial expectation was that the Nazis would soon fail, and hence he temporized about any substantial change of pattern in interacting with the state. No explicit confrontation with the regime was necessary because it was assumed that interactions between the physics community and the broader society would soon return to a more "normal" state. As Cassidy points out, this was not an unreasonable conclusion given Germany's political history. But to have looked more deeply into the situation would have required a "political sensibility and commitment to democracy that did not exist" with Heisenberg or, by implication, with many of his colleagues.[17] It should be noted that this point of view was by no means absent among "non-Aryans" and that even some of those driven from their jobs in 1933 for several months thereafter apparently maintained some hope that the status quo ante would soon return.[18]

Helmuth Albrecht argues that the loyalty to the state of a figure such as Max Planck allowed, or even demanded, a continued collaboration with the state under National Socialism in a way that was beneficial to scientific institutions by most conventional measures.[19] Albrecht's views

[15] John L. Heilbron, *Dilemmas of an Upright Man: Max Planck as Spokesman for German Science* (Berkeley: University of California Press, 1986).

[16] See John L. Heilbron, "The Earliest Missionaries of the Copenhagen Spirit," in Edna Ullmann-Margalit (ed.), *Science in Reflection* (Dordrecht: Kluwer Academic, 1988), 201–233.

[17] David C. Cassidy, *Uncertainty: The Life and Science of Werner Heisenberg* (New York: W. H. Freeman, 1992), here 305; also see David C. Cassidy, *Beyond Uncertainty: Heisenberg, Quantum Physics, and the Bomb* (New York: Bellevue Literary Press, 2009).

[18] See Skúli Sigurdsson, "Physics, Life, and Contingency: Born, Schrödinger and Weyl in Exile," in Mitchell Ash and Alfons Söllner (eds.), *Forced Migration and Scientific Change* (Washington: German Historical Institute, 1996), 48–70.

[19] Helmuth Albrecht, "'Max Planck: Mein Besuch bei Adolf Hitler'–Anmerkungen zum Wert einer historischen Quelle," in Helmuth Albrecht (ed.), *Naturwissenschaft und Technik in der Geschichte: 25 Jahre Lehrstuhl für Geschichte der Naturwissenschaft und Technik am historischen Institut der Universität Stuttgart* (Stuttgart: Verlag für Geschichte der Naturwissenschaften und der Technik, 1993), 41–63, here 53–54.

come close to the "capitulation" perspective – about which more in a moment – but they are informed by a skepticism toward the assumption that science was ever noncollaborative to begin with.

HISTORIOGRAPHY OF CAPITULATION

The historiographical perspectives considered so far thus proceed from the assumption that a well-defined domain of free, socially autonomous science was illusory or, at best, highly problematic in its translation from ideal to reality. By contrast, the negative argument that science did not preserve its freedom under Nazism presupposes that this freedom was there to begin with. In this narrative, science appears at best a hapless passenger and at worst a willing co-pilot of the Nazi juggernaut. Once innocent, science had either been co-opted by or deliberately aligned itself with the Nazi regime. Those who have taken this position have argued for one or both of two possible outcomes that are, strictly speaking, mutually incompatible. On the one hand, some authors have pointed to ways in which highly competent scientists made contributions of knowledge and technique to the Nazi cause; a powerful example from the early post-war years is Max Weinreich's *Hitler's Professors*.[20] On the other hand, others have argued that the result of this capitulation to Nazi ideology was a pervasive incompetence, or a diversion into pseudoscience. Perhaps the most notorious example of the incompetence thesis has been Samuel Goudsmit's interpretation of German atomic research in *Alsos*.[21] The Aryan physics affair, to be discussed later, has sometimes been interpreted as an example of the pseudoscience argument, but events surrounding the careers of Philipp Lenard, Johannes Stark, and their followers also have provided material for a narrative of resistance to Nazism.

HISTORIOGRAPHY OF RESISTANCE

Typically, the argument that science preserved its freedom or its autonomy under Nazism has functioned within a heroic narrative. The narrative rests on (at least) three generally unspoken assumptions: first, that "freedom," however it is defined, constitutes an integral part of the scientific

[20] Max Weinreich, *Hitler's Professors* (New York: YIVO, 1946; reprinted New Haven: Yale University Press, 1999).

[21] Samuel A. Goudsmit, *Alsos* (New York: H. Schuman, 1947; reprinted Woodbury, NY: AIP Press, 1996); see Mark Walker, "Heisenberg, Goudsmit, and the German Atomic Bomb," *Physics Today* (May 1991), 13, 15, 90–92, 94–95.

method, so that loss of "freedom" would mean the demise of science; second, that science represents a good in itself, so that actions taken to "advance" or "preserve" science can unproblematically be given moral approbation; third, that scientific autonomy constituted, ipso facto, resistance to or at least noncompliance with the Nazi regime.

The self-representations of many German scientists after 1945 express this narrative of resistance, particularly so for members of the DPG. In this discourse, the Nazi era represented an attempt on the part of outside forces or agencies to take over a community in which distinct values and norms prevailed that were at odds with those of the outsiders. Against these intrusions, the boundaries of the community were successfully defended – perhaps not everywhere and always, but at least sometimes and in some places, and for this the members of the community have reasons to be proud. Thus Ernst Brüche, the editor of the DPG organ *Physikalische Blätter*, quoted a report by Wolfgang Finkelnburg:

> The Executive board of the German Physical Society... did everything in its power, despite all difficulties and with a great deal of courage... in order to maintain a clean and decent scientific physics in opposition to the Party and Ministry and... to avoid worse things. I believe that this struggle against the Party physics may certainly be designated as a glorious chapter of the true German physics... [22]

Elsewhere, the journal described the work of the DPG undertaken "to save physics in Germany."[23] As Michael Eckert emphasizes, these early examples of the historiography of resistance cannot be understood without reference to the context of denazification, wherein the DPG members in general, and several figures in the DPG leadership in particular, were anxious to ensure their acceptability in the post-war political climate.[24] In a similar vein, theoretical physicist Fritz Bopp said of figures such as Heisenberg and Planck that even if they did little openly to confront the regime, they did create "islands of freedom" in which the norms and values of science could be preserved for future generations, echoing the notion of a community whose boundaries, even if put under pressure or

[22] Ernst Brüche, "'Deutsche Physik' und die deutschen Physiker," *Physikalische Blätter*, 2 (1946), 232–36, on 232; Finkelnburg's report is translated in Klaus Hentschel (ed.), *Physics and National Socialism: An Anthology of Primary Sources*, trans. Ann M. Hentschel (Basel: Birkhäuser, 1996), 339–345.

[23] Editor's introduction to "Eingabe an Rust," *Physikalische Blätter*, 3/2 (1947), 43.

[24] See the essay by Michael Eckert in this volume.

even shrunk, nevertheless remained coherent.[25] Although Heisenberg and Planck were not in much danger of being labeled pro-Nazi, there were nevertheless powerful reasons to bolster their reputations as anti-Nazis as Planck (after his death in 1947) became a potent symbol of the intellectual integrity of science and as Heisenberg increasingly took on the role of a public intellectual in the Federal Republic.[26]

These claims for the freedom of science in the Third Reich were offered by the ideological and professional opponents of Stark, but he also, as noted earlier, asserted after 1945 that he always acted to protect scientific freedom. Stark's memoirs describe confrontations with Nazi "party bigwigs" and claim that though he had alliances with competent administrators in the Reich Interior Ministry (*Reichsinnenministerium*), he sought to fend off interference by incompetents in the Reich Ministry for Science, Education and Culture (*Reichsministerium für Wissenschaft, Erziehung und Volksbildung*, REM).[27] The biases in these post-war statements are close to the surface: Stark put the label of "politician" on those Nazi functionaries with whom he had conflict, but for other party members he downplayed their political affiliations and emphasized their "objectivity" and "sound judgment." Stark also obscured his own political activism and overt anti-Semitism, which dated back to well before 1933.[28]

In the first few years of the National Socialist regime, Stark translated this agenda into a program to build a kind of empire of science, largely under the authority of himself and his close colleagues, to include the presidency of the Imperial Physical-Technical Institute (*Physikalisch-Technische Reichsanstalt*, PTR), control of the German Research Foundation (*Deutsche Forschungsgemeinschaft*, DFG), and the chairmanship of the DPG. He was at least partially successful in that he did become head of the PTR and the DFG, but ongoing conflicts with the education ministry and, most fatally to his career, new conflicts with the *Schutz-Staffel* (SS)

[25] Quoted in Hermann, *Werner Heisenberg*, 49; Hermann himself reaches a more nuanced conclusion, for example in "Die Deutsche Physikalische Gesellschaft 1899–1945," *Physikalische Blätter*, 51 (1995), F61-F106, here F104.

[26] See Albrecht, "Max Planck: Mein Besuch," 41–63; Gerhard Oexle, "Wie in Göttingen die Max-Planck-Gesellschaft entstand," *Max-Planck-Gesellschaft Jahrbuch*, (1994), 43–60; Cathryn Carson, "New Models for Science in Politics: Heisenberg in West Germany," *Historical Studies in the Physical and Biological Sciences*, 31 (1999), 115–171.

[27] Stark, *Erinnerungen*, 118–122.

[28] On Stark's professional relations before 1933, see Beyerchen, *Scientists*, 103–115; Mark Walker, *Nazi Science: Myth, Truth, and the German Atomic Bomb* (New York: Perseus Press, 1995), 6–16.

FIGURE 1. Johannes Stark (*Source*: MPGA)

eventually resulted in his being forced out from these posts.[29] He was, however, unsuccessful in his bid to become chair of the DPG, due in large part to resistance from Max von Laue. This confrontation has sometimes been described as a struggle for the autonomy of science against outside interference, but it can also be seen as a struggle for authority within the physics community in the context of pressures on the external boundaries of that community.

THE VON LAUE–STARK CONFRONTATION

Although its roots reached back well into the Weimar era, the Stark imbroglio became overt at the annual physicists' conference in September 1933 in Würzburg. Stark presented here his bid to add the DPG to his growing empire – if not by becoming chair himself then through election of a Berlin-based candidate who would work closely with Stark in the

[29] Beyerchen, *Scientists*, 118–122; Walker, *Nazi Science*, 31–59.

FIGURE 2. Max von Laue (*Source*: MPGA)

latter's capacity as PTR president. As part of his platform, he proposed a centralized coordination of the major physics journals and of research funding in physics. Although certainly protecting his freedom as defined in his memoirs, the plan just as certainly threatened the authority of other physicists. Notoriously, Stark concluded his lecture to the assemblage with the expression that if they (the physicists) were not willing to accept this plan, he (Stark) would "use force." A group from the extant DPG leadership (i.e., the executive committee) led by Max von Laue had worked behind the scenes to produce a candidate to outflank Stark. They found one in the industrial physicist Karl Mey, who was a Berliner and hence met one of Stark's criteria, but who could certainly be expected to maintain the extant patterns of authority within the society rather than cooperate with Stark's vision of top-down governance.[30]

[30] See Clemens Schaefer to Johannes Stark, copy to Max von Laue, 3 Nov. 1934, in Archiv zur Geschichte der Max-Planck-Gesellschaft (hereafter MPGA), III. Abt., Rep. 50, Nr. 2386, 30. For discussion of the von Laue–Stark confrontation at Würzburg, see Beyerchen, *Scientists*, 115–116; Walker, *Nazi Science*, 18–20; Hermann, "Deutsche Physikalische Gesellschaft," F94–95; Dieter Hoffmann, "Between Autonomy and Accommodation: The German Physical Society during the Third Reich," *Physics in Perspective*, 7/3 (2005), 293–329.

FIGURE 3. Karl Mey (*Source*: DPGA)

Mey's emergence as an alternative candidate was not the result of a grassroots movement but instead the product of a deliberate campaign on the part of the society's leadership. That leadership here proved to be a self-selecting elite group; however, it was precisely this pattern of elite self-selection that guaranteed scientific freedom in their eyes. Indeed, by no means did all DPG members sympathize with von Laue's viewpoint. Jonathan Zenneck wrote to von Laue that some colleagues had expressed to him (i.e., Zenneck) – apparently at the DPG meeting – the view that von Laue had brought a localized concern into the national forum; Zenneck rather pointedly asked them in return if they had "taken leave of their senses" in even countenancing Stark as a candidate for chair.[31] In the end, it was not so difficult for the extant leadership to outmaneuver Stark. Stark, of course, saw in the self-selecting elite everything that was wrong with the profession and presented his own adherence to the leadership principle (*Führerprinzip*) as the surest chance to maximize the freedom

[31] Jonathan Zenneck to Max von Laue, 27 Sep. 1933, MPGA, III. Abt., Rep. 50, Nr. 2203, 1–2.

of the discipline of the whole in the context of the new National Socialist state. In fact, Stark's aim was to substitute for the previous leadership elite an even more centralized form of authority: his own.

Von Laue made his case to the physicists in an address that concluded with an extended reference to the 300th anniversary of Galileo's trial by the Inquisition, clearly a metaphor for contemporary events. Von Laue retold the legend of Galileo's statement: "And yet it moves." While recognizing that this statement was not necessarily part of the factual historical record, von Laue noted that it was "indelible in the vernacular." According to von Laue, Galileo must have been bolstered throughout the trial by the conviction that "Whether I, or any other person now claims it or not, whether political or ecclesiastic power is for or against it, that changes nothing about the facts." And, von Laue remarked, the facts were in the end accepted even by the church that had condemned Galileo.[32]

Note how von Laue describes the confrontation: not between two rival theories that have better or worse value in explaining scientific observations, nor between two rival political systems that are perceived to be either more admirable or more obnoxious, but between facts as such and "political and ecclesiastical" power. Von Laue's specific objection here is thus not Nazi policies as such, as obnoxious as these may have seemed to him. Rather, von Laue was concerned about any external powers whatsoever impinging upon the prerogative of the competent scientist to reach his own conclusions. But who judges competence? Precisely the community of professionals, just the group von Laue was addressing. Moreover, this same group had already, well before the Nazis' arrival in power, passed collective judgment against Stark's competence and marginalized him within the physics community.[33] In their view, it was only Stark's alliance with the Nazis that had led to his PTR appointment, and now he was seeking to become to an even greater degree an agent of this external authority within the community. Von Laue's speech was thus, at least in part, an assertion of the community's right to decide for itself the internal status of its members.

An irony of von Laue's Galileo anecdote might well have escaped him and his audience, but it should not escape later historians of science.

[32] Max von Laue, "Ansprache bei Eröffnung der Physikertagung in Würzburg am 18. September 1933," *Physikalische Zeitschrift*, 34 (1933), 889–890; translated in Hentschel, *Physics*, 67–71.

[33] On Stark's relations with his colleagues before 1933, see Beyerchen, *Scientists*, 103–115; Walker, *Nazi Science*, 6–16; Stark, *Erinnerungen*.

Thanks to recent investigations of Galileo's career, we now have a greater appreciation of the extent to which he was an active participant in the court culture of early modern Italy. Galileo was no political innocent, but he was unable to play the political game successfully to the end.[34] So one moral of Galileo's story might well be that it pays for the scientist to be politically astute, not to be apolitical. Von Laue, however, probably chose the metaphor precisely because he could safely assume his audience would associate the Galileo story with the concept of the apolitical scholar.

Von Laue's lecture was also clearly intended as, and understood as, a rebuke to Stark as someone who had associated himself with these external authorities; namely, in this instance, the Nazi movement. As such, it was deprecated by Stark, who began his own lecture at the Würzburg meeting by saying that he found von Laue's reference to Galileo in this context obscure and incomprehensible, but that there was no need in any event to say in regard to German science "nevertheless, it still moves," as German science was "absolutely free" to begin with.[35] Conversely, it was appreciated by other members of the DPG leadership, notably the successful candidate Mey as well as Zenneck, who had been von Laue's original first choice but who moved aside in favor of Mey, as having the Berliner Mey as chair would deflate the force of Stark's centralization proposals.

Indeed, part of Stark's image as a melodramatic villain depends on the fact that, in the long run, he proved to be an inept politician. He presented a plan to convert the PTR to the "Central Organ" of German physics research, which would offer "advice or help" to the various individual physics research institutes and, more significantly, "mediate" between them. In essence, Stark sought to make his PTR the clearinghouse for physical information in Germany as well as the de facto arbiter of the direction of physics research. The paper presenting the plan is conspicuous in its use of the first-person pronoun. It begins: "On the 1st of May of this year Reich Minister Frick appointed me head of the Reich Institute of Physics and Technology. Thus a great responsibility has been transferred into my hands.... " Besides making this obeisance to Frick, Stark also ostentatiously cites the "genius" of Hitler in expressing a hope

34 See Mario Biagioli, *Galileo, Courtier: The Practice of Science in the Culture of Absolutism* (Chicago: University of Chicago Press, 1993).

35 Johannes Stark, typescript of lecture, MPGA, III. Abt., Rep. 50, Nr. 2387, 1. The published version of this lecture, "Organization der physikalischen Forschung," *Zeitschrift für technische Physik*, 14/11 (1933), 433–435, does not contain this opening passage; translated in Hentschel, *Physics*, 71–76.

for his support.[36] In all these respects, Stark showed an open disregard for traditional community authority and indeed his desire to displace those patterns of authority altogether.

Stark's attempts to bring the weight of outside authority to bear on the internal affairs of physics might have been more effective had he chosen his extramural authorities more astutely. By 1936, Stark had lost his position as head of the DFG and along with it much of his influence. He was pushed from power not so much by his opponents within the physics community but rather by rivals within the National Socialist power structure. There is a consensus among historians that the education ministry, headed by Bernhard Rust and his deputies Rudolf Mentzel and Erich Schumann, was one of the weakest, least dynamic agencies in the Nazi administrative structure. Nevertheless, even Rust and his department were able to outmaneuver Stark when the latter tried to expand his empire in science policy in another direction to include the Kaiser Wilhelm Society. Stark picked as his main ally Alfred Rosenberg, head of the National Socialist German Workers Party (*Nationalsozialistische Deutsche Arbeiterpartei*, NSDAP) Foreign Policy Office and (from 1934 on) the official supervisor of ideological indoctrination for the party.[37] Rosenberg's fulsome statements, however, received relatively little attention among the Nazi hierarchy. Consequently, by the mid-1930s, Stark himself became a negligible factor in the struggle for the direction of German physics and German science generally.[38]

As noted above, in 1934 Stark insisted that his plans for the reorganization of research did not in any way compromise the ideal of freedom of research. And in post-war apologias, Stark interpreted his actions in the early years of the Nazi era as an effort to preserve the autonomy of the profession by using the "leadership principle" (*Führerprinzip*) to fend off extremists from the education ministry, such as Mentzel, who wanted to steer physics toward military research.[39] This apologia was laughed out of court, for example by von Laue in the pages of *Physikalische*

[36] Stark, typescript of lecture, MPGA III. Abt., Rep. 50, Nr. 2387; Stark, "Organization," in Hentschel, *Physics*, 71–76, quotations on 71–72, 75.

[37] Rosenberg's official title was *Beauftragter des Führers für die gesamte geistige und weltanschauliche Schulung und Erziehung der NSDAP*. For simplicity's sake, his position is often referred to simply as the "Rosenberg Office" (*Amt Rosenberg*).

[38] Beyerchen, *Scientists*, 118–122; Walker, *Nazi Science*, 31–59; Hermann, "Deutsche Physikalische Gesellschaft," F96.

[39] Johannes Stark, "Zu den Kämpfen in der Physik während der Hitler-Zeit," *Physikalische Blätter*, 3 (1947), 271–272; Stark, *Erinnerungen*, especially 123–131.

Blätter.[40] But as we shall see, it was not Stark's tactic per se that was objectionable, as other physicists successfully played off one sector of the state or NSDAP apparatus against another and, moreover, successfully sold the idea of physics contributing to the military effort of the Reich. Rather, Stark counted as a villainous failure because in his ambitions of 1933, he revealed himself as a transgressor against extant community authority patterns. Von Laue spelled out his objections to Stark's actions later, in December, in a speech to the Prussian Academy of Sciences opposing Stark's election: Stark sought to set himself up as a kind of "dictator of physics."[41]

Direct confrontation between von Laue and Stark if anything only escalated in the succeeding months. Von Laue's reference, in an obituary for Fritz Haber, to the ancient Greek tale of Themistocles led Stark to ask the DPG board to expel von Laue for libeling the government; after hearing von Laue's strong denials, the board declined to do so and endorsed von Laue's statement.[42] On May 26, 1934, Stark wrote to the board again, claiming that Mey's election had been predicated on some power-sharing arrangement with Stark; after assembling its evidence, the board eventually wrote back denying this assertion vigorously.[43] Finally, Stark engaged in a personal correspondence with von Laue, culminating in a letter of August 8 expressing "my strongest regret" that von Laue's speech in Würzburg had been "for the benefit of Albert Einstein, this traitor and verbal abuser of the National Socialist government, applauded by the Jews and comrades of the Jews present."[44] After consulting a lawyer, von Laue wrote to Stark that he would refuse any further private correspondence.[45]

THE ARYAN PHYSICS STRUGGLE AS COMMUNITY POLITICS

The energy and attention that the physics community expended on the Aryan physics debate in the 1930s have, understandably, been the focus

40 Max von Laue, "Bemerkung zu der vorstehenden Veröffentlichung von J. Stark," *Physikalische Blätter*, 3 (1947), 272–273.

41 Laue, "Bemerkung."

42 Karl Mey to Johannes Stark, 5 Apr. 1934, MPGA III. Abt., Rep. 50, Nr. 2386, 18; discussed in Hermann, "Deutsche Physikalische Gesellschaft," F98.

43 Karl Scheel to Johannes Stark, 11 Oct. 1934, MPGA III. Abt., Rep. 50, Nr. 2386, 28.

44 Johannes Stark to Max von Laue, 21 Aug. 1934, MPGA III. Abt., Rep. 50, Nr. 2386, 19.

45 Max von Laue to Johannes Stark, 1 Sep. 1934, and Max von Laue to Karl Mey, n.d., MPGA III. Abt., Rep. 50, Nr. 2386, 21.

of many historical accounts.[46] The story seems almost like a scripted melodrama, with a clear-cut moral and a corps of villains who would be comical if they were not so filled with menace. The apparently straightforward moral of the story is this: Science as the independent, disinterested pursuit of truth was directly thwarted by ideological interference. The result seems to be an object lesson in the danger of political interference in science and in favor of granting the highest possible degree of autonomy to the scientific community to regulate its own affairs. The chief villain of the piece is, once again, Stark, who combined personal unpleasantness with shortcomings as a physicist in not being able to come to terms with the relativistic and quantum revolutions. The story apparently also has a happy ending: Despite the ideological assault on its integrity, the community was able to preserve itself and its patterns of authority, at least enough to maintain continuity to the end of the Nazi era and, crucially, through the tumult of the end of the war and the immediately following period of reconstruction.

This rendering of the Aryan physics struggle is not as such incorrect, and it has its appeal. However, it is worth exploring several ironies concealed beneath the surface. At first glance, the story is one of an asymmetrical interference by a political entity in an essentially nonpolitical domain, thereby transgressing the boundaries of the physics community. But the Aryan physics struggle was *not* in the first instance the interference of outsiders in the business of the physics community, but rather the attempt by marginalized members of that community to subvert its established patterns of authority while seeking out and relying on allies from the state, party, and so forth. (Nota bene: In seeking out such external allies, their conduct was not formally different from that of the established leaders of the community.) The distinction is crucial from the perspective of community boundary-keeping. Philipp Lenard and Johannes Stark had long since deviated from the physical mainstream, as historians of the episode such as Alan Beyerchen, Steffen Richter, Mark Walker, and (in most detail) Freddy Litten have made abundantly clear. In the 1920s, Stark and Lenard had backed the "wrong" theories, but even worse, had done so ungracefully and in violation of rules of engagement in professional politics. The Aryan physics struggle was in the first place an

46 Beyerchen, *Scientists*; Steffen Richter, "Die 'Deutsche Physik,'" in Mehrtens and Richter, *Naturwissenschaft*, 116–139; Walker, *Nazi Science*; Freddy Litten, *Mechanik und Antisemitismus: Wilhelm Müller (1880–1968)* (Munich: Institut für Geschichte der Naturwissenschaften, 2000); see also the essay by Michael Eckert in this volume.

intramural struggle, in which marginalized members of the physics community sought to regain a place in the status hierarchy, and the professional leadership sought to reaffirm their role as defenders of community norms.

I would hence render more precise Walker's judgment that the movement "was first and foremost political, not scientific."[47] It was indeed political, but it was above all local politics; that is, those of the community of physicists. In the abruptly changed political situation on the national level, both sides in the local struggle sought to advance their cause by appeal to external authorities: the Aryan physicists did so in an unsubtle way that ended up backfiring; their opponents, the traditional community authorities, did so in a more subtle and ultimately more successful way.

Returning to the historiographical frameworks sketched above, the initial response of German physicists to the Aryan physics struggle has conventionally been interpreted by the historical actors and by subsequent historical accounts as an issue of freedom. To have given in to the "Aryan physicists," it is asserted, would have meant abdication by the leaders of physics of the ability to discipline the physics community. But as Litten's account suggests, a takeover by the "Aryan physicists" was never in the cards, not least because they did not compose a coherent movement. Insofar as "Aryan physics" did resemble a movement, its actions were focused on very localized situations.[48] Likewise, Michael Eckert has shown that, apart from the controversies over particular professorial appointments, work in theoretical physics went on "almost as usual."[49] If Litten's and Eckert's readings are correct, the reactions to the Aryan physics struggle can be understood as a generally successful attempt to preserve extant patterns of authority within community boundaries; the picture of a systematic anti-Nazi "struggle" against a coherent pro-Nazi "movement" appeared in the post-war context.

RAMSAUER AS CHAIR OF THE DPG

By the late 1930s, another group of what we might call activist physicists had begun to exert pressure from within the DPG on its leadership.

[47] Walker, *Nazi Science*, 64.

[48] Litten, *Mechanik*.

[49] Michael Eckert, "Theoretical Physics at War: Sommerfeld Students in Germany and as Emigrants," in Paul Forman and José M. Sánchez-Ron (eds.), *National Military Establishments and the Advancement of Science and Technology: Studies in 20th Century History* (Dordrecht: Kluwer Academic Publishers, 1996), 69–86, here 80.

These new pressures were more effective in forcing the DPG to respond for two main reasons. First, these physicists, unlike Stark, could not be faulted for backing unconventional physics – in other words, they had not previously been marginalized within the community. Their concern was primarily a stricter adherence to anti-Semitic policies rather than new scientific practice or organization. They hence left the professional norms of the community, apart from matters of personal inclusion and exclusion, intact. Second, in contrast to Stark, they appeared willing to rely on existing patterns of authority with their proposal of Abraham Esau, a reasonably qualified candidate in terms of his status within the community, for president.

The response from within the DPG was to offer an alternative pathway for alignment with the Nazi state, acceding to certain key demands of the regime but nevertheless maintaining continuity in the society's leadership. Under the renewed pressure from within the society, chair Peter Debye and the DPG leadership circulated letters asking for the resignation of the society's remaining Jewish members. Thereby, the main reformatory demand of the activist group was rendered moot. Likewise, the mainstream leadership offered their own candidate to succeed Debye – he left for a professorship in the United States in 1940 – whose qualifications both locally and nationally were very strong. Carl Ramsauer was an eminent experimental physicist who as chief of the German General Electricity Company (*Allgemeine Elektrizitäts-Gesellschaft*, AEG) research laboratory possessed intimate ties to a major branch of industry. [50] This latter factor was decisive because by the late 1930s, the role of scientific research in bolstering technological capacity was becomingly increasingly clear to figures such as Hermann Göring and Heinrich Himmler. Conversely, the merely ideological offerings of the Aryan physics adherents had become less interesting, or rather, those Nazi leaders who showed most interest in them were losing out in internecine party power struggles.[51]

[50] For further discussion of Ramsauer's chairmanship, see the essays by Stefan L. Wolff and Dieter Hoffmann in this volume.

[51] See Dieter Hoffmann, "Carl Ramsauer, die Deutsche Physikalische Gesellschaft und die Selbstmobilisierung der Physikerschaft im 'Dritten Reich,'" in Helmut Maier (ed.), *Rüstungsforschung im Nationalsozialismus: Organisation, Mobilisierung und Entgrenzung der Technikwissenschaften* (Göttingen: Wallstein, 2002), 273–304; on the context of Ramsauer's election as chair, see also Dieter Hoffmann and Rüdiger Stutz, "Grenzgänger der Wissenschaft: Abraham Esau als Industriephysiker, Universitätsrektor und Forschungsmanager," in Uwe Hossfeld, Jürgen John, Oliver Lemuth, and Rüdiger Stutz (eds.), *Kämpferische Wissenschaft: Studien zur Universität Jena im Dritten Reich* (Cologne: Böhlau, 2003), 136–179.

Ramsauer appointed as deputy Wolfgang Finkelnburg, a party member who had participated in the defense of modern theoretical physics in discussions with members of the Aryan physics group and hence united qualities expected by all major sides of the local political struggle. Ramsauer commented explicitly on the political strategy behind his choice of Finkelnburg in a letter to Ludwig Prandtl: "I considered it necessary to have a modern party member on the board of trustees, in order to be able to cope better with the extreme-minded colleagues."[52]

Ramsauer's rationale was frequently echoed in post facto reflections on the Nazi era, but commonly with a new nuance. Thus Otto Scherzer, in a 1965 memoir in a symposium at the University of Tübingen, wrote:

> Every larger institute needed . . . someone who could reassure the Party about the political acceptability of the members of the institute. . . . From the colleagues, who had come to the Party in such a way, soon came the call: Help us make the Party decent from the inside. We will not be rid of the Party. If you stay off to the side, then it will be your fault if rowdies and ignoramuses dominate the powerful Party apparatus.[53]

In this view of the development of relations between science and the state under Nazism, those individuals who mediated between the scientific community and the party were thus making a kind of self-sacrifice, dirtying themselves politically in order to grant to other community members a relatively greater degree of freedom.

But was this sense of contamination necessarily part of the interaction between the DPG and the state under Ramsauer's leadership? We have no access to his or Finkelnburg's inner feelings. Ramsauer's representations to the regime of the value of a self-governing physics community were grounded predominantly in the usefulness of science to the state, especially in the context of war and, more broadly, international competition. Simply put, the modern state at war needed physics and its applications to survive. But this approach cannot be taken as an extraordinary, novel set of actions evoked by the vagaries of the Nazis; rather, it continued assumptions about the social role of science present long before 1933. Ramsauer had worked extensively on military-related research at various points in his career. His approach to the Nazis was thus not a novel

[52] Carl Ramsauer to Ludwig Prandtl, 4 Jun. 1941, quoted in Hentschel, *Physics*, 268.

[53] Otto Scherzer, "Physik im Totalitären Staat," in Andreas Flitner (ed.), *Deutsches Geistesleben und Nationalsozialismus: Eine Vortragsreihe der Universität Tübingen* (Tübingen: Rainer Wunderlich, 1965), 47–57, on 50.

attitude compelled by a Nazi regime putatively incapable of dealing rationally with science, but rather a familiar pattern of interaction.[54] And as we now know, despite his post-war representations, Ramsauer was even willing to concede, pragmatically, that the limitation of the so-called Jewish influence in physics could be accepted as long as the basis for the instrumental relationship with the state was not adversely affected.[55] It was not the fate of individual scientists that was of concern, but rather the stability of the physics community taken as a social institution with its professionally sanctioned leadership intact.

THE HISTORY OF THE DPG AND OF LOCAL COMMUNITIES IN THE THIRD REICH

If we pursue the implications of seeing the scientific community *qua* community, with its more or less well-defined boundaries and within them its more or less well-defined patterns of authority, then these episodes in the history of the DPG appear less as examples of capitulation or resistance than as examples of a kind of local conservatism. Perhaps useful here is a heuristic simile: the physics profession, or rather the DPG as its institutional embodiment, as like a kind of small provincial town. To be sure, there are many ways in which this simile is imperfect, and in some ways it is perhaps even misleading. But as a historiographical model, it does provide some illuminating comparisons of the history of the DPG with the history of other communities – in a more literal sense – under National Socialism.

What similarities between a professional society and local communities make the comparison at least plausible? Both are large enough to have complex internal social networks but small enough that many if not most of their members are personally known to each other. At any rate, if complete objectivity depends on impersonality, this is practically impossible given the size of the organization. Such mutual personal knowledge is not synonymous with egalitarianism; we may well find in small towns as well as professional societies a powerful, if only implicitly expressed, social hierarchy. At any rate, whether formal political structures are democratic

54 On Ramsauer's earlier career and path to the DPG chairmanship, see especially Hoffmann, "Carl Ramsauer."

55 Carl Ramsauer, "Letter to Bernhard Rust," translated in Hentschel, *Physics*, 278–281, especially 279; see Gerhard Simonsohn, "Physiker in Deutschland," *Physikalische Blätter*, 48 (1992), 23–28, on 28.

or authoritarian, important decisions are often made on the basis of informal relationships and personal commitments, and de facto leadership is sometimes determined on this basis with the community as a whole only ratifying decisions taken behind the scenes. Newcomers may be treated with politeness, but if they lack personal connections with the locals, it may take some time for them to feel really at home. There is typically a strong set of shared values and norms. If one adheres to these values, there follows a strong sense of identity and loyalty to the community. Deviation from these norms, on the other hand, can lead to an intense ostracism.

There is, of course, a vast literature on the fate of local communities and their institutions in the Third Reich. Surveying this literature, the late Detlev Peukert wrote the following about the coordination (*Gleichschaltung*) of social institutions with the Nazi power system in the context of small towns and rural communities:

> The Nazi seizure of power in the countryside made use of the local traditional structures in many ways, but it also came up against many stubborn obstacles.... The seizure of power was made easier by the fact that as a rule the prominent members of provincial communities, and the small-town and rural middle classes on which their positions were mainly based, mistrusted the republican, democratic state, [and] rejected the Weimar "system." ... The selfsame rural tradition ... however, and the dominant role played by prominent local citizens, formed a fairly sizeable barrier to the formulation, dissemination and ... implementation of the NSDAP's aim of a political monopoly.[56]

This passage seems in several respects, mutatis mutandi, an apt description of several aspects of the history of the DPG in that era. Peukert's description suggests several corollaries of the small, provincial town metaphor, assuming it is useful; these and others can be elaborated by looking at some of the historical literature on the subject.[57]

The metaphorical comparison with community histories highlights several aspects of the relationship between the DPG and the state that we

[56] Peukert, *Inside*, 86.

[57] An incomplete list of this literature includes: William Sheridan Allen, *The Nazi Seizure of Power: The Experience of a Single German Town 1922–1945*, rev. ed. (New York: Franklin Watts, 1984; orig. 1966); Martin Broszat et al., *Bayern in der NS-Zeit*, 6 vols. (Munich: R. Oldenbourg, 1977–1983); Ian Kershaw, *Popular Opinion and Political Dissent in the Third Reich: Bavaria 1933–1945* (Oxford: Clarendon Press, 1983); Rudy Koshar, *Social Life, Local Politics, and Nazism: Marburg, 1880–1935* (Chapel Hill: University of North Carolina Press, 1986); Gerhard Wilke, "Village Life in Nazi Germany," in Richard Bessell (ed.), *Life in the Third Reich* (New York: Oxford University Press, 1987), 17–24.

have already alluded to. One is the gradual nature of the process of *Gleichschaltung*: It did not happen overnight, but rather was a protracted process that relied on a variety of methods and agents, varying from case to case. A second point relates to the background to the selection of leadership positions in the DPG, as we have seen in the case of Mey and Ramsauer. Many of the most important decisions in both local communities and in professional organizations are de facto taken through informal consensus or personal interaction of key authority figures and only later ratified formally. Unfortunately for the historian, this means that some of the most crucial turning points in such decisions are precisely those that are *not* committed to paper. Thus, alongside specific documentary evidence, we also need to be alert to overarching patterns of interaction that point to implicit or explicit agreements or disagreements behind the scenes.

On a more general level, we might look at the question of if, or how, Nazism related to and possibly changed the internal social structures of towns. As several historical studies suggest, insofar as communities already had relatively stable political structures and community members already had a strong sense of group identification, their established patterns of interaction and structures of authority were less amenable to incorporation within the Nazi power system. Local authority presented a kind of inertia against the Nazis' quest for a monopolized organization of power. In some cases, Nazi Party cells at the town level preached the displacement of traditional local authorities. Coming to the fore in local leadership in such situations were individuals who had previously experienced marginalization and who hence were at odds with the local elites. Where this happened, social structures were indeed profoundly transformed.[58] But in other specific cases, *Gleichschaltung* depended not on the displacement but instead on the co-optation of traditional local authorities. Despite the revolutionary rhetoric of Nazism, success in these cases came not through contravention, but rather through an appeal to traditional values.[59] If we make a historiographical comparison to the

[58] A now classic description and analysis of this kind of scenario is Allen, *Nazi Seizure*; see also Wilke, "Village Life."

[59] For example, see, various essays in the volumes of Broszat et al., *Bayern*, including: Martin Broszat, "Ein Landkreis in der Fränkischen Schweiz: Der Bezirk Ebermannstadt 1929–1945" and Elke Fröhlich, "Stimmung und Verhalten der Bevölkerung unter den Bedingungen des Krieges," in Vol. I (1977), 21–192 and 571–688, respectively; Helmut M. Hanko, "Kommunalpolitik in der 'Hauptstadt der Bewegung' 1933–1935: Zwischen 'revolutionärer' Umgestaltung und Verwaltungskontinuität," in Martin Broszat,

history of the DPG, then the actions of Stark and the Aryan physics group seem in many ways to be an example of the first scenario. In the end, however, the DPG proved more analogous to the second type of effect of Nazism on local communities; that is, the co-optation of extant authority patterns.[60] This should not be surprising, as the DPG had something to bargain with that most literal communities lacked: technical–instrumental expertise that was increasingly desired by the state.

The historical literature on communities in the Third Reich also speaks to the point, raised by Peukert, that the monopolizing demands of Nazism often faced a local inertia that made their complete implementation difficult. Thus Ian Kershaw describes the opposition of Protestant communities in Bavaria to the incarceration of their bishop and the reaction of Catholic communities to the removal of crucifixes from schoolrooms. In both of these cases, the religious communities' protests led to reversals of the regime's actions. This success was due not least, Kershaw believes, to the active role of the clergy in organizing and channeling these efforts. But also contributing to success were the facts that the protests were limited in scope to specific issues and that they did not challenge loyalty to the system as a whole. Thus Kershaw recounts an episode in which Protestant protesters defused a confrontation with police by singing the national anthem and the *Horst-Wessel-Lied* (the Nazi Party anthem). In the end, one is left with a sense of tragedy that the attempts to deviate from Nazism's totalizing claims, surprisingly successful in some situations, were so weak in others. Concerning the Catholic schools protest, Kershaw writes: "On matters directly affecting the Catholic traditional 'way of life' detestation of Nazi interference prompted spectacular opposition. But . . . the Catholic response on matters of less than immediate concern to the Church was muted in the extreme."[61] Without making strong claims about a close analogy, a similar assessment might apply

Elke Fröhlich, and Anton Grossmann (eds.), *Bayern in der NS-Zeit III. Herrschaft und Gesellschaft im Konflikt: Teil B*, Vol. III (Munich: R. Oldenbourg, 1981), 329–441; Zdenek Zofka, "Dorfeliten und NSDAP: Fallbeispiele der Gleichschaltung aus dem Bezirk Günzberg," in Martin Broszat, Elke Fröhlich, and Anton Grossmann (eds.), *Bayern in der NS-Zeit IV, Herrschaft und Gesellschaft im Konflikt: Teil C* (Munich: R. Oldenbourg, 1981), 383–433.

60 I explore the relationship between scientific institutions and the Nazi state as the co-optation of extant local authority patterns – namely, through the suppression of more potentially radical models – in examples from the Kaiser Wilhelm Society in "Maintaining Discipline in the Kaiser Wilhelm Society During the National Socialist Regime," *Minerva*, 44 (2006), 251–266.

61 Kershaw, *Popular Opinion*, especially 177, 223.

to the DPG's responses to the state. Did maintaining the "freedom of science" in the sense discussed above constitute resistance to Nazism? On one level, it is important to recognize any kind of deviance from the Nazi context, be it the prerogatives of Protestant clergy, the defense of the "Catholic way of life," or the institutional integrity of the DPG. On another level, however, it is also important to recognize the limits of such examples and the conformity in other areas that made them possible. In contrast with what other forms of opposition to the Nazi regime could and did accomplish, it is fairly difficult to label such inertial nonconformity "resistance," even if it did effectively maintain "islands of freedom" and sustain the social boundaries of communities.[62]

Thus, the history of the DPG in the Third Reich cannot easily be seen as either resistance or capitulation. If we see this history as a community dynamically seeking to maintain its own social boundaries and to maintain the extant patterns of authority within those boundaries counterposed to a regime claiming a monopolistic, centralized control of power, then this history, like other examples of the attempted *Gleichschaltung* of local communities in the Third Reich, features a complex range of mixed responses and strange bedfellows. In the end, the physics community was able to pursue and advance its own agenda under its own self-selected leadership: not always and everywhere, but nevertheless sometimes and in some ways. If this is what the contemporary and post-war responses meant by "freedom," then the DPG was indeed responsible for maintaining a boundary in which freedom could prevail. However, as comparison with the history of communities generally in the Third Reich suggests, notions of freedom and autonomy in this historical context are fraught with ambivalences and difficulties.

[62] See Simonsohn, "Physiker"; Kershaw, *Popular Opinion*, 223; Martin Broszat, "Resistenz und Widerstand: Eine Zwischenbilanz des Forschungsprojekts," in Broszat, et al., *Bayern in der NS-Zeit IV*, 691–709; Peukert, *Inside*.

3

Marginalization and Expulsion of Physicists under National Socialism

What Was the German Physical Society's Role?

Stefan L. Wolff

The transfer of power to National Socialism, a process of several steps, among them the Enabling Act ("*Ermächtigungsgesetz*") from March 24, 1933, also quickly had severe consequences for German cultural and intellectual life. On April 7th the coalition government of NSDAP (*Nationalsozialistische Deutsche Arbeiterpartei*), DNVP (*Deutschnationale Volkspartei*), and other independent right-wing politicians began to "reorganize" the civil service through a new law and a subsequent series of regulatory statutes. Despite the euphemistical title, the "Law for the Restoration of the Professional Civil Service" did not intend to restore professional standards but instead to remove so-called non-Aryans as well as political opponents from all positions financed by the state. This included universities as well as research institutes like those of Kaiser Wilhelm Society or the Imperial Physical-Technical Institute (Physikalisch-Technische Reichsanstalt, PTR) and led to a large number of suspensions from office and dismissals. Quite soon, many of those affected – the younger individuals in particular – thus deprived of their livelihoods, were effectively driven out of the country. In 1933, it was not yet necessarily evident that these new laws and statutes were not merely temporary measures instead of just the beginning of a series of discriminatory laws and ordinances. The subsequent development, including the anti-Semitic Nuremberg Laws of 1935, led to a complete exclusion of the "non-Aryans," respectively the somewhat more narrowly defined "Jews," from the civil service and subsequently from all of public life.

Among the academic disciplines, physics was affected more than the average.[1] Against this backdrop the question arises: To what extent was the professional organization for physicists, the German Physical Society (*Deutsche Physikalische Gesellschaft*, DPG), willing and able to use whatever maneuvering room it had to represent the interests of its marginalized members?

THE DPG'S REACTIONS TO THE DISMISSALS

The DPG was active in two central spheres. The first was in publications; that is, managing the association's own periodicals (its proceedings, *Verhandlungen der Deutschen Physikalischen Gesellschaft*, and reports, *Physikalische Berichte*), in addition to certain engagements with other professional journals. The second was the organization of gatherings such as regional meetings and the annual fall national convention. The DPG also sought to promote research and instruction in physics. Its statutes otherwise declared its aim "of protecting the professional interests of physicists."[2] In 1929, the awarding of the Max Planck Medal for exceptional achievement in theoretical physics became another one of the society's obligations.[3] These activities were performed independently from any state institution, which is why initially, in 1933, the DPG did not seem to be directly affected as an organization by the new laws. In view of the many suspensions and dismissals among its membership, however, its set aim of guarding "professional interests" should nevertheless have made clear that the DPG was unavoidably involved in the political processes under way in 1933.

Expressions of public protest against these measures were few and far between. The society's acting chairman, Max von Laue, who largely repudiated this government policy, regarded such activism as more harm than help. Thus, he was dismayed at Albert Einstein's critical statements to the press abroad and wrote him reproachfully: "But why did you have

[1] For example, see Stefan L. Wolff, "Vertreibung und Emigration in der Physik," *Physik in unserer Zeit*, 24 (1993), 267–273.

[2] DPG statutes, implementation ordinances of §2, *Verhandlungen der Deutschen Physikalischen Gesellschaft*, 3rd series, 6, (1925), 59–68, especially 64.

[3] The nominations lay with the Max Planck Medal Committee composed of former awardees; the DPG's board made the final decision. See the contribution to this volume by Richard H. Beyler, Michael Eckert, and Dieter Hoffmann.

to come forward *politically* as well?!"[4] On 10 May, the same day as the Nazi-orchestrated book burnings, von Laue wrote a letter to all the directors of Germany's physics departments.[5] He asked them for the names and addresses of all physicists active in research, including advanced students, who were affected by the Law for the "Restoration" of the Professional Civil Service of 7 April.[6] Together with its implementation ordinances from 11 April, the law stipulated the exclusion of so-called non-Aryans. Anyone with at least one grandparent of the Jewish faith was relegated to this group. James Franck and Fritz Haber were among the very few to react by resigning their academic, state-paid, tenured positions in protest on 17 and 30 April, respectively. At that time, as war veterans they were exempted from the general rule, thanks to a proviso clause attributable to Reich President Paul von Hindenburg. The rector of the Stuttgart Polytechnic, Paul Ewald, also tendered his resignation from this position on 20 April because he found it impossible "to share the national government's standpoint on the race issue."[7]

The first leaves of absence and dismissals were implemented as the summer term was already getting started. More appeared to be imminent. Von Laue was wondering who might be next and whether anyone was suffering material hardship as a result. "As far as I can organise support for colleagues, it will be given," he promised.[8] Previous discussions with Paul Ehrenfest, a faculty member at the University of Leiden in the Netherlands, apparently had prompted him to launch this campaign. Von Laue had invited Ehrenfest to Berlin at the end of April for outside consultations about the state of affairs.[9] Ehrenfest stayed from 5 to 8 May to assess the situation, talking to numerous people. Not even fundamental

[4] Von Laue to Einstein, 14 May 1933, The Albert Einstein Archive of the Jewish National and University Library at the Hebrew University of Jerusalem, Israel, hereafter AEA, no. 16–088; English translation in Klaus Hentschel, *The Mental Aftermath: The Mentality of German Physicists 1945–1949* (Oxford: Oxford University Press, 2007), 138.

[5] Von Laue to Sommerfeld, 10 May 1933 (circular to all universities), Archives of the Deutsches Museum, Munich (*Deutsches Museum Archiv*, hereafter DMA), Sommerfeld papers.

[6] The quotation marks have been added by the author.

[7] Ewald to the senate of the Stuttgart Polytechnic, 20 April 1933, DMA, HS 1977–28/A, 88. See Michael Eckert and Karl Märker (eds.), *Arnold Sommerfeld. Wissenschaftliche Briefwechsel, Vol. 2: 1919–1951* (Berlin: Verlag für Geschichte der Naturwissenschaften und der Technik, 2004), 357.

[8] Von Laue to Sommerfeld, 10 May 1933, DMA, Sommerfeld papers, 024.

[9] Ehrenfest to von Laue, 26 April 1933, Ehrenfest to Kapitza, 25 April 1933. Archive for the History of Quantum Physics (hereafter AHQP), Ehrenfest Scientific Correspondence (hereafter EHR) (7, 2) and (6, 4).

differences of opinion were an obstacle to a 3-hour-long meeting with Johannes Stark, Philipp Lenard's fellow protagonist of "Aryan physics"; Stark's membership in the NSDAP dated back to 1930.[10] In the end, Ehrenfest's evaluation of the situation for physicists inside Germany was bleak. He began to write many letters in search of temporary positions or invitations as guest lecturers for these dismissed physicists. Shortly before his visit to Berlin, Ehrenfest had learned that Ernest Rutherford was ready to organize aid in England for his "German-Jewish" colleagues.[11] One important precondition was to receive as accurate and specific information as possible about the affected group of people. It was this information that von Laue attempted to obtain by his circular letter.

The society's letterhead on which von Laue's letter was written lent it an official character; and the attribute "chairman" attached to his signature also indicated that von Laue was acting in the name of the society. To him it was a campaign on behalf of the German physics community, to which, in his view, the affected colleagues continued to belong. With reference to the relief committees formed in other countries in aid of German émigrés, he wrote to Einstein how grim it was that, on the heels of the Armenians and other "half-wild peoples," now "we" too need help like that.[12] Von Laue did not criticize the legal regulations directly, however, restricting himself to practical aid for its victims. He encountered support even from the holder of the physics chair at the Munich Polytechnic, Jonathan Zenneck, a registered member of the DNVP since 1929,[13] who considered the measures against "non-Aryans" entirely legitimate.[14]

10 Ehrenfest to Kapitza, 19 May 1933, EHR (6, 4). On Stark, see Alan D. Beyerchen, *Scientists under Hitler. Politics and the Physics Community in the Third Reich* (New Haven: Yale University Press, 1977), 103–122; Mark Walker, *Nazi Science* (New York: Perseus Publishing, 1995), 41–63.

11 Ehrenfest to Kapitza, 4 May 1933, EHR (6, 4).

12 Von Laue to Einstein, 26 June 1933, DMA, copy.

13 Certificate of party membership in the Bavarian association for the Munich district, Office of the Deutschnationale Front, 18 July 1933, DMA, Zenneck papers, box 15. See Stefan L. Wolff, "Jonathan Zenneck als Vorstand im Deutschen Museum," in Elisabeth Vaupel and Stefan L. Wolff (eds.), *Das Deutsche Museum in der Zeit des Nationalsozialismus* (Göttingen: Wallstein, 2010), 78–126. Stefan L. Wolff, "The Establishment of a network of Reactionary Physicists in the Weimar Republic," in Cathryn Carson, Alexei Kojevnikov, and Helmuth Trischler (eds.), *Weimar Culture and Quantum Mechanics* (London: Imperial College Press, 2011), 293–318.

14 Zenneck to Karl Kiesel, 16 August 1933. He reports there that many of the English were faulting Germany for the measures against "non-Aryans." In an effort to explain, he offered them some examples, "about an institute in which all the assistants were Jews, entire faculties in which the majority of the professors, and companies in which the greatest number of directors and engineers were Jews." Wolff, "Zenneck."

Although he had nobody to report about from his own institution, Zenneck replied, "if anything should change in there, I will inform the Physical Society about it."[15] Von Laue forwarded the staffing information he had gathered in this way to Ehrenfest, who planned to relay it to international relief organizations. He also passed it on to the banker Carl Melchior, who was a member of the board of the Central Committee of German Jews for Aid and Development, founded that April.[16]

Whether, as a professional association, the DPG did or did not have more options available to stand up for its suspended and dismissed members will have to remain an open question. The answer may well be negative, if one takes as a gauge the individual efforts by prominent physicists to apply, carefully and discreetly, what influence they deemed they had. Ultimately, they did not achieve anything. After his visit to Berlin, Ehrenfest reported that some "top-ranking" colleagues were trying "to plead for their Jewish fellow professionals with those in power, as discreetly as possible, making use of all their connections."[17] The prevailing view held by these German physicists was that government supporters were far more radical than the government itself, which was accordingly rated as comparatively moderate.[18] The appeals by National Socialist students and their organized lecture boycotts did, in fact, expose a particularly aggressive strain of anti-Semitism. But this very extremism and a perceived dilution of individual provisions in the new education laws for schools apparently raised the impression that responsible policy makers would soon adopt a more moderate course.[19] Moreover, it was believed that these legal measures were only temporary. That, in any case, is what physicists such as von Laue, Max Planck, and Werner Heisenberg believed.[20] They even hoped that after things had calmed down, a few of their suspended colleagues would perhaps return to their old posts.

15 Zenneck to von Laue, 18 May 1933, Wolff, "Zenneck."

16 Von Laue to Sommerfeld, 21 May 1933, DMA, Sommerfeld papers, 024: published in Eckert and Märker, *Arnold Sommerfeld*, 385–386. Notice about the *Zentralausschuss der deutschen Juden für Hilfe und Aufbau* and its board in the *CV Zeitung*, no. 16, 20 April 1933, 5. In this context, this central committee's importance may have been rather minor. Besides, only some of those affected regarded themselves as Jewish.

17 Report by Ehrenfest, 13 May 1933, EHR (3, 8).

18 See Hans Kopfermann to Niels Bohr, 23 May 1933, AHQP, Bohr Scientific Correspondence (hereafter BSC) (22, 2).

19 See Finn Aaserud, *Redirecting Science* (Cambridge: Cambridge University Press, 1990), 112–113.

20 Heisenberg to Max Born, 2 June 1933, cited with commentary in Karl von Meyenn (ed.), *Wolfgang Pauli. Wissenschaftlicher Briefwechsel.* Volume 2 (Berlin: Springer, 1985), 168.

Thus, to them it seemed that public protests would be counterproductive. It was not a form of expression they were accustomed to using anyway. They were afraid it would only make the situation more precarious for their vulnerable colleagues, and they thought they could achieve more by engaging their exclusive contacts.[21] Planck asked to see Adolf Hitler to pay his respects to the new Reich chancellor in his capacity as presiding chairman of the Kaiser Wilhelm Society. The visit occurred on 16 May. There are conflicting reports about the conversation Planck had with Hitler; in the end, it had no impact on changing the policy of dismissals.[22] Therefore, during the subsequent period, efforts concentrated on helping jeopardized individuals, avoiding all the while any open criticism of the government and its policy. Objections raised confidentially in a few special cases remained unsuccessful, however, such as the attempt to claim war-veteran status for Peter Pringsheim on the grounds of his internment during the First World War or the plan to preserve the right of Lise Meitner to give lectures (*venia legendi*) at the University of Berlin. She had worked as an X-ray assistant at the Eastern Front during the war.[23] The only practical use von Laue's circular letter had was to pass information on to relief organizations. Any campaigns beyond that appeared not to make much sense to von Laue and the DPG; neither would it have suited their preference for staying out of any public political debates.

After the usual greetings, the ninth physicists' national convention in Würzburg from 17 to 22 September 1933 began with the opening speeches of Karl Mey in his role as chairman of the German Society for Technical Physics (Deutsche Gesellschaft für technische Physik, DGtP) and von Laue for the DPG. Whereas Mey declared that German physicists were prepared to support the new chancellor Hitler in his efforts to overcome the economic crisis, von Laue, who was vacating his position as DPG chairman after a statutory 2-year term, spoke first about the physics tradition in Würzburg and subsequently about the Inquisition against

21 Report by Ehrenfest, 13 May 1933, AHQP, EHR (3,8).

22 Helmuth Albrecht, "Max Planck: Mein Besuch bei Adolf Hitler – Anmerkungen zum Wert einer historischen Quelle," in Helmuth Albrecht (ed.), *Naturwissenschaft und Technik in der Geschichte. 25 Jahre Lehrstuhl für Geschichte der Naturwissenschaft und Technik am Historischen Institut der Universität Stuttgart* (Stuttgart: GNT Verlag, 1993), 41–63.

23 Letter by von Laue, Planck, and Erwin Schrödinger to Verwaltungsdirektor der Friedrich-Wilhelms-Universität, 1 June 1933, published in Rudolf Schottlaender (ed.), *Verfolgte Berliner Wissenschaft* (Berlin: Edition Hentrich, 1988), 91.

Galileo. This latter subject must have been understood as an allusion to recent events revolving around Einstein, whose name was not mentioned explicitly.[24] Following von Laue, the new president of the PTR, Johannes Stark, spoke on the "Organization of Physical Research."[25] Although his offensive bid at this meeting to chair the DPG as von Laue's successor encountered clear resistance, the minutes provide no information on this. There is just the note: "Mr. K. Mey is elected as chairman." No quantitative result of the vote is given.[26] This supposedly "ideologically neutral" industrial physicist Mey was director of the Osram Lighting Company and had presided over the DGtP since 1931.

In the years that followed, the DPG evidently wanted to avoid any mention of the expulsion of "non-Aryan" scientists. Little indication can be found in the society's official proceedings, the *Verhandlungen der Deutschen Physikalischen Gesellschaft*, related to the political changes and emigrations. This information bulletin mainly reported about meetings of the regional associations, sometimes with brief outlines of the subjects of talks. It additionally recorded the minutes of the administrative meetings along with obituaries of important physicists and listings of recently registered members. A few staffing changes were mentioned casually at the regular administrative meeting of the members of the Berlin association in December 1933 with the addition "instead of Mr. P. Pringsheim, who has moved away from Berlin," Mr. Ebert would become his substitute.[27] So much for the not inaccurate, yet not entirely complete, information about the emigration of one colleague, who had meanwhile found a position at the Université Libre in Brussels. Once again, the forced migrations did not receive the slightest mention in this official publication. Even von Laue's obituary of Haber in the proceedings of 1934 announced his death in a Swiss spa while on travels without mentioning that he had left Germany and was looking for a suitable place for his emigration.[28]

[24] Mey's speech was published in *Zeitschrift für technische Physik*, 14 (1933), 567–568; von Laue's speech in: *Physikalische Zeitschrift*, 34 (1933), 889–890. For an interpretation, see Beyerchen, *Scientists*, 64f.

[25] *Zeitschrift für technische Physik*, 14 (1933), 433–435.

[26] *Verhandlungen der Deutschen Physikalischen Gesellschaft*, 3rd series, 14 (1933), 32. Many years later von Laue mentioned that Stark got only two votes: Von Laue to Meitner, October 7, 1946, in Jost Lemmerich (ed.), *Lise Meitner – Max von Laue. Briefwechsel 1938–1948* (Berlin: ERS Verlag, 1998), 464.

[27] *Verhandlungen der Deutschen Physikalischen Gesellschaft*, 3rd series, 14 (1933), 36.

[28] *Verhandlungen der Deutschen Physikalischen Gesellschaft*, 3rd series, 15 (1934), 7–9. Haber had left Germany in August 1933. The negotiations on the Reich Evasion Penalty Tax were still in progress, and although his household had already been dissolved,

James Franck and Max Born, also mentioned in this article, likewise lack any such reference.[29]

In February 1934, the regional associations and local chapters of the two physical societies finally received an official reference to the "emigration of German scholars" in a circular letter from their chairman Mey after the journals *Science* and *Nature* had criticized these events. In Mey's words, they engendered an unfavorable mood against Germany and gave Anglo-Saxon countries the appearance of being "refuges for intellectual and scientific freedom." This evidently posed a problem for German physicists because their relations abroad, such as attendance at organized scientific events, generally became an issue. Mey contacted the "influential Reich authorities" and received their approval, not to stay away from such meetings, but instead to make active use of them to represent the German point of view. According to Mey, the members of the two societies should regard the societies' offer to grant travel subsidies in such cases as active support of the government.[30] Mey was able to maintain a certain continuity concerning the international contacts because he could convince the government that they had common interests in this area.

Karl Scheel's historical sketch at the festive meeting for the Berlin Physical Society's 90th anniversary in January 1935 ignored the recent past.[31] One fact, however, likewise directly identifiable on the membership lists, did slip out: that one-third of the membership "lives beyond the Reich border."[32] This share even grew a bit, as surprisingly many émigrés initially remained members of the DPG; the addresses in the membership lists trace their paths in exile.

On 29 January 1935, the DPG and the Society of German Chemists (*Gesellschaft Deutscher Chemiker*, GDCh) jointly organized a memorial meeting under the auspices of the Kaiser Wilhelm Society on the occasion

an official residence still existed in Berlin. Thus, Haber's emigration had not yet been completed. But von Laue undoubtedly knew that Haber did not want to return to Germany. For example, see Dietrich Stoltzenberg, *Fritz Haber* (Weinheim: VCH, 1994), 602–616.

29 *Verhandlungen der Deutschen Physikalischen Gesellschaft*, 3rd series, 15 (1934), 7–9.

30 Mey to the regional associations (*Gauvereine*) and local chapters (*Ortsgruppen*) of the DPG and DGtP, 27 February 1934, Archives of the Max Planck Society (*Archiv der Max-Planck-Gesellschaft*, hereafter MPGA), div. III, rep. 19, Debye papers, no. 1011, sheets 1–3.

31 *Verhandlungen der Deutschen Physikalischen Gesellschaft*, 3rd series, 16 (1935), 1–11.

32 *Verhandlungen der Deutschen Physikalischen Gesellschaft*, 3rd series, 16 (1935), 11. See also the statistics by Beyerchen, *Scientists*, 74f.

FIGURE 4. Jonathan Zenneck (*Source:* Deutsches Museum)

of the first anniversary of Haber's death, who had been one of their former presidents as well. This memorial acquired an unintentionally demonstrative tone, primarily from a last-minute ministerial decree on 15 January banning all members of the civil service from attending it.[33] In a letter to the minister, Mey energetically objected as chairman of the DPG to any contention that the planned event was a critique of the government, which he tried to disqualify as slander.[34] Retrospectively, however, this event has been repeatedly evaluated as an act of opposition.[35]

Zenneck was elected the DPG's new chairman in 1935. The belief within the society apparently still was that the political demands could essentially be appeased by certain rituals and turns of phrase. Seeking instructions on the correct salutation to use during the fall convention,

[33] See the contribution to this volume by Deichmann.

[34] Mey to Reichsminister Rust, 24 January 1935, Archives of the German Physical Society (*Archiv der Deutschen Physikalischen Gesellschaft*, hereafter DPGA), no. 10011. See the documentation reproduced in Dieter Hoffmann and Mark Walker (eds.), *Physiker zwischen Autonomie und Anpassung* (Weinheim: Wiley-VCH, 2007), 557–561.

[35] On this, see, for example, Beyerchen, *Scientists*, 66–69.

Zenneck wrote to his predecessor Mey in September 1936: "Likewise, it should probably be agreed who should raise the 'Sieg-Heil' cheer for the Führer. Besides, a laudatory telegram to the Führer should probably be drafted and sent out. I have no experience in these matters and do not know in what manner this affair has hitherto been settled."[36] The DPG purposely did not want to participate in the ideological controversy about "Aryan physics."[37] When Friedrich Hund contacted Zenneck about Stark's attacks in the *Schutz-Staffel* (SS) weekly *Das Schwarze Korps*, he, just as von Laue, thought it would be senseless for the society to intervene.[38]

The society clearly followed an evasive strategy whenever confrontation with political authorities was anticipated. Hence, its members were left to wage such battles individually. The problematic state of physics in academia, particularly in the area of theory, became the subject of a personal petition by Max Wien in 1934, for example. He cosigned another one with Heisenberg and Hans Geiger in 1936 in the form of a memorandum appending the signatures of 75 physics professors.[39] Zenneck's signature was among them, along with those of many other members of the society's board. However, despite the "professional interests" undoubtedly at stake here, the DPG did not step forward as an organization.

In response to an inquiry in November 1935, Zenneck stated his support of accepting "non-Aryan" members in principle, provided they could exclude the possibility that "any kind of drawbacks would follow from it."[40] Chairman Zenneck belonged to that conservative wing of the German physics community who frequently spied unwelcome "Jewish forces" at work among the academic competition and in the ideological debates of the Weimar period, such as after Gustav Hertz received the professorship at the University of Halle in 1926.[41] Zenneck did not think

36 Zenneck to Mey, 2 September 1936, DMA, Zenneck papers, box 12. See also Wolff, "Zenneck," 90.

37 See the contribution to this volume by Michael Eckert.

38 Hund to Zenneck, 21 July and 2 August 1937, DMA, Zenneck papers, box 38; Zenneck to Hund, 2 August and 5 August 1937 (carbon copy), DMA, Zenneck papers, box 38; von Laue to Zenneck, 31 July 1937, Zenneck papers, box 50. On the article in *Das Schwarze Korps* see also Klaus Hentschel (ed.), *Physics and National Socialism. An Anthology of Primary Sources* (Science Networks. Historical Studies, vol. 18) (Basel: Birkhäuser, 1996), 152–161.

39 Dieter Hoffmann, "Die Physikerdenkschriften von 1934/36 und zur Situation im faschistischen Deutschland, Wissenschaft und Staat," *itw-kolloquien*, no. 68 (1989), 185–211; also see Hentschel, *Physics*, 91–95 and 137–140.

40 Zenneck to Ernst Brüche, 23 November 1935, DMA, Zenneck papers, box 12.

41 Zenneck to Himstedt, 4 April 1927, DMA, Zenneck papers, box 38.

it beneath himself to instrumentalize anti-Semitic biases either, such as when he tried to prevent Arnold Sommerfeld's candidacy for the editorship of the journal *Zeitschrift für Physik* in 1936. Zenneck argued that one could justifiably blame Sommerfeld for having "formerly developed a strong pro-Semitic activity."[42] Therefore, Zenneck's support for accepting "non-Aryan" members shows that anti-Semitism played no role in this context. One very decisive aspect was surely the financial difficulties that the society was experiencing. In 1937, the national physics convention originally scheduled to take place in Salzburg had to be relocated to Bad Kreuznach owing to the lack of foreign currency.[43] Thus, tight budgetary constraints alone made it advantageous for the DPG to maintain the highest total number of members possible. The proportion living beyond the Reich borders, amounting to about one-third, gained special importance, as they provided the society with the very desirable foreign currency. It was suspected that these members felt particular solidarity with the Jewish members living within the country, so their treatment and status were always situated within a financial context.[44]

To be accepted into the DPG, nomination by one of the members was needed, after which point the board had to pass judgment. There were some isolated approvals of "non-Aryan" scientists even after 1933, at least one of which predated Zenneck's answer mentioned earlier.[45] Thus, it was not a matter of principle but rather something that could be settled on a case-by-case basis. An émigré, Wolfgang Berg, became a new member of the DPG in 1937 at Treasurer Walter Schottky's proposal. Formerly an assistant at the University of Berlin, Berg had been dismissed in summer 1933 on the grounds of the law affecting the professional civil service. Initially supported by a stipend of the Imperial Chemical Industries, Berg was working at the time for the Kodak Company in Harrow, England.[46]

42 Zenneck to Mey, 4 January 1936 (carbon copy), DMA, Zenneck papers, box 12.

43 Zenneck to Ministerium, 17 June 1937, according to Georg Schmucker, "Jonathan Zenneck, 1871–1959. Eine technisch-wissenschaftliche Biographie" (Stuttgart: Dissertation Universität Stuttgart, 1999), 445.

44 Some statements by Treasurer Schottky speak for this assessment, for example, Walter Schottky to Samuel A. Goudsmit, 3 January 1938, American Institute of Physics, Niels Bohr Library, College Park, MD (hereafter AIP), Goudsmit papers, box 3, folder 45 (reproduced in Hoffmann and Walker, *Physiker*, 554), as well as Schottky to Peter Debye, 3 December 1938, MPGA, div. III, rep. 19, Debye papers, no. 1014, sheet 36.

45 Such as, 1934 Walter Deutsch, 1935 Rolf Landshoff and Hans Beutler. See *Verhandlungen der Deutschen Physikalischen Gesellschaft*, 3rd series, 15 (1934), 27; 16 (1935), 24 and 31.

46 On the Imperial Chemical Industries stipends, see Stefan L. Wolff, "Frederick Lindemanns Rolle bei der Emigration der aus Deutschland vertriebenen Physiker," *Yearbook of the Research Center for German and Austrian Exile Studies*, 2 (2000), 25–58.

Zenneck's own nominee, the Italian physicist Emilio Segrè, counted not only as a "non-Aryan" but also, pursuant to the narrower definition of the Nuremberg Laws, as a Jew.[47] By contrast, the Max Planck Medal had an entirely different quality.[48] It attracted the public eye. After a pause of a number of years, the awards were resumed in 1937. Approval by the Ministry for Science, Education and Culture ("*Reichsministerium für Wissenschaft, Erziehung und Volksbildung*," REM) about the chosen winner was obtained each time in advance. When in spring 1938 the potential winner Enrico Fermi prompted "objections of a racial kind," therefore making it seem questionable that approval would be granted, the idea was soon dropped.[49]

At the initiative of its chairman, von Laue, in 1933, the DPG tried to assist fellow physicists affected by the dismissals by passing along their personal data. Otherwise, it avoided any mention of the emigration in its official publications. All in all, the DPG did what it could to keep out of conflicts, even where "professional interests" were at stake. The status of "non-Aryan" members was not a matter of any discussion. On one hand, they did not excite much notice, and on the other hand, it was feared that their leaving could lead to the loss of the high proportion of foreign members who provided the society with an influx of the all-important foreign currency.

THE RELATIONSHIP BETWEEN ÉMIGRÉS AND THE DPG

Only a small minority of the dismissed émigrés considered boycotting German institutions altogether, the DPG included, because of the way the German state had treated them. On the contrary, many retained their memberships for a long time precisely because they did not want to let themselves be excluded. Another reason was that they regarded the DPG as independent of the state. Just as frequently, they had pragmatic reasons for continuing membership. Some exiles were continuing to publish their work in German journals,[50] and the revised address lists kept them traceable for their fellow professionals, including other émigrés.

47 *Verhandlungen der Deutschen Physikalischen Gesellschaft*, 3rd series, 18 (1937), 36.

48 See the contribution to this volume by Beyler, Eckert, and Hoffmann.

49 Report on the board meeting of the Berlin Physical Society, 30 March 1938, MPGA, div. III, rep. 19, Debye papers, Nr. 1176, sheet 1. In fact, it had been a mistake to suppose that Fermi was Jewish. His wife was of Jewish origin. C. F. von Weizsäcker tried to procure the pertinent information for the DPG through the German embassy.

50 See the remarks on Cornelius Lanczos in this chapter as well as the contribution to this volume by Simonsohn.

TABLE 1. *Withdrawals from the DPG*

Year	Total Withdrawals (Including an Unknown Number of Cancellations for Unpaid Dues)	Documentable Withdrawals of Members Known To Be Vulnerable to "Racial" or Political Discrimination
1932	69	Not Available
1933	113	19
1934	74	4
1935	108	27
1936	32	4
1937	51	8
1938	121	47
Total as of 1938	568	109

Since its founding, the DPG had never been an exclusive professional association just for physicists. As an organization of the leading discipline of the sciences, it always welcomed a considerable number of chemists, meteorologists, astronomers, mineralogists, industrialists, and technicians among its membership. The high number of foreigners lent the society a certain internationality.

The total number of members varied between 1,225 (early 1933) and 1,079 (early 1939), excluding firms and institutions, which totaled 212 and 239, respectively. The individual membership cancellations listed in the second column of Table 1 attained a level of up to 10 percent of the total membership.[51] The figure for 1932 roughly indicates the scale unaffected by political events. The third column of the table presents the statistics for those known not to be able or willing to work in Germany because of their or their spouses' origins or because of their political convictions. (Details are listed in the table in the appendix to this volume.) These figures are essentially based on the *International Biographical Dictionary of Central European Émigrés 1933–1945* (*Biographisches Handbuch der deutschprachigen Emigration nach 1933*) and the lists of two relief organizations: first, the printed *List of Displaced German Scholars* by the Emergency Association of German Science Abroad (*Notgemeinschaft deutscher Wissenschaftler im Ausland*) from 1936, with a

[51] Based on comparative counting of the available membership lists printed in the *Verhandlungen der Deutschen Physikalischen Gesellschaft* or a separately printed list of members for 1937. For 1938/1939 see also Hentschel, *Physics*, lxx, and the Internet list at http://www.uni-stuttgart.de/hi/gnt/hentschel/Dpg38–39.htm, accessed 20 December 2010.

supplement from 1937; second, the files of the British Society for the Protection of Science and Learning (SPSL), founded in 1933 as the Academic Assistance Council.[52] Not all the scientists appearing on these two lists eventually managed to emigrate, but their mention here serves as evidence of their falling among the discriminated group of people under National Socialism or that they had made their political opposition known in some form. Some survived inside Germany, some fell victim to the organized mass murder of the National Socialist state, and the fates of some individuals could not be clarified. Among the migrants, that is, scientists who had already accepted a position abroad before 1933, those are here included who were denied the possibility of returning to positions in Germany as a result of the changed political conditions.

The majority of membership withdrawals cannot be ascertained in the above sense, however. The instruments available here do not permit such a categorization of all these persons. Younger, not yet established members are not listed in reference works, for instance. The same applies to many who were employed outside of universities or research establishments. Even in cases of existing career information, it is often not possible to state the reasons for their departures. The only exceptions are the deceased. Also, unpaid membership dues were a cause for deletion of memberships following the society's statutes. In turn, it was accordingly possible to terminate the relationship passively by ceasing to make the payments. One characteristic of the statistics is prominent, however. The individually identifiable exits in the third column of Table 1 show three sharp peaks in the years 1933, 1935, and particularly 1938.

This is not very surprising, as it very clearly mirrors the effects of the political events. The retracted memberships can be related to several developments: the *Machtergreifung* in 1933 with the Law for the Restoration of the Professional Civil Service; the 1935 Nuremberg Laws, one consequence of which ultimately led to the cancellation of the exemption clauses for war veterans as well as for senior civil servants (*Altbeamten*); and in 1938, many events occurred, from the incorporation of Austria in March to the November pogroms. Added to these was the decision by the DPG in December 1938 (described later in more detail) to urge

[52] Werner Röder (ed.), *Biographisches Handbuch der deutschsprachigen Emigration nach 1933, 3 Vol.* (Munich: Saur, 1980-1983); *List of Displaced German Scholars* (London: Speedee Press Services, 1936); Herbert Strauss (ed.), *Emigration. Deutsche Wissenschaftler nach 1933. Entlassung und Vertreibung* (Berlin: Technische Universität, 1987) (contains a reprinting of the London list of 1936 and the supplements of 1937).

the remaining Jewish resident members to withdraw. This was the end of a longer development that, statistically speaking, did not have any further trenchant consequences.[53] Altogether, these figures document that even though terminations by affected members rose considerably in some years, it was nonetheless a continuous process. The surprisingly high number of more than 30 émigrés or migrants still retaining membership in the DPG even in 1939 is remarkable (see the table in the documentary appendix to Hoffmann and Walker, *Physiker*, 555–556). However, three of them had not paid their membership dues since 1937, and it was still an open question whether they would keep their memberships.[54]

Although the DPG encompassed a broad spectrum of professions, there were, conversely, many physicists who were not registered members. The lists compiled by the relief organizations reveal that only roughly half of the affected group of persons were members of the DPG.[55] Such prominent names as Hans Bethe, Felix Bloch, Otto Robert Frisch, Victor Weisskopf, or Edward Teller are missing. Yet, before 1933 some of them had nevertheless participated in the society's activities. For example, Bethe and Rudolf Peierls had originally met during one of the DPG's conventions.[56]

An account of individual withdrawals or, as the case may be, decisions not to take this step will serve to illustrate the situations determined by individual personal circumstances, as rendered chronologically in the third column of Table 1. Among at least 19 withdrawals in 1933 that can be correlated in the above-defined sense with racial or political discrimination, we find the names Albert Einstein, who himself had served as chairman of the DPG between 1916 and 1918, and Alfred Landé. Having made some critical statements about the new German government,

[53] Existing analyses mostly focus on this decision of 1938 forced upon the society from the outside and rather attribute too much importance to it. For example, see Helmut Rechenberg, "Vor fünfzig Jahren," *Physikalische Blätter*, 44 (1988), 418, and also Klaus Hentschel on his Internet list: http://www.uni-stuttgart.de/hi/gnt/hentschel/Dpg38–39.htm, accessed 20 December 2010.

[54] Hans Beutler, Robert Emden, and Heinz Kallmann were among a list of 15 persons and one institution (Physics Institute in Bucharest) with long-outstanding dues. See Schottky's letter to Debye, 16 March 1939, MPGA, div. III, rep. 19, Debye papers, no. 1016, sheet 31.

[55] We can take a sampling of the 158 persons listed under the column "physics" by the Society for the Protection of Science and Learning (hereafter SPSL), for example. If this figure is reduced to physicists working in Germany, Austria, and Prague, there remain about 135, among which 63, hence almost only half, belonged to the DPG. BLO, SPSL archives. Likewise, only half of the physicists listed by the *Notgemeinschaft* were DPG members (69 of 139, including the supplement).

[56] Charles Weiner's interview with Rudolf Peierls in August 1969. AIP, transcription, 4.

Einstein had caused a furor by publicly announcing his withdrawal from membership in the Prussian Academy of Sciences on 28 March 1933.[57] It had to be expected that any other organization of which he was a member would become the target of political attack as a result. He wrote to von Laue to this effect:

> Dear [von] Laue, I heard that my unsettled relationship with German corporate bodies in which my name still appears on the membership lists could cause difficulties for friends of mine in Germany. That is why I ask you to make sure sometime that my name be struck from the lists of these bodies. They include, e.g., the German Physical Society and the Society of the Ordre pour le mérite. I explicitly empower you to arrange this for me. This is probably the right way, because thus new theatrical effects are avoided. Amicable greetings, yours, A. Einstein.[58]

Einstein regarded his withdrawals as a discreet retreat to spare his friends from trouble. He thus certainly did not want to set a symbolic example for others to follow. He had done this already in the Prussian Academy. By contrast, Landé, one of the migrants, attached to such a step a concrete political message that reached beyond his personal circumstances. He tried to motivate other foreign members to withdraw their memberships together with him and wanted to add emphasis to this demonstrative drive with a public statement in protest against the treatment of Jews inside Germany. Landé did not regard himself as Jewish but had a Jewish family background.[59] Nevertheless, the time he had spent at the University of Tübingen had evidently strongly sensitized him to this issue. First appointed there in 1921, he soon felt driven into the role of an outsider, inappropriately characterized by the labels "Jewish" and "communist."

Such surroundings offered him no chance to integrate socially, so he was only too glad to be able to accept an offer in 1929 to teach at Ohio State University for one-half year. A second invitation in 1930 subsequently developed into a permanent position. Memories of the

[57] Christa Kirsten and Hans-Jürgen Treder (eds.), *Albert Einstein in Berlin, Part I: Darstellung und Dokumente* (Berlin: Akademie Verlag, 1979), 246.

[58] Einstein to von Laue, 5 June 1933, MPGA, div. III, rep. 19, Laue papers, no. 536, sheet 1. Transcribed in the documentary appendix to Hoffmann and Walker, *Physiker*, 532–533; English translation in Siegfried Grundmann, *The Einstein Dossiers* (Berlin: Springer, 2005), 288.

[59] Landé to Epstein, 23 March 1933, Archives of the California Institute of Technology, Pasadena, CA (hereafter CITA), Epstein papers, cited in Beyerchen, *Scientists*, 75. Concerning the family of Landé: Elke Brychta, Anna-Maria Reinhold, and Arno Mersmann (eds.), *Mutig, streitbar, reformerisch. Die Landés – Sechs Biografien 1859–1977* (Essen: Klartext-Verlag, 2004).

atmosphere at Tübingen, as the National Socialist activists were just getting going, were probably responsible for the special concern motivating Landé to call for a coordinated exodus from the society.[60] But he evidently did not find much support. Paul Epstein, for instance, who had already been working in Pasadena since 1921 and to whom he had written, initially declined. When von Laue wrote to a few émigrés in 1935 with a reminder about their unpaid dues, he tried to argue against such drives, appealing to them to stay in the DPG specifically given the situation. Theodore von Kármán, another migrant also employed in Pasadena who was behind with his membership dues since 1933, was one of the few to respond positively. In his reply, Kármán used a physical metaphor to characterize the politically triggered migration of scientists: "Undoubtedly, the developments of the past few years has set off a true Brownian motion of physicists. It seems to me that the diffusion related to it can be quite useful to many countries." The weak response to his efforts disappointed von Laue; he borrowed Kármán's metaphor in his somewhat resigned reply: "But tell the German physicists who are still in Brownian motion or have already become sedimentary that the man who beat the pack because he wanted the donkey to move has never seemed to me to be a model of particular cleverness."[61]

The remaining retractors from among this group in 1933 otherwise mainly included people who had been dismissed or suspended on the basis of the law for the "restoration" of the professional civil service and had already left to seek their fortunes in exile. One of these was the mathematician Richard Courant, who initially left for England before moving on to the United States in 1934; another was the chemist Friedrich Paneth, who never returned to Germany from a visit to England in 1933. The DPG likewise lost the theoretical physicist Paul Hertz from Göttingen and Karl Weissenberg from Berlin. Hertz was employed in Geneva from 1934 on; Weissenberg, who was stripped of his position at the Kaiser Wilhelm Institute and soon afterward of his extraordinary professorship at the University of Berlin, was hired by the University of Southampton in 1934.[62]

[60] Charles Weiner's interview with Alfred Landé on 3 October 1973, AIP.

[61] Von Laue to von Kármán, 4 May 1935, von Kármán to von Laue, 3 June 1935, and von Laue to von Kármán, 15 June 1936, CITA, Kármán papers. The last of these three letters is also cited in Beyerchen, *Scientists*, 76. Kármán personifies how the concepts of "migrant" and "émigré" could overlap. He was still dismissed from the Aachen Polytechnic: Ulrich Kalkmann, *Die Technische Hochschule Aachen im Dritten Reich (1933–1945)* (Aachen: Aachener Studien zu Technik und Gesellschaft, 2003), 130–132.

[62] Weissenberg's personnel file, Bodleian Library, Oxford (hereafter BLO), SPSL archives.

The year 1934 offered no new prospects. A few other émigrés left the society, such as Heinrich Kuhn, who had been dismissed from the University of Göttingen in 1933 and got a position at Oxford University. The following year 1935 exhibited a marked rise in numbers but they were mainly unrelated to the emigrations. Such cancellations could occur years in advance. The theoretical physicist Friedrich Kottler from Vienna was one example, and Erich Marx from Leipzig, who had been working in a private institution for years before emigrating to the United States in 1939 and 1941, was another. Yet others, such as Max Born, Leo Szilard, and Hans Reichenbach, who had lost their positions in 1933 and had felt compelled to leave Germany, only now, after an interval of some 2 years, dispensed with their memberships – this, on the assumption that they were not struck from the rolls for not keeping up with their membership fees. Paul Epstein's exit in 1935 – who had refused to join Landé's boycott in 1933 – speaks for a more critical assessment of the circumstances in Germany and the DPG's role. One could take the exodus of nine members from English-speaking countries as another indication in this regard, which members possibly considered this step as an act of solidarity.

The year 1936 brought comparatively fewer losses in numbers for the DPG. Nevertheless, it would be a mistake to conclude that stability therefore prevailed in the attitude of the remaining members. Samuel Goudsmit, originally from the Netherlands, who had studied in Germany and still had very good contacts there despite his work in the United States since 1927, struggled with his scruples in this regard for more than a year. Notwithstanding the fact that Goudsmit was Jewish, Walther Gerlach called him his "old institute comrade" and "a friend and coworker" in 1936 and, prompted by Goudsmit's tardy payments, tried to keep him as a member of the DPG.[63] On 24 June 1936, Goudsmit replied: "Sometimes I cannot see at all what purpose there is to support the German Physical Society any longer. The inhumane treatment of many excellent German scientists saddens me deeply."[64] Despite all this, his old attachment prevented him from dropping his membership, so he sent Treasurer Schottky his annual fee.[65] However, at the end of 1937, Goudsmit did

[63] Gerlach to Goudsmit, 10 February 1936, AIP, Goudsmit papers, box 3, folder 45; transcription in Hoffmann and Walker, *Physiker*, 552f.

[64] Goudsmit to Gerlach, 24 June 1936, published in Rudolf Heinrich and Hans-Reinhard Bachmann, *Walther Gerlach*, catalogue (Munich: Deutsches Museum, 1989), 75.

[65] Goudsmit to Schottky, 24 June 1936, with acknowledgment in Schottky to Goudsmit, 14 July 1936, AIP, Goudsmit papers, box 3, folder 45.

FIGURE 5. Walter Schottky (*Source*: DPGA)

finally break with the DPG and cancelled his membership: "I am disappointed that the society as a whole never protested against the vicious attacks on some of its most prominent members."[66] He evidently had an understanding of the tasks and the range of maneuverability of a professional association that was incompatible with the passive role of the DPG. Schottky expressed his dismay at this step, also because he wanted to emphasize to Goudsmit, in the name of the society, the special importance it placed on relations with colleagues abroad.[67] He could have pointed out that, in spite of all the political circumstances described above, many émigrés were still keeping their memberships. In 1937 the DPG lost more of them, however, including Walter Heitler and Rudolf Peierls, who had both emigrated to England years ago.

This series of withdrawals was probably not in direct reaction to actions taken by the DPG itself. As the Goudsmit example shows, it was

66 Goudsmit to Schottky, 17 December 1937, AIP, Goudsmit papers, box 4, folder 48; transcription in Hoffmann and Walker, *Physiker*, 554.

67 Schottky to Goudsmit, 3 January 1938, AIP, Goudsmit papers, box 3, folder 45; transcription in Hoffmann and Walker, *Physiker*, 554.

TABLE 2. *Reviewers (Later) Affected by Dismissals, Early Retirements, or "Racial" or Political Discrimination (Definition as in Column 3 of Table 1)*

Year	Number of Thus-Defined Reviewers
1933	25
1934	9
1935	7
1936	6
1937	4
1938	2
1939	1

more a matter of making a professional organization partially responsible for the effects of the policy on its own comparatively narrow area of professional expertise. Isolated efforts on behalf of "non-Aryans" could not conceal that the DPG members still inside Germany who were being marginalized by the professional civil service and Nuremberg Laws and other measures were by no means safely settled in a reserve of collegial solidarity. By 1938 at the latest, painful breaks awaited them here, too. But gradual changes had been registered earlier on as well. Most members of this group simply inconspicuously vanished one after another from the *Physikalische Berichte* in their capacities as reviewers.[68]

The preparations for Sommerfeld's 70th birthday on 5 December 1938 finally exposed a chasm that presumably had existed for a while but only then became visible. A special issue of the *Annalen der Physik* that the society was planning to honor Sommerfeld would not include contributions by "non-Aryans." The group of Sommerfeld pupils in charge, including, among others, Peter Debye, the DPG's chairman since 1937, and Otto Scherzer, who had special responsibility for this volume of the *Annalen*, simply accepted this stipulation by the publisher. When Paul Ewald and Wolfgang Pauli heard about it, they were appalled.[69] Pauli hoped he could rely on the solidarity of fellow physicists in boycotting

[68] In 1938 there are contributions only by Karl Przibram from Vienna and Reinhold Fürth from Prague. The latter appears again in 1939 as the sole member of this group. This information comes from the contributor lists in the *Physikalische Berichte* for the pertinent years. See *Physikalische Berichte*, 14 (1933) to 20 (1939).

[69] Debye to Ludwig Hopf, 18 October 1938, MPGA, div. III, rep. 19, Debye papers, no. 377, sheet 6.

FIGURE 6. Peter Debye (*Source*: MPGA)

the periodicals of such publishers in the future.[70] He regarded it as an unacceptable precedent.[71] This was the first instance of a new, discriminatory rule being introduced without physicists in Germany posing any opposition – insofar as they personally were unaffected.

In a letter that October, Debye wrote that he suspected that Jews might even be prohibited from attending the meeting planned to take place in Munich in celebration of Sommerfeld's birthday.[72] The Upper Bavarian District of the DPG had taken over the organization of the meeting; hence, it was an official event for the society.[73] The consequences beginning to emerge from a few incidents at the society's fall meeting of the foregoing

[70] Pauli to Heisenberg, 15 August 1938, von Meyenn, *Wolfgang Pauli*, 593.

[71] Pauli to Epstein, 21 August 1938, von Meyenn, *Wolfgang Pauli*, 595–596.

[72] Heisenberg to Pauli, 15 July 1938, von Meyenn, *Wolfgang Pauli*, 587 (see also von Meyenn's commentary on 572). Debye to Hopf, 18 October 1938, MPGA, div. III, rep. 19, Debye papers, no. 377, sheet 6.

[73] See the report in *Verhandlungen der Deutschen Physikalischen Gesellschaft*, 20 (1939), 7–9.

month, which will be discussed in the next section, may have bred Debye's suspicion.

Ludwig Hopf, who had lost his professorship at the Aachen Polytechnic because of the 1933 Law for the Restoration of the Professional Civil Service, belonged to the group of Sommerfeld's former students.[74] When details about this celebration had still not arrived, which he was naturally expecting to receive, he contacted Debye. Having abstained from participating in any of the recent events of the past years, he had no illusions about his own status and inquired whether his participation was perhaps unwelcome "under the present circumstances."[75] The "pure Aryan" issue of the *Annalen* had been no surprise for him, but he did not want to accept being barred from an event of the DPG: "as long as I am a member and pay so-and-so many marks, I have the right to be treated like any other member.... If I am not going to be invited, then I will, of course, withdraw."[76] He underscored this point in another letter from 10 November: "In any case, I place importance on participating in the festive meeting, etc., with all the rights of a member of the Physical Society and definitely expect an invitation."[77]

The "Aryan" *Annalen* issue and the Sommerfeld birthday celebration organized by the DPG's Bavarian section quite openly terminated any collegial solidarity. Such restrictions on the rights of "non-Aryan," or (in the narrower definition of the Nuremberg Laws) Jewish, members in Germany reduced their memberships to a mere formality in the already rather loose association of common professional interests. Besides Hopf, three former Sommerfeld students, Fritz London, Wolfgang Pauli, and Arthur Rosenthal, also left the DPG in 1938.[78] There is no documentary evidence of any causal connection; nevertheless, it can be assumed. Rosenthal had remained in Germany and in the context of the events of the pogrom of

74 For Hopf's dismissal, see his retroactively compiled personnel file, university archive of the Rheinisch-Westfälische Technische Hochschule Aachen, PA 2013.

75 Hopf to Debye, 16 October 1938, MPGA, div. III, rep. 19, Debye papers, no. 377, sheet 5.

76 Hopf to Debye, 19 October 1938, MPGA, div. III, rep. 19, Debye papers, no. 377, sheet 7.

77 Debye to Hopf, 9 November 1938, MPGA, div. III, rep. 19, Debye papers, no. 377, sheet 8, and Hopf to Debye, 10 November 1938, MPGA, div. III, rep. 19, Debye papers, no. 377, sheet 9.

78 A list of Sommerfeld's students supplements Debye's letter to Pauli, 23 April 1928, compiled on the occasion of Sommerfeld's 60th birthday, in von Meyenn, *Wolfgang Pauli*, 704–710.

November 9 – known as the "Night of Broken Glass" (*Reichskristallnacht*) – was arrested and spent four weeks in the Dachau concentration camp. London had already emigrated. Pauli, who held a professorship in Zurich, likewise belonged to the potential victims, because his ancestry prevented him from being allowed to work in Germany anymore.[79]

There were a number of components leading to the sharp rise in membership cancellations in 1938. More émigrés or migrants left the DPG. Von Kármán was one of these, having previously responded positively to von Laue's appeal in 1935 to stay a member of the society; Lise Meitner was another, having just fled from Germany in 1938. However, a much larger number of withdrawals came from the German Reich, to which Austria belonged since the *Anschluss*[80] in March. From among these 30 members, one-third came from Vienna alone. Most of them had been working in the Radium Institute and had been dismissed through enforcement of the Nazi civil service laws in Austria. With the exception of the director Stefan Meyer, who was able to survive in Bad Ischl, all these affected members emigrated soon afterward.[81]

More factors led to a rise in the membership cancellations in 1938. The political and social situations for Jewish citizens of the German Reich were aggravated across the board. The pogrom of 9 November marked an escalation beyond anything that had preceded it. Arbitrary arrests followed, touching, among others, Rosenthal (as mentioned above) and Hans Baerwald, professor of physics at the Darmstadt Polytechnic. Baerwald spent a month in the Buchenwald concentration camp and upon his release had to commit himself to leaving the Reich very soon.[82] Hopf escaped the same fate only because his son endured the month of

[79] For Rosenthal, see: Dorothee Mussgnug, *Die vertriebenen Heidelberger Dozenten* (Heidelberg: Winter, 1988), 155. Pauli describes his own status as "75 per cent Jewish" in Pauli to Aydelotte, 29 May 1940, Library of Congress, Washington, DC, Manuscript Division, Neumann, box 8.

[80] Literally translated as "connection," better understood as annexation.

[81] Besides Meyer, there were six other members of the DPG: Karl Przibaum, Marietta Blau, Elisabeth Rona, Franz Urbach, Gustav Kürti, and Stefan Pelz. See Wolfgang Reiter, "Die Vertreibung der jüdischen Intelligenz: Verdopplung eines Verlustes – 1938/1945," *Internationale Mathematische Nachrichten*, 187 (2001), 1–20, especially 16. On Meyer himself, see Wolfgang Reiter, "Stefan Meyer: Pioneer in radioactivity," *Physics in Perspective*, 3 (2001), 106–127, especially 122f.

[82] Autobiographical account in Baerwald to Rektorat der Technischen Hochschule Darmstadt, 30 January 1946, archive of the Technische Universität Darmstadt, personnel file TH 25/01–19/8. Baerwald still appeared on the membership list for 1939.

confinement in Buchenwald in his father's stead.[83] Among the measures designed to restrict academic research was the instruction issued by the administration of Berlin University, on 14 November, prohibiting Jews from attending colloquia, likewise the ban on use of university facilities in general, including the libraries, dated 8 December.[84] Planck's attitude toward these two new regulations were symptomatic of the apolitical self-image of the elite. He regretted it, yet accepted that in the given situation it had been simply unavoidable in the long run.[85]

Ever since its fall meeting in September 1938, the society had been attracting the critical attention of its superior ministry, the REM. But it was mostly because of the mounting pressure applied by its own members that the board of the DPG felt obliged to urge its resident Jewish members (in the sense of the Nuremberg Laws) on 9 December to leave the society. Details about these circumstances are discussed in the following section. The documents reveal that merely seven such members accordingly submitted written notice by the beginning of January 1939. There were probably very few left who considered themselves affected. The statutes of the society initially did not formally permit the board to expel these members.

Richard Gans was the first to answer. The 84-year-old Emil Cohn along with Georg Jaffé and Leo Graetz declared their withdrawals before 13 December. Walter Kaufmann from Freiburg and Hans Adolf Boas from Berlin followed a few days later. Hartmut Kallmann cancelled his membership in the beginning of January 1939.[86] Gans, Kallmann, and Kaufmann were able to stay inside the country and survive, owing to

83 Dietmar Müller-Arends and Ulrich Kalkmann, "Ludwig Hopf 1884–1919," in Klaus Habetha (ed.), *Wissenschaft zwischen technischer und gesellschaftlicher Herausforderung. Die Rheinisch-Westfälische Technische Hochschule Aachen 1970–1995* (Aachen: Einhard, 1995), 208–215, especially 213.

84 Decree by the rector (Hoppe) of the University of Berlin, 14 November 1938, MPGA, div. III, rep. 19, Debye papers, no. 1268, sheet 55; Saul Friedländer, *Das Dritte Reich und die Juden* (Munich: Beck, 1998), 308; Joseph Walk, *Das Sonderrecht für die Juden im NS-Staat* (Heidelberg: Müller, 1981), 264.

85 Planck to von Laue, 17 November 1938, DMA, von Laue papers; also see Wolff, "Vertreibung," 270.

86 Letters by Debye's office to Grotrian, 10 December, 13 December, 17 December 1938, and 14 January 1939. MPGA, div. III, rep. 19, Debye papers, no. 1914, sheets 39, 40, and 43 and MPGA, div. III, rep. 19, Debye papers, no. 1016, sheet 1. Therein are reported the already effectuated withdrawals by the persons involved. The number of thus documented withdrawals is relatively low; however, there is no indication that the files are incomplete. Also see Rechenberg, "Vor fünfzig Jahren," 418; Stefan L. Wolff, "Hartmut Kallmann (1896–1978) – ein während des Nationalsozialismus verhinderter Emigrant verlässt Deutschland nach dem Krieg," in Dieter Hoffmann and Mark Walker (eds.),

special conditions and circumstances (e.g., "privileged mixed marriage").[87] Graetz died at 85 years of age in 1941 in Munich. Cohn, Jaffé, and Hopf emigrated to Switzerland, the United States, and England/Ireland, respectively, before the war broke out. The total number of withdrawals (Table 1, column 2) also reached its peak in 1938. It may be that some of the 22 foreigners among these did so as an act of solidarity. As the membership list for 1939, from which their names were struck, was published in February, it is possible that it happened in reaction to the events of the final months of 1938 just described.[88]

Relatively few members of the DPG were subjected to the deportations and killings. The statistics surely remain incomplete, because not all are identifiable. Compared with other refugees, the scientists under examination here, some of them very distinguished, had above-average possibilities for finding a new position abroad. Those residing in Germany who lost their posts as a result of the civil service law and received no pension could only continue to conduct research in some niche in the private sector, at best for a limited time. Only emigration offered some way to continue a scientific career. Consequently, we find most of the émigré former members in Great Britain and the United States (see the documentation in the appendix to this volume). It was virtually only there that the necessary infrastructure capable of absorbing them existed. Thanking Pauli for his birthday greetings, Sommerfeld wrote him in this sense on 1 January 1939, that "perhaps ... the future of civilization does not lie 'on the water' but beyond the great water."[89] A few former members managed to escape to the United States (still neutral until

"Fremde" Wissenschaftler im Dritten Reich. Die Debye-Affäre im Kontext (Göttingen: Wallstein, 2011), 310–334.

87 On Gans, see Edgar Swinne, *Richard Gans. Hochschullehrer in Deutschland und Argentinien* (Berlin: ERS-Verlag, 1992). Kallmann mentions his "privilegierte Mischehe" in a letter to Federal President Heuss, 1 April 1954, MPGA, Kallmann's personnel file. On Kaufmann, see his papers, Archives of the University of Freiburg (*Archiv der Universität Freiburg i.Br.*, hereafter UFA), Kaufmann, C139.

88 Letter by Reinhold Mecke to Bothe, 8 May 1939, MPGA, div. III, rep. 19, Debye papers, no. 1016, sheet 43. He refers there to the membership list from February. However, Kallmann's son remembers that his father was disappointed about the lack of solidarity by foreign members: "He always complained throughout his life that no foreign member resigned!" Unpublished manuscript of Klaus Kallmann, 2005. Cf. Stefan L. Wolff, "Das Vorgehen von Debye bei dem Ausschluss der jüdischen Mitglieder aus der DPG," in Dieter Hoffmann and Mark Walker (eds.), *"Fremde" Wissenschaftler im Dritten Reich. Die Debye-Affäre im Kontext* (Göttingen: Wallstein, 2011), 106–130.

89 Sommerfeld to Pauli, 1 January 1939, reprinted in von Meyenn, *Wolfgang Pauli*, 615–616.

December 1941) even after the war in Europe had begun. In 1940 Hedwig Kohn succeeded, after many failed attempts, and Fritz Reiche as well as Erich Marx got out in 1941.[90]

The expansion of the area under German control also turned some of the émigrés, once again, into victims. As the war progressed, the Nazi persecution had mutated from marginalization to ultimate elimination. As a result, both Herbert Jehle, who was no longer willing to work in Germany for political reasons and out of religious conviction as a Quaker, and Pringsheim were extradited from Brussels to different internment camps in France. In both these cases, external intervention (by Arthur Stanley Eddingten and Franck, respectively) achieved their release. Jehle and Pringsheim then obtained permits to leave for the United States in 1941.[91] Karl Przibram, who had likewise emigrated to Brussels, did not make it to his original destination in England but managed to survive in the Belgian Underground. A worse destiny befell Franz Pollitzer, who had fled to France. He was deported to Auschwitz in 1942 and perished. Alfred Byk belonged to the numerous victims who were – as it was euphemistically called – evacuated from Berlin to the east. His deportation on 13 June 1942 brought him to the Majdanek death camp. Arnold Berliner committed suicide to avoid the imminent threat of such a transport to Poland. The Prague manufacturer Emil Kolben was sent to Theresienstadt, where he died in 1943. Kurt Sitte survived the Buchenwald concentration camp as a political prisoner.[92]

90 See Brenda Winnewisser, "Hedwig Kohn – eine Physikerin des zwanzigsten Jahrhunderts," *Physik Journal*, 2 (2003), 51–55; Valentin Wehefritz, *Verwehte Spuren – Prof. Dr. phil. Fritz Reiche 1883–1969: ein deutsches Schicksal im 20. Jahrhundert* (Dortmund: Univ.-Bibliothek, 2002).

91 See Wolfgang Drechsler and Helmut Rechenberg, "Herbert Jehle (5.3.1907–14.1.1983)," *Physikalische Blätter*, 39 (1983), 71; and on Pringsheim: Valentin Wehefritz, *Gefangener zweier Welten* (Dortmund: Univ.-Bibliothek, 1999), 33–35.

92 See Max von Laue, "Arnold Berliner (26.12.1862–22.3.1942)," *Die Naturwissenschaften*, 33 (1946), 257–258. On Pollitzer, see his widow's statement in *Poggendorff*, vol. 7a, 607, as well as Liste des Convoi No 30 en Date du 9 Septembre 1942, in Serge Klarsfeld, *Le Memorial de la Déportation des Juifs de France* (Paris: Klarsfeld, 1978), unpaginated, or *Memorial to the Jews Deported from France 1942–44* (New York: B. Klarsfeld Foundation, 1983), 265; on Byk: Freie Universität Zentralinstitut für sozialwissenschaftliche Forschung (ed.), *Gedenkbuch Berlins der jüdischen Opfer des Nationalsozialismus* (Berlin: Edition Hentrich, 1995), 181, and Bundesarchiv Koblenz (ed.), *Gedenkbuch: Opfer der Verfolgung der Juden unter der nationalsozialistischen Gewaltherrschaft in Deutschland 1933–45*, Vol. 1 (Frankfurt am Main: Weisbecker, 1995), 190. Heinz Kolben, "Dr. h.c. Ing. Emil Kolben zum Gedächtnis," *Bohemia*, 26 (1985), 111–121, only mentions the fact that Kolben fell victim to the Third Reich. See also: Sitte's personnel file, BLO, SPSL archives.

Against this backdrop, the fact that 30 émigrés and migrants remained members of the DPG as late as 1939 is an astonishing phenomenon (see the table in the documentary appendix to Hoffmann and Walker, *Physiker*, 555–556). This situation provided grounds for objections within the society as well. Viewing this "great number of émigrés," Professor Reinhold Mecke from Freiburg set as a condition for his election on the board of the regional association of Baden-Palatinate that this breach of the "existing regulations of the Reich" had to be rectified.[93]

Thus arises the question: What motivated these émigrés to retain their memberships despite all this? Specific examples may not be able to answer this completely, but they can offer some indication. Rudolf Ladenburg, for instance, was one member of this group whose biography cannot be faulted for either a lack of solidarity with émigrés or any supposed political naïveté. He had left the country early on for Princeton University in 1931, initially just for a year, but his sojourn turned into a permanent position in 1932. Calling to mind the events in Germany, he and Eugene Wigner cosigned a circular to 27 other physicists in the United States, almost all of whom were likewise of European origin, suggesting they make available 2 to 4 percent of their salaries over a period of 2 years to come to the aid of a number of named destitute fellow professionals.[94] They closed with the statement: "We hope that you, too, feel the importance of such assistance and will not deny your support and cooperation."[95]

Ladenburg had no illusions about the conditions in Germany. In 1938, he had come face-to-face with "racial" discrimination by German law in his capacity as a foreign member of the Kaiser Wilhelm Society. Its managerial board informed him on 16 June 1938 that "all Foreign Members, who pursuant to the German racial legislation are in fact not in a position to exercise their memberships," would be "struck from the membership rolls." On 3 July 1938, Ladenburg felt challenged to respond by pointing out that his American citizenship relieved him of any obligations toward

93 Mecke to Bothe, 8 May 1939, MPGA, div. III, rep. 19, Debye papers, no. 1016, sheet 43; Grotrian to Debye, 13 May 1939, MPGA, div. III, rep. 19, Debye papers, no. 1016, sheet 41; Grotrian to Bothe, 13 May 1939, MPGA, div. III, rep. 19, Debye papers, no. 1016, sheet 44; Walter Grotrian to Eduard Steinke and Mecke, 13 May 1939, MPGA, div. III, rep. 19, Debye papers, no. 1016, sheet 45–46.

94 Charles Weiner, "A new site for the seminar. The refugees and American physics in the Thirties," in Donald Fleming and Bernhard Bailyn (eds.), *The Intellectual Migration: Europe and America 1930–1960* (Cambridge, MA: Belknap Press, 1969), 190–234, especially 215–216.

95 Quoted with English translation, Weiner, 229–232.

Germany. As the Kaiser Wilhelm Society was nevertheless applying the German racial laws to foreign members, he explicitly requested his name be deleted from the membership rolls.[96] The DPG's own proposal in December 1938, however, discussed later, only applied these "racial criteria" to members residing within the German Reich. In the light of these events, it might have been a quite deliberate decision by Ladenburg to stay within the DPG.

The situation of another migrant shows that there might have been entirely practical reasons for not leaving the DPG and to remain in contact with former German colleagues. Cornelius Lanczos had likewise gone to the United States in 1931, first as visiting professor; then, in view of the political changes in Germany, he decided to stay there at Purdue University in Lafayette, Indiana. His focus on mathematical physics was a problem, however, for it was not appreciated within the American academic landscape of scientific research. Not even Einstein, with whom he collaborated for a year, could help him very much: "The directly applicable result is simply valued more in this country than quality and matters of lasting worth."[97]

Flat rejections by a few American journals moved Lanczos to reengage the contacts he still had in Germany, particularly the chief editor of the *Zeitschrift für Physik*, Karl Scheel, to publish there. Einstein could not "understand that you as a Jew still want to publish in Germany. This is a kind of treason. On the whole, German intellectuals behaved disgracefully, with all those horrendous injustices and thoroughly deserve being boycotted."[98] But Lanczos thought he could differentiate between state institutions and a professional association. He emphasized in his reply to Einstein that from various sources he had gained the impression that:

> [A] larger proportion of German physicists (and also of mathematicians) certainly have no connection with the present criminal administration and are not involved

[96] Vorstand der KWG to Ladenburg, 16 June 1938, Ladenburg to Vorstand, 3 July 1938, Telschow to the ministry, 2 August 1938, MPGA, Ladenburg's personnel file. On the changes to the statutes, see Helmuth Albrecht and Armin Hermann, "Die Kaiser-Wilhelm-Gesellschaft im Dritten Reich (1933–1945)," in Rudolf Vierhaus and Bernhard vom Brocke (eds.), *Forschung im Spannungsfeld von Politik und Gesellschaft* (Stuttgart: DVA, 1990), 356–406, especially 385–386.

[97] Einstein to Lanczos, undated, beginning February 1938, AEA, no. 15–271 and 15–272.

[98] Einstein to Lanczos, 11 September 1935, AEA, no. 15–246 and 15–247. See John Stachel, "Lanczos's early contributions to relativity and his relationship with Einstein," in J. David Brown et al. (eds.), *Proceedings of the Cornelius Lanczos International Centenary Conference* (Philadelphia: SIAM, 1993), 201–221, especially 218–219.

in any way with it, although it is, of course, practically impossible to say actively anything *against* it. Since I certainly do consider the Zeits. f. Phys. as an organ of German physicists, and not as a journal of Germany, I saw no reason barring me from placing my paper there.[99]

So here we encounter one émigré with the view, or perhaps the wishful belief, that his fellow physicists back in Germany were a professional community independent of the regime, a community he still believed he belonged to through personal ties. Given his predicament of lacking recognition in the United States, this sense of belonging helped him quite concretely with publication alternatives.

Thus, the exodus of members since 1933 proceeded in stages, almost never connected with open protest against the conditions inside Germany. Landé and Goudsmit appear here rather as exceptions. For some, emigration provided a reason to withdraw. A considerable number of others, however, wanted to maintain their ties, which – as Lanczos's example demonstrates – could have quite practical reasons. Moreover, some of them even encouraged new memberships in exile.[100] The DPG appreciated this internationality as well. Shortly after the outbreak of war, the board circulated within the German Reich a list of names of all members living in other – neutral – countries. The intention was to find out who among their resident members was in contact with these colleagues. The list included émigrés as well.[101] Inside Germany, belonging to the DPG afforded no protection from the growing discrimination, not even within the small professional sphere. By 1938, it no longer even ensured free access to the DPG's own activities.

CHANGES TO THE STATUTES

Like other scientific societies with a high percentage of foreign members, including the Society of German Chemists and the German Mathematical Association (*Deutsche Mathematiker-Vereinigung*, DMV), the DPG also did not expel its Jewish members until 1938. They all saw a connection to the memberships of their foreign members, which they were afraid to lose. The foreign memberships were not only important for the

99 Lanczos to Einstein, 14 September 1935, AEA, no. 15-248.

100 Examples in *Verhandlungen der Deutschen Physikalischen Gesellschaft*, 15 (1934), 27; 16 (1935), 52; 17 (1936), 14; and 18 (1937), 36.

101 Circular letter by Grotrian (managing director) with a list of names and addresses dated 1 November 1939, MPGA, div. III, rep. 19, Debye papers, no. 1017, sheets 27–30.

reputations of these societies but also played an important financial role. They provided the societies with treasured foreign currency. The REM had also accepted this point. However, during the national physics convention in fall 1938, an omitted toast to the *Führer* excited the ministry's displeasure and also directed its attention to the fact that Jews could still continue to be members of the DPG and publish in the society's periodicals.[102] The general political atmosphere was also changing by that time. Regulations prohibiting Jews (here always used in the sense of the definition in the Nuremberg Laws) from membership in scientific academies were being promulgated. Max Planck took upon himself the task of writing a letter to all the Jewish members of the Prussian Academy of Sciences on 10 October 1938 to ask them to relinquish their memberships voluntarily in view of the modifications to the statutes required by the ministry.[103] On 1 December he was obliged to announce an intensification of this measure, as members of "mixed blood" (*Mischlinge*) or "Jewish by marriage" (*jüdisch Versippte*, i.e., those with a Jewish spouse) henceforth likewise had to leave.[104] In this situation, not pressure from the ministry but rather a letter by the enthusiastic Nazi DPG members Wilhelm Orthmann and Herbert Arthur Stuart (both members of NSDAP as well as of the *Sturmabteilung* [SA] since 1933) might have been the reason that Debye took the initiative in this direction.[105] In consultation with the rest of the board, Debye composed a circular letter to all the German members on 9 December, a variant of its first draft of 2 December. The opening words, "Under the compelling prevailing circumstances," implying a reluctant act by the society, preceded the announcement that "continued membership in the German Physical Society by German-Reich Jews in the sense of the Nuremberg Laws"

[102] REM to State Secretary Otto Wacker, Memorandum dated 3 October 1938, Hentschel, *Physics*, 178–181.

[103] Max-Planck-Gesellschaft zur Förderung der Wissenschaften (ed.), *Max Planck. Vorträge und Ausstellung zum 50. Todestag* (Munich: Max-Planck-Gesellschaft, 1997), 87.

[104] Planck to all members within the Reich, regular, foreign, corresponding, and honorary members, 1 December 1938, official copy to Debye, MPGA, div. III, rep. 19, Debye papers, no. 1191, sheet 16.

[105] The letter itself is unknown. It is mentioned in the minutes of the meeting of the Berlin Physical Society from 14 December 1938. MPGA, div. III, rep. 19, Debye papers, no. 1176, sheets 8–10; see Hentschel, *Physics*, 182–183. About Stuart personally, see his handwritten résumé dated 17 January 1939, Federal German Archives, Berlin (*Bundesarchiv*, including the former Berlin Document Center, hereafter BA), RS, Herbert Stuart born 27 March 1899. For the NSDAP and SA membership of Orthmann and Stuart, see their files in "Kartei aller Hochschullehrer," 1934, BA R 4901.

could "no longer be maintained." The members concerned were asked to give notice of their cancellations.[106] Von Laue expressly approved of this procedure as a member of the board.[107] Only Treasurer Schottky, who had evidently not been fully briefed about the current situation, raised any objections, fearing a chain reaction of withdrawals by émigrés and foreigners, who at that time still amounted to 25 percent of the membership. He warned about a consequential loss of "not insignificant foreign exchange receipts." But this was by no means the only aspect worrying Schottky. Relations abroad were an important pillar of the DPG's tradition and he – again not being fully in the picture – regarded the measure as regrettable even from the point of view of state policy.[108] In other respects, Schottky's view certainly agreed with the self-image that the DPG had long been propagating. The society's prestige partly depended on it. The officiating chairman, Zenneck, even pointed this out at the end of his opening speech at the national physics convention during the war year 1940: "And if prior to the present war the number of memberships in the German Physical Society among our foreign colleagues amounted to more than 20% . . . then that is a sign of the reputation our society and German physics have been enjoying abroad as well."[109]

On 15 December, the DGtP followed suit, choosing the same approach of writing a similar letter to its members.[110] The comparatively moderate tone the two societies chose for these requests seems to correspond with the awareness that, at this point, by dispensing with their memberships the affected persons were, in practice, losing little more than what had already been taken away from them. The DPG's existing statutes only had provisions for rescinding memberships in cases of delinquent dues or

[106] Draft by Debye, 2 December 1938, MPGA, div. III, rep. 19, Debye papers, no. 1014, sheet 34. Debye to the German members of the DPG, 9 December 1938, MPGA, div. III, rep. 19, Debye papers, no. 1014, sheet 38; in English in Hentschel, *Physics*, 181–182; a transcription and facsimile of both versions are in the appendix to Hoffmann and Walker, *Physiker*, 511, 564. See also Helmut Rechenberg, "Vor fünfzig Jahren," 418; Wolff, "Vorgehen."

[107] Handwritten note under Debye's draft of 2 December 1938, MPGA, div. III, rep. 19, Debye papers, no. 1014, sheet 34 (see the facsimile in Hoffmann and Walker, *Physiker*, 511).

[108] Schottky to Debye, 3 December 1938, MPGA, div. III, rep. 19, Debye papers, no. 1014, sheet 36.

[109] Speech by Zenneck, 1 September 1940 at the convention of German physicists in Berlin, in *Verhandlungen der Deutschen Physikalischen Gesellschaft*, 3rd series, 21 (1940), 31–34, especially 34.

[110] Mey to domestic members of the DGtP, 15 December 1938, MPGA, div. III, rep. 19, Debye papers, no. 1006, sheet 17.

revoked citizenship rights. To that extent, to abide by its statutes the DPG could not just rid itself of its Jewish members by crossing out their names; it relied on their "cooperation."[111] For that reason, the statutes also did not provide any possibility for retracting honorary memberships of the deceased. Consequently, although Emil Warburg and Eugen Goldstein were by definition Jewish according to the circular letter, they continued to be named on the list of honorary members for 1939.

At the board meeting of the Berlin Physical Society on 14 December 1938, Debye reported about the "developments of the non-Aryan issue" and announced as the next step after the circular letter a modification of the statutes.[112] On 2 March 1939, the DPG's board met in Berlin. During the discussions about the statutes issue, Chairman Debye reported that, because his circular letter had settled the "Jewish question," no Jews resident in the German Reich were members of the DPG anymore.[113] Regarding the problem of changes to the statutes along the lines of membership restrictions, he had made inquiries at REM.[114] From there he had obtained the information that no uniform guidelines existed for formulating the statutes of scientific societies, but he had received the corresponding details for scientific academies. Furthermore the ministry declared that "an immediate clearing-up of the Jewish question in the DPG would be appreciated very much."[115] Since REM was not aware of the circular letter in December, Debye could now report that "because of a 'Sonderaktion' (special action) already executed at the end of the last year no Jews living in Germany were members of the DPG any more.[116]

Given this situation, Debye supported the view that the statutes ought to be modified but that it was neither necessary nor advisable to take

[111] Statutes of the German Physical Society, article 6, in *Verhandlungen der Deutschen Physikalischen Gesellschaft*, 3rd series, 6 (1925), 59–60. The situation was different for academies. There, too, voluntary withdrawals were the preferred course; but in negative cases, article 11 of their statutes could be applied, which provided for expulsions. Appendix to a letter by Dames (government advisor at the REM) to Debye, 1 March 1939, MPGA, div. III, rep. 19, Debye papers, no. 1201, sheets 40–41.

[112] Minutes of the board meeting of the Berlin Physical Society on 14 December 1938, MPGA, div. III, rep. 19, Debye papers, no. 1176, sheets 8–9; see Hentschel, *Physics*, 182–183.

[113] Minutes of the board meeting of the DPG on 2 March 1939, MPGA, div. III, rep. 19, Debye papers, no. 1016, sheets 24–30, especially sheets 28–29.

[114] Debye to Dames, 18 February 1939, MPGA, div. III, rep. 19, Debye papers, no. 1201, sheet 39.

[115] Dames to Debye, 1 March 1939, MPGA, div. III, rep. 19, Debye papers, no. 1201, sheet 40.

[116] Debye to Dames, 11 March 1939, MPGA, div. III, rep.19, Debye papers, no. 1201, sheet 42.

immediate action. The board then agreed, to be on the safe side, to first of all appoint a committee to handle the topic should it become necessary. Aside from Debye, it consisted of von Laue, Schottky, Stuart, and the managing director, Walter Grotrian.[117] Stuart considered it important not to be the only "Pg" (*Parteigenosse*, member of the NSDAP) on the committee, which was secured by including Grotrian. An alteration to the statutes was of less interest to Stuart because he favored political alignment of the DPG: "I have long been thinking that the societies ought to take steps at their own initiative, like the Chemical Society and the VDI [Union of German Engineers, *Verein Deutscher Ingenieure*] to become somehow subordinate to Todt."[118] With this view, however, he stayed in the minority.

Contrary to Debye's expectations, the committee had to become active very soon thereafter, because the matter was deemed urgent after all.[119] In the following months there were at least two more meetings in Berlin.[120] The committee found it difficult, not only from the linguistic point of view, how to define who could be a member of the DPG. The treatment of foreign members posed a special problem. The REM did not want to admit any essential difference between them and resident members.[121] That could only be disadvantageous for the DPG. So Debye tried to use the interests "of the Foreign Office and the Ministry of Finance (foreign exchange authorities)" to make a counterargument.[122] To orient itself, the committee requested that the statutes of the other scientific and technical associations be sent to them that already contained all the formulations satisfying the new political requirements.[123] Debye probably wanted to

[117] Minutes of the board meeting, 2 March 1939, MPGA, div. III, rep. 19, Debye papers, no. 1201, sheet 29.

[118] Stuart to Georg Stetter, 17 March 1939, AIP, Goudsmit papers, box 28, folder 53; transcription in the appendix to Hoffmann and Walker, *Physiker*, 566–567. See also Stuart to Grotrian, 20 March 1939, MPGA, div. III, rep. 19, Debye papers, no. 794, sheet 11. Todt is meant here as the head of the roof organization National Socialist League of German Engineers (*Nationalsozialistischer Bund Deutscher Technik*, NSBDT).

[119] With reference to the urgency: Grotrian to Stuart, 25 May 1939, MPGA, div. III, rep. 19, Debye papers, no. 794, sheet 15.

[120] Minutes exist dated 21 June and 7 August: handwritten note on the verso of Stuart's letter to Grotrian, 26 May 1939, MPGA, div. III, rep. 19, Debye papers, no. 1018, sheets 10 and 25–28.

[121] Dames to Debye, 1 March 1939, MPGA, div. III, rep. 19, Debye papers, no. 1201, sheet 40.

[122] Debye to Dames, 5 August 1939, MPGA, div. III, rep. 19, Debye papers, no. 1201, sheet 49.

[123] Grotrian to GDCh and Deutsche Bunsengesellschaft, 4 June 1939, MPGA, div. III, rep. 19, Debye papers, no. 1201, no. 1018, sheets 11–12; GDCh and Deutsche Bunsengesellschaft to Grotrian, 6 June 1939, ibid., sheets 14 and 16; DMV to Grotrian, 23 June

FIGURE 7. Walter Grotrian (*Source*: Archiv der Berlin-Brandenburgischen Akademie der Wissenschaften)

relieve the DPG of some of the responsibility when he also asked REM on 1 July for instructions on what they should contain.[124] Such a request was repulsed there, however, and the DPG was asked to present a draft for subsequent approval. The ministry then pointed out additionally, in no uncertain terms, that any future support of scientific organizations depended on the "purging of the societies and associations of Jewish or mix-blooded members, as well as those related by marriage to Jews or persons of mixed blood."[125] The committee thereupon agreed on draft statutes at the following meeting on 7 August, which were submitted to

1939, MPGA, div. III, rep. 19, Debye papers, no. 1201, no. 1018, sheet 18; the file also contains the statutes of the Deutsche Lichttechnische Gesellschaft without any letter of request, MPGA, div. III, rep. 19, Debye papers, no. 1201, no. 1018, sheet 21.

124 Debye to Dames, 1 July 1939, MPGA, div. III, rep. 19, Debye papers, no. 1201, sheet 45.

125 Dames to Debye, 4 August 1939, MPGA, div. III, rep. 19, Debye papers, no. 1018, sheet 23.

the ministry on 19 August for its approval.[126] The outbreak of war on 1 September 1939 delayed completion of this matter.

After Debye left for what was initially scheduled as a temporary research stay in the United States in January 1940, Zenneck returned to the presiding chair. At the board meeting of 1 June 1940, he reported about the changes that the minister had made to the committee's draft statutes in the interim. The board accepted the revised draft and ratified it at the following administrative session held in connection with the fall convention.[127] The final statutes recorded in the association registry in 1941 limited memberships to resident citizens (*Reichsbürger*), which meant to those of "German or kindred blood." Foreigners residing beyond the Reich borders were exempted from that rule; but the state of war now made that of minor practical significance.

The changes included a passage concerning spouses:

> Only Reich Germans who, along with their wives [female members were accordingly clearly not foreseen] possess the rights of a Reich citizen, can become regular members, just as foreigners residing within the German Reich who fulfill the requirement for acquisition of Reich citizenship pursuant to German law. . . .

Foreigners residing abroad could become extraordinary members.[128]

Erich Regener's reaction illustrates the consequences of the new regulations affecting wives. He wrote to Zenneck in October 1941:

> I would like to inform you, pro forma, that my wife was listed earlier as a Jewess, but according to a *preliminary* decision by the Reich Office of Family Research can now count as of first-degree mixed blood.[129]

Later, in 1943, a questionnaire for new members was prepared in consultation with the ministry.[130]

THE POST-WAR PERIOD: VARIOUS SENSIBILITIES

After the war, communication problems, as well as much incomprehension, generally existed between Germans who had stayed and those who

[126] Debye to Dames, 19 August 1939, MPGA, div. III, rep. 19, Debye papers, no. 1201, sheet 50.

[127] Minutes of the board meeting of the German Physical Society, 1 June 1940, DPGA, no. 10014 (excerpted transcription in the appendix to Hoffmann and Walker, *Physiker*, 570).

[128] Statutes of the Deutsche Physikalische Gesellschaft, registered on 15 July 1941 (brochure attached to the society's proceedings).

[129] Regener to Zenneck, October 1941, DMA, Zenneck papers, box 52.

[130] Report on the meeting of the DPG's council, 31 May 1943, DPGA, no. 10023.

had been forced to leave the country. The root cause was a clash between various situations and experiences. The émigrés had been driven away from their homes and places of work, and their emotions were shaped by the crimes committed by Nazi Germany. Not a few of the émigrés had lost members of their very own families. Several of Einstein's cousins, for example, were murdered in Theresienstadt and Auschwitz. Most of his colleagues who had remained in Germany refused to assume any personal responsibility for these events, preferring rather to focus on their own sufferings.[131]

Such is the background to many quotes from this period. In June 1945, Lise Meitner reprimanded Otto Hahn and other members of the profession for working for Nazi Germany. In her view, they had not even attempted to exert passive resistance. "Certainly, to buy off your consciences you have helped a person in distress here and there, but have allowed millions of innocent people to be slaughtered without making the least protest."[132] Albert Einstein thought similarly, which is why he could not muster up any sympathy for an appeal his colleague Franck wanted to direct to the American government, within the context of the Morgenthau Plan, to better its treatment of Germany. In December 1945, Einstein refused point blank to sign such a petition:

> The Germans slaughtered millions of civilians according to a carefully conceived plan in order to capture their places.... From the few letters I have received from there and reports by a few reliable persons who had recently been sent there, I see that among the Germans there isn't any trace of remorse.[133]

When almost at that very time Sommerfeld asked Einstein to agree to rejoin the Bavarian Academy, he reacted quite similarly but not without stressing his appreciation of a few personal contacts:

> After the Germans slaughtered my Jewish brothers in Europe, I want to have nothing to do with the Germans ever again, neither with a relatively harmless Academy. It is different for the couple of individuals who remained steadfast, within the range of feasibility. I was pleased to hear that you were among them.[134]

[131] See the contribution to this volume by Hentschel.

[132] Meitner to Hahn, 27 June 1945, transcribed in Fritz Krafft, *Im Schatten der Sensation* (Weinheim: VCH, 1981), 181f.; English translation in Hentschel, *Physics*, 332–335, especially 333.

[133] Einstein to Franck, 11 December 1945, reprinted in Jost Lemmerich (ed.), *Max Born, James Franck. Physiker in ihrer Zeit. Der Luxus des Gewissens*, exhibition catalogue (Wiesbaden: Reichert, 1982), 141–142. English translation in Hentschel, *Mental*, 165.

[134] Einstein to Sommerfeld, 14 December 1946, DMA, HS 1977–28/A, 78. Reprinted in Eckert and Märker, *Arnold Sommerfeld*, 602–603.

Kasimir Fajans, who had emigrated from Munich in 1936, wrote to Sommerfeld on 1 October 1946 from the United States about his feelings in this regard:

> I would like to mention that in the various European countries during the German occupation, many of my friends and relatives were killed violently and that specifically in Poland two of my wife's brothers together with their families, one of my sisters together with all those close to her, as well as numerous other relatives were murdered. It will always remain incomprehensible to me how the German nation let itself be "led" into such a policy, which ultimately caused the collapse as well.[135]

In this scenario, Heisenberg found some of his contacts with émigrés during a visit to England in 1948 pleasant: "Born was as friendly and nice as in the old days, Simon and Peierls, too, were very hospitable but do seem to have a harder time getting over the injustice done to them."[136]

The dominant sentiment among those who had remained in Germany soon was that most had likewise just been victims. Even von Laue took this view. This leveling left no more room for moral distinctions.[137] From such a perspective, von Laue thought he could argue away special compensation rights to émigrés and those persecuted under National Socialism for the losses they had suffered. He wrote to Otto Hahn in May 1954: "Kallmann is complaining that the National Socialists robbed him of his wealth. Well, ours has been robbed by the Allies, which in practice comes down to the same thing. In times *like those*, pretty much everyone becomes impoverished."[138]

During the first few years after the war, Germans found themselves once again in a political environment that to them seemed needlessly to hamper their work. In addition to problems with the economy, there were, on one hand, political controls that Otto Hahn also referred to as the "denazification evil"; on the other hand, there were compulsory services of Germans in the countries of the victorious powers.[139] The émigrés' need to keep their distance from such Germans is alluded to in an unsigned letter published in the *Physikalische Blätter* in 1948, written by a physicist working in the United States under such constraints:

[135] Fajans to Sommerfeld, 1 October 1946, DMA, no. 89, Sommerfeld papers, 008.

[136] Heisenberg to Sommerfeld, 31 March 1948, Max Planck Institute for Physics (*Max-Planck-Institut für Physik*, hereafter MPIfP), Heisenberg papers.

[137] See the contributions to this volume by Hentschel and Rammer.

[138] Von Laue to Hahn, 8 June 1954, MPGA, Kallmann's personnel file.

[139] Ernst Brüche, "Rückblick auf 1947," *Physikalische Blätter*, 4 (1948), 45–49, especially 45–46.

The only people who are sometimes pointedly reserved (but not one-hundred percent either) are the former German émigrés from the period after 1933, who had partly suffered severely in the past system and who therefore are not happy to see us here, primarily trying more or less obviously to keep us away from their universities in which they are established.[140]

The émigré Max Born returned as a guest to the national physics convention in September 1948 in Clausthal-Zellerfeld, at that time within the British zone. He was to be awarded the first Max Planck Medal since the war.[141] On that occasion, Born apparently did not consider himself a guest of the Germans, however: "We took part and were given a friendly reception, but we were at that time still regarded as visitors from England, watched over and taken care of by the occupying power."[142] In the following year, Lise Meitner received this award (together with Otto Hahn). In her letter to von Laue from April 1949, she expressed her delight, emphasizing her special connection to Germany:

I would most heartily like to thank you as chairman (and as an old friend) and naturally also the board of the German Physical Society for this great and so entirely unexpected distinction. Every bond tying me to the former – my dearly beloved – Germany, the Germany to which I cannot be grateful enough for the profound joy of doing scientific research during the decisive years of my academic development and for a very dear group of friends, is a very valuable gift for me.[143]

After the medal had gone out to Peter Debye in 1950, whose transfer to the United States in 1940 had an entirely different quality to the emigrations otherwise discussed here, as it took place with the approval of the pertinent Reich ministry,[144] Gustav Hertz and James Franck shared the prize for 1951. Einstein, who otherwise continued to refuse any personal contact with German institutions, supported this decision.[145] At that time, professional aspects were not the only factor in selecting the winner. Von

[140] "Anonymous letter 'from USA,'" in *Physikalische Blätter*, 4 (1948), 268–269. English translation in Hentschel, *Mental*, 156f.

[141] See the circular letter by Sommerfeld, 24 February 1948, in which he nominates Born and Debye, DPGA, no. 20961.

[142] *The Born–Einstein Letters. Correspondence between Albert Einstein and Max and Hedwig Born from 1916 to 1955, with commentaries by Max Born* (New York: Walker, 1971), 200.

[143] Meitner to von Laue, 25 April 1949, in Lemmerich, *Meitner–Laue*, 528–529.

[144] See Dieter Hoffmann, "Peter Debye (1884–1966): Ein Dossier," *Max Planck Institut für Wissenschaftsgeschichte, Preprint* 314 (Berlin: MPIfWG, 2006).

[145] Einstein authorized von Laue to represent him in this regard: Einstein to von Laue, 5 January 1951, DPGA, no. 20963.

Laue had already mentioned this in his circular of 29 December 1950. The 1951 award had previously been considered for Dirac. But von Laue, who was able to refer to similar opinions by former awardees, argued that it was important to give James Franck some moral encouragement at that time. The latter's research achievements offered von Laue and the other supporters ample arguments in his favor.[146] Franck gratefully accepted the distinction.[147]

In fall 1950, the regional societies in West Germany had been reconsolidated into the registered association: *Verband deutscher physikalischer Gesellschaften*. Zenneck again assumed the function of chairman. He exercised this office for just 1 year. Prominent names such as Walter Weizel, Hans Kopfermann, and Walther Gerlach were initially mentioned as his potential successor. But the choice in 1951 fell, not on any of these internationally known candidates, but rather on Karl Wolf, chairman of one of the regional associations. He was director of the metrology and testing department of the chemical concern BASF, where he had worked ever since taking his doctorate under Wilhelm Wien in 1927. Along with his administrative skills, the general wish to stay in close contact with industry may also have been a decisive point for Wolf's election.[148] He remained the association's chairman until 1954.

THE INVITATION TO REJOIN THE PROFESSIONAL ORGANIZATION OF GERMAN PHYSICISTS

In 1952, the new chairman was busy with an initiative intended to encourage refugee former members who were living abroad to rejoin the society. Writing personal letters was preferable to making a general public announcement. At its meeting in September 1952, the board charged Wolfgang Gentner with drafting a suitable letter to be forwarded to von Laue and Karl Wolf.[149] Political circumstances in the Federal Republic of Germany may have promoted such an initiative at that point in time.

[146] Von Laue to his "Dear Colleagues" (Einstein, Bohr, Sommerfeld, Heisenberg, Schrödinger, de Broglie, Jordan, Hund, Kossel, Born, Hahn, Meitner, Debye), 29 December 1950, DPGA, no. 20964. Positive responses by Born on 11 January 1951, Meitner on 2 April (?) 1951, and Hahn on 4 January 1951, DPGA, no. 20964.

[147] Franck to Zenneck, 25 May 1951, reprinted in Lemmerich, *Max Born*, 148.

[148] Wolf to the board members of the Physical Society in Württemberg-Baden-Palatinate, 24 August 1951, DPGA, no. 40208. I refer to information by Gerhard Wolf, Heidelberg, son of Karl Wolf.

[149] Minutes of the board meeting, 28 September 1952, DPGA, no. 20004.

The law regulating compensation for the National Socialist injustices to public servants, passed in the foregoing year, was extended to apply to émigrés in 1952. Just 18 days before the above-mentioned resolution by the board, the Federal Republic and Israel had signed a reparations treaty ("*Wiedergutmachungsabkommen*").

What Gentner had seen and experienced under National Socialism equipped him for a particularly sensitive treatment of the consequences of this past. In occupied Paris, he had been commissioned to direct a group assigned with setting the local cyclotron in operation. His intercession for his French fellow scientists – he was able, among other things, to arrange the release of Frédéric Joliot and Paul Langevin from prison – ultimately was the reason why he was sent back to Germany.[150]

Although the first draft of the letter apparently came from Max von Laue,[151] the final version ultimately emerged out of a discussion with Gentner and Wolf. Gentner understood much better than von Laue what psychological barriers existed on the side of the émigrés. He perceived how inappropriate von Laue's formulation was that the board "may emphasize, however, that both societies [the DGtP included] retained [their purged members] as long as ever possible."[152] In Gentner's view, it was inadmissible for von Laue to extend his doubtlessly honorable personal attitude at that time to the entire DPG. Gentner rather missed a confession of partial guilt: "We should regret that we were not in a position, perhaps because we lacked the courage, to prevent these laws by the Nazi government."[153] He thought it important to stress the injustice done to the refugee members. At first, "after a double exchange back and forth," as he disappointedly informed Wolf, he could only wrest out of von Laue the formulation: "we have always deeply regretted."[154] Wolf then made a few more corrections to soften what in Gentner's opinion were "still extant harsh formulations."[155] These included crossing out the expectation in von Laue's version that it should be possible "to draw a line under ('*Schlussstrich setzen*') the disastrous occurrences of the Hitler

[150] See Dieter Hoffmann and Ulrich Schmidt-Rohr (eds.), *Wolfgang Gentner (1906–1980). Festschrift zum 100. Geburtstag* (Heidelberg: Springer, 2006), 37f.

[151] Forwarding of a draft, see von Laue to Gentner, 3 November 1952, DPGA, no. 40209.

[152] Max von Laue's draft, November 1952; Gentner to von Laue, 15 November 1952, DPGA, no. 40209.

[153] Gentner to von Laue, 15 November 1952, DPGA, no. 40209.

[154] Gentner to Wolf, 4 December 1952, DPGA, no. 40209.

[155] Wolf to Gentner, 3 December 1952; Gentner to Wolf, 4 December 1952, DPGA, no. 40209.

period insofar as this is due to us."[156] The revised draft was completed in December 1952; the final version was ready on 12 March 1953.[157] Because Wolf was largely unknown outside the country, whereas von Laue had an untarnished reputation of political integrity, particularly among the émigrés, they both signed it.[158] Initially, 112 such letters were sent out, later followed by some to émigrés who had not been members before (P. Bergmann, H. Bethe, J. von Neumann, V. Weisskopf). Not only émigrés but also foreigners who had formerly belonged to the DPG were included in this group (see the table in the documentary appendix to Hoffmann and Walker, *Physiker*, 655–658). The invitation extended in this circular letter thus also expressed compassion for the injuries inflicted during the National Socialist era. With reference to the alleged relatively late membership expulsions, emphasis was placed on the fact that both societies have "tried as long as possible to avoid the injustice." Against this backdrop of a supposedly rather irreproachable position by the organization, the initiative was presented "as an expression of our sincere effort to repair the ignoble actions of the Hitler period, insofar as this is possible for us."[159]

With this in mind, among the physicists contacted, only the émigrés from the National Socialist period would have felt properly addressed. More than 40 percent of the addressees, however, were former foreign members who had never lived under Nazi rule; and in some cases these were even Germans, such as Frank Matossi, who had left Germany only after the war under completely different circumstances and whose social or political backgrounds did not fit the group of persecuted or refugee members at all. Ernst Stuhlinger, for instance, had even formerly collaborated at Peenemünde. Indeed, the official drafter of the request, that Jewish members should leave the society, Peter Debye, was contacted in this way as well. Suitable variations in such individual letters would have been easily possible. Besides a lack of care, for certain addressees ought to have been excluded, there is also a certain "shop-window effect" as regards the foreigners. They, too, were offered an apology, although they

[156] Comparison of the version from 9 December 1952 against von Laue's draft of November 1952, DPGA, no. 40209.

[157] Verband Deutscher Physikalischer Gesellschaften e.V. (Wolf and von Laue) to former members residing abroad, draft of the circular, 12 March 1953, DPGA, no. 20437.

[158] Following Gentner's suggestion; see Gentner's letter to von Laue, 26 November 1952, DPGA, no. 40209.

[159] Circular letter, 12 March 1953, DPGA, no. 20437.

had not actually been affected. This was apparently supposed to prop up the society's reputation and thereby favor its return into the international community. Purposefully omitted from this action had been those who were living "in countries on the other side of the iron curtain" – which conformed to the general political situation – and those who had belonged solely to the Austrian regional association.[160] Nevertheless, the names of some émigrés were still missing, which was mostly due to the lack of knowledge of their addresses. In the period between March and August 1953, relatively few replies arrived (somewhat less than 20 percent), but all of them were friendly and in agreement. Among this total of 20 positive reactions, just 12 originated from the group of actually affected persons.

Richard Gans, who, thanks to special circumstances, had been able to survive inside Germany and had since emigrated to Argentina, only too willingly accepted the idea that the community of German scientists had merely bowed to inescapable external pressures: "I can give you my assurance that I never felt any bitterness about my elimination from the German Physical Society, because I knew that it involved an act enforced from outside against the society's will."[161] Already intending to spend part of the year in Germany after his imminent retirement, Max Born blocked out the symbolical significance of this matter a little by broaching the question whether membership could still make any practical sense for him, as he would soon no longer be an active researcher. The society attached to this type of reaction the hope that Born could soon be won over, after all.[162] Born did, in fact, rejoin in the following year.[163]

The former treasurer, Marcello Pirani, perceived the circular letter as friendly and was "sympathetically" disposed to the invitation to become a member again. As he had been addressed with the title of doctor, he did not fail to point out that until 1933 he had been professor at the Berlin Polytechnic, which was the appropriate title to use.[164] Brief replies arrived from Friedrich Dessauer, Andreas Gemant, Gerhard Herzberg,

160 Schoch to Wolf, 7 March 1953, DPGA, no. 20437.

161 Gans to Wolf, 29 April 1953, DPGA, no. 20437.

162 Max Born to "Sehr geehrter Kollege" (Wolf), 24 March 1953; Wolf to Schoch, 31 March 1953, Schoch to Wolf, 31 March 1953, Schoch to Wolf, 2 April 1953, Schoch to Born, 13 April 1953; all DPGA, no. 20437.

163 Ebert (managing director) to the board members of the reg. association, 12 April 1954, DPGA, no. 20437. He wanted to confirm Born's readmission.

164 Pirani to Wolf, 30 March 1953; Schoch to Pirani, 13 April 1953; DPGA, no. 20437.

and Hans Reinheimer announcing their willingness to join the association again.[165] Paul Rosbaud in London was so touched by the gesture that he even volunteered to advertise the DPG further. He emphasized the personal credibility of both signers, Wolf and von Laue.[166] A positive signal likewise came from the political émigré Herbert Jehle.[167] Just as in the case of the Australian immigrant Ilse Rosenthal-Schneider, approval in principle was the general response, with perhaps added questions and doubts about the technicalities of paying the membership fees.[168]

Wigner recruited his colleague John von Neumann, Kurt Sitte, and Peter Bergmann.[169] Neither von Neumann nor Bergmann had been members before the war, but such additions were regarded as legitimate, as it was assumed that only the political circumstances had formerly prevented them from entering the society. Along these lines, Victor Weisskopf and Hans Bethe were also approached. A positive answer only arrived from the latter. Bethe wanted his approval to be interpreted "as a sign of my friendship with many of my colleagues in Germany."[170] Thus, with Bethe, the society had gained another first-ranking émigré scientist besides Wigner and von Neumann. Lise Meitner, who was likewise on the contact list, had already been made an honorary member in 1948.

At the subsequent 18th convention of physicists in Innsbruck in September 1953, Chairman Wolf reported about the success of this drive, consequently being in a position

> to welcome into our association those physicists, former members of the German Physical Society and the German Society of Technical Physics now living abroad, who announced their entry into the association in the past few weeks in response to our recent invitation. We are especially pleased that those among these new members, who had been subjected to great injustice during the Hitler period, did

[165] Dessauer to "Sehr geehrte Herren," 1 April 1953; Gemant to Wolf, 4 April 1953, Herzberg to Wolf, 20 April 1953, Reinheimer to Wolf, 1 May 1953; all DPGA, no. 20437.

[166] Rosbaud to Wolf, 8 April 1953, DPGA, no. 20437.

[167] A. L. Johnson replied also on behalf of Jehle: Johnson to von Laue, 27 April 1953, DPGA, no. 20437.

[168] Rosenthal-Schneider to "Sehr geehrter Herr Doktor" (Wolf), 22 August 1953; Schoch to Rosenthal-Schneider, 25 November 1953; DPGA, no. 20437.

[169] Wigner to Wolf, 14 April 1953; Sitte to Wolf, 15 April 1953, Schoch to von Neumann, 4 May 1953, Schoch to Bergmann, 4 May 1953, Bergmann to Schoch (secretary), 26 May 1953; all DPGA, no. 20437.

[170] Schoch to Weisskopf, 4 May 1953, Schoch to Bethe, 4 May 1953, Bethe to Schoch, 27 May 1953; all DPGA, no. 20437.

not let their so understandable bitter feelings dominate, preferring rather to tie a hopeful new bond with their former homeland.[171]

However, another voice audible within the DPG did note that the chosen point in time had certainly not been the earliest possible. It came in the form of Georg Joos's commentary: "this decision, which should have been taken years ago. . . . "[172] His more critical assessment was presumably connected with a guest professorship that Joos had meanwhile had in the United States.

Just as before the war, the list of members henceforth included a separate column for those residing abroad. What counted was evidently not the quite slim response – 80 percent of the fellow scientists addressed did not reply at all – but the acceptances, however low in number. They could prove helpful for reintegration into the international scientific community. In fall 1952, the association had filed an application for readmission into the International Union of Pure and Applied Physics, originally founded in 1922.[173] Bethe's remarks about his personal friendships show that colleagues who had been purged were certainly capable of serving as bridgeheads in the international rapprochement with German physicists.

At the end of his life, even Einstein used a more reconciliatory choice of words in his relations with the DPG. On the 50th anniversary of his important publications in the *Annalen der Physik*, he was invited by both Max von Laue and Gustav Hertz to a joint East–West celebration in Berlin. His letter in February 1955, addressed to his "dear colleagues," did not reject it categorically, rather graciously declined owing to his bad state of health: "It would be a great pleasure for me if I could be present there. But the 50 years in the interim have left only decrepit remnants of myself so I cannot undertake any major trips. This nonetheless does not dampen my delight about this proof of good will."[174]

171 *Physikalische Verhandlungen*, 4 (1953), 128.

172 Joos to Wolf, 22 February 1954, DPGA, no. 20437.

173 Minutes of the meeting of the members of the Verband Deutscher Physikalischer Gesellschaften on 29 September 1952, DPGA, no. 40209. See also Ebert to Gentner, 29 October 1952, DPGA, no. 40209, in which Gentner is informed about having been elected into the "National Committee," which was supposed to represent the DPG in the International Union, which they were intending to join.

174 Einstein to the Physical Society of Berlin, 10 February 1955, DPGA, no. 10683; reprinted in Horst Nelkowski, "Die Physikalische Gesellschaft zu Berlin in den Jahren nach dem Zweiten Weltkrieg," in Theo Mayer-Kuckuk (ed.), *Festschrift 150 Jahre Deutsche Physikalische Gesellschaft, special issue of Physikalische Blätter*, 51 (1995), F-143–F-156, especially F-148.

CONCLUDING REMARKS

The DPG took pains to evade conflicts of any kind during the first years of National Socialist rule. This restraint may have been a strategy to expose as few flanks as possible to political attack and thereby not awaken any critical interest among the public. Obviously, the DPG had the illusion of preserving a kind of autonomy in this way but by doing so restricted its range of action, the limits of which were circumscribed much more closely than would have been called for in "preserving the professional interests," as claimed in its statutes. During the first wave of dismissals in 1933, the DPG did not take a political stance; rather, its chairman tried personally to provide concrete help for those affected fellow professionals on a humanitarian level. In the following years, similar to other scientific societies with a high percentage of foreign members, there were no initiatives to question the memberships of "non-Aryans." There was the fear that, if "non-Aryans" in Germany were excluded, then most of the foreigners would also quit their memberships. This concerned not simply the self-image of an association that always built its reputation on international contacts, for keeping the foreign members also meant receipts of valuable foreign exchange. However, the "non-Aryans," respectively the Jewish members, did not and subsequently could not participate in the activities of the DPG.

This exclusion became generally visible in a jubilee volume of the *Annalen* to honor Sommerfeld's 70th birthday in 1938. Therefore, these Jewish members did not lose anything when Debye's circular letter of 9 December 1938 requested their withdrawals, for this was not the beginning but rather the formal conclusion of a development already under way. This was the DPG's own initiative, because at that time there had been no direct pressure of the corresponding ministry to act in this direction. However, it would have become unavoidable only a few months later.

Some years after the war had ended, the DPG sought to retrieve its persecuted and expelled former members. The back and forth between Gentner and von Laue about the vocabulary to use in the letters to these colleagues, such as "partial guilt" or "drawing a line," certainly reflects a quarrel that would occupy German society as a whole for many more years to come. The group of people written to, including many foreigners who had not been affected at all by the Nazi injustices, shows that it was not purely a matter of reconciliation but also involved the issue of the

DPG's reentry into the international community. Thus, under different circumstances, those formerly excluded or marginalized members were supposed to become useful for the DPG again. Up until 1938, they had been kept to maintain the memberships of foreigners; after 1953, they helped to reestablish international contacts.

4

The German Physical Society and "Aryan Physics"

Michael Eckert

"Aryan physics" (*Deutsche Physik*, literally translated as "German physics"), as it was called after the title of Philipp Lenard's textbook, is a prominent example of ideologized science under National Socialism. Alan Beyerchen drew an impressive picture of it in his 1977 study, *Scientists under Hitler*.[1] Although Aryan physics was confined to a few fanatics gathered around the two Nobel laureates Lenard and Johannes Stark, its influence was strong enough to secure professorships for some of their followers. The most striking incident took place at the University of Munich, where the aerodynamicist Wilhelm Müller, a completely unknown figure among physicists, was appointed to Arnold Sommerfeld's renowned chair for theoretical physics.

How did physicists in Germany respond to Aryan physics – more specifically, how did their representative professional association, the German Physical Society (*Deutsche Physikalische Gesellschaft*, DPG), act? This is the answer the editor of the *Physikalische Blätter*, Ernst Brüche, offered to this question in 1946:

> I believe that the physics community has a right to know how the chairman of the German Physical Society did everything within his power in the years since the last convention of physicists of 1940, despite all the difficulties, and with a lot of courage, to represent the matter of a proper and decent physical science at odds with the party and the ministry, and to prevent worse from happening than what already has. I believe that this battle against party politics may surely be described as the true "German physics," because – although actively conducted

[1] Alan D. Beyerchen, *Scientists under Hitler: Politics and the Physics Community in the Third Reich* (New Haven: Yale University Press, 1977).

FIGURE 8. Ernst Brüche around 1940 (*Source*: DPGA)

by just a few – it has been effectively and morally supported by the overwhelming majority of physicists.[2]

In 1947 Carl Ramsauer, who had chaired the DPG from 1940 to 1945, published in the *Physikalische Blätter* excerpted documents regarding the history of the DPG during the Hitler period. One passage reads:

Among the steps that the German Physical Society took to save physics in Germany, the Society's petition that had been sent by the Society's chairman

[2] Ernst Brüche, "'Deutsche Physik' und die deutschen Physiker," *Physikalische Blätter*, 2 (1946), 232. See the comparative transcription of this post-war article against the original submission in the appendix to Dieter Hoffmann and Mark Walker (eds.), *Physiker zwischen Autonomie und Anpassung. Die Deutsche Physikalische Gesellschaft im Dritten Reich* (Weinheim: VCH, 2007), 594–600. For Brüche's role as editor of the *Physikalische Blätter*, see Swantje Middeldorff, "Ernst Brüche und die Geschichte der Physikalischen Blätter 1944–1974" (Diplom-Thesis: University of Hamburg, 1993).

Prof. Ramsauer on 20 January 1942 to Education Minister Rust plays a special role. In this petition, drafted by the two board heads Prof. Ramsauer and Prof. Finkelnburg, the responsible ministry was told in no uncertain terms what needed to be said.[3]

By using expressions like "battle against party politics," or "to save physics in Germany," the representatives of the DPG gave their organization the appearance after 1945 of having been a professional association in resistance. Beyerchen's historical analysis also acknowledges the "offensive against Aryan physics" as evidence of the DPG's struggle for the autonomy of its profession. Ramsauer's petition particularly denounced the choice of Müller as Sommerfeld's successor as a scandalously poor appointment. As Beyerchen put it: "In a sense, Aryan physics had been *too* successful" in Munich: it was "a tactical victory but a strategic defeat for Aryan physics." While acknowledging that the DPG sought "adherence to professional values" in the face of ideologically motivated Nazi influences, Beyerchen left open whether this ought to be considered as resistance against National Socialism.[4]

Allusions to representatives of Aryan physics are retrospective categorizations. The study on Sommerfeld's successor, Wilhelm Müller, objects to this common usage of Aryan physics. Rather than a movement, Aryan physics comprehended "a very loose set of little groups only temporarily apparently coherent through practical alliances or through – mostly false – assumptions about having found like-minded fellows," some of whom flocked around "self-assigned and self-centered 'leaders' like Stark or Lenard, as their students or potential beneficiaries."[5] There was no consistency, either, in the bias either for experiment or against relativity theory. Müller, for example, was a mathematician by training; his manner of conducting aerodynamics was ranked by Ludwig Prandtl, the doyen of this discipline in Germany, as "amply formalistic," a quality otherwise rather attributed to so-called Jewish physics. And if one blindly chooses opposition to relativity theory as a unifying criterion, one overlooks the at times considerable differences concerning the causes of this opposition. It could have been motivated by racial or political, if not scientific, arguments. If the following analysis nevertheless uses the label

[3] Carl Ramsauer, "Zur Geschichte der DPG in der Hitlerzeit," *Physikalische Blätter*, 3 (1947), 110–114.

[4] Beyerchen, *Scientists*, 166f, 175f, and 198.

[5] Freddy Litten, *Mechanik und Antisemitismus: Wilhelm Müller (1880–1968)* (Munich: Institut für Geschichte der Naturwissenschaften, 2000), 380.

"Aryan physics," this should not be mistaken as an acknowledgment that it represented a movement but rather taken as a descriptive term for a phenomenon that is analyzed here only with regard to the conduct of the DPG. On one hand, historical accounts of this topic since the end of the Second World War themselves deserve historical scrutiny, so use of the term "Aryan physics" has become unavoidable. On the other hand, the above-cited critique by Freddy Litten has arguably revealed it as undifferentiated black or white oversimplification.

The DPG has hitherto not been the subject of special inquiry, so despite the numerous studies on Aryan physics, it is not clear how the society's representatives behaved toward this phenomenon. The business at hand is to indicate how the DPG's board responded to situations of conflict. The scandal surrounding Sommerfeld's succession, which the society itself put forward as proof of its resistance, is a starting point in this chapter. That section is followed by an analysis of the debate within the DPG's leadership during those years (1935–1940). The next section will subject to critical test the society's involvement in the downfall of Aryan physics. The subsequent two sections are devoted to the emergence of the myth of resistance: The way the DPG's representatives came to terms with the past is contrasted with statements by former "Aryan physicists." The "polyphony"[6] of various readings shows how controversial the topic Aryan physics was after the Third Reich came to an end.

TEST CASE: SOMMERFELD'S SUCCESSOR

The precise wording Ramsauer chose as DPG chairman in his petition of 1942 to the Reich Minister of Education in protest against ideologically motivated, incompetent appointments to professorships in physics was made public in the *Physikalische Blätter* in 1947 as follows:

> Professorship appointments in physics do not always follow the principle of preference according to ability, which has proven successful both in the past and the present. I do not want to go into the details of well-known and obviously inappropriate appointments, as this would not change the circumstances and would only greatly annoy individuals. But I am prepared to support my opinion more specifically upon request. Nevertheless, in order to elucidate the absolute seriousness of the state of affairs, I would like to make an exception, all the

[6] Klaus Hentschel, "Finally, some historical polyphony!" in Matthias Dörries (ed.), *Michael Frayn's "Copenhagen" in Debate. Historical Essays and Documents on the 1941 Meeting Between Niels Bohr and Werner Heisenberg* (Berkeley: Office for History of Science and Technology, 2005), 31–37.

more so since the case in point is of symptomatic significance precisely to the state of theoretical physics in Germany. I enclose an opinion on the successor to Sommerfeld, Prof. W. M. (Munich), by Prof. L. Prandtl (Göttingen), which the latter had submitted to me in another connection but had placed at my disposal.[7]

By this, the DPG itself called its reaction to the Sommerfeld succession a test case for its attitude toward the Nazi regime. With the following quote from this petition, the editor of the *Physikalische Blätter* documented the society's advocacy of the cause of modern theoretical physics against the animosities of Aryan physics:

In our country one main branch of physics, theoretical physics, is being pushed more and more into the background.... In contrast to this, it must be observed that it is impossible for physics as a whole to thrive unless theoretical physics thrives. Modern theoretical physics in particular has a whole series of the greatest positive achievements to offer that could also be of vital importance to the economy and the armed forces; and the very general accusations made against the advocates of modern theoretical physics of being pioneers of the Jewish spirit are as unsubstantiated as they are unjustified.[8]

Accordingly, there is no question about *whether* Chairman Ramsauer, on behalf of the DPG, took exception to the Munich appointment and the attacks on modern theoretical physics. But one has to fill in the missing passages of the petition to gain an authentic picture of *how* this criticism had been presented. The ellipses between the words "into the background.... In contrast to this," contract the sentence:

The legitimate struggle against the Jew Einstein and against the excrescences of his speculative physics has spread to the whole of modern theoretical physics and has brought it largely into disrepute as a product of the Jewish spirit (see Attachment II).[9]

Attachment II lists a total of 17 "Publications against modern theoretical physics" – albeit without naming the major work whose title, *Deutsche Physik*, was adopted as the name of the ideology of Aryan physics. Its author, Lenard, happened to be Ramsauer's former teacher. In the printed version in the *Physikalische Blätter*, the original wording of the petition was tampered with, shifting the reference to attachment II elsewhere while

[7] Carl Ramsauer, "Eingabe an Rust," in *Physikalische Blätter*, 3 (1947), 44 (reprinted in Hoffmann and Walker, *Physiker*, 516, 594–617). A translation of the original submission is provided in Klaus Hentschel (ed.), *Physics and National Socialism: An Anthology of Primary Sources* (Basel: Birkhäuser, 1996), docs. 90–93, the present quote on 280.

[8] Hentschel, *Physics*, 279–280.

[9] Ramsauer to Rust, 20 January 1942. Carbon copy with enclosure letter to Prandtl, 23 January 1942, Archives of the Max Planck Society (*Archiv der Max-Planck-Gesellschaft*, hereafter MPGA), div. III, rep. 61, no. 1413; translation in Hentschel, *Physics*, 279.

otherwise raising the impression of complete authenticity. So one should at least bear these omissions in mind with the reiterated suggestion that the DPG "openly presented reservations about Rust's science policy."[10]

But what measures did the DPG take against Aryan physics before 1940; that is, at a time when the subsequently so regretted "inappropriate appointments" could still have been prevented? It had been clear since 1937 that the Sommerfeld succession in particular had become far more of a political issue than usual in the bureaucratic process of university appointments in the civil service. Others besides the selection committee officially appointed by the faculty were working in the background: the university president, the Bavarian Culture Ministry and the Reich Ministry for Science, Education and Culture (*Reichsministerium für Wissenschaft, Erziehung und Volksbildung*), the National Socialist German University Lecturers League (*Nationalsozialistischer Deutscher Dozentenbund*, NSDDB), the Nazi party (*Nationalsozialistische Deutsche Arbeiterpartei*, NSDAP), the SS (*Schutz-Staffel*, translated as "Protection Squadron"), and various influential individuals each pushed their favorite among the short-listed candidates. Thus, intervention by the DPG would certainly not have been unusual. This politicization caused the selection of Sommerfeld's successor to become the subject of a number of detailed studies,[11] without it being possible to pinpoint what exactly had ultimately triggered the scandal surrounding Müller's appointment. With an eye to possible influence by the DPG, I will review the most important stages of this appointment controversy.

In January 1935, a new law concerning the retirement of university teachers reduced the maximum age of professors from 68 to 65 years. As a result, Sommerfeld became an emeritus as of the end of the semester then in progress. He was asked to continue teaching, however, until his successor had been appointed. In April 1935, the selection committee produced the following list of finalists for the position:

1. Werner Heisenberg
2. Peter Debye
3. Richard Becker

[10] Wilhelm Walcher, "Fünfzigste Physikertagung," in *Physikalische Blätter*, 42 (1986), 218.

[11] Beyerchen, *Scientists*; Litten, *Mechanik*; David Cassidy, *Uncertainty: The Life and Science of Werner Heisenberg* (New York: Freeman, 1991), revised edition David Cassidy, *Beyond Uncertainty: Heisenberg, Quantum Physics, and the Bomb* (New York: Bellevue Literary Press, 2009); Michael Eckert and Karl Märker (eds.), *Arnold Sommerfeld. Wissenschaftlicher Briefwechsel, vol. 2: 1919–1951* (Berlin: Verlag für Geschichte der Naturwissenschaften und der Technik, 2004).

The second and third in line were included for the sake of appearance, since Heisenberg had long been known to be the candidate of choice for Sommerfeld, who was the committee's guiding voice. This choice found general approval among the faculty, and there was no notable objection even by the SS member and representative of the Nazi party and NSDDB, Wilhelm Führer. That was the situation at the outset.

Signs that Heisenberg's call would encounter opposition soon emerged, but the first protests did not come from "Aryan physicists." Problems are to be expected, Peter Debye wrote to Sommerfeld in September 1935 after a visit to REM in Berlin, because the responsible REM official (Franz Bachér) "wants 'the Munichites' to conjure up a new list, because he wants to have Heisenberg go to Göttingen."[12] At the University of Göttingen, Max Born's chair for theoretical physics had been vacant since 1933, and there, too, Heisenberg was deemed the most suitable successor. The "Munichites" satisfied this wish in November 1935 and named other eligible candidates (Friedrich Hund, Gregor Wentzel, Ralph Kronig, Ernst Stückelberg, Erwin Fues, Fritz Sauter, Albrecht Unsöld, and Pascual Jordan), every one of them qualified theorists. At the same time, however, the appointment committee made it unmistakably clear that Heisenberg was the man they wanted. The matter remained up in the air until April 1936, when the representative for the NSDDB, Wilhelm Führer, informed the university president that Debye's candidacy could no longer be considered because he had received an appointment to Berlin at the Kaiser Wilhelm Institute of Physics. Führer had written to Stark and Rudolf Tomaschek to obtain further recommendations. The REM dismissed the names they submitted (Hans Falkenhagen and Sauter), however, and Sommerfeld was asked to substitute for his chair in the coming semester as well.

In the interim, the first attack was made on Heisenberg. In December 1935, Stark delivered a speech during the renaming ceremony of the physics department at the University of Heidelberg. Henceforth it was to be called the Philipp Lenard Institute. In this speech, Stark raised polemical objections to modern theoretical physics. Excerpts of it were published on 29 January 1936 in the *Völkischer Beobachter*, the main Nazi newspaper, under the headline "German and Jewish Physics" ("*Deutsche und jüdische Physik*"). Heisenberg immediately defended himself in another

[12] Debye to Sommerfeld, 20 September 1935, Archives of the Deutsches Museum, Munich (Deutsches Museum Archiv, hereafter DMA), HS 1977–28/A, 61, transcribed in Eckert and Märker, *Sommerfeld*, 425–426.

article in the *Völkischer Beobachter*. The abstractness for which Aryan physicists faulted modern physics struck a chord with many people, so Sommerfeld drafted a detailed list of helpful arguments for the dean of his faculty to use during the negotiations to convince the ministerial officials that Heisenberg's research was also "of the greatest practical import."[13] Heisenberg did likewise for the REM. He was asked to submit to the minister a memorandum that the minister could use as a basis in forming his own opinion for the anticipated debate. The memorandum written by Heisenberg and the experimental physicists Max Wien and Hans Geiger[14] put in words the broad consensus among German physicists and strengthened Heisenberg's position against Stark's hostilities.

But another year went by without a decision. Then, in March 1937, the REM completely unexpectedly offered Heisenberg the Munich chair, requesting his immediate response. Heisenberg did not feel sufficiently prepared and asked to postpone his decision until the end of the summer term. During this delay in July 1937, Stark launched a new attack, this time in the SS publication *Das Schwarze Korps*. The article itself was unsigned, but Stark appended a postscript in which he applauded the substance of its arguments. Its title was "'White Jews' in science" and its vehemence outdid the foregoing attack on Heisenberg.

This article caused displeasure at the DPG, but it did not move the society to ply its influence on behalf of the attacked theorist. Jonathan Zenneck wrote in his capacity as then-chairman of the society to Max Planck that he had read the article in the *Das Schwarze Korps* "with the greatest indignation" and that these were "really shameful circumstances."[15] But nothing more was forthcoming. Max von Laue recommended that "the board of the German Physical Society should do nothing in this affair," and Zenneck declared he completely agreed: The "flouted gentlemen" should go to court about the attacks, "not only against the responsible chief editors of *Das Schwarze Korps* but also against Stark."[16] So the DPG did not consider the quarrel an incident concerning physics as a whole but a matter between *Das Schwarze Korps* and the specifically named theorists Heisenberg, Sommerfeld, and Planck. Each of these

[13] Sommerfeld to Kölbl, 17 February 1936, DMA, NL 89, 004.

[14] See Dieter Hoffmann, "Die Physikerdenkschriften von 1934/36 und zur Situation im faschistischen Deutschland, Wissenschaft und Staat," *itw-kolloquien*, no. 68 (1989), 185–211.

[15] Zenneck to Planck, 28 July 1937, DMA, *Nachlässe* (hereafter NL) 53, 012.

[16] Zenneck to Laue, 3 August 1937, MPGA, div. III, rep. 50 (excerpt transcribed in Hoffmann and Walker, *Physiker*, 563).

reacted differently: Heisenberg sought his rehabilitation through contacts his mother had with the family of the SS head, Heinrich Himmler.[17] Sommerfeld directed a formal protest to the president of the university with the request that it be forwarded to the Bavarian Ministry of Culture.[18] It is not known how Planck reacted.

As a result, the SS initiated a lengthy investigation of Heisenberg. Almost the same time that the article in *Das Schwarze Korps* appeared, the leader of the NSDDB declared that appointing Heisenberg to Munich was out of the question. Sommerfeld was again called upon to continue teaching until the investigation of Heisenberg had been concluded. That was when Heisenberg's and Sommerfeld's opponents at the University of Munich became active. Notable among them was the astronomer Bruno Thüring, who had been representing the local Lecturers League since 1936. In a rival short list to the appointing committee, Thüring named as their candidates Johannes Malsch, Wilhelm Müller, and Hans Falkenhagen. Malsch, at the top of the list, was a technical physicist who had not as yet stood out as an ideological promoter of Aryan physics. Neither could Falkenhagen be considered as belonging to that milieu. Only Müller, an aerodynamicist, had earned a reputation among Nazi ideologues for anti-Semitic writings reaching beyond the bounds of his profession. But his ideology had scarcely any points in common with Lenard's Aryan physics. Yet Malsch, not Müller or Falkenhagen, was the candidate of choice of Heisenberg's opponents at Munich. The other two names were included because they had to provide three options on such short lists.[19]

In summer 1938, the SS restored Heisenberg's reputation. Himmler assured Heisenberg that he did not approve of the article against him in *Das Schwarze Korps* and had "put a stop to any further attack on you."[20] Heisenberg promptly informed Sommerfeld and the responsible authorities at the University of Munich.[21] Even so, the affair remained in limbo. That autumn, it was still unclear whether the successor to Sommerfeld would be picked from the selection committee's list or from Thüring's

[17] Cassidy, *Heisenberg*, 472–485.

[18] Sommerfeld to the *Rektorat* of the University of Munich, 26 July 1937, Archives of the University of Munich (*Universitätsarchiv, Ludwigs Maximilian Universität München*, hereafter UAM), E-II-N (Sommerfeld), transcribed in Eckert and Märker, *Sommerfeld*, 442f.

[19] Litten, *Mechanik*, 70–95.

[20] Himmler to Heisenberg, 21 July 1938, transcribed in Samuel Goudsmit, *Alsos*, 2nd ed. (Los Angeles: Tomash, 1983), 119; translated in Hentschel, *Physics*, doc. 64.

[21] Heisenberg to Sommerfeld, 23 July 1938, DMA, *Handschriften* (hereafter HS) 1977–28/A, 136, transcribed in Eckert and Märker, *Sommerfeld*, 451f.

list. Heisenberg wrote to Walther Gerlach, who directed experimental physics at the University of Munich:

> As concerns the question of the Sommerfeld succession, you shouldn't think that I was becoming nervous at all. I'm thankful to you and Sommerfeld from the bottom of my heart for all the trouble and all the unpleasantness you are taking upon yourselves with this and I can still very well imagine that you will be successful. Besides, I do also have an utterly thick skin and would not become dismal even about an adverse outcome. I cannot deny, of course, that I would be annoyed at a victory of Messrs. Thüring and cohorts [*der Herren Thüring und Genossen*] and their methods. But we happen to be living in the midst of a revolution and thus one oughtn't regard one's own private fate as all that important.[22]

There is no reliable information from contemporary archival documentation – despite an intense search – about the final selection of Sommerfeld's successor. Despite Heisenberg's rehabilitation, the choice at the REM did not fall on him or on any of the other candidates on the selection committee's list but on number two on the list submitted by the NSDDB: Wilhelm Müller. The decision was presumably taken in spring 1939. Why the ministry did not choose number one on the rival list, Malsch, taking instead the aerodynamicist Wilhelm Müller, who was so remote from physics proper, remains unclear (see later). But one thing can be said with certainty: The DPG did not bring its influence to bear either in favor of Heisenberg or against Müller.

Another professorship went to the Aryan physics camp in spring 1939 at the Munich Polytechnic. Lenard's pupil Rudolf Tomaschek became Jonathan Zenneck's successor as professor for experimental physics. Heisenberg wrote to Sommerfeld that thus the "other side (Thüring, etc.)" would be strengthened; after this appointment it would be "even more difficult to overcome the resistance by this orientation."[23] The decision to refill Zenneck's position – like the Sommerfeld succession – was only made after years of preliminary deliberations. "Only genuine or masked Einstein people are critically set against Tomaschek. Tomaschek, as a pupil of Lenard, comes from a school of the founder of Aryan physics. This obviates further justification, if only because Tomaschek numbers among Lenard's most astute pupils." Thus the dean of the Faculty for the General Sciences implied to the president of the Munich Polytechnic in 1936 that Tomaschek's appointment had the potential of becoming

[22] Heisenberg to Gerlach, 16 November (1938), DMA, NL 80, 092-01.

[23] Heisenberg to Sommerfeld, 13 May 1939, DMA, HS 1977-;28/A, 136, transcribed in Eckert and Märker, *Sommerfeld*, 465f.

a political issue.[24] The NSDAP likewise took sides with Tomaschek and stated to the REM that it "would be very welcome if Tomaschek were called to Munich, not merely with regard to the polytechnic but also to the university; for it is absolutely necessary that a man come to Munich who originates from a different physicist school (Lenard) than the current Munich physicists."[25]

But no special side-taking was necessary to ensure that Tomaschek receive an appointment to the polytechnic. Unlike the situation with respect to the Sommerfeld succession at the university, the faculty, the president, and the ministries were not divided up into opposing camps with regard to the Zenneck succession. Tomaschek was called to the polytechnic, effective 1 April 1939. With the arrival of Müller and Tomaschek, Munich had apparently become "the capital of the physical reactionaries," as Sommerfeld's pupil Karl Bechert wrote to his former teacher in December 1939.[26] Such remarks remained confined to private correspondence, however. No protest is known to have been made about Tomaschek's appointment, as far as the DPG is concerned, even though Zenneck himself had chaired the society from 1935 to 1937.

THE DEBATE ABOUT ARYAN PHYSICS WITHIN THE DPG, 1935–1940

The published reports about the DPG's conferences do not mention any reactions to the Munich scandal. Zenneck, who became the society's chairman again in 1939 after Debye's emigration, did not utter a single word about the outcome of the Munich appointment process at the opening of the DPG's annual conference in early September 1940. He rather praised his organization for its "sense of pan-German cohesion"

[24] Boas to Rektor der TH München, 20 July 1936, Historical Archives of the Technical University of Munich (*Historisches Archiv der Technischen Universität München*, hereafter HATUM), professorship search files, 1938–1943, Registratur div. II 1a, vol. 4, no. 6. For the history of physics in general at the Technical University in Munich during the Nazi period, see Ulrich Wengenroth, "Zwischen Aufruhr und Diktatur. Die Technische Hochschule 1918–1945," in Ulrich Wengenroth (ed.), *Technische Universität München. Annäherungen an ihre Geschichte* (Munich: Technische Universität, 1993), 215–260.

[25] NSDAP to Wacker, 21 July 1938, HATUM, professorship call files, 1938–1943, Registratur div. II 1a, vol. 4, no. 6.

[26] Bechert to Sommerfeld, 30 December 1939, DMA, HS 1977-28/A, 12. On Zenneck's attitude, see Stefan L. Wolff, "Jonathan Zenneck als Vorstand im Deutschen Museum," in Elisabeth Vaupel and Stefan L. Wolff (eds.), *Das Deutsche Museum in der Zeit des Nationalsozialismus* (Göttingen: Wallstein, 2010), 78–126.

and enthused about the "collaboration of physicists in the great tasks that any war presents to a nation."[27] Not a hint of reproach directed at the Aryan physics! If even party factions, the SS, and other groups remote from physics chose to take sides, why not the DPG as the self-professed organization representing physicists in Germany? Did the DPG even see itself as the guardian of the interests of its members, as stipulated in its own statutes?[28] Did the Munich events trigger any internal debate among its members, at least? How did the society's representatives assess the threat posed by the ideologues of Aryan physics to physics in Germany?

In contrast to the revisionism in the *Physikalische Blätter* after 1945, the following reconstruction of the discussions held within the society relies exclusively on contemporary sources. The relevant time frame falls within the chairmanships of Zenneck (1935–1937) and Debye (1937–1939). Prior to 1935, Aryan physics did not yet pose a problem for the DPG. Stark's attempt to secure the position of "leader" (*Führer*) of physics in Germany in 1933 signaled the germination of ideologically motivated influences, but it was not yet part of any concerted Aryan physics campaign. Stark's move was entirely autonomous. Quite irrespective of the views he now presented in the dress of Nazi ideology, Stark had earned a reputation as an unsavory troublemaker long before 1933 from the frequent polemics he engaged in with his fellow physicists. The trigger event usually taken to mark the beginning of Aryan physics is the dedication ceremony of the Philipp Lenard Institute in Heidelberg on 13 and 14 December 1935. Stark used this occasion to launch an attack on modern theoretical physics, and this battle continued to be waged in the *Völkischer Beobachter*. When the champions of Aryan physics were forced to concede the groundlessness of their positions after 1940 during a meeting known as the physicists' "synod" (*Religionsgespräch*) organized by the NSDDB, the movement lost virtually all its impact. By the time Ramsauer submitted his petition in January 1942, Aryan physics had already become a thing of the past. Thus, Zenneck's and Debye's terms of office from 1935 to 1940 can be considered as the crucial period for this ideological debate.

Johannes Stark's ambitions to seize leadership of physics in Germany had been successfully stymied long ago during von Laue's term of office: "We have thwarted Stark's attack on the Physical Society's chair

[27] *Verhandlungen der Deutschen Physikalischen Gesellschaft*, 1940, no. 2, 31–34.
[28] See the contribution by Wolff in this volume.

brilliantly," von Laue wrote to Einstein after the society's annual meeting in Würzburg in 1933.[29] During Mey's tenure, an argument had arisen with the REM about a memorial celebration for Fritz Haber on 29 January 1935. The ministry initially regarded it as a provocation and forbade physicists from attending it. But Max Planck defended this veneration of Haber as apolitical. In his words, the memorial ceremony for Haber was "a customary, natural act of piety prescribed by old tradition toward a deceased member who, without ever making himself politically conspicuous in any way, had earned undying merit in Germany's science, economy and war technology." Not to do so would "generally cause a highly embarrassing sensation abroad as well, and the attitude of our government toward science would be unfavorably interpreted."[30]

Planck had written this to the Reich Minister of Education in his capacity as president of the Kaiser Wilhelm Society for the Advancement of the Sciences (*Kaiser-Wilhelm-Gesellschaft zur Förderung der Wissenschaften*, KWG), of which Haber had been a member. As Haber had also been a member of the DPG, having even served as its chairman for the term 1914–1915, Mey was able to adopt Planck's arguments in a letter of his own to the minister. The Haber memorial could proceed as planned, without its organizers (the German Chemical Society, the KWG, and the DPG) receiving any further complaint from the ministry. As it turned out, the decree forbidding attendance at the Haber ceremony had been issued by a mere subordinate. The minister himself had been completely unaware of it. Even "high authorities of the Reich and other agencies" had announced their intention to attend.[31]

Thus the waves from earlier differences had smoothed over again when Zenneck relieved his predecessor Karl Mey of the chairmanship of the DPG in 1935. The first conflict Zenneck had to face was only indirectly related to Aryan physics and would probably not have occurred in this form without Stark's onslaught against Sommerfeld, Heisenberg, and modern theoretical physics in December 1935. It revolved around

29 Cited after: Armin Hermann, "Die Deutsche Physikalische Gesellschaft 1899–1945," in Theo Mayer-Kuckuk (ed.), *Festschrift 150 Jahre Deutsche Physikalische Gesellschaft, special issue of the Physikalische Blätter*, 51 (1995), F95.

30 Planck to Rust, 18 January 1935. MPGA, div. V, rep. 13, no. 1850; transcribed in Hoffmann and Walker, *Physiker*, 558.

31 Mey to Rust, 24 January 1935. DPG to the members of the board, 25 January 1935. Archives of the German Physical Society (*Archiv der Deutschen Physikalischen Gesellschaft*, hereafter DPGA), no. 10011; see the transcription in Hoffmann and Walker, *Physiker*, 559–561.

the issue of the editorship of the *Zeitschrift für Physik*. Max von Laue had proposed that it be offered to Sommerfeld. Zenneck considered this suggestion "not very auspicious," as he wrote to Mey, because

> Sommerfeld is deemed an advocate of the theoretical physics orientation, which in some regards has certainly not proved itself. Besides this, he is – not illegitimately – being accused of having hitherto developed a strongly pro-Semitic working approach. So I fear all parties would kick up a storm if one were to consider making Sommerfeld the chief editor of "*Zeitschrift für Physik*."[32]

The negative sentiment about modern theoretical physics that shines through here and in other letters certainly gives the appearance that on certain points Stark's attacks struck a chord with experimental physicists.[33] Even so, Zenneck had no sympathy for Aryan physics. At the end of January 1936 he wrote to Mey, "squabbles within the German Physical Society are particularly misplaced under present conditions."[34] His primary aim was to keep himself and the DPG out of trouble. The same way of handling matters emerged with the issue of political etiquette at the annual physics convention. During the course of preparing for the autumn meeting of 1936 to be held at Bad Salzbrunn jointly with the German Society for Technical Physics (*Deutsche Gesellschaft für technische Physik*, DGtP), Zenneck was faced with deciding on whether to introduce the Hitler salute. He wrote to his counterpart Mey at the other society: "I don't have any experience about these matters and don't know how this issue has hitherto been settled."[35]

This same fear of difficulties is visible during the search for a successor to the DPG's managing director Karl Scheel, who died on 8 November 1936. At the suggestion to entrust von Laue with this office, Zenneck wrote to Mey that although he thought "extraordinarily highly" of von Laue as a physicist and as a person, he was not suitable for the managerial position, on the following arguments:

> 1. He is very much disliked by circles we cannot avoid. 2. He has very much time. But I'm more afraid of people who have lots of time than of bad people. They can make one's life thoroughly unpleasant. I fear that, based on my experience

[32] Zenneck to Mey, 4 January 1936, DMA, NL papers 53, 012.

[33] For example, Walther Gerlach wrote to Samuel Goudsmit on 7 November 1936: "The developments in both fields [theory of metals, ferromagnetism], just as what is happening in wave mechanics, do not appeal to me at all (nor do I understand it – that's precisely why I don't like it!)." American Institute of Physics, Niels Bohr Library, College Park, MD (hereafter AIP), Goudsmit papers, box 5, folder 45.

[34] Zenneck to Mey, 23 January 1936, DMA, NL 53, 012.

[35] Zenneck to Mey, 2 September 1936, DMA, NL 53, 012.

with von Laue up to now, there would be constant restlessness. 3. Von Laue is unpredictably jumpy, so one wouldn't know for sure whether he wouldn't do something completely inept.[36]

Other considerations played a part in the choice of the new managerial director besides these rather personal characteristics. Zenneck and Mey, both of whom had been chairing the DPG and the DGtP, respectively, since 1933, were technical physicists. Max von Laue wanted to assume the management, as he wrote Sommerfeld, "so that the German Physical Society doesn't get so entirely caught up in the tow line of the German Society for Technical Physics. Certain dangers exist in this regard, less due to the persons falling under consideration than due to the general pressures of the times."[37]

This avoidance strategy obviously did not always work. The Planck medal, the highest distinction that the society had been awarding since 1929 for exceptional achievements in theoretical physics, became a particular bone of contention.[38] But even then a conflict between the DPG and Aryan physics never came to a head. Whenever there was a fear that the ministry, Stark, or other Nazi followers were making accusations against the DPG, the medal was not awarded unless the ministry's approval had been obtained ahead of time.

THE DPG'S PART IN THE DOWNFALL OF ARYAN PHYSICS

There is no sign of any struggle with Aryan physics proponents in the proceedings of the society's board meetings throughout the course of Zenneck's and Debye's terms of office. Alfred Rosenberg's public declaration of neutrality in 1937 clarified the stance that the Nazi party itself assumed: "The NSDAP cannot take a dogmatic position on these ideological questions; so no party comrade may be compelled to acknowledge a position on these problems of experimental and theoretical science as being the official party line."[39] Since the polemical assault in *Das Schwarze Korps*, Stark found himself fighting increasingly on the defensive. In April 1938, he justified himself to Lenard about why he had taken refuge in the "Jewish magazine" *Nature* to publicize his conceptions

[36] Zenneck to Mey, 7 November 1936, DMA, NL 53, 012.

[37] Laue to Sommerfeld, 13 February 1936, DMA, HS 1977-28/A, 197.

[38] See the contribution by Beyler, Eckert, and Hoffmann in this volume.

[39] Rosenberg on the freedom of research, memorandum of 9 December (1937); also see Beyerchen, *Scientists*, 175.

about Jewish science. It was "no longer possible" to disseminate his views in Germany, ever since Rosenberg had been preventing "the appearance of any more articles against the Jewish spirit in the V[ölkischer] B[eobachter] since 1936, of course," adding: "*Das Schwarze Korps* doesn't accept any more articles that follow the lines of my earlier one, either."[40] By 1936, Stark had given up his attempt to become the national leader of German science when he was forced to resign his presidency of the German Research Foundation (*Deutsche Forschungsgemeinschaft*, DFG). In 1939, he relinquished his office as president of the Imperial Physical-Technical Institute (*Physikalisch-Technische Reichsanstalt*, PTR), as well. Stark and Lenard may have had sympathizers within the Nazi party, but certainly by the end of the 1930s, it was out of the question that their views could be implemented as official science policy.

Following the party's lead, the NSDDB also declared its neutrality in the "synod" of 1940. After that, the activists began to turn on each other. Wilhelm Führer was irritated by the "bilious crap attitude of so-called National Socialist physicists," after seeing "the product of this meeting of physicists."[41] Thüring was shocked at the "lack of instinct of the leaders of the Lecturers' League," when they planned "to allow the relativists to appear here with equal rights" at a follow-up debate.[42] When this debate took place in November 1942 in Seefeld in the Tyrol, Thüring felt left in the lurch: "Here stands a large, quite closed front against us, whose sole representative here is me! 1 against 30."[43]

The "synod" and the debate in Seefeld fell within Ramsauer's chairmanship of the DPG. The "synod" took place essentially at the initiative of Wolfgang Finkelnburg, who was Ramsauer's deputy on the DPG's board. May we thus recognize in at least these two events an autonomous measure taken by the DPG against Aryan physics? Is there any connection between the submission to Rust of January 1942, in which, among other things, another "debate" was called for to settle "the internal conflicts within German physics"?[44]

40 Cited from: Andreas Kleinert, "Der Briefwechsel zwischen Philipp Lenard (1862–1947) und Johannes Stark (1874–1957)," *Jahrbuch 2000 der Deutschen Akademie der Naturforscher Leopoldina*, 3rd series, 46 (2001), 259.

41 "schleimscheißerisches Benehmen": Führer to Thüring, 8 January 1941, Archive of the Viennese Observatory. I thank Freddy Litten for the perusal of this material.

42 "Instinktlosigkeit": Thüring to Dingler, 1 February 1941, Hofbibliothek Aschaffenburg (hereafter HBA), Dingler papers.

43 Thüring to Dingler, 2 November 1942, HBA, Dingler papers.

44 Ramsauer to Rust, 20 January 1942 (translation in Hentschel, *Physics*, doc. 90, 281). Carbon copy with enclosure letter to Prandtl, 23 January 1942, MPGA, div. III, rep. 61.

The Ramsauer era and the consequences of the submission to Rust are analyzed in detail elsewhere.[45] For our purposes here, it suffices to point out that the submission's main purpose had *not* been to wrestle with Aryan physics. Ramsauer's primary concern had been Germany's loss of ground to the United States as the leader in physical research. The first cause of this development that he listed was a lack of funding for physics departments at universities. Other causes he then mentioned were a neglect of theoretical physics, bad appointments in filling vacant professorships, and problems with raising the next generation of physicists. Aryan physics was only partly held to blame for these problems.

In 1939, as a board member of the DPG, Ramsauer had already demanded: "steps definitely have to be taken to achieve better funding for physics institutes." The trigger for this had been a similar initiative by chemists, "who had set very far-reaching demands in this regard for chemistry institutes." So a committee was formed "whose task should be to assemble documents about the financial situation of physics institutes and to prepare a petition to the minister." It is noteworthy that Ramsauer did not attach any blame to Aryan physics in this demand, recognizing the bad state of physics departments as due instead to "a deplorable legacy of the *Systemzeit* [Weimar Republic]."[46] The fact that the subsequent submission to Rust set its sights on Aryan physics in assigning blame was the result of an initiative unattached to the society. The person responsible was the aerodynamicist Ludwig Prandtl, who stood in high regard at Göring's Reich Aviation Ministry and in July 1938 lent his support in clearing Heisenberg's name with Himmler. In this submission, Prandtl argued:

> The difficulties encountered by this field are brought about primarily by the fact that a relatively small group of experimental physicists, who were unable to keep pace with the theorists' research, has vehemently resisted the newer developments in theoretical physics, basing themselves chiefly on the fact that significant parts included in the conceptual framework of current theoretical physics stem from non-Aryan scientists. It must be conceded that among these non-Aryan scientists were also those of inferior quality, who called out their sham wares [*Talmudwaren*] with the industry characteristic of their race. It is only just and good that such articles disappear; but among the non-Aryans there are also prime scientists who strive ardently to advance science and who have actually promoted it in the past.[47]

45 See the contribution by Dieter Hoffmann to this volume; only the aspects of relevance to the downfall of "Aryan physics" will be treated here.

46 Minutes, board meeting of the DPG, 2 March 1939, DMA, NL 53, 012.

47 Prandtl to Himmler, 12 July 1938, MPGA, div. III, rep. 61; translation of quote in Hentschel, *Physics*, doc. 62, 173. See also Beyerchen, *Scientists*, 162f.

In spring 1941, Prandtl again rose as spokesman for theoretical physics. The occasion was provided by an incident at Sommerfeld's former institute, which was by then directed by Müller: Contrary to custom, Sommerfeld was refused the right usually granted an emeritus of having a room at his disposal in which he could peruse the library books of his former institute. "You perhaps don't yet know that Müller threw me out of my institute," he wrote afterward to Prandtl. This behavior by Sommerfeld's successor shocked Prandtl so much that he resolved "to do something about this matter."[48] He immediately sent a submission to Göring to complain about the "sabotage" of theoretical physics by Lenard and his followers.[49] The Ministry of Aviation, then busy with drawing up the plans for the war against Russia, replied that "for obvious reasons," it was "currently unfortunately unable to look into this matter." Prandtl was given the recommendation "to approach the Education Minister directly and confidentially."[50]

Prandtl had written his letter of complaint together with the Göttingen physicists Georg Joos and Robert Wichard Pohl. They may well have suggested he contact Ramsauer because they knew about his earlier plans to promote a strengthening of physics departments. That was why Prandtl sent a copy of his complaint to Ramsauer and kept him informed about how things were progressing. Prandtl and Ramsauer initially doubted whether Rust's ministry was the best place to address their complaint, but they decided to do so because Ramsauer should "make an appearance at the Education Ministry *qua* the Physical Society... hence following 'official channels.'"[51] On 31 October 1941, Ramsauer asked Prandtl for permission to use his letter to Göring for the planned submission by the DPG: "Formally we could then take the position that you had made this material available to me as chairman of the German Physical Society."[52]

It is clear from this prehistory that the DPG as an organization was not the driving force behind the submission to Rust. It was rather the result of a fusion of two different initiatives: the one by Prandtl, who was concerned about the Munich scandal, and the one by Ramsauer, whose

48 Sommerfeld to Prandtl, 1 March 1941, and Prandtl to Sommerfeld, 22 March 1941, transcribed in Eckert and Märker, *Sommerfeld*, 538f.

49 Prandtl to Göring, 28 April 1941, MPGA, div. III, rep. 61, transcribed in Johanna Vogel-Prandtl, *Ludwig Prandtl. Ein Lebensbild. Erinnerungen. Dokumente* (Göttingen: Mitteilungen aus dem Max-Planck-Institute für Strömungsforschung, no. 107, 1993), 210–214; translated in Hentschel, *Physics*, docs. 84f.

50 2nd State Secretary to Prandtl, 19 May 1941, MPGA, div. III, rep. 61.

51 Prandtl to Pohl, 18 June 1941, MPGA, div. III, rep. 61.

52 Ramsauer to Prandtl, 31 October 1941, MPGA, div. III, rep. 61; transcribed in Hoffmann and Walker, *Physiker*, 592–593.

FIGURE 9. Wolfgang Finkelnburg (*Source*: DPGA)

primary concern was bolstering university departments of physics. Nor was the "synod" in November 1940 the work of the DPG. Finkelnburg, in his capacity as representative of the NSDDB in Darmstadt, wanted to dissuade the league leaders in Munich from supporting Aryan physics. He was not acting at the behest of the DPG when he set out on his campaign against Aryan physics in autumn 1940. Ramsauer only entrusted him with the vice-chairmanship of the DPG in May 1941.[53] This does not reduce Finkelnburg's personal investment against Aryan physics, but the DPG had clearly not been the instigator either of his or of Prandtl's actions. Both Prandtl's complaint to Göring and Finkelnburg's "synod" arrangements were adopted by the society *after* the fact, in order to lend greater weight to Ramsauer's submission to Rust.

COMING TO TERMS WITH THE PAST, I: THE VERSION BY THE DPG

An article by the Munich mathematician Oskar Perron initiated the reckoning with Aryan physics after the demise of the Third Reich. The editor of the *Physikalische Blätter* reprinted excerpts of it under the heading

[53] Beyerchen, *Scientists*, 184.

"'*Deutsche Physik' und die deutsche Physiker*," arguing that it was of interest to know "how and with what success scientists defended themselves against spoon-feeding and political coercion."[54] From this introduction was born the DPG's rendition of a resistance to "political physics" (*Parteiphysik*). Subsequent articles in the *Physikalische Blätter* advanced further "evidence," as we observed earlier. According to this version, the society demonstrated its opposition to Aryan physics in the form of Finkelnburg's "synod" and the submission to Rust by its chairman, Ramsauer. Equating Aryan physics here with political physics served to certify the society's general opposition to the Nazi regime.

Although most historical accounts have questioned this equation, this "battle against political physics" has not been examined more closely as a myth in the making. It is striking that none of the targeted modern theorists were among the publicists of such resistance to Aryan physics after 1945: not Heisenberg, for instance, but Finkelnburg, Ramsauer, and Brüche. The rhetoric of resistance that developed in the *Physikalische Blätter* must be seen in the context of denazification. By presenting Ramsauer's and Finkelnburg's actions as resistance to the Third Reich in their capacity as representatives of the DPG, and therefore in the name of the majority of physicists, Brüche's periodical attempted collective exoneration of the physics community from complicity with the Nazi system. But individual exonerations in the form of "whitewash certificates" (*Persilscheine*) were as much a part of it. Just a few months after the end of the war, Finkelnburg approached Sommerfeld with the following request:

> It involves the fact, now held against me, that in August 1940 I declared my willingness to accept the commission of acting administrator for the leadership of the Lecturers League at the Darmstadt Polytechnic, on the explicit condition that I could then take up the fight with the party against the Lenard-Stark-Müller-Bühl group. You know that despite all warnings about personal disadvantage I then engaged in the fight and brought about the Munich debate of November 1940, in which for the first time a stand was energetically taken against the discrimination of theoretical physics and its most prominent representatives in Germany.[55]

[54] Ernst Brüche, "Deutsche Physik," 232. Perron's article, "Verfälschung der Wissenschaften," originally appeared in *Neue Zeitung*, 7 November 1946. As a close colleague of Arnold Sommerfeld, Perron was fully informed about the events at the University of Munich. He also took a strong personal and active interest in other incidents during the Nazi period; for instance, about the attitude that the Bavarian Academy of Sciences assumed toward émigrés. See Perron to Sommerfeld, January 1951, DMA, NL 89, 012; transcribed in Eckert and Märker, *Sommerfeld*, 647–649.

[55] Finkelnburg to Sommerfeld, 14 October 1945, DMA, NL 89, 020, folder 8, 3, transcribed in Eckert and Märker, *Sommerfeld*, 574f.

Ramsauer, too, contacted Sommerfeld in search of a mitigating character reference:

I, without having been a Pg [*Parteigenosse*, i.e., party member], am being politically attacked because I had been chairman of the German Physical Society during the Nazi period. It is being assumed that I could only have received and held the chair under the decisive influence of the National Socialist party. The contrary is, in fact, the case. At the time, I had been suggested by our colleague Zenneck as his successor and had then been duly elected by the assembly in a secret ballot, because it was hoped that I would provide the society with a more independent leadership than our colleagues in the civil service. I believe I met these expectations fully. Despite many more or less clear warnings by the party, I carried forward, among other things, the festivities in honor of Planck's birthday, the memorial for Nernst, and a number of conferrals of the Planck medal on *our* candidates. But most particularly, I defended the existential interests of German physics at the National Socialist Ministry in a major submission, in which I primarily fought the defamation of German theoretical physics as a Jewish sham and the filling of important professorships with National Socialist incompetents.[56]

Denazification was also at the focus of Brüche's correspondence in the immediate post-war years. In May 1946, Finkelnburg wrote to Brüche, as he previously had to Sommerfeld, that the American Military Government had denied him permission to assume a substitute professorship at Karlsruhe on the argument that he had served as acting head of the NSDDB in Darmstadt. "If you know what could be done in this regard," he wrote Brüche, "I would be grateful. You are perhaps able to attract the attention of some influential and reasonable American to this case, or are able to obtain some personal reference." In the very same breath, he went on to propose that the DPG also be cleared of suspicion of complicity with the Nazi regime: "It also applies to our DPG that she has always kept well away from Nazism, thus is also 'not implicated.'"[57] Brüche complied and wrote the affidavit Finkelnburg wanted but doubted its effectiveness because, as he put it: "I myself could be in need of such a thing in the foreseeable future."[58]

In May 1946, the DPG's "extraordinarily energetic battle against the science policy of the Third Reich" had thus been carefully prepared.[59]

[56] Ramsauer to Sommerfeld, 8 March 1946, DMA, NL 89, 020, folder 8, 3.

[57] Finkelnburg to Brüche, 16 May 1946, Archives of the State Museum for Technology and Work in Mannheim (*Archiv des Landesmuseums für Technik und Arbeit in Mannheim*, hereafter LTAMA), Brüche papers, folder 106.

[58] Brüche to Finkelnburg, 20 May 1946, LTAMA, Brüche papers, folder 106.

[59] Brüche, certificate concerning Prof. Dr. W. Finkelnburg for presentation at the tribunal, 29 May 1946, LTAMA, Brüche papers, folder 106.

Finkelnburg's correspondence in pursuit of his own exoneration soon afterward contains the suggestion: "wouldn't it now be suitable, for the sake of physicists as well as of foreign opinion, to write candidly sometime about the battle the DPG had waged against 'party physics'?" But he did not want his enclosed "short account" to appear under his name so as not to appear as if he were defending his own case.

I do not deny that I also attach a personal purpose to this plan: My role in this battle must become better known than it now is among the physicist public. Additionally, some sad news: but please, in confidentiality! Just as I was writing down my account, I received the information from the Karlsruhe rector that Hiedemann has been appointed to take Bühl's chair, because the Karlsruhe Military Government has marked me down as "not to be employed as teacher or in any other capacity by the Ministry of Education." So, it's shattering! Ramsauer affidavit, etc., notwithstanding! You do understand that I would not like to have this broadcast about, e.g., so as not to prevent Springer from printing my atomic physics, etc.[60]

"That's horrendous," Brüche exclaimed in reply and assured him that he would do everything possible to arrange the publication of Aryan physics. "I wrote to Heisenberg, because it seems best that one of the Göttingen men sign the report." As for himself, Brüche considered himself "not competent" regarding this question. Besides, he had to respect Ramsauer's wish that he "not deal with the affairs of the Physical Society, because that was the business of the Göttingers."[61]

That is how the affair turned into a political issue for the physics profession as a whole, because the DPG had yet to be re-founded as an organization uniting all physicists in Germany. At Göttingen, where Heisenberg was serving as the president of the just recently inaugurated German Physical Society in the British Zone, Brüche's inquiry was not well received. Indeed, Göttingen had no control over such independent initiatives that the *Physikalische Blätter* was proposing. "Weizsäcker wrote me that he has great reservations," Finkelnburg informed Brüche. "I would now propose that my report be incorporated into your 'archive' as material for a possible future publication."[62] Brüche responded:

Heisenberg is, no doubt, cross with me. Now I have just recently unintentionally nettled Weizsäcker as well. I fear we aren't going to be able to get on with these gentlemen. I would very much appreciate it if you could carefully investigate the

60 Finkelnburg to Brüche, 14 June 1946, LTAMA, Brüche papers, folder 106.
61 Brüche to Finkelnburg, 17 June 1946, LTAMA, Brüche papers, folder 106.
62 Finkelnburg to Brüche, 18 July 1946, LTAMA, Brüche papers, folder 106.

true reasons. It is undoubtedly important for the *Neue Physikalische Blätter* that the Göttingen problem be solved. It's just a very finicky business dealing with mimosas.[63]

At the end of September 1946, Finkelnburg informed Brüche that he had "now fortunately become a 'fellow-traveler.'" In the same letter he made a new attempt to have the "attitude of the Phys. Soc. during the Nazi period" be made into the topic of a publication: Ramsauer was supposed to author the article, without putting too strong an accent on Finkelnburg's activity. "Furthermore, this publication ought, in my view, not happen without the approval of the Göttingen leaders of the Phys. Soc., even if they don't have to participate directly in it."[64] What followed was the already mentioned series of articles in the *Physikalische Blätter*.[65]

This tension between Brüche's *Physikalische Blätter* and the "Göttingen leaders" was not just a tiff about some article or other. In some respects it resembles the former squabbling between Berlin and the rest of Germany. Frustrated about his continuing difficulties with obtaining a professorship, Finkelnburg later bitterly complained to Sommerfeld that "one only has prospects if one has been ordained at Göttingen."[66] In a letter to Ramsauer, he wrote that he was still of the opinion, "irrespective of all the know-alls today," that during the war he "had acted rightly on the question of physics policy." What triggered this outburst, Ramsauer told Brüche, was von Laue's criticism of the account about the battle against Aryan physics that had appeared in his periodical. The explanation Ramsauer offered for this criticism was as follows:

Laue feels ashamed of himself for not having come to the defense of German physicists like Planck, Heisenberg, Sommerfeld (in the W. Müller case!), and is probably also irritated that I, an outsider, at least tried to follow an active policy as chairman of the German Physical Society and to give the society the position and the influence that befits it, in contrast to my predecessors (including Laue), who let things continue on as they had been for years.[67]

[63] Brüche to Finkelnburg, 10 September 1946, LTAMA, Brüche papers, folder 106. Concerning the objections raised by the "Göttinger Herren" with the *Physikalische Blätter*, see Middeldorff, Brüche, 22.

[64] Finkelnburg to Brüche, 23 September 1946, LTAMA, Brüche papers, folder 106.

[65] Ernst Brüche, "'Deutsche Physik' und die deutschen Physiker," *Physikalische Blätter*, 2 (1946), 232–236; Carl Ramsauer, "Eingabe an Rust," *Physikalische Blätter*, 3 (1947), 43–46; translated in Hentschel, *Physics*, docs. 90–93; and Carl Ramsauer, "Zur Geschichte der Deutschen Physikalischen Gesellschaft in der Hitlerzeit," *Physikalische Blätter*, 3 (1947), 110–114.

[66] Finkelnburg to Sommerfeld, 8 October 1947, DMA, NL 89, 008.

[67] Ramsauer to Brüche, 26 January 1948, LTAMA, Brüche papers, folder 104.

COMING TO TERMS WITH THE PAST, II: THE "OTHER SIDE'S" VERSION

It was pointed out earlier that the term "Aryan physics" is a retrospective categorization. The contemporaries themselves used such designations as those in Prandtl's complaint to Göring, "Lenard circle" or "Lenard group"; on one occasion Zenneck spoke of the "group of psychopathic physicists"[68]; Heisenberg called those responsible for the rival candidate list at Munich the "workers' and soldiers' soviets à la Thüring,"[69] the "opposing side (Thüring, etc.)," or the "Thüring group."[70] Prandtl regarded them merely as a "Munich clique,"[71] and Ramsauer described them in his submission as a "numerically small group of extreme-minded physicists, astronomers, and philosophers."[72] The advocates of Aryan physics themselves simply used the term "Aryan physics" when referring to Lenard's textbook of that title (1935) or took it as the headline of a pamphlet by Stark and Müller (1941).

When Thüring,[73] Dingler,[74] and other champions of the "other side" found themselves being pilloried after the war as "Aryan physicists" and representatives of a "political physics," their initial reaction was amazement. Thüring wrote to Dingler:

> Haven't you read the pamphlet that Perron, in his total shamelessness, published in the "*Neue Zeitung*" of October 7th 1946 under the title "Distortion of the sciences"? Contra Lenard, Stark, Müller, Dingler, Thüring, Vahlen, Bieberbach. In it he calls me a "sinister being, buoyed up" by the 3rd Reich, and you, as far as I can recall, "Hugo Dingler, the glittering, 'grand' natural philosopher of every stripe of the knighthood bonanza." The article's "factual" topic is the theory of rel[ativity] with all its trappings. Now, the essential parts of precisely this pamphlet has been reprinted by Prof. Brüche, editor of the "*Neue Physik. Blätter*," in the March issue (I believe)! Added to it, stupid comments about "political physics," etc. (This was the physics we were supposedly propagating!)[75]

68 Zenneck to Grotrian, 19 August 1937, DMA, NL 53, 012.

69 Heisenberg to Sommerfeld, 9 April 1939, DMA, HS 1977–28/A, 136.

70 Heisenberg to Sommerfeld, 13 May 1939, DMA, HS 1977–28/A, 136.

71 Prandtl to Ramsauer, 28 January 1942, MPGA, div. III, rep. 61.

72 Ramsauer to Rust, 20 January 1942. Carbon copy with enclosure letter to Prandtl, 23 January 1942; translation of quote in Hentschel, *Physics*, doc. 92, 285.

73 About Thüring see: Freddy Litten, *Astronomie in Bayern 1914–1945* (Stuttgart: Steiner, 1992), 225–228.

74 For Dingler see Gereon Wolters, "Opportunismus als Naturanlage. Hugo Dingler und das 'Dritte Reich,'" in Peter Janich (ed.), *Entwicklungen der methodischen Philosophie* (Frankfurt Main: Suhrkamp, 1992), 257–327.

75 Thüring to Dingler, 6 May 1947, HBA, Dingler papers.

This sarcasm quickly turned into dead earnest, however. Denazification was at least as much of a threat to the "other side" as to the DPG officials, and the charge of having promoted "political physics" could have extremely harmful consequences on any hopes of an academic future. In anticipation of his tribunal proceedings, Thüring drafted an affidavit that was supposed to purge him and Dingler of the accusation of this charge. He sent a copy of this text to Dingler for his signature. It included the statement:

> There never was such a thing as "political physics." But it is absolutely impossible to assign this description to the scientific orientation that Prof. Thüring represented. The contrary would rather be correct. He tried to find recognition for his scientific arguments, against the strong resistance of positions of authority among the Lecturers League leadership.[76]

In reviewing his own conduct during the Nazi period one more time, just before his hearing in October 1948, Thüring characterized himself to Dingler as someone persecuted by the "clique of physicists." Besides his antirelativistic stance and his "support for Dingler's philosophy," he was "being held to account" for Müller's appointment, as a "particularly grave crime against the Sommerfeld spirit."[77] The charge that was sent out to him shortly afterward, ranking him in the second category on a scale of 5 was, as Thüring saw it, just "a repeat in different words of a part of the long-winded, mean and hateful denunciation full of lies that Perron had poured forth in Nov. 1946."[78] In the end, Thüring's tribunal issued a very mild verdict: "End result: group III, 1 year period of probation, 500 DM 'atonement,' perhaps remunerable by special labor (25 DM per day), assumption of the suit costs."[79] It fell to Thüring's favor that the tribunal deemed the debate around Aryan physics a "scientific controversy": "The tribunal lacks the necessary expertise to be able to pass judgment here," the grounds of the tribunal's ruling read. "If the defendant as representative of the Lecturers League has a seat and a voice in the faculty and has rejected Prof. Heisenberg for scientific reasons, then that is his good right." As it has been proved "that Prof. Heisenberg was certainly not an opponent of the NSDAP," the claim that Heisenberg's rejection came as a result of his anti–National Socialist attitude was therefore unjustified.[80]

[76] Dingler to Thüring, 30 April 1947, HBA, Dingler papers.
[77] Thüring to Dingler, 19 October 1948, HBA, Dingler papers.
[78] Thüring to Dingler, 21 January 1949, HBA, Dingler papers.
[79] Thüring to Dingler, 16 March 1949, HBA, Dingler papers.
[80] Excerpts from the tribunal's ruling, HBA, Dingler papers.

Despite this mild outcome, Thüring never succeeded in regaining a foothold in science. At the end of 1949 he wrote bitterly to Dingler:

> Who else is there to prevent, or even just to try to prevent people like Perron, Brüche, Heckmann, et al. from spraying the venom of their calumny and defamations? What I (and others) have gone through these past 5 years in this regard has opened my eyes about friend and foe, as experiences of my entire former life never could.[81]

At almost 70 years of age, Dingler was not directly affected by denazification. But his commiseration with Thüring about his fate stemmed from a shared hatred for their former opponents. His words upon hearing about Sommerfeld's death were as follows: "With his death the terroristic dictate of atonal physics has lost the strongest bastion it still had."[82]

How strongly Thüring and Dingler believed that they were victims of intrigue resurfaced in 1954 when Dingler happened to encounter Wilhelm Führer, who had headed the NSDDB at the University of Munich until 1936 and distinguished himself as the most politically active among the "Munich clique." Afterward, he was employed as an expert official at the Bavarian Ministry of Culture, transferring in 1939 in that capacity to the REM and ending up working as a company commander of the SS at its Ancestral Heritage Office (*Ahnenerbe*). By the end of the war, Führer had taken on a senior command in the ordnance branch *Waffen-SS*. During this encounter, as Dingler recounted, Führer recalled from his days at the Bavarian Ministry of Culture how Gerlach had come to him "with the demand that under no condition may Heisenberg be appointed. (It was evidently clear to Gerlach that next to this celebrity he would cut a very poor figure)."[83] Thüring confirmed this:

> It is surely the truth that Gerlach spoke out against Heisenberg at the time and surely also that his opinion was more effective at the ministry than mine. But that doesn't redraw the current front, as Gerlach is an intriguant of entirely like caliber, as you know from personal experience.[84]

Führer was so certain about this recollection that he even mentioned it to Gerlach in the hope of getting help for his denazification. He wrote in a letter to Gerlach in 1947:

> I had the impression that we concurred in our views on science policy, even though I was a "Nazi" and I certainly knew that you had other alliances. If I

81 Thüring to Dingler, 27 December 1949, HBA, Dingler papers.
82 Dingler to Thüring, 8 May 1951, HBA, Dingler papers.
83 Dingler to Thüring, 2 June 1954, HBA, Dingler papers.
84 Thüring to Dingler, 23 June 1954, HBA, Dingler papers.

acted against the Sommerfeld school and its influence, it was just that I wanted to help experimental physics along; and thereby I concurred with every physicist that had come to see me. Or were they trying to mislead me? I don't believe so. None of the Sommerfeld pupils suffered by it and, in the end, commissioning Heisenberg with the directorship of the K.W.I. was not a reprehensible deed.[85]

As concerned the succession to Sommerfeld's chair, Führer alleged: "I was completely uninvolved in the Müller appointment. That appointment was carried out by my predecessor."[86]

But Gerlach declined to exonerate Führer.[87] The ministerial advisor, Wilhelm Dames, had been the expert official responsible for Müller's appointment. Rudolf Mentzel was his superior. After the war, Mentzel also applied to Gerlach for a whitewash certificate.[88] He passed the buck, however, as far as the Sommerfeld succession was concerned:

Besides the former Lecturers League, Johannes Stark had been most particularly involved in the Sommerfeld-Müller matter. At the time, St[ark] had circumvented all the referees and obtained directly from Rust a decision in favor of Müller, who I believe was a pupil of Stark's.[89]

Gerlach did not refuse Mentzel's request for an affidavit. He confirmed that Mentzel had "disapproved" of Müller's appointment in his presence and had even advised him to file a complaint against Müller, because "one really ought to demonstrate once and for all by this case what consequences the administrative meddling by the Lecturers League had."[90]

How did Müller judge these events after the war? He saw himself as a victim in a double sense. He, too, decidedly distanced himself from the accusation of having championed "political physics." His own activities "had not been determined by any partisan activism, but solely by my scientific work," he declared before the tribunal. He described his hate-ridden anti-Semitic tirades of the time as the result of an apolitical philosophy of life. His appointment to Munich had appeared, "from my point of view, at least, to have been conducted entirely normally." The Munich faculty had failed "to warn me in time and to put me into the

85 Führer to Gerlach, 4 May 1947, DMA, NL 80, 410.

86 Führer to Gerlach, 6 February 1948, DMA, NL 80, 332–01.

87 Gerlach to Führer, 20 February 1948, DMA, NL 80, 332–01.

88 Mentzel to Gerlach, 12 October 1948, DMA, NL 80, 290.

89 Mentzel to Gerlach, 20 December 1948, DMA, NL 80, 290. Müller had not been a student of Stark. But he confirmed in a "Report about my experiences at the University of Munich" from March 1943 that the appointment had "come about mainly through the intercession of the president, Prof. Joh. Stark and the university's Lecturers League." Quoted from Litten, *Mechanik*, 399.

90 Gerlach, Eidesstattliche Erklärung, 13 December 1948, DMA, NL 80, 290.

picture about the anticipated problems." So he saw himself as having been "completely impartially and unsuspectingly planted into a hostile and malicious atmosphere."[91] In the end, Müller was also ranked in category III among lesser offenders. He was not held to account for his advocacy of Aryan physics because that had involved "a scientific controversy" fought "between Einstein's followers, that is, theoretical physicists, and practical physicists." In a subsequent appellate proceeding, the category was lowered even further to IV, so that Müller was considered only a "Fellow Traveler."[92]

UPSHOT

Even despite all the finger-pointing in the correspondence by the belligerents, the search for clear answers continues. The contradictory sources only let us surmise how the decision in this test case concerning Sommerfeld's succession fell in favor of Müller. Mentzel's version that Stark "had circumvented all the referees and obtained directly from Rust a decision in favor of Müller" appears entirely plausible.[93] But it does not explain *why* Rust gave in to Stark's meddling. At that time (1938–1939), Stark had already lost his influence on science policy. Neither did his recommendation of Müller agree with the vote of the "Munich clique," whose favorite candidate had been Malsch. Müller was a completely unknown figure to them before he accepted office at Munich.[94] The ministry's recommendation to the physicists at Munich to file a complaint about Müller speaks for Mentzel's expressed plan to expose the damaging "administrative meddling by the Lecturers League." As Gerlach had been summoned by Mentzel, Sommerfeld also had an interview with Dames at the REM on 16 July 1940 from which he received the same message: in the event that Müller proved incapable of meeting his obligations, complaints should be filed and sent to the ministry.[95]

91 Wilhelm Müller, statements before the tribunal, 20 November 1948, quoted from Litten, *Mechanik*, 439–450.

92 Litten, *Mechanik*, 218 and 223.

93 There is no relevant documentation concerning Müller's appointment among the REM's files. The only record of interference by Stark regards a rather trivial argument between the president of the University of Munich (Kölbl) and Stark from December 1937 to November 1938. Stark states his position on "tendentiously concocted information" that former "collaboration between me and Jews should draw my legitimacy for a battle against Heisenberg's appointment to the University of Munich" into question. Federal Archives in Berlin (Bundesarchiv Berlin, hereafter BA), holdings R4901, PA, St. 47.

94 Dingler to Thüring, 8 December 1939, HBA, Dingler papers.

95 Ulrich Benz, *Arnold Sommerfeld. Lehrer und Forscher an der Schwelle zum Atomzeitalter 1868-1951*. (Stuttgart: Wissenschaftliche Verlagsgesellschaft, 1975), 183–184.

This fits with Heisenberg's analysis of the ministerial tactics:

> I find it very understandable that Dames is working toward Müller's appointment: He wants to mobilize the counter forces to the Lenard clique by it and – if that doesn't work out – to have Müller disgrace this group. The purely political issue of Lenard's influence among the leadership of the Lecturers League is naturally more important to him than the Munich professorship.[96]

There is other evidence that Heisenberg was right about the mood at REM. On 6 August 1938, Stark had written to Rust to complain that Dames "was under the influence of the Jewish-spirited group around Heisenberg, as he had been long-time assistant to the now emigrated full Jew James Franck at Göttingen." Dames responded to this charge with a memorandum to Rust in which he pointed out that the characterization Stark had chosen "would be a construction of his ideas by which he believes he must describe the representatives of modern theoretical physics." Dames regarded himself less a pupil of Franck than "of W. Wien and experimental physicists" and hence "did not think much of relativity theory and quantum mechanics." But he could "not bear the responsibility of contributing toward having modern theoretical physics excluded from the intellectual activity at our universities."[97] As early as 1936, Mentzel and Rust had sharply curtailed Stark's ambitions as a science policy maker. First Stark had to cede the presidency of the DFG to Mentzel. Then Rust intervened personally with Hitler to prevent Stark from becoming president of the KWG. Rust's arguments against Stark's presidency of the society were that he was being "rejected by so many reputable leading figures and high authorities of the state."[98] In other words, Stark had earned the reputation of a hot-head among all persons of influence at the REM concerned with the Sommerfeld succession, from Rust to Mentzel to Dames. If his vote for Müller got approved, it was certainly not because the ministry sympathized with Stark in any way. The hope rather was to let the fanatics clustered around Lenard and Stark "disgrace" themselves thoroughly.

[96] Heisenberg to Sommerfeld, 13 May 1939, DMA, HS 1977–28/A, 136, transcribed in: Eckert and Märker, *Sommerfeld*, 465–466.

[97] Quoted in Jost Lemmerich, "Ein Angriff von Johannes Stark auf Werner Heisenberg über das Reichsministerium für Wissenschaft, Erziehung und Volksbildung (REM)," in Christian Kleint, Helmut Rechenberg, and Gerald Wiemers, *Werner Heisenberg 1901–1976, Beiträge, Berichte, Briefe. Festschrift zu seinem 100. Geburtstag* (Leipzig: Verlag der Sächsische Akademie der Wissenschaften, 2005), 213–221.

[98] Quoted in Walter Stöcker, *Der Nobelpreisträger Johannes Stark (1874–1957). Eine politische Biographie* (Tübingen: MVK, 2001), 78.

Another clue lending plausibility to this version is an offhand remark in a letter Prandtl had written to Ramsauer. Rust had purportedly "bitterly complained" to Pohl that "professorship appointments were being forced upon his ministry by the *Braune Haus* [the Nazi Party headquarters in Munich] (NSDDB leadership) that he personally thought were very damaging, but he was completely powerless against this business."[99]

If this series of indicators accurately describes the developments and the forces working behind the scenes in the refilling of Sommerfeld's chair, then Müller had become a plaything tossed between rival forces within the Nazi regime – but then the DPG's submission to Rust, too, had been drafted entirely in line with the minister. We can conclude that, as far as the Ramsauer era is concerned, the DPG's reaction to Aryan physics was *not* an act of opposition to the Nazi regime. It was, in fact, an act of cooperation to correct adverse developments also noticed by other Nazi authorities. During the 1930s, when these unfavorable developments could still have been countered, the DPG avoided conflicts that would have brought it on a confrontational course with the regime. Where such danger emerged, such as the Max Planck Medal awards, official approval by the responsible authority, the REM, was first secured. Critical analysis in other studies about mathematics and science during the Third Reich has also revealed supposed opposition as conformance with the system. After close scrutiny of Max Planck's "My audience with Hitler" along the lines of Herbert Mehrtens's study about applied mathematicians,[100] Helmuth Albrecht described the opposition to Aryan physics as a maneuver to distract attention "away from the true collaborative relations with those in power and the full scale of the realignment with the ideology of National Socialism."[101] This same résumé would likewise apply to the conduct of the DPG.

99 Prandtl to Ramsauer, 8 June 1941, MPGA, div. III, rep. 61.

100 Herbert Mehrtens, "Angewandte Mathematik und Anwendungen der Mathematik im nationalsozialistischen Deutschland," *Geschichte und Gesellschaft*, 12 (1986), 317–347. For a translation of Planck's article, see Hentschel, *Physics*, doc. 114.

101 Helmuth Albrecht, "'Max Planck: Mein Besuch bei Adolf Hitler' – Anmerkungen zum Wert einer historischen Quelle," in Helmuth Albrecht (ed.), *Naturwissenschaft und Technik in der Geschichte. 25 Jahre Lehrstuhl für Geschichte der Naturwissenschaft und Technik, am Historischen Institut der Universität Stuttgart* (Stuttgart: GNT Verlag, 1993), 61.

5

The Ramsauer Era and Self-Mobilization of the German Physical Society

Dieter Hoffmann

> Final adjustment by the D.P.G. within the Third Reich is, without a doubt, urgently needed but a delicate affair. I was not very pleased at the time, when Debye became chairman of the society in Kreuznach. But to my knowledge, the election had taken place with the Ministry's approval and, as far as I could observe, Debye always did take his cues for his official leadership from the Ministry as well. The D.P.G.'s handling of the Jewish question shows, however, that he lacks the necessary understanding for the political issues, as would not otherwise be expected.... All this together makes supporting Debye's reelection impossible.... I would very much welcome it if Esau became chairman [and] that this autumn the Society would get a leader who guides her fortunes in a positive and uninhibited manner toward the Third Reich. Esau probably offers the best guarantee for this.[1]

The Königsberg physicist Wilhelm Schütz wrote this in spring 1939 to his colleague Herbert Stuart in Berlin. He was speaking for a group of young physicists seeking to deepen the German Physical Society's (*Deutsche Physikalische Gesellschaft*, DPG) commitment to the political system of the Third Reich. They wanted to put a stop to the refractory delay tactics used by the society's leadership on this issue. These tactics had made it possible for the DPG to withstand the political pressure to conform and maintain a certain amount of autonomy during the early years of National Socialist rule. As the Nazis consolidated their power in

[1] Schütz to Stuart, 4 April 1939, American Institute of Physics, Niels Bohr Library, College Park, MD (hereafter AIP), Goudsmit papers, box 28, folder 53. Transcribed in Dieter Hoffmann and Mark Walker (eds.), *Physiker zwischen Autonomie und Anpassung – Die DPG im Dritten Reich* (Weinheim: Wiley-VCH, 2007), 568–569.

FIGURE 10. Wilhelm Schütz (second from left) and Werner Heisenberg (second from right) in 1952 (*Source:* DPGA)

the second half of the 1930s, there was mounting pressure for the society to fall properly into line within the National Socialist system of control. The Reich Research Council (*Reichsforschungsrat*, RFR) was founded in 1937, and one of its central tasks was to seek the collaboration of scientific associations in organizing research and development within an economy focused on military defense and autarky.[2]

THE BELATED POLITICAL REALIGNMENT OF THE DPG

According to the *Führerprinzip* ("Leader principle"), the position of chairman was the gateway to the society's total capitulation to Nazi science policy. The previous chairmen of the DPG – Karl Mey, Jonathan Zenneck, and Peter Debye – had basically been conservative nationalists by persuasion and even demonstrated a partial affinity for some of the goals and conceptions of National Socialism. Yet this willingness to reach compromise and conform to the Nazi system ended where the society's

[2] Inaugural speech by the general Prof. Dr. Becker, in *Ein Ehrentag der deutschen Wissenschaft. Zur Eröffnung des Reichsforschungsrats am 25. Mai 1937* (Berlin: Press Office of the Reich Education Ministry, 1937), 26.

political independence began. The chairmen jealously defended the society against party interference in an effort to guard its integrity. Recurrent conflicts with representatives of the National Socialist regime were the result. In the eyes of these representatives – but also of the young guard of Nazis within the DPG – these leading scientists only incompletely satisfied the expectations of a *Führer* (leader) of one of the most important scientific professional associations in the Third Reich. During the physics convention at Bad Kreuznach in 1937, Peter Debye had been unanimously elected as chairman of the DPG.[3] He had just been awarded the Nobel Prize in Chemistry in the previous year and had become head of the new Kaiser Wilhelm Institute (*Kaiser-Wilhelm-Institut*, KWI) of Physics in spring 1937. His scientific renown and prominent position as director of a KWI were thought sufficient guarantee that this independence would be maintained and the society would remain safe from overly stiff injunctions or other interference by National Socialist policy and party activists. The first test of this came with the official function in honor of Max Planck's 80th birthday and the resumption of awarding of the Max Planck Medal in April 1938.

Ministerial bureaucrats and other Nazi authorities viewed both events with great reservation because a type of scholar and tradition was being celebrated that the Third Reich did not condone. Indeed, it was even considered taboo, among Nazis at least. In the previous year, questions had been raised about the renaming of the new KWI of Physics as the Max Planck Institute (*Max-Planck-Institut*). The Aryan science protagonists Philipp Lenard and Johannes Stark had argued: "Planck had not done enough for physics to be worthy of such a distinction."[4] Thus the Planck festivities had taken place in an atmosphere of dissent, even if it was not – as frequently alleged after the war – against the expressed wish of the Reich Ministry for Science, Education and Culture (*Reichsministerium*

3 Debye and his relationship with the DPG are explored by Horst Kant, "Peter Debye und die Deutsche Physikalische Gesellschaft," in Dieter Hoffman, Fabio Bevilacqua, and Roger H. Stuewer (eds.), *The Emergence of Modern Physics* (Pavia: Università degli Studi di Pavia, 1996), 507–520, and most recently, Horst Kant, "Peter Debye als Direktor des Kaiser-Wilhelm-Instituts für Physik in Berlin," in Dieter Hoffmann and Mark Walker (eds.), *"Fremde" Wissenschaftler im Dritten Reich. Die Debye-Affäre im Kontext* (Göttingen: Wallstein, 2011), 76–109. For Debye's overall role during the Third Reich, see Dieter Hoffmann, *Peter Debye. Ein Dossier*, Preprint No. 314 (Berlin: Max Planck Institute for History of Science, 2006).

4 Debye to Heisenberg, 29 May 1937, Archives of the Max Planck Society (*Archiv der Max-Planck-Gesellschaft*, hereafter MPGA), div. III, rep. 19, Debye papers, no. 331, sheet 41.

FIGURE 11. Banquet in honor of Max Planck's 80th birthday (*Source:* Landesmuseum Technik & Arbeit Mannheim)

für Wissenschaft, Erziehung und Volksbildung, REM). The minister opposed its character and scale and therefore made do with issuing an official birthday greeting but sent no high-ranking representatives of the government. This was quite in contrast to the pompous inauguration of the Philipp Lenard Institute at Heidelberg in December 1935 or the celebration of Lenard's 80th birthday, at which numerous Nazi notables made a glamorous appearance.[5] All the same, or perhaps precisely because of this, Planck's birthday celebration in the *Harnack-Haus* turned into an impressive eulogy of Germany's doyen of physics and signified the culmination of the DPG's scientific activities.[6] It was naturally also interpreted as an unmistakable demonstration of scientific autonomy and tradition.[7]

[5] August Becker (ed.), *Naturforschung im Aufbruch. Reden und Vorträge zur Einweihungsfeier des Philipp-Lenard-Instituts der Universität Heidelberg am 13. und 14. Dezember 1935* (Munich: Lehmann, 1936).

[6] Dieter Hoffmann, Hole Rößler, and Gerald Reuther, "'Lachkabinett' und 'großes Fest' der Physiker. Walter Grotrians 'physikalischer Einakter' zu Max Plancks 80. Geburtstag," *Berichte zur Wissenschaftsgeschichte*, 33/1 (2010), 30–53.

[7] "Feier des 80. Geburtstages des Ehrenmitglieds der Deutschen Physikalischen Gesellschaft, Herrn Geheimrat Professor Dr. Max Planck, am Sonnabend, dem 23. April 1938 im Harnack-Haus in Berlin-Dahlem," *Verhandlungen der Deutschen Physikalischen*

This niche existence had its rock-solid boundaries, however, and it was essentially impossible to detach oneself from the social reality of the Nazi state. The expulsions of the society's Jewish members in fall 1938 underscored this.[8] After repeated calls that the DPG conform its statutes to National Socialist precepts and, in particular, solve the problem of its non-Aryan membership, the society eventually buckled under state pressure. It issued a circular letter dated 9 December 1938 requesting that all "German Reich Jews in the sense of the Nuremberg Laws" send notification of "their withdrawal from the Society."[9] The DPG's circular formally obeyed the ministry's instructions but abstained from taking any public position on it or issuing any individual declarations of allegiance. In this case, this was not much for the DPG to brag about, of course, but neither can it be taken for granted, as similar circulars and activities at other institutions demonstrate.[10] The group of young Nazi activists noticed this as well. That was why there were further discussions about the "non-Aryan question" at the board meeting of 14 December. The result was an altercation between Debye and Wilhelm Orthmann, who "points out that the first sentence of the letter addressed to the German members of the Society is formulated in such a way that the letter could be misunderstood. Mr. Debye asks that this sentence be taken just as it is meant and assumes responsibility for the wording selected."[11] The letter to Stuart quoted at the beginning of this chapter was much more direct. Schütz effectively denounced Debye:

> The DPG's handling of the Jew problem shows, however, that he lacks the necessary understanding for the political issues, as would not otherwise be expected. At the time, I tried in vain to elicit from the chairman a clear statement and thus set in motion final resolution of the problem.[12]

Gesellschaft, 19 (1938), 57–76. See also: Ernst Brüche, "Vom großen Fest der Physiker im Jahre 1938." Accompanying text to the photographic record "Stimme der Wissenschaft," Frankfurt Main, undated; see the contribution by Beyler, Eckert, and Hoffmann to this volume.

8 See the contribution by Wolff to this volume.

9 An die deutschen Mitglieder der Deutschen Physikalischen Gesellschaft, Berlin, 9 December 1938, MPGA, div. III, rep. 19, Debye papers, no. 1014, sheet 38. Transcribed in Hoffmann and Walker, *Physiker*, 511, 565, translated in Klaus Hentschel (ed.), *Physics and National Socialism. An Anthology of Primary Sources* (Basel–Boston: Birkhäuser, 1996), 182.

10 See the contributions by Deichmann and by Remmert to this volume.

11 Minutes of the meeting of the board of 14 December 1938, Archives of the German Physical Society (*Archiv der Deutschen Physikalischen Gesellschaft*, hereafter DPGA), no. 10012, translated in Hentschel, *Physics*, doc. 68, 183.

12 Schütz to Stuart, 4 April 1939, AIP, Goudsmit papers, box 28, folder 53.

The Information Service of the National Socialist German University Lecturers League (*Nationalsozialistischer Deutscher Dozentenbund*), of which Herbert Stuart was a functionary, likewise sarcastically censured the affair:

One evidently still seems to be very far behind at the German Physical Society and still very much attached to the beloved Jews. It is indeed remarkable that only "under the prevailing compelling circumstances" can the membership of Jews no longer be upheld.[13]

The DPG was also very far behind as regarded its statutes.[14] It stalled in response to inquiries made by REM and did not take the initiative to adjust its statutes to the new political situation in Germany. When the ministry started its offensive in science policy and more or less gave an ultimatum to politically delinquent science institutions to conform their "internal activities" – as it reads in a relevant letter to the Berlin Academy[15] – "in accordance with the basic views upon which the state and intellectual life of the German present day are grounded," the DPG's leadership was forced to react. At the board meeting of 2 March 1939, a proposal was made to form a panel for the "preparation of the statutory modifications by the German Physical Society." But Debye stated for the record that although "the statutes should be modified, it was neither necessary nor practical at the moment to hurry through changes to the statutes." He suggested that a panel be immediately formed anyway, so as to conduct preliminary discussions on the question of statutory changes. There would then be a functioning panel when the decisive moment arrived.[16] The members appointed to this panel were Debye himself as well as Max von Laue, Walter Schottky, Herbert Stuart, and Walter Grotrian. In a letter to Georg Stetter, Stuart mocked the board's stalling tactics and the composition of the panel. With the exception of Stuart, its members had been entirely recruited from the board – hardly determined followers of National Socialism. Stuart mentioned there his intention

that at least one more party comrade ... get on this panel.... It is actually a disgrace that we have to deal with such trifles and these bourgeois, while our *Führer* is making history and pointing out to us one great mission after the next.[17]

[13] *Informationsdienst der Reichsdozentenführung*, 2 (1939), 27.
[14] See the contribution by Wolff to this volume.
[15] Rust to the Prussian Academy of the Sciences, archive II, Ia, vol. 14, sheet 16.
[16] Minutes of the meeting of the board on 2 March 1939, MPGA, div. III, rep. 19, Debye papers, no. 1016, sheet 24.
[17] Stuart to Stetter, 17 March 1939, AIP, Goudsmit papers, box 28, folder 53.

Stuart and his group were not the only ones applying pressure on the DPG and its board about this issue. Eduard Steinke, full professor of physics at the University of Freiburg, wanted to set as a condition for his election to the board of the society's Regional Association (*Gauverein*) for Baden and the Palatinate that "the adaptation of the Society to the existing regulations of the Reich" take place.[18] In his reply letter, the managing director pointed out on behalf of Debye that the DPG had actually done everything within its power on this question and the required statutory changes were being held up solely by the fact that the ministry had not yet provided any binding guidelines; that is, the buck was passed on to the ministry responsible for scientific affairs.

The first thing the panel did was to assign the DPG's managing director the task of sending out a circular letter requesting information about the valid statutes of other scientific associations.[19] It also told the ministry it confidently anticipated receiving "the final guidelines by the Ministry . . . before the conclusion of this task."[20] So the assumption was to treat the matter in the same dilatory manner as before. At the beginning of 1940, influential representatives of the DPG like Max von Laue and Jonathan Zenneck still held the view (privately, of course) "that today really is not the time to change the statutes. That can be discussed when normal conditions have returned again, but not during wartime."[21] This strategy did not work, however. Spring 1939 had not even gone by before the ministry instructed Debye to go ahead with the statutory changes, even without concrete prerequisites, and "to submit the modified statutes of the German Physical Society as soon as they are finished in draft form, for explicit approval" as soon as possible.[22] Subsequently, the ministry also advised the DPG that

> closer ties with scientific societies and associations are only fostered when they are capable, suitable and willing to collaborate on current problems of science and are deployable by the State for the furtherance of scientific purposes. A purge by societies and associations from members related by descent or marriage to those of Jewish or mixed blood is a precondition for this collaboration.

[18] Grotrian to Steinke, 13 May 1939, MPGA, div. III, rep. 19, Debye papers, no. 1016, sheet 45.

[19] Minutes of the board meeting of 2 March 1939, MPGA, div. III, rep. 19, Debye papers, no. 1016, sheet 24.

[20] Debye to Dames, 11 March 1939, MPGA, div. III, rep. 19, Debye papers, no. 1016, sheet 34.

[21] Zenneck to von Laue, Munich, 12 February 1940, DPGA no. 10014.

[22] Wacker to Debye, 25 March 1939, MPGA div. III, rep. 19, Debye papers, no. 1016, sheet 35.

Only in individual personal cases are exceptions in this regard possible, under the condition that special merit or circumstances exist.... The Ministry does not intend to act pedantically on individual personal cases, whether domestic or foreign, if such personalities are of value to science and are positively disposed to the New Germany as well as at least satisfying the basic civil laws of the Reich also as regards wives.[23]

During the meeting of the statutory panel on 7 August 1939, which von Laue, Ramsauer, Schottky, and Grotrian also attended, Debye jotted down the essentials of the planned modifications to the society's statutes: "Reich civil law, *Führer* principle. Chair is elected, then has *Führer* qualities."[24] At this meeting, concrete suggestions for changes to individual articles of the statutes were then formulated, and finally the new draft statutes were handed over to the ministry for confirmation. The outbreak of war apparently left the draft lying around in some ministerial office for a number of months.[25] On 1 June 1940, the board was at last able to conduct the final deliberations on the new statutes (without Debye, who had left Germany for America[26]) and then "accept the form approved by the Minister after brief discussion."[27] The ministry had made changes to a few points on the draft, as a comparison of Debye's notes against the statutes that ultimately came into force in 1941 shows. They primarily concerned section 4, which regulated membership. The original draft by the DPG's board reads, according to Debye's notes:

Only Reich Germans can be regular members who possess the right of Reich citizenship as well as foreigners residing in the Ger. Reich who meet the condition for acquiring the right of Reich citizenship according to German law.[28]

This article of the statutes was made considerably more severe in the version corrected by the ministry and subsequently confirmed by the DPG in fall 1940. Accordingly, *Reichsdeutsche* (German citizens according to the 1935 Nuremberg Laws) could only make a claim for membership in

[23] Dames to Debye, 4 August 1939, MPGA, div. III, rep. 19, Debye papers, no. 1018, sheet 23.

[24] Handwritten notes; minutes of the board meeting of 2 March 1939, MPGA, div. III, rep. 19, Debye papers, no. 1018, sheet 25.

[25] Debye to Grotrian, 8 January 1940, MPGA, div. III, rep. 19, Debye papers, no. 1019, sheet 1.

[26] See Hoffmann and Walker, "*Fremde*."

[27] Minutes of the board meeting of the DPG on 1 June 1940, DPGA no. 10014.

[28] Notes by Debye from the statutory panel meeting on 7 August 1939, MPGA, div. III, rep. 19, Debye papers, no. 1018, sheet 26.

the DPG if their wives likewise possessed citizenship rights.[29] The consequences and humiliation that this article caused for individual members is reflected in the following excerpt from a letter by Erich Regener in October 1941. He considered himself obligated to inform Jonathan Zenneck that

> now that the Physical Society, too, has extended the Aryan article to wives of its members,... to set the record straight [I would like] to note that my wife, formerly recorded as Jewess by *provisional* decision of the Reich Office of Race Research, can now, however, count as being of mixed blood to first degree and accordingly has an identity card.[30]

A few members of the DPG demanded that the statutes even exceed the stipulations of the civil law of the Reich and banish Jewish authors and reviewers from the pages of the society's professional publications. But these radical calls could be averted. The panel only introduced into the statutes National Socialist stipulations that formed part of the indispensable political requirements imposed by the Nazi state. Besides excluding Jewish members and introducing the *Führerprinzip*, the new statutes committed the society to conceding supervisory rights to REM. From then on, all important decisions and personnel questions had to be submitted there for confirmation. The new statutes thus did not go beyond what had already become everyday practice during the Third Reich. They were more moderate than revisions by other scientific societies and institutions.[31] After final confirmation at the administrative meetings of the individual regional associations, the revised statutes received the society's official stamp of approval at the DPG's administrative meeting that took place during the physicists' convention in Berlin in September 1940. Formal legal recording of the society in the Association Registry took place on 15 July 1941.[32]

Debye's style of leadership and cautious cross-checking with the Reich ministry responsible for the DPG before making any decisions may have signaled a willingness for compromise and cooperation with Nazi superiors. Even so, his term of office is characterized by a certain restraint, not just with regard to the statutory revisions and ultimate realignment of

[29] See Satzungen der Deutschen Physikalischen Gesellschaft (E.V.), 1941.

[30] Regener to Zenneck, undated (around October 1941), Archives of the Deutsches Museum, Munich (*Deutsches Museum Archiv*, hereafter DMA), papers 053 (Zenneck), no. 52.

[31] See the contributions by Deichmann and by Remmert to this volume.

[32] *Verhandlungen der DPG*, 22 (31 December 1941) 2, p. 29.

FIGURE 12. Abraham Esau (*Source:* DPGA)

the society within the National Socialist system. The overall impression it left was decidedly apolitical. For instance, Debye delivered only one official address as chairman of the DPG and that happened to be during the above-mentioned Planck celebration. It contained none of the embarrassing rhetoric that studded the speeches of his predecessors Karl Mey and Jonathan Zenneck. For this reason, Nazi activists inside and outside of the society faulted Debye for a lack of the necessary understanding of political issues of the day.

A more suitable candidate had to be found for the upcoming reelections in fall 1939 to fit the political profile expected in the Third Reich. The letter quoted at the beginning of this chapter reveals their favorite candidate for chairman of the DPG as Abraham Esau. This specialist in high-frequency physics who hailed from Jena was sure to "guide her fortunes in a positive and uninhibited manner toward the Third Reich."[33] Esau

[33] Schütz to Stuart, 4 April 1939, AIP, Goudsmit papers, box 28, folder 53.

was an experienced, accredited physicist who had, moreover, demonstrated his political loyalty on repeated occasions. He joined the Nazi party in 1933, and his career benefited from the implementation of the Four-Year Plan, culminating in a position as leading science administrator of the Third Reich.[34] At the founding of the RFR in 1937, he was appointed director of its special physics division (which included mathematics, astronomy, and meteorology). The council was supposed to harness scientific and technological research in Germany to the goals of the Four-Year Plan through a concerted distribution of tasks and resources. It was thus an important steering mechanism for the National Socialist policy of autarky and rearmament.[35] Two years after this appointment, Esau received a professorship in military signal engineering at the Faculty of Defense Technology of the Berlin Polytechnic in Charlottenburg. At the same time, he was vested in the acting presidency of the Imperial Physical-Technical Institute (*Physikalisch-Technische Reichsanstalt*, PTR), the renowned bureau of standards and national research center in the capital.[36] The DPG's chairmanship was supposed to add the finishing touch to this array of influential offices, making him a similar guiding figure that Johannes Stark had been in the opening years of the Third Reich. This move from Stark to Esau was more than a mere generational change. It is symptomatic of the transformation in the policies pursued by the Nazi system on science and research. As the end of the 1930s neared, the focus turned increasingly away from ideological battles about principle and the fanaticism of the world-view elite exemplified by the Aryan physics movement. The Four-Year Plan encouraged scientists and engineers that were able to guarantee an interlinking between academic research and the defense and armament industries to advance official science policy.[37]

34 See Dieter Hoffmann and Rüdiger Stutz, "Grenzgänger der Wissenschaft: Abraham Esau als Industriephysiker, Universitätsrektor und Forschungsmanager," in Uwe Hossfeld, Jürgen John, Oliver Lemuth, and Rüdiger Stutz (eds.), *Kämpferische Wissenschaft: Studien zur Universität Jena im Dritten Reich*, (Cologne: Böhlau, 2003), 136–179.

35 See Notker Hammerstein, *Die Deutsche Forschungsgemeinschaft in der Weimarer Republik und im Dritten Reich* (Munich: Beck, 1999), 205ff.

36 See Ulrich Kern, *Forschung und Präzisionsmessung. Die Physikalisch-Technische Reichsanstalt zwischen 1918 und 1948* (Weinheim: VCH, 1994).

37 According to Adolf Fry at the beginning of 1941 about the connection between National Socialism and rearmament. See Hans Ebert and Hermann Rupieper, "Technische Wissenschaft und nationalsozialistische Rüstungspolitik: Die Wehrtechnische Fakultät der TH Berlin 1933–1945," in Reinhard Rürup (ed.), *Wissenschaft und Gesellschaft. Beiträge zur Geschichte der Technischen Universität Berlin 1879–1979*, Vol. 1 (Berlin: Springer, 1979), 490.

Esau's rise signified this change. His name became very specifically connected with the chairmanship of the DPG at the physics convention in Baden-Baden. When Karl Mey, chairman of the German Society for Technical Physics (*Deutsche Gesellschaft für technische Physik*, DGtP), offered the first toast to the German Physical Society and not to the *Führer* at an evening banquet, it almost became a political scandal. Esau supposedly saved the day by adding the officially expected toast to Hitler immediately afterward.[38] Another episode during the general assembly of the DGtP also illustrates Esau's adept political correctness. Esau proposed that in the future, the annual meetings only be convened after the 20th of September so as to exclude scheduling conflicts with the national conventions of the National Socialist German Workers Party (*Nationalsozialistische Deutsche Arbeiterpartei*, NSDAP). According to the minutes of the meeting, Mey as chairman of the DGtP simply responded by tersely setting forth "the guiding aspects for the timing of the annual conventions." But no concrete resolutions were passed.[39] One may assume that Esau felt somehow slighted or ridiculed by Mey. In any case, just a few weeks later Esau moved quickly to carry out Otto Wacker's plans at REM to relieve Mey because of his political blunder. New statutes were imposed upon the DGtP as well. Esau's informal interactions with Wacker at the ministry reveal that more was involved than simply injured pride. Esau was generally more willing than his fellow physicists actively to carry forward the Nazi rulers' policies of political realignment and anti-Semitic discrimination. In this connection, Esau explicitly approved setting down in the DGtP's statutes REM as its supervisory authority and henceforth not just to exclude Jewish members from regular membership but also to refuse granting any more space to Jewish authors and reviewers in the society's professional publications. Other physicists besides Wacker had supposedly also raised these demands at the annual convention during which the evening banquet had taken place.[40]

Despite, or perhaps precisely because of, this political context, the alliance between REM and the party fraction within the DPG did not manage to push their candidate through. The physics convention with the election of the DPG's chairman on the agenda was scheduled to take

[38] See Hentschel, *Physics*, 179.

[39] "Bericht über die 20. Hauptversammlung der Deutschen Gesellschaft für technische Physik e.V. am Mittwoch dem 14. September 1938 in Baden-Baden," *Zeitschrift für technische Physik*, 19 (1938), 616.

[40] See Hentschel, *Physics*, 179.

FIGURE 13. Carl Ramsauer (*Source:* MPGA)

place at the end of September 1939 in Marienbad. But the outbreak of war led to postponement. At the board meeting on 1 June 1940, a proposal by Stetter that this office be "suggested to Mr. Esau" was officially discussed and promptly rejected on the argument made by the officiating chairman Jonathan Zenneck (Debye was then already living in the United States)[41] "that Mr. Esau was being weighed down by so many disparate obligations that he would not be able to devote himself sufficiently to this task."[42] The board seconded this objection and nominated Carl Ramsauer instead. In September 1940, Ramsauer was unanimously elected chairman of the society at the physics convention in Berlin and with some delay was finally confirmed by REM at the beginning of April 1941.

CARL RAMSAUER

This election yielded a physicist with managerial experience. Ramsauer was already directing the DPG's most important member association, the

[41] Hoffmann, *Peter Debye*.

[42] Minutes of the board meeting of 1 June 1940, DPGA no. 10014.

Physical Society for Berlin, with great circumspection. As an acknowledged experimental physicist and head of the German General Electricity Company (*Allgemeine Elektrizitäts-Gesellschaft*, AEG) research laboratory, he enjoyed high regard among scientists. Born in 1879, the same year as Albert Einstein and Max von Laue, Ramsauer was a member of a particularly illustrious class of physicists. He attended preparatory school in his native town Oldenburg until Easter 1897, thereafter studying mathematics and physics at the Universities of Munich, Tübingen, Berlin, and Kiel. Philipp Lenard was one of his professors at Kiel. In 1902, Ramsauer defended his thesis there on projectile analysis (*Über den Ricochetschuß*) and became research assistant at the local Imperial Torpedo Laboratory where he worked as a researcher for a number of years. After a half-year in London to study and having completed his year's military service, he decided to pursue a university career and acquire academic certification. For this purpose, he followed Lenard to Heidelberg to work as his assistant and took his habilitation degree there 2 years later with a thesis on experimental and theoretical bases of elastic and mechanical collisions. For the following years, Ramsauer was engaged as a staff researcher at the newly founded Radiological Institute of the University of Heidelberg. This phase of his professional life was cut short for many years by his deployment as artillery officer during the First World War. He was also a part of the Artillery Testing Commission for the Army Ordnance Office (*Heereswaffenamt*). He subsequently made an important discovery that earned him worldwide fame, the so-called Ramsauer effect: an anomalous scattering cross-section of slow electrons traveling through a gas. Classical theory was unable to explain this effect, and it was only solved years later by means of wave or quantum mechanics. The Ramsauer effect is one of the first experimental indicators of the wave-like nature of electrons or matter. In 1921, Ramsauer accepted a professorship at the polytechnic in Danzig, where he directed the local physics department until 1927 and immersed himself in his investigations on the Ramsauer effect.

In the mid-1920s, Berlin's major electrical combine AEG offered the physicist a position as director of a new "large-scale industrial research institute." Ramsauer accepted the post in 1928 and within a short time transformed the AEG laboratory into one of the country's leading and most productive industrial laboratories.[43] With this move into the

[43] Burghard Weiss, "Forschung zwischen Industrie und Militär. Carl Ramsauer und die Rüstungsforschung am Forschungsinstitut der AEG," *Physik Journal*, 4 (2005), 53–57.

industrial sector, this physicist's engagement as a science administrator and industrial manager grew steadily. He soon joined the AEG's board, in which capacity Ramsauer earned the trust and support of its chairman and general director Hermann Bücher. On the side, he also coedited the *Zeitschrift für Physik* and became increasingly active in the DGtP, becoming its treasurer in 1935. The DPG's most important regional affiliate in Berlin elected him to its board in 1937 and a year later as its chairman. In this capacity, he shared the responsibility with Peter Debye for the previously mentioned "physicists' great festival" in Berlin in honor of Max Planck's 80th birthday on 23 April 1938.[44]

These merits, a sensitive feel for what was politically feasible, and, not least of all, his position in the industrial sector made Ramsauer the most suitable candidate for the DPG's chair in the eyes of the majority of its members. As Arnold Sommerfeld put it: "he can and will assume a more independent stance toward the government than the other state-employed professors."[45] Jonathan Zenneck wrote in this sense to his designated successor:

> We definitely need a chairman today who enjoys scientific & personal prestige & possesses the necessary independence. After the recent election for the Berlin Physical Society, the danger is particularly great that a chairman would then be impressed upon us.[46]

As earlier, when in fall 1933 the society declined the party's favored representative of so-called Aryan physics, Johannes Stark, in favor of the industrial physicist Karl Mey for its new chairman, the preference in 1940, too, was to demonstrate relative independence and not slavishly follow the party line. In addition, Ramsauer's scientific profile touched only lightly on a few points in the ideologically sensitive areas of modern physics. His former assistantship under the protagonist of Aryan physics, Philipp Lenard, made him virtually impervious to attack by overzealous ideologues and other Nazi activists. Although conservatively and nationalistically minded, it also fell in his favor that – contrary to Esau – he did not become a direct adherent of the Nazi party or join as a member after

[44] Dieter Hoffmann, Hole Rößler, and Gerald Reuther, "'Lachkabinett' und 'großes Fest' der Physiker. Walter Grotrians 'physikalischer Einakter' zu Max Plancks 80. Geburtstag," *Berichte zur Wissenschaftsgeschichte*, 33/1 (2010), 30–53.

[45] Sommerfeld to Ramsauer, Munich, 22 March 1946, DMA, papers 89 (Sommerfeld), 020, folder 8.3.

[46] Zenneck to Ramsauer, Munich, 20 March 1941, DPGA no. 10018.

1933. Neither did he take on any prominent political functions. In his memoirs published after the war, Ramsauer states in this regard:

> I could not befriend myself with political Hitlerism, nor was I pressured in any way by the overall attitude at the AEG to improve my and my institute's public image substantially through an entry by me into the party, which in and of itself would have been very possible.[47]

Such postwar testimony cannot simply be taken at face value, of course, because it conceals how much Ramsauer and other members of the technocratic elite generally conformed to the Nazi system; it is a part of their self-stylizations. In contrast, it reflects important reasons for Ramsauer's acceptance among the physicist membership. His position in industry and ostensible political abstention made Ramsauer the embodiment of a certain independence. As an influential technocrat, he was a broker acceptable to all sides, a mediator between the interests of the scientific and technical communities and the pragmatically imbued elements among the Nazi leadership. Technocrats found broad acceptance among professionals, politicians, and military circles alike. Thus, they provided the basis for the collaborative relations[48] typical of the dealings by these elite and were an influential part of the reality of the Third Reich. For this reason, such conduct had nothing to do with opposition or renitency, not to mention active political resistance.

RAMSAUER AS A LEADING FIGURE FOR SELF-MOBILIZATION

In this sense, Ramsauer's "election success" and what was thought the "more independent stance" of the DPG also had its price. Instead of total political realignment imposed on the society from the outside, a partial self-realignment took place during the Ramsauer era with DPG's forced engagement in the armament efforts and war economy of Nazi Germany. This change happened less in passive reaction to related pressure by the National Socialist leadership than as a result of self-imposed initiatives. The DPG leadership and circles close to the German community of scientists were highly active in these dealings. The motive behind such

[47] Carl Ramsauer, *Physik–Technik–Pädagogik. Erfahrungen und Erinnerungen* (Karlsruhe: Braun, 1949), 128.

[48] Herbert Mehrtens, "Kollaborationsverhältnisse: Natur- und Technikwissenschaften im NS-Staat und ihre Historie," in Christoph Meinel and Peter Voswinckel (eds.), *Medizin, Naturwissenschaft, Technik und Nationalsozialismus*, (Stuttgart: GNT-Verlag, 1994), 13–31.

self-mobilization lay, as Ramsauer saw it, least of all in politics or even in an emphatic act of solidarity with the dominant Nazi regime. Rather, the attempt was to pawn off the accelerated integration of the natural and technological sciences into the German war and defense economies in exchange for a lasting strengthening of influence not just for the society but also for the professional field and an appropriate reaffirmation of the "right of experts to have a say, which in their opinion had unjustifiably been denied them by the state."[49] Because conventional behavioral models considered engagement for the fatherland a national or patriotic duty, such actions were still regarded apolitical.

The failure of Germany's "Lightning War" (*Blitzkrieg*) strategy exposed the severe deficits in the country's material and technological resources in 1941–1942 and even more so the developmental gaps in weapons technology for the German forces. This crisis situation offered critical but politically loyal scientists like Ramsauer the opportunity to present in a "factual" light professional criticism and status-conditioned grievances to the political authorities in charge. That this meant entering into an unholy alliance with the military, ultimately extending the duration of the war and acting as a stabilizing force for the system as well, hardly disturbed scientists like Ramsauer, if it ever even occurred to them. Ramsauer had been at home in military contexts since the beginning of his career as a research engineer. It has already been mentioned that he had been employed as a young physicist in the Imperial Torpedo Laboratory. During the First World War, he had placed his competence in physics at the disposal of the Army Ordnance Office out of a sense of patriotic duty. This explains why practically no personal inhibitions existed in this regard. Such research contexts were seen as more or less natural and rather confirmed an apolitical self-understanding of the work. In a mixture of traditional patriotism, political opportunism, and pragmatic calculation, Ramsauer was also open to giving his research a military and technical profile during the Third Reich. Within the framework of the Four-Year Plan, this led among other things to a noticeable increase in the number of important war-related research projects for his institute at the AEG.[50]

Conversant as this research and industrial manager of the AEG was in the advantages and necessities of such an image within the Nazi state

[49] Karl-Heinz Ludwig, *Technik und Ingenieure im Dritten Reich* (Düsseldorf: Droste, 1979), 241.
[50] Weiss, *Forschung*.

(if not overall), he also knew how to put it to good use. The same attitude also guided his activity as chairman of the DPG. Thus, it was not coincidence that Ramsauer became the standard-bearer of initiatives that integrated the DPG more strongly than before in the science and research policy of the National Socialist state right after he took up his new function. The society strove to overcome its sociopolitically marginal position and sought to strengthen the position of physics generally within the overall canon of research. The expulsion of reputable colleagues in the profession, an exaggerated politicization and ideologization of physical content, as well as a general decline in physics training and a permanent underfunding of basic research after 1933 threw central parts of physical research, notably modern theoretical physics, into crisis. Physics in Germany fell severely behind, particularly against English-speaking countries.

These deficits became the central topic of a submission by Ramsauer and his fellow board members on behalf of the DPG to the Reich Ministry of Science, Education, and Culture in the fall of 1941.[51] The immediate occasion for the submission was the serious predicament of theoretical physics in Germany and the ongoing defamation by Aryan physics activists who were referring to its modern branches (relativity and quantum theory) as "Jewish physics." This attempt to rehabilitate modern theoretical physics, with its associated public attacks on unsavory Aryan physics, had a greater aim, however. The petition emphasized the general role and importance of physics for modern society, particularly specifying the relevance of modern physics to the German war economy. It pointed out, with detailed documentation, the severe shortfall of German physics against the recent advances made by America and its allies. The result of this petition was that Aryan physics was permanently driven into the defensive and marginalized, ultimately falling into complete insignificance in science policy.[52] Political physicists, stale or second-rate though they were, may well have been good implementers of a National Socialist revolution, especially when Aryan physics fit so nicely within their own ideological conceptions. But a war and the development of modern weapons systems were only partly served – particularly considering the

[51] Carl Ramsauer, "Eingabe an Rust," *Physikalische Blätter*, 3 (1947), 43–46; see the reprint in Hoffmann and Walker, *Physiker*, 594–617; partial translation in Hentschel, *Physics*, docs. 90–93.

[52] Alan D. Beyerchen, *Scientists under Hitler: Politics and the Physics Community in the Third Reich* (New Haven: Yale University Press, 1977), 191f.

extremely small size of the Aryan physics group and that it included hardly any other prominent or established physicists besides Lenard, Stark, and perhaps Rudolf Tomaschek. The typical ideological and political conformance politically expected at the beginning of the Third Reich was only necessary in a very basic sense by the beginning of the 1940s. The technocratic elite was not called upon to go beyond what is generally describable as "political loyalty." The policy of the Four-Year Plan, and more so the Second World War, called for practical service to the *Führer, Volk, und Vaterland.*

This is also reflected in the submission by the DPG, delivered to Reichsminister Rust by registered mail on 20 January 1942. The submission names "with refreshing clarity" – as Ernst Brüche characterized it in his post-war retrospective, not entirely without a gush of apologetics[53] – the fears that the DPG harbored "about the future of German physics as a science and as a factor of power."[54] The petition's preamble immediately makes the unmistakable diagnosis:

> German physics has lost its former supremacy to American physics and is in danger of continuing to lag behind.... The Americans have made very great progress indeed. This is not solely due to the fact that the Americans employ far larger amounts of material resources than we do. It is due at least as much to the fact that they have succeeded in attracting a large new generation of carefree and motivated young scientists, the individual achievements of whom are equivalent to those of our own at the best of times with the advantage that they are able to work in teams.[55]

The reasons for the deplorable state of physics in Germany were then individually set forth:

1. The normal tangible budgets of physics departments at universities and polytechnics receive only a fraction of the funds that are absolutely necessary in our times of advanced technology for physical research, teaching, and education....
2. In our country one main branch of physics, theoretical physics, is being pushed more and more into the background.... Modern theoretical physics in particular has a whole series of the greatest positive achievements to offer that could also be of vital importance to the economy and the armed forces; and the very general accusations made against the advocates of modern theoretical physics of

[53] "mit erquickender Deutlichkeit": Brüche's introduction to Ramsauer, Eingabe, 43.
[54] Ramsauer, Eingabe, 43; partial translation in Hentschel, *Physics*, doc. 90, 278.
[55] Ramsauer, Eingabe, 43; partial translation in Hentschel, *Physics*, doc. 90, 278.

being pioneers of the Jewish spirit, are as unsubstantiated as they are unjustified . . .

3. Professorship appointments in physics do not always follow the principle of preference according to ability that has proven successful both in the past and the present . . .
4. An academic career in physics is rapidly losing its former appeal among our best students . . . transfers into this career from positions in industry are hardly attempted anymore, quite in contrast to the past . . . [56]

The remedies to address these demonstrable deficiencies were then separately listed:

Re 1: The tangible budgets of physics chairs should be raised substantially in accordance with modern requirements.

Re 2: The internal conflicts within German physics must be settled . . .

Re 3 & 4: The deficiencies described here should be eliminated in the future or mitigated as far as possible.[57]

Various attachments were enclosed with the submission. These offprints, speech scripts, publications, and expert opinions were supposed to underscore or illustrate the individual arguments demanding resolution. They were a four-page analysis on "American physics outdoes German physics" (attachment I), "Publications against modern theoretical physics" (attachment II), "The crucial importance of theoretical physics and particularly of modern theoretical physics" (attachment III), "Refuting allegations that modern theoretical physics is a product of the Jewish spirit" (attachment IV), an excerpt "From Professor L[udwig] Prandtl's submission" (attachment V), as well as the proceedings of "The Munich conciliation and pacification attempt [with regard to Aryan physics]."[58]

Ramsauer gave the following report about the effect this submission had after the war:

I had expected at least to have been heard, but had also counted on a rough rebuff in response or, in view of the sharpness of the criticism, severe personal reprisals. What I was not prepared for, however, was that I received no reply at all. I heard indirectly that the submission had aroused the greatest disapproval and that it had been brought to the attention of influential authorities in the Reich, and I believe

56 Ramsauer, Eingabe, 44; translated in Hentschel, *Physics*, doc. 90, 279–280.

57 Ramsauer, Eingabe, 44; translated in Hentschel, *Physics*, doc. 90, 280–281.

58 Ramsauer, Eingabe, 45; translated in Hentschel, *Physics*, 85, 91–93.

I can also attribute some harassments by the Ministry against Mr. Finkelnburg and me for our proceeding. But otherwise nothing, nothing![59]

The lack of response undoubtedly reflects the notorious weakness in the leadership of Rust's ministry. Josef Goebbels himself repeatedly complained about it in his diaries: Rust "does absolutely nothing and lets all things run down to the very worst."[60] About the same time that the DPG was working on its submission, Victor Klemperer wrote in September 1941 the following humorous corroborating anecdote: In Germany "one Rust" counts as a new unit of time for the period between two decrees and their retractions.[61] Weak leadership and loss of prestige for Rust's ministry within the hierarchy of power of the Nazi state were just one angle responsible for the purported dead silence about the DPG's resolution. The other angle is that by then, the lot of physics had already experienced some improvement. The Four-Year Plan and the outbreak of the Second World War offered physicists much more powerful alliance partners, particularly among the technocratic and military elites of the Nazi state, than Rust's Reich Ministry of Science, Education, and Culture. Notwithstanding the statements to the contrary after the war, Ramsauer and his fellow physicists certainly had been heard – if not by Rust and his ministry then certainly by the meanwhile much more influential circles of the Nazi polycracy of power: Josef Goebbels, Hermann Göring, Albert Speer, and above all the upper echelons of the German military. We find, for instance, this note in Goebbels' diary from 1943:

The famous physicist Professor Ramsauer... submits a memorandum to me about the state of German... physics. This memorandum is very depressing for us.... We notice this in the air war as well as in the submarine war.... In any case, Professor Ramsauer is also of the opinion that we can win the advantage back from Anglo-Saxon physicists.... But it will require quite some time. Even so, it is better to start with it... than to let things continue on.[62]

Ramsauer's statement quoted above, that "nothing, nothing!" had happened, must thus be described as one of the only too familiar postwar euphemisms or myths. The facts speak for themselves.

59 Carl Ramsauer, "Zur Geschichte der Deutschen Physikalischen Gesellschaft in der Hitlerzeit," *Physikalische Blätter*, 3 (1947), 113.

60 Elke Fröhlich (ed.), *Die Tagebücher von Joseph Goebbels*, part I, vol. 4, 1 January 1940–8 July 1941 (Munich: Saur, 1987), 335.

61 Victor Klemperer, *Ich will Zeugnis ablegen. Tagebücher*, vol. 1 (Berlin: Aufbau-Verlag, 1997), 666; English translation as Victor Klemperer, *I Will Bear Witness* (New York: Random House, 1998).

62 Ludwig Lochner (ed.), *Goebbels Tagebücher 1942/43*, (Zurich: Atlantis Verlag, 1948), 347.

RAMSAUER AS REPRESENTATIVE OF THE INDUSTRIAL–MILITARY COMPLEX

A comparison between the rendition of Ramsauer's memorandum published in the *Physikalische Blätter* and its full text is revealing.[63] In 1942, REM, the authority formally in charge of the DPG, was evidently regarded by the society as but one contact, and not even the most important one, at that. The ellipses preceding Ramsauer's signature subsume omissions that certainly were not marginal.[64] The partners in league with the society are explicitly named there. This exposes Ramsauer's post-war contentions as conscious understatement. After "present my concerns to you in person" the original text continues:

> I would like to mention in this regard that I have been exchanging thoughts on these questions with Colonel-General Fromm for a while now already, and that furthermore the authorized persons at the Reich Aviation Ministry (*Reichsluftfahrtministerium*, RLM) have approached me regarding these issues. I have therefore forwarded a transcript of this submission to these two agencies upon their request. I am convinced that the whole Armed Forces would readily apply their influence in order to help obtain approval of the required funds from the Finance Ministry. In addition, the Reich University Lecturers leadership, which has been very interested in these questions for some time now, has received a transcript of this submission. Heil Hitler! Yours very sincerely, sig[ned] C. Ramsauer.[65]

Other omissions also reveal adjustments and diction now considered the mark of the spurned representatives of Aryan physics. For example, we read under No. 2 of the submission after the sentence ending with "into the background":

> The legitimate struggle against the Jew Einstein and against the excrescences of his speculative physics has spread to the whole of modern theoretical physics and has brought it largely into disrepute as a product of the Jewish spirit (see Attachment II)...[66]

Gerhard Simonsohn pointed out this omission two decades ago and asked whether this had been "out of conviction or just the exercise of duty in

[63] There is a carbon copy of the submission, together with an enclosure letter by Ramsauer to Prandtl, Berlin, 23 January 1942, among Prandtl's papers in the MPGA, div. III, rep. 61, Prandtl papers, no. 1413. See the comparative transcription in the German version of this volume: Hoffmann and Walker, *Physiker*, cited above: Ramsauer, "Eingabe."

[64] Ramsauer, "Eingabe," 44.

[65] Ramsauer to Reichsminister Rust, Berlin, 20 January 1942. MPGA, div. III, rep. 61, Prandtl papers, no. 1413; translated in Hentschel, *Physics*, doc. 90, quote on 281.

[66] Ramsauer to Reichsminister Rust, Berlin, 20 January 1942. MPGA, div. III, rep. 61, Prandtl papers, no. 1413, sheet 3; translated in Hentschel, *Physics*, doc. 90, quote on 279.

order to get a hearing at all?"[67] Even a favorable response to this difficult historical question, after an interim of 60 years, leaves one "sobered." Ramsauer never alluded to such statements, either in the published version of the submission or in his "physics memoirs" from 1949,[68] not even to differentiate his views about these things or to express any criticism. He rather joined the mighty ranks of those who preferred to remain silent and minimize their National Socialist pasts; he is yet another, more specific example of the German unwillingness – if not inability – to mourn about his or her personal legacy from the Nazi era.[69] Such an emotional outpouring or critical assessment of one's own conduct during the Third Reich would, of course, have defused the myth of an unrelenting battle against "party physics." It would also have attracted attention to the collaborative relations or arrangements made with the National Socialist rulers that the reality of living within the Third Reich had imposed, also on Ramsauer.

The reality was not just opposition to "party physics" and therefore Aryan physics. It was also political action in furtherance of special interests. Ramsauer and the DPG thereby willingly made a pact with the military – or to use current terminology, with the industrial–military complex. With such powerful backing, it was a relatively speedy affair to regain direct influence on the filling of professorship vacancies. The ideologically motivated disparagement and public discrimination against modern theoretical physics had been put to a stop when Aryan physics suffered its dramatic loss of prestige after the "synods" of 1940 and 1942. Fritz Sauter's appointment in 1942 to the polytechnic in Munich is an illustration. He was chosen against the opposition of the "side" backed by Wilhelm Müller, the former successful Aryan physics candidate for Sommerfeld's chair. Pascual Jordan's appointment to the University of Berlin (1944) or Siegfried Flügge's to the University of Königsberg (1944), as well as the appointments of Wolfgang Finkelnburg, Carl Friedrich von Weizsäcker, and Rudolf Fleischmann – all representatives of the professional "DPG-fraction" – to the University of Strasbourg (1942), may be regarded as the outcome of the DPG's submission to the ministry. It was a victory by "professionally oriented" physicists over Aryan physicists. Another fact demonstrates that the DPG perceived an upswing in its

67 Gerhard Simonsohn, "Physiker in Deutschland 1933–1945," *Physikalische Blätter*, 48/1 (1992), 13–18.

68 Ramsauer, *Physik*, 99–130.

69 Alexander Mitscherlich and Margarete Mitscherlich, *Die Unfähigkeit zu trauern* (Frankfurt: Piper, 1967).

fortunes after the filing of its submission: The society resumed conferring its highest scientific award, the Max Planck Medal, in 1943.[70] It was by no means coincidental that Friedrich Hund and Pascual Jordan were thus distinguished. These two theoreticians were leading representatives of modern physics. The award to Jordan, a pioneer of the oft-targeted quantum mechanics during the 1930s, was perhaps even a careful calculation by the DPG's prize committee and board, as he was also a resolute supporter of the Third Reich.[71] Ramsauer likewise carried through an official ceremony in honor of Walther Nernst, who had died in 1941.[72] It did not have the same political impact that the Haber festivities had had 10 years earlier; nor did it attract the same degree of attention as Planck's birthday celebration in 1938. But the official authorities were certainly not pleased about it – the ministry made a point of not sending any representative to attend.[73] It could not afford to make such a public demonstration in favor of an intellectual tradition and type of scholar that the Third Reich was no longer supposed to be promoting.

In the months that followed, Ramsauer delivered numerous talks to inform the leadership of the Third Reich about the problems set forth in the submission and to publicize his message as widely as possible. In summer 1942, he presented a review at a NSDDB training camp as well as before the Lilienthal Society for Aviation Research about "The key position of physics in science, technology and armament."[74] In February 1943, he spoke at the University of Cologne "On the state of physics in Germany and the training of the next generation,"[75] and 2 months later, he spoke before the German Academy for Aviation Research (*Deutsche*

[70] See the contribution by Beyler, Eckert, and Hoffmann to this volume.

[71] Richard H. Beyler, "Targeting the Organism: The Scientific and Cultural Context of Pascual Jordan's Quantum Biology, 1932–1947," *Isis*, 87 (1996), 248–273; Norton Wise, "Pascual Jordan: Quantum Mechanics, Psychology, National Socialism," in Monika Renneberg and Mark Walker (eds.), *Science, Technology and National Socialism* (Cambridge: Cambridge University Press, 1994), 224–254. See also Dieter Hoffmann and Mark Walker, "Der gute Nazi: Pascual Jordan und das Dritte Reich," in *Pascual Jordan (1902–1980), Mainzer Symposium zum 100. Geburtstag. Preprint 329* (Berlin: Max Planck Institute for History of Science, 2007), 83–112.

[72] *Verhandlungen der Deutschen Physikalischen Gesellschaft*, 23 (1942), 2–34.

[73] See von Laue to Meitner, Berlin, 26 April 1942, in Jost Lemmerich (ed.), *Lise Meitner – Max von Laue. Briefwechsel 1938–1948* (Berlin: ERS-Verlag, 1998), 183.

[74] An expanded version was published as Carl Ramsauer, "Die Schlüsselstellung der Physik für Naturwissenschaft, Technik und Rüstung," *Naturwissenschaften*, 31 (1943), 285–288. Reprinted in Hoffmann and Walker, *Physiker*, 618–625; translated in Hentschel, *Physics*, doc. 102.

[75] Carl Ramsauer, "Über die Lage der Physik in Deutschland und die Nachwuchsausbildung." See the notes taken during the talk among the papers of P. Neubert. DPGA, no call number.

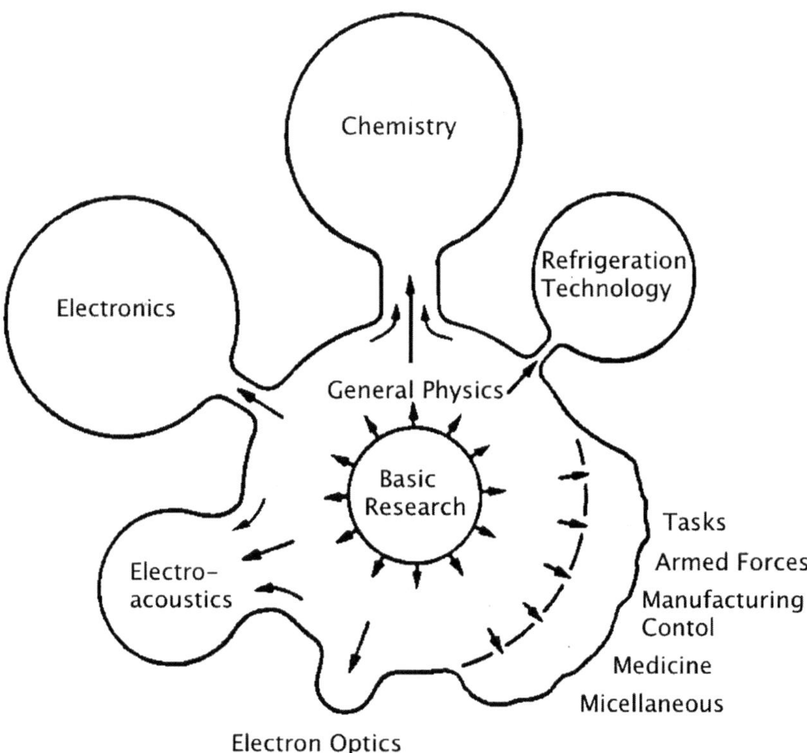

FIGURE 14. Diagram of the "Key Position of Physics for Science, Technology, and Armaments," 1943 (*Source:* Adapted from [captions translated]: Carl Ramsauer, "Die Schlüsselstellung der Physik für Naturwissenschaft, Technik und Rüstung," *Die Naturwissenschaften*, 31 (1943), 285–288)

Akademie für Luftfahrtforschung, DAL) "On the efficiency and organization of Anglosaxon physics with perspectives on German physics."[76]

A number of other essays and studies from this period built upon these topics: "The physics institutes at universities and polytechnics as centers of German physics, situation and future of German physics with suggestions" and "Proposals for the recovery of physics instruction at schools of higher education."[77]

[76] Carl Ramsauer, "Über Leistung und Organisation der angelsächsischen Physik mit Ausblicken auf die deutsche Physik," *Schriften der Deutschen Akademie für Luftfahrtforschung* (Berlin: Deutsche Akademie der Luftfahrtforschung, 1943).

[77] Ramsauer, *Physik*, 81.

The basic tenor of these talks accentuated the central importance of physics. According to Ramsauer, it extended far beyond the boundaries of an individual science; and its role

> will influence future technology to at least the same degree that it has influenced it in the past and the present, but probably even more so. Thus, physics is one of the most essential foundations of our economy and our military strength. This applies especially to all competition with other nations on the economic and military levels. . . . The State must also make allowance for the key position of physics in its decisions on funding physical research or on the optimal organization of physical education.[78]

In consequence of this central role, physics was "a weapon of the greatest, perhaps one day of pivotal importance in the economic and military contest among nations."[79] Ramsauer concluded from this the necessity that teaching and research in physics at German universities undergo extensive development. He also demanded preferential promotion of basic research and general physical disciplines over individual subdisciplines. The former had been grievously neglected and at times even vilified. Ramsauer repeatedly bolstered his arguments with examples from American physics: German leadership in physics had been lost to the Americans, who had mobilized it on such a grand scale that "they will ultimately draw from their scientific superiority everything of military value that can be drawn from it."[80] In exasperation about this dismal outlook, Ramsauer exclaimed: "as a German, [I] do not know whether my sense of awe about the foreign accomplishments should outweigh my sense of worry about Germany's future!"[81] From the results of his report, he derived the demands already presented in the submission to Rust: strengthening physical research at German universities and improving the position of modern theoretical physics, as well as generally implementing professional criteria in hiring policy. In view of the immediate demands posed by the war, he suggested that a central registry of available physicists in Germany be established to obtain a concrete picture of the situation. He proposed a "most efficient utilization of available physicists, in particular the establishment of new criteria for military deployment of physicists." Compared with the enemy forces, they made up a comparatively small

[78] Ramsauer, Schlüsselstellung, 288; translated in Hentschel, *Physics*, 320.
[79] Ramsauer, "Leistung."
[80] Ramsauer, "Leistung," 16.
[81] Ramsauer, "Leistung," 12.

FIGURE 15. Ludwig Prandtl (*Source:* MPGA)

"capacity," but should "at least be set fully on scientific tasks for the purposes of the war."[82]

ALLIES AMONG THE HIGH COMMAND AND RLM

Ramsauer and the DPG were not acting in isolation. They had a particularly influential ally in the aerodynamicist Ludwig Prandtl.[83]

Both Prandtl and Ramsauer delivered many talks that helped interweave the ideas contained in the society's submission within the network of ideas and plans of official research policy. Prandtl was personally involved in plans to strengthen the clout of research in the field of aviation,

[82] Ramsauer, "Leistung," 20f.

[83] About Prandtl, see Moritz Epple, "Rechnen, Messen, Führen. Kriegsforschung am Kaiser-Wilhelm-Institut für Strömungsforschung 1937–1945," in Helmut Maier (ed.), *Rüstungsforschung im Nationalsozialismus. Organisation, Mobilisierung und Entgrenzung der Technikwissenschaften* (Göttingen: Wallstein, 2002), 305–356.

which generally tended to fall short when compared with development.[84] Thus, the efforts by Ramsauer and Prandtl complemented each other very well. Thanks to the central role of aviation technology in German defense research, Prandtl was not coincidentally very much more influential than Ramsauer. He was the driving force behind linking the DPG's submission with other initiatives of science policy and making them generally known. There are even indications that Prandtl could have been the true instigator of the society's submission, at least having served as a stimulating player in its drafting. In a letter from April 1941, he informed Ramsauer that

> at the urgings of the Göttinger physicists... [he had] contacted Reich Marshal Göring in the matter of sabotage of theoretical physics by the Lenard group. It is to be expected that you will be asked about the situation. That is why I forward to you in the enclosed a carbon copy of my submission and an attachment in commentary of it.... It would, naturally, be good if you presented examples about how important a thorough understanding of theoretical physics is for the aims of industrial development.[85]

Ramsauer promptly informed Prandtl about the steps that followed in this matter and in such detail that one could even speak of an interactive coordination between the two.[86] Prandtl received a copy of the DPG's submission when it was sent out to REM.[87] His reply deserves quoting at length because it reveals the tactics that they used to implement their research initiatives:

> Esteemed Colleague,
>
> I have read with the greatest interest the transcript of the submission to Reich Minister Rust that you kindly forwarded to me regarding the plight of theoretical physics in Germany; and I find it exceptionally impressive and convincing precisely because of its impartial style. The military agencies will doubtlessly make sure that the matter does not get stuck in the sand, which from previous experience with the Education Ministry would otherwise certainly have been irrecoverably the case, since various lower-ranking positions seem to pull together with the Munich clique.
>
> First and foremost I think that the military agencies should enforce through official channels a ban on any publishing activity against theoretical physics by

[84] Helmuth Trischler, *Luft- und Raumfahrtforschung in Deutschland 1900–1970* (Frankfurt Main: Campus-Verlag, 1992), 250ff.

[85] Prandtl to Ramsauer, Göttingen, 28 April 1941, MPGA, div. III, rep. 61, Prandtl papers, no. 1302, sheet 6 (transcribed in Hoffmann and Walker, *Physiker*, 592).

[86] See the correspondence in MPGA, div. III, rep. 61, Prandtl papers, no. 1302, sheets 6 ff.

[87] Ramsauer to Prandtl, Berlin, 2 January 1942, MPGA, div. III, rep. 61, Prandtl papers, no. 1302, sheet 12.

FIGURE 16. Hermann Göring opens a conference of the German Academy for Aviation Research in 1938; Ludwig Prandtl is in the front row, third from the right (*Source:* Ullstein)

the sabotaging Munich group; and then the essential new appointments would just have to be made, where necessary by creating new positions. This internal war is naturally complicated by the fact that it is also a war against the "Brown House"; and the military agencies would have to be made particularly aware of this as well.[88]

We have already seen how quickly the wishes and demands mentioned in this letter were granted. This letter also shows that from the beginning, Ramsauer apparently did not consider either the party or civilian authorities like Rust's REM and Education suitable and potent allies for carrying through his demands for a strengthening and general appreciation of physical research in Germany. His confidence lay instead in the influence and power of the military. In fall 1941, while the submission to Rust was still on the drawing board, he reported to Prandtl along these lines:

I have succeeded today after a long discussion with Colonel-General Fromm, Commander of the Reserve Army and Chief of Defense, in convincing him of the

[88] Prandtl to Ramsauer, Göttingen, 28 January 1942, MPGA, div. III, rep. 61, Prandtl papers, no. 1302, sheet 13; translated in Hentschel, *Physics*, doc. 94, 292.

serious menace to German physics. He promised to plead for these issues from the standpoint of the Army and is of the opinion that his resources of power are large enough to carry his intentions through the Reich Ministry of Science, Education and Culture. In addition, he will assign a suitable high-ranking military man to summarize the Army's interests in physics and to work together with the German Physical Society.[89]

That spring, other influential military figures were similarly informed. They included people like Admiral-General Carl Witzell, who directed the Navy Ordnance Headquarters at the Supreme Command of the Armed Forces (*Oberkommando der Wehrmacht*, OKW); General Field Marshal Erhard Milch, who directed the Technical Office of the Reich Aviation Ministry; or General Carl Thomas, head of the Defense Economy and Armaments Office at the OKW. They, too, received copies of the submission. At a meeting with General Thomas on 26 March 1942, Ramsauer mentioned how backward physical research was in Germany against that in the United States, clairvoyantly pointing out:

The large-scale research there in the area of atom smashing could become a great danger for us, one day. Ramsauer asked for support and suggested creating a small advisory board within the Armed Forces to summarize the physical interests in the area of the military.[90]

The decisive support Ramsauer obtained did not come from the OKW, however, but from the Air Force. Milch lent a certain "official tone" to the DPG's submission. Prandtl had Milch ask the German Academy for Aviation Research, "as an impeccable organization of science policy," for an expert opinion on the situation. This would – in Ramsauer's words[91] – "be of the greatest importance for the further course of the entire affair." Prandtl plied his influence on Milch, taking the trouble of personally drafting the appropriate letter to the chancellor of the academy and enclosing it with his letter of recommendation. This letter mentions that Ramsauer's submission had been

sent in transcript also to Messrs. Colonel-General Fromm and Admiral-General Witzell with the request that the necessary weight be attached. From a statement

89 Ramsauer to Prandtl, Berlin, 31 October 1941, MPGA, div. III, rep. 61, Prandtl papers, no. 1302, sheet 10; transcribed in Hoffmann and Walker, *Physiker*, 592f.; translated in Hentschel, *Physics*, doc. 88, 276f.

90 Quoted from Bernhard R. Kroener, Rolf-Dieter Müller, and Hand Umbreit, *Das Deutsche Reich und der Zweite Weltkrieg* (Stuttgart: DVA, 1999), vol. 5/II, 733.

91 Ramsauer to Prandtl, Berlin, 12 May 1942, MPGA, div. III, rep. 61, Prandtl papers, no. 1302, sheet 15.

by Prof. Ramsauer I gather that neither area of command has any scientific agency of sufficient authority at its disposition.

In order to carry this, indeed, very urgent matter forward, I therefore approach the German Academy for Aviation Research for such an expert opinion, to be drawn up by representatives of physics from among their membership for presentation to me.[92]

The fact that suitable experts had already been agreed upon in advance demonstrates how concerted this action really was. Another letter names in this context Jonathan Zenneck, Robert Pohl, Walther Gerlach, Georg Joos, and Richard Becker[93] – all of them equally reputable physicists who fully supported the DPG's submission. It has not been possible to ascertain whether the opinion in question was ever in fact issued. But Ramsauer was invited to speak at the DAL in spring 1943. In the presence of Milch, the academy's vice-president, he gave a detailed sketch of the crisis situation in which German physical research found itself, making the focus of his argument the so unfavorable comparison with research under way in English-speaking countries.[94] It was quite an achievement to have attracted the interest of Milch at the DAL. This was a key administrator of research in defense technology during the Third Reich, and the academy was a leading institution.[95]

Aviation research was, at that time, the largest, most progressive and extensive field of research in the Third Reich, ranked well ahead of the Kaiser Wilhelm Society and other establishments for the development of science and technology.[96] There was an additional factor. Aviation was by nature based on a significant proportion of physics-based research, and Ramsauer's analysis was particularly relevant to this area of research. The reception the DPG's submission received from this key area of military technology undoubtedly strengthened Ramsauer's and the society's position. The abundance of invitations that Ramsauer accepted as a speaker in 1942–1943 is good documentation of this. Other evidence is that Ramsauer's submission was welcomed in other areas as well, becoming the theme of official statements.

[92] Draft of a letter to Kanzler der Deutschen Akademie der Luftfahrtforschung, MPGA, div. III, rep. 61, Prandtl papers, no. 1302, sheet 17.

[93] Ramsauer to Prandtl, Berlin, 12 May 1942, MPGA, div. III, rep. 61, Prandtl papers, no. 1302, sheet 15.

[94] Ramsauer, "Leistung"; there is a facsimile of the title page of this paper in Hoffmann and Walker, *Physiker*, 517.

[95] Trischler, *Luft- und Raumfahrtforschung*, 183ff. and 236ff.

[96] Trischler, *Luft- und Raumfahrtforschung*, 246.

In June 1943, a discussion took place at Speer's ministry with Ramsauer on "the state of German physics,"[97] and in the following month, another meeting was held on this issue with Rudolf Mentzel. Abraham Esau was called upon as head of the physics section of the RFR to give an assessment that basically agreed with Ramsauer's suggestions for raising the performance and reputation of physics in Germany.[98] General consensus was not the submission's only impact. It also aroused rivals. Walter Schieber, department head at Speer's ministry and director of the chemistry division of the National Socialist League of German Engineers (*Nationalsozialistischer Bund Deutscher Technik*, NSBDT), regarded as far too one-sided Ramsauer's emphasis on physics as a key discipline for German armaments technology and the war economy. He feared a promotion of the needs of physical research would come at the expense of chemistry. His suggestion for solving the problems in physics was to have physicists join the NSBDT (at last).[99]

In view of such a flurry of activity, the ready reception and ultimate success triggered by the petition in the months after its submission to Rust, Ramsauer's declaration that in this regard "nothing, nothing!" had happened should be fundamentally questioned. Clearly, it should be interpreted as a retrospective attempt to shroud the alliance between the DPG and the industrial–military complex of Nazi Germany. Pacifism happened to be what was wanted in post-war Germany, and the discovery of any such alliances and collaborative relations could only have harmed the rebuilding of German physics and a resumption of international contacts among scientists – not to mention the tarnish it would have left on the society's luminous historical image.[100]

MOBILIZING THE LAST RESERVES

At the tail end of the Second World War, things were seen differently, of course, and the preference was for making the most of the growing acceptance among those in power, particularly among the military. Between 1942 and 1943, there apparently was an interesting shift in perception and accent regarding Ramsauer's submission. The petition itself and more

97 "Über die Lage der deutschen Physik," Federal German Archives, Berlin (*Bundesarchiv*, hereafter BA), R 26/III/126 (unpaginated).

98 Abraham Esau's statement of 10 June 1943, BA, R 26/III/126 (unpaginated).

99 Schieber to R. Mentzel, Berlin, 25 August 1943, BA R 26/III/126 (unpaginated).

100 See Ramsauer, "Geschichte," 113–115.

specifically the majority of its attachments can be read as a directed appeal on behalf of modern theoretical physics, ultimately pitted against the fatal influence of Aryan physics. In subsequent discussions, this latter aspect recedes into the background to the point of receiving no mention at all in 1943. The topics determining these more or less internal discussions emphasize the key position of physics in the nation's armaments technology and defense economy. Ramsauer's talk before the German Academy for Aviation Research in April 1943 is a case in point.[101] There is no mention at all of Aryan physics – since the Seefeld "synod" of November 1942, it could legitimately be regarded as a dead issue. This talk by Ramsauer is what is explicitly alluded to in the discussions and opinions referred to earlier, becoming more or less synonymous with the true DPG submission. There are also historical mentions of a memorandum on the state and future of German physics (*Lage und Zukunft der deutschen Physik*) that Ramsauer supposedly had written after delivering his talk in May 1943.[102] Goebbels' diary entry as well as the discussion at Speer's ministry and the resulting opinion by Esau in summer 1943 could also be references to it. This memorandum has yet to be explicitly verified or located, however.

The shift in perception and accent given to the submission from 1942 transformed Ramsauer and the DPG into the mouthpiece of a campaign that probably demonstrates most clearly the ambivalence of Ramsauer's initiatives. In his discussions with the military and the responsible political authorities, Ramsauer specifically advocated "taking special steps for the deferment [from direct military service, *uk-Stellung*] of physicists,"[103] also demanding new criteria for their deployment: "The Armed Forces will have to be able to make do with '3,000 fewer soldiers.' '3,000 more physicists' could perhaps decide the war."[104] This was Ramsauer's persuasive argument, and his demands did not go unanswered. They materialized in what is known as the Osenberg initiative. By secret decree of the OKW, about 5,000 scientists were released from combat duty for integration into a weapons research program.[105] This initiative may have preserved the health and lives of many a young (German) scientist. Nonetheless, it was an element of the mobilization of the nation's last reserves. It was thus the specific contribution of German science to total

101 Ramsauer, "Leistung."

102 See Ludwig, *Technik*, 242; Kroener, Müller, and Umbreit, *Deutsche Reich*, 738.

103 Ramsauer to Schieber, 3 September 1943, BA R 26/II/126, Draft.

104 Ramsauer, "Leistung," 21.

105 Ruth Federspiel, "Mobilisierung der Rüstungsforschung," in Maier, *Rüstungsforschung*, 72–105.

war. As we know, it did not decide the war, but it surely helped extend it for a while.

The obvious receptivity to Ramsauer's initiatives was used to full advantage in 1943–1944. Their basic premises were worked into a "Program by the German Physical Society for the development of physics in Greater Germany."[106] This agenda was drawn up as the war was taking a catastrophic course, basic material and technical bottlenecks were tightening on the supply side of a war economy, deficits in the area of military engineering were mounting, and all the nation's forces were being concentrated on total mobilization. The DPG's governing board passed its own program in August 1944, and it was published directly afterward in the first issue of the society's proceedings for that year. It likewise is characterizable as the special contribution of the German physics community, or its professional association, toward full realization of the German war potential. But in this case, too, it was more than this. In the twilight, the *Götterdämmerung*, of the Third Reich it had probably become just as important for the realist Ramsauer to secure the best possible options for physics *after* the war. This ultimately agreed with the strategies being followed in the upper echelons of German industry,[107] and from his position as a business leader, Ramsauer was thoroughly informed about them. This is even euphemistically circumscribed in the DPG's program. It notes in passing: "implementation of this program only comes into question after the victory. If it is so soon sent to press, the reason is to mark an initial close to year-long preliminary work by the Society."[108]

In this sense, the program was not just an act of self-mobilization by physicists doing their bit for the supposed ultimate victory. It was simultaneously also an act of self-preservation in view of a military defeat in the making. Written in light of this salvage strategy of framing better conditions for physical research for the purposes of the war, just as for "the future of Germany," the program primarily emphasizes four points:

1. Providing better funding and staffing of physics departments at universities and polytechnics as anticipated core cells of a necessary revival of physics.

[106] Programm der Deutschen Physikalischen Gesellschaft für den Ausbau der Physik in Großdeutschland. *Verhandlungen der Deutschen Physikalischen Gesellschaft*, 25/1 (1944), 1–6. Reprinted in Hoffmann and Walker, *Physiker*, 626–631.

[107] Paul Erker, *Deutsche Unternehmer zwischen Kriegswirtschaft und Wiederaufbau: Studien zur Erfahrungsbildung von Industrie-Eliten* (Munich: Oldenbourg, 1999).

[108] Programm der Deutschen Physikalischen Gesellschaft für den Ausbau der Physik in Großdeutschland. *Verhandlungen der Deutschen Physikalischen Gesellschaft*, 25/1 (1944) 1–3.

2. Strengthening theoretical as well as technical physics.
3. Raising the prestige of physics at schools.
4. Establishing binding guidelines for the deployment of physicists in the war effort.

In view of Germany's dramatic situation at the time, it is not surprising that hardly any of these points could be realized in the remaining months before the fighting ended. More important was the signal sent out by the program. The chaotic situation during the last year of the war and the progressive destruction obliterated the conditions necessary for conducting research at universities and raising coming generations of physicists. But one proposal in the DPG's reform program could be immediately implemented. In summer 1943, an Information Agency of German Physicists (*Informationsstelle Deutscher Physiker*) was founded. Its purpose was to "enlighten the public about the work and importance of physical research."[109] This was to happen through organized events, lectures, and press and radio reports. A journal, the *Physikalische Blätter*, was even established for the purpose. Its aim was to inform the public about the accomplishments and potential of physics. Its first issue appeared in January 1944 – no small achievement in a period of major shortages, including paper. It is one more trace of the regard and favor of the governors in power. Indeed, almost at this very time the *Zeitschrift für die gesamte Naturwissenschaft*, the main publication of the Aryan physics movement and its allies among the various "Aryan sciences," had to stop its presses, together with other scientific journals, owing to the general shortage of paper.

"Research in distress!" – in peace as at (total) war, one might cynically add – is the guiding theme of the first issue of the *Physikalische Blätter*. This was propaganda for the sake of scientific and technological research in Germany. The neglect of basic research as an important option for the future had to be stopped. Whoever wanted to know how important – scientifically just as socially – fundamental physical research was for later applications could read in the *Physikalische Blätter*:

> and if today a farmer living on the most out-of-the-way farm can hear the voice of the *Führer* on the radio, this is, in the end, thanks to that Maxwellian system of partial differential equations.[110]

109 "Aufklärung der Öffentlichkeit über Arbeit und Bedeutung der physikalischen Forschung," *Physikalische Blätter*, 1 (1944), 1.

110 *Physikalische Blätter*, 1 (1944) 51. See the facsimile of this first issue with its motto: "Forschung tut Not!" in Hoffmann and Walker, *Physiker*, 519.

PHYSIKALISCHE BLÄTTER

IM AUFTRAGE DER DEUTSCHEN PHYSIKALISCHEN GESELLSCHAFT
HERAUSGEGEBEN VON ERNST BRÜCHE

Forschung tut not!

AUS DEM INHALT
Gründung der Informationsstelle / Zur Einführung / Forschung tut not / J. Zenneck: Die Bedeutung der Forschung / S. R. Cajal: Über die Forschung / Unsere Physikerbilder / J. Goebbels: Wissenschaft und Forschung in ihrer Stellung im Dritten Reich / A. Speer: Einschätzung der Grundlagenforschung / Anekdoten

FRIEDR. VIEWEG & SOHN BRAUNSCHWEIG

JAHRGANG 1 JANUAR-FEBRUAR 1944 HEFT 1-2

FIGURE 17. The *Physikalische Blätter*: "*Forschung tut Not*" ("Research in Danger") (*Source:* Cover of *Physikalische Blätter*, 1–2 (1944))

Certainly, it is an extreme example of blatant propaganda that should be passed off as one contemporary concession to the going ideology. Much in the few issues appearing by the end of the war is certainly readable, nevertheless.[111] The focus of the articles is not just on the role of science for war-making but also for modern industrial society in general. This topic continued to retain a high place in the journal's columns during the post-war period under the conditions of reconstruction; this (official) organ of the DPG was not forced to close down. One had thus "found the basis" – as Ramsauer, in true technocratic innocence, put it in his

[111] See the contribution by Simonsohn to this volume.

retrospective article of 1947 – "upon which we can and will rebuild German physics."[112]

Two other initiatives besides the founding of the *Physikalische Blätter* deserve mention. They were among the few points of the DPG's program that could still be addressed with some success. They also document how strongly the DPG's activities of these years were already oriented toward the post-war period. One focus of the DPG's program of 1943–1944 was "raising the prestige of physics at schools." The campaigns "Prinz Eugen" and "Blücher" launched by the Reich Youth leadership in 1944 in cooperation with military agencies did not just happen to pursue an (unstated) double strategy. They were seen as a way to help counter the deplored neglect of mathematical and scientific education at schools. At the same time, scientifically gifted schoolboys could be spared from the draft or deployment on the "home front" during the final stage of the Second World War and consequently senseless heroic sacrifice.[113]

The "Prinz Eugen" campaign assembled in special training camps boys gifted in science who were preparing for their school-leaving examinations. There they were introduced to the elements of high-frequency engineering for later deployment as aids in radar direction-finding units of the Armed Forces. A number of these 6-month courses took place at Stegskopf in Westerwald, a wooded range along the southeastern border of Westphalia. The courses were held under the patronage of the Hitler Youth and the science director for the Plenipotentiary of High-Frequency Research at the RFR. Five hundred of such graduating pupils were prepared for radar-unit deployment.[114] Younger schoolboys (born between 1927 and 1929) were assigned to a similar sister organization called "Aktion Blücher," also launched in 1944. It was probably during Speer's visit at the AEG's research laboratory that Ramsauer and Brüche mentioned the problem of future generations of scientists and engineers and obtained from him permission to establish schools for gifted pupils, which were supposed to be operated together with the Reich Youth leadership.[115] The plan was to create a nationwide network of such elite schools for pupils talented in mathematics and science. The war only permitted one such school actually

[112] Ramsauer, "Geschichte," 114.

[113] Dieter Hoffmann, "Die DPG-Initiativen "Prinz Eugen" und "Aktion Blücher" in der Endzeit des Dritten Reiches" (forthcoming).

[114] *Wir Stegskopfer. Die Funkmeß-Einheiten Prinz Eugen – Tegetthoff 1943–1945. Eine Chronik* (Marbach: Private publication, 1989).

[115] Interview with Heinrich Hartmann, Reutlingen, 16 March 2004. See also Artur Axmann, "*Das kann doch nicht das Ende sein*": *Hitlers letzter Reichsjugendführer erinnert sich* (Koblenz: Bublies, 1995), 374.

to be founded, at Weidenhof Castle near Breslau. The good connections between the Reich Youth leadership and the regional leader of Lower Silesia, Karl Hanke, explain the selection of this particular location. Breslau was particularly favorable because it offered the possibility of enlisting professors at the local university for these pupils' schooling. Such classes were subsequently held by Ludwig Bergmann and other teachers at the University of Breslau. This is what the DPG and Brüche, in particular, had applied their influence for, not – as the latter contended after the war – "in order to help the party win a long since lost war, but to salvage what was still salvageable for the future after the war."[116] Brüche was the *spiritus rector* behind this initiative at the DPG. He was not only in direct communication with Heinrich Hartmann of the Reich Youth leadership but also personally took care of concrete organizational questions regarding the school. In addition, he granted the Reich Youth leadership access to his *Physikalische Blätter* as a platform for promoting their goals among the scientific public. In an article on "Youth and engineering," Hartmann let his readers know that the Reich Youth leadership was aware "how decisive the knowledge and expertise of our engineers and the indefatigable research of our scientists are, besides tough front-line bravery and feats by our specialists."[117] The boarding school at Weidenhof opened its doors to about 100 pupils in August 1944 under this guiding idea. But a few months later, at the beginning of February 1945, it had to be evacuated to southern Germany as the frontline of fighting neared. After the war, the school's pupils described it as a generally ideology-free environment that had saved them from deployment on the "home front" as anti-aircraft aids or as peons of the "People's Storm" (*Volksturm*), the last-stand militia composed of old men and boys.[118]

THE DILEMMA OF SELF-MOBILIZATION

Overall, Carl Ramsauer's activity as chairman of the DPG leaves a clear trace. Under his aegis, the DPG stepped out of its niche existence as a representative professional association of physicists and became increasingly involved in the sociopolitical processes of the Third Reich. Ramsauer

[116] Circular letter by Brüche to the "Weidenhofer," 18 December 1946, Archives of the State Museum for Technology and Work in Mannheim (*Archiv des Landesmuseums für Technik und Arbeit in Mannheim*, hereafter LTAMA), Brüche's papers, no. 320.

[117] Heinrich Hartmann, "Jugend und Technik," *Physikalische Blätter*, 1 (1944) 75.

[118] See Karlheinz Koppe, *Das Unternehmen Blücher 1944/45. Wie 40 Schüler vor dem Fronteinsatz gerettet wurden* (unpublished manuscript, 2001); see also the collection of letters at LTAMA, Brüche papers, no. 320.

used his position as a commercial physicist and his good ties with military circles to forge an alliance with the industrial–military complex of Nazi Germany. The DPG's prestige, position, and influence gained lasting profit by this alliance, but it irretrievably turned the society into a well-functioning association of scientific specialists, who, although not in every case enthusiastic about the Nazi state, taken together became its loyal and effective servant. The society lost its political innocence – if there ever had been such innocence to lose – because this alliance became the basis of a variety of collaborative relations with political power. These relations undoubtedly had a stabilizing effect on the system and surely also extended the war. Thus, they can hardly be interpreted as opposition, let alone resistance.

When compared with the conduct of some other science epigoni of the Third Reich, Ramsauer's and the DPG's battle against "party physics" does stand out favorably. Nevertheless, these quarrels remained within the purely professional or academic realm and should therefore, at best, be characterized by the word *Resistenz*[119] only in the sense of shielding or safeguarding professional autonomy and handed-down standards within small, discipline-oriented areas.[120] Ramsauer made comparisons between the state of physical research conducted in Germany and in the United States, but he did not venture further to include in any way the political orders concerned – either before or after 1945! Ramsauer's later frequent exonerating formula that all had been done for the sake of science (implying a just cause) suppresses or entirely neglects that it involved science within and for an unjust system that had been willingly served to entirely mutual advantage and that, at times, had even been enthusiastically courted. Although as a skillful and successful science policy maker and manager, Ramsauer knew how to protect the special interests of the DPG and of physics as a whole, the Nazi dictatorship profited considerably by his competence as an ostensibly apolitical expert – whether or not Ramsauer subjectively concurred.

GÖTTERDÄMMERUNG

Before the DPG fell apart in the whirl of destruction of the Third Reich, it convened one more time in the afternoon of 18 January 1945 to celebrate

[119] This German term is taken from biology and is intended to remind the reader of "resistance" in the sense of resisting a disease.

[120] Gerhard Simonsohn, "Physiker im Dritten Reich – Resistenz und Anpassung." Talk delivered at the Freie Universität Berlin on 27 January 1999.

FIGURE 18. University of Berlin Physics Institute on the Reichstagsufer (*Source:* DPGA)

the centennial of its founding. The place chosen for the celebration was the great auditorium full of tradition at the Physical Institute on the Reichstagsufer.

This hall – as a contemporary report reads – "had witnessed many a memorable and festive meeting for physics."[121] These included the famous Wednesday colloquium and the DPG's Friday gatherings, where "all those significant scientific festivals in the history of Berliners and therefore of German physics took place."[122] The centennial event was jointly organized by the boards of the regional society of Berlin and the DPG, with the additional support of the newly founded Information Agency of German Physicists. That meant Ernst Brüche played an important role. He was not only responsible for providing information and publicity but also influential in designing the program.

[121] Vorbericht zur 100-Jahr-Feier der Deutschen Physikalischen Gesellschaft. LTAMA, Brüche papers, no. 376.

[122] "Hundertjahrfeier der Physikalischen Gesellschaft," *Physikalische Blätter*, 3 (1947), 3.

FIGURE 19. Centennial Celebration of the German Physical Society in 1945 (*Source:* Bundesarchiv Koblenz)

The opening talk by the DPG's chairman Carl Ramsauer was followed by speeches by the physicist from Danzig, Eberhard Buchwald,[123] and the physics historian from Hamburg, Hans Schimank.[124] The second part of the event was taken up by documentary film presentations from the society's "peace-time conventions" – the Bad Nauheim meeting (1932) until Baden-Baden (1938) – as well as a slide show from the history of physics.

Buchwald's speech provided a brief historical outline of the society, borrowing its five focal subheadings from Goethe's poem *Urworte Orphisch.*[125] In this classical dress, his historical review could brush more modern issues in euphemistic terms, using the terminological archetype *anangke*:

> *Anangke*, hard destiny, the fourth among these words, signifies the society being swept up in temporal events that encumber it with duties and obligations.... During the '30s and then in wartime *anangke* appeared in two forms. It

[123] Eberhard Buchwald, "Die Deutsche Physikalische Gesellschaft an der Schwelle ihres zweiten Jahrhunderts," *Physikalische Blätter*, 2/5 (1946), 97–106.

[124] Hans Schimank, "Die Physik des 19. Jahrhundert," *Die Naturwissenschaften*, 34/1 (1947) 2–10.

[125] Johann Wolfgang Goethe, *Gesammelte Werke* (Berlin: Tillgner, 1981), vol. 2, 158.

forced us to do battle for the recognition of theory and mathematics.... This battle seems to have been fortuitously fought out; how painful that it was necessary! We are in the throws of a harder *anangke*, with temporal events dictating the society's actions. We can summarize these tribulations under the name "Ramsauer plan," very serious proposals for how German physics should secure its merited place in the world, a program regarding scientific research, staffing and professorship appointments, the coming generation of physicists, the problem of senior schoolteachers, etc. This carefully considered and energetic intervention in a fateful situation, an *anangke* in the truest sense of the word, is the society's main accomplishment in our times.[126]

That the orators of this centennial celebration confine their rhetoric within the apolitical sphere was not the intention. Ramsauer explicitly urged Schimank to "touch on German/foreign relations."[127] Ramsauer's own inaugural speech was unabashedly political. In accordance with the "Ramsauer plan," he explained,

modern physics, for all its importance, [is] no longer a narrowly limited science; it has become a techno-military factor of the first order whose decisive importance for the present day and the future is becoming increasingly clear not just to us but also to our enemies...; the importance of physical research in deciding future peace and war thus offers us a platform for our decisions to further the development of German physics.[128]

In this sense, the centennial celebration became a forum for the DPG's new program. It propagated the "Ramsauer plan's" call for adequate recognition of the key position of physical research in all science, technology, and the economy, and not least in the ongoing war. This celebration, with Reich Youth Leader Artur Axmann and his adjunct Heinrich Hartmann in attendance, received much press coverage. But the appeal was naturally swallowed up in the din of the last battle. The notable attendance of the Reich Youth leader as the highest ranking National Socialist official was anyway not so much the result of any elevation in standing by the DPG within the Nazi hierarchy. It rather reflected the special personal relationships of trust developed during the "Blücher" and "Prinz Eugen" initiatives of the final years of the war.[129]

The modest ceremony could not remotely compare to the glitter of other founding festivities or to the Planck celebration of 1938. It was

[126] Buchwald, "Deutsche," 105.
[127] Vorbereitung der 100-Jahrfeier. LTAMA, Brüche papers, no. 376.
[128] Carl Ramsauer, "Hundert Jahre Physik," *Deutsche Allgemeine Zeitung* (18 January 1945), 1; reprinted in Hoffmann and Walker, *Physiker*, 632–635.
[129] Interview with Heinrich Hartmann, Reutlingen, 16 March 2004.

weighed down by dismal surrounding conditions and an all-pervasive eschatological mood. Although the hall was overcrowded, it was unheated. The Physical Institute building already bore scars of the bombings, and the Red Army was marching unhindered toward the river Oder and Berlin. Not without reason did Buchwald anxiously swear "to prevail, in hope of times of peaceful labor in a safe fatherland and under one's own authority!"[130]

In a letter to Arnold Sommerfeld, who could not have reached Berlin because of the catastrophic travel conditions, Ramsauer reported: "The centennial celebration went very nicely. Privy Councillor Planck was represented by his wife. All the participants perceived this event as an intellectual oasis in the present times and followed the entire event to the end with the greatest of interest, despite the relative lengthiness on the whole and despite the so chilly temperatures of the hall."[131]

A quarter of a year later, this interest was but history, and the once so splendid physics institute at the Reichstagsufer was not the only ruin – it had been burnt to the ground in the battle over the government district in April. The DPG, too, was no more. After Germany's unconditional surrender, the Allies declared not only parties and political organizations dissolved but also all associations and societies, including the German Physical Society and the German Society for Technical Physics.

130 Buchwald, "Deutsche," 106.

131 Ramsauer to Sommerfeld, Berlin, 24 January 1945, DPGA no. 20958.

6

The Planck Medal

Richard H. Beyler, Michael Eckert, and Dieter Hoffmann

"The Planck Medal is awarded annually by the German Physical Society [*Deutsche Physikalische Gesellschaft*, DPG], as a rule on the 23rd of April, to a scholar for achievements in theoretical physics, specifically for those extending Planck's oeuvre."[1] This is article 1 of the statutes for a medal first awarded on 28 June 1929 on the occasion of the golden jubilee of Max Planck's doctorate. The prominence of its recipients is just one reason why it is still regarded as the DPG's most renowned distinction. A listing of their names reads like a *Who's Who* in modern – theoretical – physics.[2] The initiative for this award came from leading German theorists: Max Born, Albert Einstein, Max von Laue, Erwin Schrödinger, and Arnold Sommerfeld. In 1927, in anticipation of Max Planck's 70th birthday, they signed an appeal for contributions to an endowment for a gold medal. The response to the appeal was great, reflecting the overwhelming acclaim that Max Planck enjoyed not just from the scientific community of his time as Germany's doyen of theoretical physics but also from the general public as a spokesman for science.[3] German industry shouldered

[1] Statutes of the Planck Medal, Archives of the Max Planck Society (*Archiv der Max-Planck-Gesellschaft*, hereafter MPGA), div. III, rep. 50, Laue papers, no. 2388, sheet 1.

[2] See the full list heading the documents pertaining to the Planck medal in the appendix to Dieter Hoffmann and Mark Walker (eds.), *Physiker zwischen Autonomie und Anpassung – Die DPG im Dritten Reich* (Weinheim: Wiley-VCH, 2007), 579–591.

[3] See Dieter Hoffmann, *Max Planck. Die Entstehung der modernen Physik* (Munich: Beck, 2008); John L. Heilbron, *The Dilemmas of an Upright Man: Max Planck as Spokesman for German Science* (Berkeley: University of California Press, 1986).

FIGURE 20. The Max Planck Medal (*Source*: DPGA)

a large portion of the total endowment capital.[4] The fundraising itself is evidence of the high rank to which theoretical physics had ascended in the first decades of the 20th century, making it the prestigious guiding subdiscipline of physics.[5] The initiators of the appeal were able to hand to Planck on his 70th birthday a letter that contained

> besides the text of the appeal and a listing of all the individuals, societies and companies by name that the endowment had brought together, also the request for his approval of the purpose of the endowment, prompt naming of the scientific body to make the awards, and finally his acquiescence to the necessary sittings with an acknowledged artist so that the medal face might bear his portrait.[6]

The scientific body designated to award the medal was – as had probably been envisioned from the outset – the German Physical Society. It

[4] A. Vögler to Petersen, 31 December 1927, Archives of the German Iron & Steel Institute (*Verein Deutscher Eisenhüttenleute*, hereafter VDEh), Düsseldorf, Helmholtz-Gesellschaft holdings, file HG5, vol. 1.

[5] See Michael Eckert, *Die Atomphysiker. Eine Geschichte der theoretischen Physik am Beispiel der Sommerfeldschule* (Wiesbaden: Vieweg, 1993).

[6] "Planck-Medaille," *Die Naturwissenschaften*, 16 (1928), 368.

Nummer 27
7. Juli 1929

Zeitbilder

Beilage zur Vossischen Zeitung

Deutschlands große Physiker.

[illegible] Planck überreicht am Tage seines goldenen Doktorjubiläums die für Fortschritte auf dem Gebiet der theoretischen Physik geschaffene Planck-Medaille seinem Fachgenossen Albert Einstein.

FIGURE 21. Max Planck and Albert Einstein, the first recipient of the Max Planck Medal in 1929 (*Source*: MPGA)

was instructed to form a medal committee responsible for informing the society's board about proposed recipients. According to the statutes, this committee was composed of Max Planck as chairman and the medalists. After Planck's resignation, the "oldest bearer of the medal" was supposed to take over the committee chairmanship. In the annual selection process, each committee member would name two candidates; then, the committee chairman would pass on to the DPG's board the names of the two with the most nominations, and the board would select the awardee. Thus, a rather exclusive structure was erected around the Max Planck Medal that turned the laureates into a self-recruiting elite. The list of the first medalists also reflects this. They were not just prominent representatives of theoretical physics but also pioneers of quantum theory in particular.

At the beginning of November 1933, Werner Heisenberg received the medal under rather modest and subdued circumstances, not at the physicists' congress but at a specially organized scientific colloquium in Berlin, and the medal committee and the DPG board decided, without offering any concrete reasons, to discontinue further awarding of the medal for the time being. A lack of worthy candidates was certainly not what lay behind this decision. By then, the list of awardees – Max Planck, Albert Einstein, Niels Bohr, Arnold Sommerfeld, Max von Laue, and, as just mentioned, Werner Heisenberg – rather displayed a group of physicists currently at the brunt of a Nazi smear campaign. Advocates of the so-called Aryan physics (*Deutsche Physik*) particularly denounced them as exponents of an overly mathematical and formalistic, speculative conception of (theoretical) physics. Their stereotyping as representatives of "the Jewish spirit" could just as well be made to fit the two candidates for the year 1934 – Max Born and Erwin Schrödinger – who had been nominated by a convincing consensus. The inquiry to the former recipients had yielded a unanimous vote in their favor.[7] But this nomination went no further, for there was no award ceremony. As the pertinent documents by the DPG's board or by Max Planck as chairman of the medal committee have not come down to us, the causes can only be surmised. But it is surely not mistaken to assume that after the embarrassments surrounding the physicists' convention in Bad Pyrmont, the DPG's board wanted to

[7] See the compilation in Planck's hand as well as Einstein's letter to Planck, 27 January 1934, Archives of the German Physical Society (*Archiv der Deutschen Physikalischen Gesellschaft*, hereafter DPGA), no. 20952; transcribed in Hoffmann and Walker, *Physiker*, 543, and Planck to Sommerfeld, 15 January 1934, Archive of the Deutsches Museum (Deutsches Museum, Archiv, hereafter DMA), NL 89, 018, folder 3.12.

avoid a further escalation of the conflict with the ruling National Socialists and their representatives in the physics profession. An award to Born and Schrödinger would undoubtedly have led to a clash of worldviews and ideologies, and it had a political dimension besides. Both these physicists had emigrated from Germany: Born because of his dismissal on the basis of the Law for the Restoration of the Professional Civil Service, and Schrödinger because – as he later once stated – he could "not tolerate being harassed by politics."[8] Schrödinger had disguised his departure as a leave of absence and officially continued to remain a member of the faculty of the University of Berlin, but what Schrödinger's absence really meant was clear to everyone. Consequently, as a letter to the Reich Ministry for Science, Education and Culture (*Reichsministerium für Wissenschaft, Erziehung und Volksbildung*, REM) phrased it,[9] the DPG's board deemed it "appropriate to postpone the award." The board was following its general line of doing its utmost to avoid public and direct confrontation with the Nazis.

Evidently at Max Planck's own urgings, in 1937 new efforts were made to resume the awarding of the Planck medal. The board deliberated on this question on 10 March 1937. Born no longer entered into the discussions as a potential awardee for the above-mentioned racist and political reasons, so it was all about Schrödinger. His nomination was held to have meanwhile become defensible because he had accepted an appointment to Graz. That meant it was no longer compelling to view him as an émigré – especially considering that, as it is put in one of Planck's letters to Jonathan Zenneck, he "is still listed as a member of the faculty in our university's most recent staff directory."[10] Schrödinger's nomination by the DPG's board became a political issue nonetheless. The minutes of the board meeting record:

> Mr. Zenneck notes that Privy Councillor Planck upholds his proposal: Born and Schrödinger for the award of the Planck medal and moves that the award go to Mr. Schrödinger. At the suggestion of Mr. Schottky it is resolved that the medal be awarded this year to Mr. Schrödinger, provided the Society's fall convention take place in Salzburg. Should the convention take place at another venue, then the award should be postponed. At the proposal of Messrs. Geiger and Mey it is

[8] According to Max Born, *Mein Leben. Die Erinnerungen des Nobelpreistägers* (Munich: Nymphenburger Verlag, 1975), 363.

[9] Zenneck to W. Dames, 4 August 1937, DPGA, no. 20953; transcribed in Hoffmann and Walker, *Physiker*, 583.

[10] Planck to Zenneck, 11 July 1937, DPGA, no. 20953; transcribed in Hoffmann and Walker, *Physiker*, 580f.

resolved: Provided the convention take place in Salzburg, an inquiry should be directed at the responsible authorities in advance about whether any reservations exist against an award of the medal to Mr. Schrödinger.[11]

The DPG's board was therefore itself divided on the matter of the Planck medal. The officiating president, Jonathan Zenneck, was not pleased. He complained: "the gentlemen of the German Physical Society's board are a trifle too fearful,"[12] also noting that "the responsible authorities are being consulted more than is absolutely necessary."[13] He assured Planck that he had been "thoroughly annoyed when in the spring the, in my view, entirely unnecessary decision was made to postpone the award of the medal if the physics meeting would not be in Salzburg and to solicit from the Reich Ministry for Science, Education and Culture whether the award to Schrödinger was amenable to them."[14] Planck had not been too pleased about the board's vote either. It could in fact be regarded as aimed at stymieing his efforts. Owing to the general control of the foreign exchange inside Germany, it was highly uncertain whether the convention and hence also the medal award could take place in Salzburg at all. With the relocation of the fall meeting from Salzburg to Bad Kreuznach definitely fixed for the early summer, Planck pointed out again in no uncertain terms:

[I]t really is time that the medal endowment, whose activities for understandable reasons had been arrested for a while, now finally return to regular operation. For, simply ignoring the statutes, which the German Physical Society had resolved and ratified in due form, while the endowment capital and the corresponding interest yields are abundantly available and all the other conditions for its implementation are satisfied, really is not acceptable. Then it would be preferable to liquidate the entire endowment and direct its funds to some scientific cause, which I would very gladly agree to. In any event, though, if this affair is not put right, I would resign the chairmanship in the medal committee – as you can believe me, I would feel very much relieved to be released from an obligation that constantly forces me to work at an affair that, after all, was also originally conceived as an honor for me personally.[15]

This threat from the pen of the universally esteemed doyen of physics in Germany had its effect. His resignation would have caused a public scandal and certainly would have marred the DPG's standing. A modification

11 Meeting of the DPG's board, 10 March 1937, DPGA, no. 20953.

12 Zenneck to von Laue, 3 August 1937, DPGA, no. 20953.

13 Zenneck to W. Grotrian, 22 March 1937, DMA, NL 53, 011.

14 Zenneck to Planck, 28 July 1937, DMA, NL 53, 012.

15 Planck to Zenneck, 11 July 1937, DPGA, no. 20953; transcribed in Hoffmann and Walker, *Physiker*, 580.

of the board's resolution was introduced by written vote that summer that removed the link between the conferral of the medal to Schrödinger and the venue, Salzburg. This vote documents the differing positions represented within the board on this issue. It yielded the following:

> [A]n unreserved "yes" was voted by Messrs. Backhaus, W. Meißner, von Laue, Westphal, Grotrian, Madelung, Stetter, Lohr, Steinke, Zenneck, Mey, Valentiner. Messrs. Schottky and Clemens Schäfer voted "yes" with the reservation that clarity exist about the Reich Ministry for Science, Education and Culture's stance.[16]

Walter Schottky even threatened to resign if his condition was not met. Walter Grotrian feared that the ministry was not capable of mustering up any kind of recommendation. Zenneck's reply and comment about the results of the vote was that Schottky had to admit that "everything feasible has happened" to find a respectable solution, and that this could not be used as a pretext for resigning "if he did not want to be counted among the group of psychopathic physicists."[17] Yet Schottky was a source of further annoyance: He intimated that an attack was in the making against the DPG in the event that Schrödinger received the Planck medal but would not divulge the source of his information. Zenneck preferred to wait and see, replying to Schottky that he did not know "what to make of such general warnings." To Grotrian he said: "What role Mr. Schottky is playing in the whole affair is not clear to me. Joos already wrote me a while ago that the future of the German Physical Society would be endangered if the medal was given to Schrödinger."[18]

A letter by the other conditional "yes" vote, professor of physics Clemens Schäfer at the University of Breslau, provides further details about the interests and prevailing emotional state:

> I had written to Grotrian that I, provided no political reservations existed, would vote yes. In the alternate case, in my opinion, no board member of the German Physical Society could justifiably vote yes. I am thus very much in favor of Zenneck having written to the Ministry. However, I am not of the opinion that silence by the Ministry should be counted as approval. Very often silence is slovenliness or – design. I can therefore, in the interest of the Physical Society, only vote yes if there is *explicit* approval by the Ministry.[19]

Schäfer alluded here to the feelers the DPG's board had extended, both along official channels as well as unofficially, about what the responsible

[16] Grotrian to Zenneck, 16 August 1937, DMA, NL 53, 012.
[17] Zenneck to Grotrian, 19 August 1937, DMA, NL 53, 012.
[18] Zenneck to Grotrian, 25 August 1937, DMA, NL 53, 012.
[19] Schäfer to unknown, 16 August 1937, DPGA, no. 20953; transcribed in Hoffmann and Walker, *Physiker*, 583.

political authorities thought about the plan of awarding the Planck medal to Schrödinger. At the beginning of August, Max von Laue had inquired of Ministerial Advisor von Rottenburg at the REM what he thought "the right thing" to do might be,[20] and a short time later Zenneck, too, officially consulted the ministry about this issue.[21] By specifically pointing out that the original nominees had been "Born and Schrödinger," whereas now the intention was to "award the medal to Professor Dr. Schrödinger of the University of Graz" alone,[22] Zenneck apparently wanted to highlight the DPG's political loyalty to the ministry and document its ability to compromise or cooperate on this matter.

Just 3 weeks later, it was clear that on the part of the ministry "no objections exist to the intended award of the Planck medal to Professor Schrödinger at the University of Graz."[23] Now – as Zenneck wrote to Sommerfeld – "the matter [is] completely clear and we can and must award the medal to Schrödinger, even at the risk of some kind of difficulties ensuing."[24] Consequently, with the ministry's backing, the Planck medal was awarded to Schrödinger 3 weeks later at the physicists' congress in Bad Kreuznach – albeit *in absentia*. This was a novelty but not in breach of the medal statutes. Planck had already expressly pointed out in his letter in summer 1937 that a personal appearance by the laureate was not absolutely required.[25]

The affair would have come to a fortunate close, had the inquiry made at the ministry not posed a troublesome precedent for the future selection process. Overeager obedience by the DPG's board and the general political atmosphere were to blame rather than any official instructions issued by the ministry. As a result, discussions about whether and how the ministry was to be informed or directly involved became one of the attendant circumstances, unwelcome and politically instrumentalized, for future awards of the Planck medal. In addition, determined opponents of the DPG seized the opportunity of the award to Schrödinger to redouble their attacks on the society and modern physics generally. Right after the fall

[20] Zenneck to Grotrian, 2 August 1937, DMA, NL 53, 012; Zenneck to von Laue, 3 August 1937, DPGA, no. 20953.

[21] Zenneck to REM, attn: Dr. Dames, 4 August 1937, DPGA, no. 20953; transcribed in Hoffmann and Walker, *Physiker*, 583.

[22] Zenneck to REM, attn: Dr. Dames, 4 August 1937, DPGA, no. 20953; transcribed in Hoffmann and Walker, *Physiker*, 583.

[23] REM to the president of the DPG, 24 August 1937, MPGA, div. III, rep. 19, no. 1013, sheet 9.

[24] Zenneck to Sommerfeld, 1 September 1937, DPGA, no. 20953.

[25] Planck to Zenneck, 11 July 1937, DPGA, no. 20953.

meeting, Johannes Stark demanded to see the statutes of the medal endowment. Zenneck advised Grotrian to deny him this wish because with his article in *Das Schwarze Korps*, titled "'White Jews' in Science,"[26] Stark had "attacked German theoretical physicists in an unheard of manner" and he was worried that Stark would "abuse the statutes of the Planck medal solely for similar purposes." He added: "In this entire affair we are, of course, in an entirely impregnable position, after the Reich Ministry for Science, Education and Culture explicitly approved the award to Schrödinger."[27]

Stark did not, in fact, simply accept this new slight by the DPG and launched another unsigned article that appeared on 18 November 1937 in *Das Schwarze Korps* under the heading "The other side of the medal," targeting the Planck award: "Professor Schrödinger demonstratively placed himself politically in opposition to Germany; a more insensitive sign of national debasement cannot be imagined than honoring such a man as well for some sort of professional achievement."[28] The author admitted that he did not know the exact circumstances leading to the award to Schrödinger, but he hastened to add, "neither the German Physical Society nor its board could be held responsible for the award."[29] The DPG's board had been "spared," Zenneck suspected, "because Stark knows that attacks against the board in the present case would have to be regarded as attacks against the Reich Ministry for Science, Education and Culture, after it had given its approval." Even so, he advised the board recently elected in Kreuznach with Peter Debye at its head: "It would, in my view, be a disgrace if we troubled ourselves with an article of our own in *Das Schwarze Korps*. We would only be doing the author a favor if we gave him occasion for a new article."[30]

This advice was followed, but Debye and his fellow board members were soon confronted nonetheless with new challenges and problems. The manner of dealing with them did not change during Peter Debye's term, either. Conflicts with the politically mighty and their representatives

[26] Johannes Stark, "'Weisse Juden' in der Wissenschaft," *Das Schwarze Korps* (15 July 1937); translated in Klaus Hentschel (ed.), *Physics and National Socialism: An Anthology of Primary Sources* (Basel–Boston: Birkhäuser, 1996), doc. 55, 152–157.

[27] Zenneck to Grotrian, 19 October 1937, DMA, NL 53, 012.

[28] Anonymous, "Kehrseite der Medaille," *Das Schwarze Korps* (18 November 1937); reprinted in Hoffmann and Walker, *Physiker*, 584–587.

[29] Anonymous, "Kehrseite der Medaille," *Das Schwarze Korps* (18 November 1937; reprinted in Hoffmann and Walker, *Physiker*, 584–587.

[30] Zenneck to Grotrian, 23 November 1937, DMA, NL 53, 012.

in the scientific community were avoided as best as possible to the point of neglecting hitherto accepted professional standards. If these frictions seemed unavoidable anyway, backing was secured at the responsible political authorities. One instance is when Planck resigned his chairmanship of the medal committee in fall 1937. According to the statutes, "the next-oldest employable holder" of the medal ought to have taken his place. (The reference quantity was not age but the designated year of the award.) This person just happened to be Einstein. He, as Planck put it, "naturally does not come into consideration"[31]; neither did Bohr, the next medalist in line. A foreigner would not be appropriate for this function during the Third Reich. So the choice finally fell on Arnold Sommerfeld – certainly not a second-rate or bad choice at all for the chairmanship of the medal committee. Nevertheless, the decision undermined the statutes laid down by the endowers 10 years before.

The award to Schrödinger had undoubtedly been made with the intention of resuming the annual awards of the Planck medal pursuant to its statutes. New conflicts emerged, however, already in 1938. The reason was not the deferral of Max Born's distinction. Although it was actually his turn to receive the medal, the political wrangling excluded him from serious consideration. As Grotrian explained in a letter to Sommerfeld, "such a proposal would only be suitable for landing the institution of the Planck medal its fatal blow."[32] In describing the political dimension of the upcoming prize, Grotrian remarked:

> It is naturally somewhat different if Mr. Born would be one of the two gentlemen nominated by the medal committee, as in the foregoing year, and the other could be the medal bearer picked by the board of the German Physical Society.... Under these circumstances it is extremely desirable that then the second be the right person, and if, of course, merit and achievement are supposed to tip the scales, it would perhaps be possible to choose from among equals whoever would be able to deliver a scientific talk in the German language for Mr. Planck's birthday celebration. In my conversation with Mr. Planck he mentioned the names de Broglie and Fermi. I would believe that based on their merits they are of equal stature but that Mr. Fermi would be very much more suitable as a speaker at the anniversary celebration.[33]

After further consultations about this affair between the DPG's new president, Peter Debye, and Planck's successor as chair of the medal committee, Arnold Sommerfeld, the committee arrived at the clearly not coincidentally virtually unanimous vote in favor of Enrico Fermi and Louis de

[31] Planck to Sommerfeld, 28 September 1937, DPGA, no. 20953.

[32] Grotrian to Sommerfeld, 2 December 1937, DPGA, no. 20954.

[33] Grotrian to Sommerfeld, 2 December 1937, DPGA, no. 20954.

Broglie for the award for 1938.[34] Erwin Schrödinger, who in faraway Graz had apparently not been included in the DPG's in-house communications, voted for Max Born and Wolfgang Pauli.[35] In the preliminary deliberations, Heisenberg had submitted an explicit vote for Pascual Jordan. It appeared to him that:

> [I]t would not be particularly advisable to award the next medal to a foreigner. Nor does an award to de Broglie appear to me to be so pressing, as de Broglie already has the Nobel Prize and for him another distinction would not mean as much. On the other hand, I know that at the time Jordan contributed as much toward the discovery of quantum mechanics as Born. Since an award to Born is not possible at the moment and since Jordan's work dates before Fermi's, I would for that reason be very pleased if an award to Jordan were possible.[36]

Heisenberg also mentioned in this letter the decision on Sommerfeld's succession at Munich, which was still up in the air. It appeared propitious to him also in this regard to suggest a colleague acceptable to all parties. Jordan would without a doubt satisfy this condition, as he was not just one of the pioneers of quantum mechanics but also a follower of National Socialism and a party member since 1933.[37]

Louis de Broglie had otherwise been Max Planck's favorite. In a letter to Sommerfeld, he had entered de Broglie's name into the running because as founder of the concept of matter waves he "ought not to be missed among the medal bearers" and among the candidates he had "the oldest claims" to it. Also, "with the good relations that we currently have towards France, even the Ministry would not be so unhappy at all to see such an honoring of a Frenchman."[38] As the medal award was supposed to take place within the framework of the planned festivities for Planck's 80th birthday in April 1938, Planck's vote gained special weight. It was shored up by the racist objections subsequently attached to Fermi's nomination because his wife was Jewish. This situation must have led to intense consultations with the REM and other government agencies during the course of March, after the DPG's board had resolved

[34] Bericht über die Sitzung der Kommission zur Vorbereitung der Feier des 80. Geburtstages von Hrn. Planck, 5 January 1938, MPGA, div. III, rep. 19, no. 1013, sheet 4.

[35] Schrödinger to Sommerfeld, 8 January 1938, MPGA, div. III, rep. 19, no. 1013, sheet 4.

[36] Heisenberg to Sommerfeld, 8 January 1938, DPGA, no. 20953; transcribed in Hoffmann and Walker, *Physiker*, 587.

[37] See Richard H. Beyler, "Targeting the organism: The scientific and cultural context of Pascual Jordan's quantum biology, 1932–1947," *Isis*, 87 (1996), 248–273, as well as Dieter Hoffmann and Mark Walker, "Der gute Nazi: Pascual Jordan und das Dritte Reich," in *Pascual Jordan (1902–1980), Mainzer Symposium zum 100. Geburtstag. Preprint 329* (Berlin: Max Planck Institute for History of Science, 2007), 83–112.

[38] Planck to Sommerfeld, 22 October 1937, DPGA, no. 20953.

on 2 March "to confer the Planck Medal for the year 1938 to Messrs. de Broglie and Fermi, provided that the Ministry raises no objection."[39]

Contacts had been made early on with the REM; in January 1938, Debye informed Wilhelm Dames, the ministerial official in charge, about the imminent medal nominations of de Broglie and Fermi.[40] At this point in time, the ministry also articulated its preference for Fermi, as "surely no objections exist" toward the friend and ally, the "foreign nation [of] Italy."[41] An inquiry at the Foreign Office then drew attention to Laura Fermi's non-Aryan origins. For National Socialist Germany, that weighed more heavily than its political ties with Italy. In any event, even intense efforts by Carl Friedrich von Weizsäcker with his direct contacts (through his father) with the Foreign Office[42] were unable to extract any clear statement by the ministry with regard to Fermi. So Fermi's delivery of the honorary scientific speech at the Planck festivities had to be dispensed with for the time being.[43] In the end, it was resolved to award the Planck medal to Louis de Broglie alone, as official permission had been granted only to him in a letter by the minister of 31 March 1938.[44]

On 23 April 1938, de Broglie was then formally distinguished as the eighth winner of the Planck medal during the celebration of Planck's 80th birthday. The laureate had to excuse himself, however, for health reasons, so the French ambassador André François-Poncet attended the festivities held in the *Harnack-Haus* in the Dahlem suburb of Berlin to receive the medal from Max Planck's hands on his behalf. Besides praising Louis de Broglie's scientific achievements, Planck also took the opportunity of this laudation to express his "genuine and yearning wish for a truly lasting peace" that "would make possible for both sides unperturbed productive research. May a kindly fate grant that France and Germany join hands before it is too late for Europe."[45] This avowal of peace and reconciliation among nations surely did not fully agree with the Nazi rulers.

39 Minutes of the board meeting of the DPG on 2 March 1938, DPGA, no. 20954; DMA, NL 53, 012.

40 Notes of a telephone conversation between Debye and Dames from 17 January 1938, MPGA, div. III, rep. 19, no. 1013, sheet 13.

41 Notes of a telephone conversation between Debye and Dames from 17 January 1938, MPGA, div. III, rep. 19, no. 1013, sheet 13.

42 Note by Debye, March 1938, MPGA, div. III, rep. 19, no. 1013, sheets 45/46.

43 Note by Debye, March 1938, MPGA, div. III, rep. 19, no. 1013, sheets 47f.

44 Note by Debye, March 1938, MPGA, div. III, rep. 19, no. 1013, sheet 50.

45 Max Planck, "Ansprache anlässlich der Verleihung der Planck-Medaille an Louis de Broglie," *Max Planck, Physikalische Abhandlungen und Vorträge* (Braunschweig: Vieweg & Sohn, 1958), vol. 3, 411.

The celebration of Planck's 80th birthday as a whole was not solely an impressive tribute to Germany's doyen of physics. It became at the same time an unmistakable demonstration in support of scientific autonomy and against ideological or politically spiked influences on physical research.[46] The event may not have taken place in outright opposition to the REM, but it was certainly contrary to the character and scale expected of an official anniversary celebration. No high-ranking officials were sent out in its congratulatory train[47] – quite in contrast to the ceremony to rename the physics institute of the University of Heidelberg as the Philipp-Lenard Institute in December 1935, for example, or the celebration of Lenard's 80th birthday in 1942, when numerous Nazi greats decked the halls.[48] This was surely meant to bruise Planck and the DPG, but that could be lived with. The society could rely on the sympathy of the overwhelming majority of German physicists and build on the support of the Kaiser Wilhelm Society, powerful industrialists, and some influential groups within the Nazi polycracy of power.

This discord with the Reich Ministry for Science, Education, and Culture and the attempt by the DPG's leadership to avoid producing any friction with the political wielders of power led to further cancellations of the Max Planck Medal awards in the years that followed. Sommerfeld, as chairman of the medal committee, had started sending around inquiries again for 1939, but – contrary to the previous nominations – they yielded no uniform proposal. The board took this as an excuse not to award the medal for 1939.[49] REM's reservations about "the award of the medal this year"[50] may well have weighed far more heavily than the medal committee's heterogenous vote, however. The DPG's board decided instead to give a physicist from Leipzig, Hans Euler,[51] a Planck Stipend, drawn from the endowment.

[46] See Ernst Brüche, "Vom großen Fest der Physiker im Jahre 1938. Accompanying text to the phonographic record 'Stimme der Wissenschaft,'" (Frankfurt am Main: unknown, undated).

[47] M. von Laue to T. von Laue, 22 April 1938. MPGA, div. III, rep. 50, supplement 7/2 (1938), sheet 42.

[48] See August Becker (ed.), *Naturforschung im Aufbruch. Reden und Vorträge zur Einweihungsfeier des Philipp-Lenard-Instituts der Universität Heidelberg am 13. und 14. Dezember 1935* (Munich: Lehmann, 1936).

[49] Debye to Sommerfeld, 11 March 1939, DPGA, no. 20955.

[50] Minutes of the board meeting on 2 March 1939, MPGA, div. III, rep. 19, no. 1016, sheet 28.

[51] Dieter Hoffmann, "Kriegsschicksale: Hans Euler," *Physikalische Blätter*, 45/9 (1989), 382–383.

The outbreak of the Second World War later in 1939 served as a convenient reason yet again in spring 1940 to forego the award of the medal and thereby circumvent potential conflict with the ministry. Max Planck hoped, though, "at the next occasion, that is hopefully next year, to award two medals so that the number of bearers of the medal grows a little faster."[52] Sommerfeld took up this suggestion in his circular to the medalists at the end of 1940 and – in conformity with Planck's proposals – named Friedrich Hund (University of Leipzig) and Peter Debye (on a leave of absence at Cornell University) as candidates. Owing to gaps in the preserved documents, it is not clear whom the medal committee finally agreed on – the vote was probably in favor of Friedrich Hund and Pascual Jordan. The lack of a medal award in 1941 was probably primarily related to the fact that the decision on the DPG's presidency was still awaiting finalization during those months. The ministry's stamp of approval of the designated successor, Carl Ramsauer, was still missing, and Jonathan Zenneck as the officiating president evidently did not want to act on such a touchy issue.[53] In any case, the proposal for 1941 was left unfinished.[54]

After officially assuming his office as the new president of the DPG in 1941, and probably also having already received the first positive reactions to the society's submission to Rust,[55] Ramsauer saw it as one of his most urgent priorities to resume the awards of the Planck medal. It suited the DPG board's new-found self-confidence. In a letter of December 1942, Ramsauer asked Sommerfeld for concrete procedural details about the medal award:

> [F]or . . . considering the Ministry's entire attitude, it is necessary that we proceed completely correctly. . . . All in all, I would like to carry out the conferral in a way that does not come off as a demonstration against the Ministry but rather as a conciliatory closure to the whole debacle about modern theoretical physics.[56]

In his reply, Sommerfeld emphasized that all had proceeded "in accordance with the statutes" and an appropriate proposal by the medal committee existed that obviated reinitiating inquiries. Thus, only the board needed to take up the matter. Otherwise, he advised Ramsauer to "rather

[52] Planck to Sommerfeld, 24 December 1939, DPGA, no. 20955.

[53] See Ramsauer to Sommerfeld, 24 January 1941, DPGA, no. 20955.

[54] Sommerfeld to Ramsauer, 19 December 1942, DPGA, no. 20956.

[55] See the contribution by Hoffmann to this volume.

[56] Ramsauer to Sommerfeld, 18 December 1942, DPGA, no. 20956; transcribed in Hoffmann and Walker, *Physiker*, 590.

approach the Ministry for advice about the material issue than formally acknowledge supervisory rights over the award, which are not provided for at all in the statutes to which we are subject."[57]

In spring 1943, Ramsauer received REM's approval of the award of the Planck medals to Pascual Jordan for 1942 and to Friedrich Hund for 1943 and informed Sommerfeld about it in a letter dated 7 April.[58] In a later letter, he made it clear that he had not

> addressed the Ministry with any inquiry about the current award, but just relayed the fact that the board of the German Physical Society, pursuant to the statutes, had arrived at the choice of Jordan and Hund in accordance with the proposals by the Planck foundation's medal committee. Dr. Fischer made some very consternated statements about this to our deputy manager Dr. Ebert, that he had expected an "inquiry," but did not take any official position on this form of notification toward me and merely wrote that the Ministry had no objections to the choice. Thus, in my view, the usage [*Usus*] that had begun to develop has been stopped and, as I believe, settled.[59]

He also informed Sommerfeld that the board had likewise resolved not to issue the medal in gold but in a substitute material during the war. The medal ceremony took place on 30 April 1943 during a special session of the DPG in the tradition-steeped lecture hall at the Physics Institute on the Reichstagsufer in Berlin. Besides orations praising the winners, scientific speeches were delivered by the recipients themselves. Pascual Jordan spoke about "The new developments in quantum mechanics" and Friedrich Hund about "Forces and their conceptualization." Max von Laue reported to Lise Meitner in Stockholm about the awards:

> Yesterday evening was a festive session of the German Physical Society in Gerthsen's institute. Ramsauer handed out the Planck medal for 1942 to P. Jordan, the one for 1943 to F. Hund. Both gave speeches about their work and they were excellent. The hall was packed full. Tomaschek had appeared from elsewhere, not Sommerfeld, however, whom I had actually expected. A smaller gathering took place afterwards in the Zentralhotel.[60]

There is no doubt that the resumption of the awards of the Planck medal can be considered a success by the DPG's leadership in reclaiming more autonomy from the National Socialist authorities for science. It also

57 Sommerfeld to Ramsauer, 19 December 1942, DPGA, no. 20956.
58 Ramsauer to Sommerfeld, 7 April 1943, DPGA, no. 20956.
59 Ramsauer to Sommerfeld, 4 June 1943, DPGA, no. 20956.
60 Von Laue to Meitner, 1 May 1943, in Jost Lemmerich (ed.) *Lise Meitner – Max von Laue. Briefwechsel 1938–1948* (Berlin: ERS Verlag, 1998), 268.

signals the new role that the society assumed within the Nazi power structure during the Ramsauer era.[61] However, partially self-imposed conformance narrowed this success, not just generally but also specifically. Ramsauer's letter made a definite claim for autonomy from the ministerial bureaucracy, displaying its newly found self-confidence, but it did so ultimately at the price of abandoning former standards previously stamped and sealed by the medal's statutes. The exchange of letters between Friedrich Hund and Max von Laue reveals, for example, that foreign physicists were no longer deemed suitable choices for the medal – as noted above, as early as 1938–1939, Werner Heisenberg had expressed doubts about "awarding the medal too frequently to foreigners."[62] Physicists viewed as controversial because of emigration or marriage to a Jew were also excluded from the pool of candidates, irrespective of their scientific achievements. As a result, Max Born, Enrico Fermi, and Paul Adrien Maurice Dirac were struck even though their names had already appeared prior to 1939.[63] This and the fact that the votes by awardees living abroad in enemy countries were more or less necessarily dispensed with – by 1943 this applied, for example, to Bohr, Einstein, and de Broglie – not only undermined the medal statutes but also resulted in a nationalization and provincialization of the awards. This certainly did not agree with the original intentions of the prize as an international reward. Not that the scientific merits of Jordan or Hund are in any way diminished or characterized as second rate. They unquestionably are on a par with other pioneers of modern physics – ranging from Born to Dirac, Debye, Fermi, and Arthur Holly Compton – who were as good as a priori eliminated from the deliberations. Even the award to Jordan, the first to receive the medal after the politically imposed interruption had been lifted, must be regarded as a sign of political loyalty. His achievements in science notwithstanding, Jordan was a professed Nazi.[64]

In the matter of the Planck medal, normality had thus returned by 1943. At the turn of 1943 to 1944, the chairman of the medal committee was able to apply more or less routinely to the six "reachable members of the medal committee" for their vote on the next award. Each had a right to nominate two candidates, whereby the Danzig physicist and Sommerfeld

[61] See Dieter Hoffmann's chapter in this volume.

[62] Minutes of the board meeting of the DPG on 2 March 1939, MPGA, div. III, rep. 19, no. 1016, sheet 27.

[63] Hund to Sommerfeld, 4 May 1943, 4 June 1943; von Laue to Sommerfeld, 17 May 1943, DPGA, no. 20956; transcribed in Hoffmann and Walker, *Physiker*, 590.

[64] Beyler, "Targeting"; Hoffmann and Walker, "Der gute Nazi."

pupil Walther Kossel was nominated by all and the Japanese physicist Hideki Yukawa by five of the jurors. One vote went to Richard Becker.[65] The board then followed the medal committee's majority vote and decided to award the Planck medal for 1944 to Kossel. Neither was this a second-rate or clearly politically motivated choice. Kossel's pioneering role in the early history of quantum theory, particularly regarding his interpretation of the chemical bond, is beyond question. But in view of the other names being discussed, his nomination does come as a surprise. It could even be described as an emergency solution. As Sommerfeld put it in a circular of January 1945, "the choice among German theoretical physicists . . . has become slim."[66] Ramsauer's letter reveals that the ministry "was again informed" about the impending award to Kossel "as a final decision by us; it politely but coolly approved."[67] Kossel was handed the medal, this time in bronze instead of gold, on 28 April 1944 at the traditional place in the large auditorium of the Physics Institute at the University of Berlin. One year later, the institute lay in ruins.

In view of the catastrophic situation in which Germany found itself in the final year of the war, it is no wonder that Kossel remained the last laureate for a while. At the turn of 1944 to 1945, Sommerfeld informed the medal committee and Ramsauer as the DPG's president that, owing to the "general political and military situation" and because "the usual honorary session in Berlin would scarcely be possible," he wanted to cancel that year's conferral of the Planck medal.[68] By 1946, the post-war situation had normalized somewhat, and a successor organization to the DPG had resumed activities in Göttingen.[69] Arnold Sommerfeld then returned to his duties as chairman of the medal committee. But the first post-war award of the Planck medal only took place in 1948, in celebration of Max Planck's 90th birthday. He had died the year before. Max Born became the new bearer of the medal. Thus, a distinction that should have taken place in 1934 was recouped. The list of awardees in the post-war years generally demonstrates a strenuous effort to make good on past omissions. Physicists whose candidacies between 1933 and 1945 had not been politically feasible or who had fallen victim to general opportunism were

65 Sommerfeld to Becker, 20 January 1944, DPGA, no. 20957.

66 Circular letter by Sommerfeld, re.: award of the Planck medal in the year 1945, DPGA, no. 20958.

67 Ramsauer to Sommerfeld, 15 April 1944, DPGA, no. 20957.

68 Circular letter by Sommerfeld, re.: award of the Planck medal in the year 1945, DPGA, no. 20958.

69 See the contribution by Rammer to this volume.

given preferential attention. Because of what Sommerfeld described to von Laue in 1951 as "some award data omitted during the Nazi period,"[70] it was decided to present double awards. Altogether, the temporary postwar successor to the German Physical Society, the Union of German Physical Societies (*Verband deutscher physikalischer Gesellschaften*), was not alone in its efforts to demonstrate an appearance of normality by resurrecting the medal as a sign of an international spirit in science. Its recipients themselves remained silent about the embarrassments involved in the conferrals during the Third Reich and by which some of them had been personally affected. For émigrés like Lise Meitner, the Planck medal was, above all, a high honor for her scientific achievements. It was also a

> bond that attaches me to the old, much beloved Germany, the Germany for which I cannot be grateful enough for the decisive years of my scientific development, the deep joy of scientific research and a very dear group of friends.[71]

[70] Sommerfeld to von Laue, 5 January 1951, DPGA, no. 20964.

[71] Meitner to von Laue, 25 April 1949, in Lemmerich, *Lise Meitner – Max von Laue*, 528.

7

The German Physical Society and Research

Gerhard Simonsohn

If one disregards its activities during the final war years, the German Physical Society (*Deutsche Physikalische Gesellschaft*, DPG) did not attempt to influence research directly during the reign of National Socialism. After 1933, the society continued to offer the conferences and journals under its purview as a stage for the presentation and discussion of the results of scientific research. It provided coverage of an unrestricted range of topics, and no rhetorical concessions were made to the "Aryan physics" movement. One can thus say that at best, the DPG exerted an indirect influence on ongoing research projects in science. The same is true of the German Society for Technical Physics (*Deutsche Gesellschaft für technische Physik*, DGtP).

Here, *research* primarily refers to fundamental research. The demands of the National Socialist totalitarian state certainly allowed some independent research oriented by professionally immanent criteria. Many physicists did not concentrate on "practical research," even if – within the context of the preparations for war and, above all, after its outbreak – more and more room had to be made for such research. Even under the changing conditions of forced emigration and the loss of Germany's "world reputation" in science, attempts were made to prevent a complete rupture of traditional lines of research.

This account of the physical research prevailing at that time focuses mainly on the conferences and periodicals officially co-issued by the professional societies. This is justifiable because other professional journals – in particular, *Die Naturwissenschaften* and *Physikalische Zeitschrift*, whose official editor was Peter Debye until 1945 (even after his emigration!) – do not yield a substantially different picture. *Ergebnisse der exakten Naturwissenschaften*, a series founded in 1922 that was allowed to resume publication after the war, provides an overview of the projects under way at home and abroad in its 10 volumes appearing between 1933 and 1945.

THE ANNUAL CONVENTIONS AND PROCEEDINGS OF THE DPG

The professional societies of physics were most publically visible during their annual conferences. Known as the "physicists' conventions" (*Physikertage*), they were jointly organized by the DPG and the DGtP, occasionally also with the participation of other associations. The conference programs were published in the society's proceedings, *Verhandlungen der Deutschen Physikalischen Gesellschaft*, together with announcements of talks and articles describing lecture series hosted by local associations (*Gauvereine*). The members' meetings were also the subject of its reports. Karl Scheel was the editor until his death; Walter Grotrian succeeded him in 1937. The full texts of the talks usually appeared in the *Zeitschrift für technische Physik*. Of interest to us here are six conventions that took place between 1933 and 1938, another that was scheduled for September 1939 in Salzburg but cancelled because of the outbreak of war, and, finally, a "German Physicists' Conference War Year 1940." The program of the cancelled event was also published in the proceedings.

The first rousing inaugural speeches oriented toward political conformity started in 1934. They were delivered by the chairman Karl Mey and his successor Jonathan Zenneck. The conference program itself, however, essentially continued to be devoted to basic research. Contributions on technical or applied physics were also included, although the distinction between the two categories was not always clear. One may assume that proponents of "pure physics" ("*reine Physik*," this term is occasionally used as a heading in the conference programs) rather viewed the opening session as just a façade in front of what really mattered.

It comes as no surprise that the conferences were viewed negatively by the "Aryan physics" camp. One of its prominent advocates, Alfons Bühl, wrote in 1939 in the *Zeitschrift für die gesamte Naturwissenschaft*:

> The bias for Einstein's Jewish way of thinking was revealed in a very conspicuous, indeed, obtrusive way at the major professional meetings (German Physical Society, Naturalists' convention [*Naturforscherversammlung*]) shortly after the seizure of power by National Socialism. There were special speeches emphasizing

the indispensability of the Jew's theories, without any sign of appropriate reaction at the general assemblies.[1]

"Einstein's way of thinking" was apparently the paragon of the whole of modern physics. Similar criticism of the conventions of scientific societies in general appear in the secret situation reports by the Security Service of the *Schutz-Staffel* (SS):

Besides demonstrating excellent German achievements, the conventions of the scientific societies presented the growing competitive abilities of foreign countries in the field of science. To some extent, the German side committed the decisive mistake of enlisting for the summarizing key-note speeches men who were neither professionally nor personally suited for giving the international audience an impression of German science.... The representatives of the German sciences (in particular the board members of the societies) often did not possess the necessary worldview and political attitude to be able to appear before foreigners as representatives of today's State.[2]

The convention in Würzburg in September 1933 formed the prelude to the reorientation within a new era. Max von Laue's appearance as departing chairman is famous. Johannes Stark held an introductory speech entitled "Organization of physical research" that exposed an obvious ambition (as documented by the subsequent board discussions)[3] to extend his power as president of the Imperial Physical-Technical Institute (*Physikalisch-Technische Reichsanstalt*, PTR) to include chairmanship of the DPG – in addition to his imminent presidency of the German Research Foundation (*Deutsche Forschungsgemeinschaft*, DFG). However, the society would have none of it. It elected Karl Mey, an industrial physicist with a leading position at the Osram Lighting Company, who was already serving as chairman of the DGtP. The same argument may have prevailed that governed the election of Carl Ramsauer as chairman 7 years later: an industrial physicist was sure to be more resistant to interference by state or party authorities than a university teacher in the civil service.

[1] *Zeitschrift für die gesamte Naturwissenschaft*, 5 (1939), 152. This sentence appears in a review of Lenard's textbook *Deutsche Physik*.

[2] Heinz Boberach (ed.), *Meldungen aus dem Reich. Die Geheimen Lageberichte des Sicherheitsdienstes der SS*, vol. 2 (Herrsching: Pawlak, 1984), annual situation report for 1938, 89.

[3] Confidential report on the meeting of the board of the DPG on 10 September 1934, TOP 10, Archives of the German Physical Society (*Archiv der Deutschen Physikalischen Gesellschaft*, hereafter DPGA), no. 10011. Johannes Stark's speech, "Über die Organisation der physikalischen Forschung," as well as von Laue's opening address at Würzburg are translated in Klaus Hentschel (ed.), *Physics and National Socialism: An Anthology of Primary Sources* (Basel: Birkhäuser, 1996), docs. 27–28.

There is nothing unusual about the scientific program of the Würzburg convention.[4] The first main theme, "Selected talks from the area of atomic research" (which at that time still incorporated nuclear research), included presentations about neutrons, atomic fission, positrons, cosmic radiation, and hydrogen isotopes by Walther Bothe, Walter Kolhörster, Erich Regener, and others. Johannes Stark even presided over one session ("Mr. Johannes Stark, later Mr. Clemens Schäfer") covering a mixture of topics – even though the second talk was about experimental verification of the quantum theory of natural line width. Only the fifth talk dealt with the Stark effect of the Lyman series of hydrogen. There was a joint session with the Heinrich Hertz Society for the Promotion of Radio Engineering under the lead theme "Limits of electrical measurement." This society had collaborated once before at the annual convention of 1930. This time, the number of technically oriented contributions was larger, despite the broad range that these 19 contributions (of a convention total of 71) covered: one was "Lateral and altitudinal measurement by means of ultrashort waves while landing an airplane"; Ernst Brüche and Otto Scherzer discussed the cathode-ray oscillograph as a problem of electron optics; and Manfred von Ardenne presented a new method for eliminating distortion from space charge in Braun tubes.

The next convention, at Bad Pyrmont in 1934, began with a lengthy address by the new chairman that would hardly have been conceivable coming from his predecessor (nor, for that matter, from many others among the society's membership). It was published in the *Zeitschrift für technische Physik*.[5] No room for doubt was left about where the society's allegiance now lay with regard to the new government: Karl Mey alluded to the "right attitude toward the state as a whole and toward the nation's welfare" and applauded "the Reich Chancellor and Führer, who under trials and tribulations seeks to lead us into a fine future." He adeptly quoted a remark about "physics as a major power" that Adolf Hitler had once surprisingly made in a speech.[6] All the same, Karl Mey appeared to consider elaborate words of conformance unnecessary:

> It has become the custom at scientific conferences to use up a good part of the time with statements about the position of the relevant science within the new

[4] *Verhandlungen der Deutschen Physikalischen Gesellschaft*, 14 (1933), 29.

[5] *Zeitschrift für technische Physik*, 15 (1934), 491.

[6] Adolf Hitler's speech at the Cultural Convention of the NSDAP on the second day of the Reich party convention in Nuremberg on 5 September 1939, *Völkischer Beobachter*, north-German ed./edition A, no. 250, 7 September 1934. In a pathos-filled account

State, with demonstrations of its importance and its allegiance to the nation; we, however, amongst ourselves, would like to preserve old custom and let the factual and professional work speak for us.

Evidently referring to the Aryan physics movement, whose purpose was to eliminate unfruitful theoretical approaches by means of experiment, he remarked:

These purification processes are thoroughly natural and would normally play themselves out, if they were not frequently turned into the hotbed of selfish ambitions, out of motivations that lie entirely beyond physical chains of arguments.

He warned against "valuing the theoretical or experimental achievements of a physicist pursuing purely fundamental research less than application of physics in technology and its present great national importance." In accordance with this advice, the conference offered two guiding themes: "Physics and materials" and "The physics of low temperatures."[7] Such a thorough treatment – 23 talks – of technical questions under a main theme like the first is a novelty. One could otherwise only call to mind the guiding theme of the convention of 1931: "Physical problems of the talking film." There were only 12 talks under the second theme, but they included contributions by such prominent authors as Peter Debye, Klaus Clusius, and Walther Meißner. Karl Mey explicitly greeted four visiting Dutch speakers in his opening speech. There were otherwise 41 talks on a variety of topics; the purely technically oriented ones remained in the minority.

The American author of a history of superconductivity, Per F. Dahl,[8] assessed the treatment of superconductivity within the framework of the second guiding theme as "taking stock" of a new situation that had arisen since the discovery 1 year before of the Meißner–Ochsenfeld effect at the PTR laboratories. "[The convention]...was to be the last important

about mankind's new venture "to expose the secrets of the world and [mankind's] own existence," there are the lines: "In the service of this errant hunt a suddenly magically unleashed genius makes inventions and discoveries.... The major power of physics and technology reaches its hand out in passing to – no less formidable – chemistry. The constantly widening knowledge of the world permits mobilization of the globe's resources for the elevation of mankind at an almost horrifying rate..." The great-power motif reappears during the convention in the telegram greeting at the end: "...extends its loyal greetings... to its Führer and Reich Chancellor, proud and motivated by his words about physics as a great power...," *Zeitschrift für technische Physik*, 15 (1934), 496.

7 *Verhandlungen der Deutschen Physikalischen Gesellschaft*, 15 (1934), 19.

8 Per F. Dahl, *Superconductivity. Its Historical Roots and Development* (New York: American Institute of Physics, 1992), 208ff. The relevant chapter bears the title: "Taking stock at Bad Pyrmont."

meeting on superconductivity in Germany, with minimal international participation, for a very long time." The contributions were published with the discussion remarks in the *Zeitschrift für technische Physik*.[9] Debye and Meißner notably cited Franz Simon and Nicholas Kürti, who had already fled Breslau for Oxford.

A Dutch speaker from the Kamerlingh Onnes Laboratory in Leyden, Cornelius Jakob Gorter, tells us a little about the tense mood at this convention. In his retrospective, "Superconductivity until 1940,"[10] he wrote: "At that time only a few German physicists had been influenced by the Nazi ideology ... and after some hesitation Keesom, as well as I, decided to attend the meeting of the German Physical Society in Bad Pyrmont in September 1934 at which low temperature physics was a main subject." In this article, the remark "though it was not hidden from me that the contacts maintained by the Netherlands scientists were hardly appreciated by Simon and his colleagues" reveals some reservations among the low-temperature group of émigrés from the University of Breslau he was visiting at the Clarendon Laboratory at the University of Oxford. All the same, one certainly cannot conclude that there was a complete break in relations. Walther Meißner visited Franz Simon at Oxford in 1935,[11] apparently in connection with his participation in a symposium hosted by the Royal Society in London on "Superconductivity and other low temperature phenomena."[12] Gorter and his colleague, Ralph de L. Kronig from Groningen, were also speakers at the annual convention of 1938 in Baden-Baden.

At the next convention in Stuttgart in 1935,[13] the first two guiding themes were clearly oriented toward fundamental science: (1) electron and ion conductivity of solids (16 presentations); (2) cosmic radiation and nuclear physics (17 presentations). The band model developed around this time was an important feature of the first theme. The sessions, presided over by Peter Paul Ewald, Robert Wichard Pohl, and Walter Schottky, included talks, among others, by Friedrich Hund, Ralph de L. Kronig, Pohl, Robert Hilsch, Bernhard Gudden, and Schottky. In his survey of

9 *Zeitschrift für technische Physik*, 15 (1934), 497: "second Pyrmont issue."

10 C. J. Gorter, "Superconductivity until 1940," *Reviews of Modern Physics*, 36 (1964), 3.

11 See the speech Walther Meißner personally delivered in honor of Franz Simon at the physics convention in (West) Berlin in 1959: *Physikalische Verhandlungen, Verbandsausgabe*, 10 (1959), 151.

12 See Walther Meißner's contribution in *Proceedings of the Royal Society A*, 152 (1936), 13.

13 *Verhandlungen der Deutschen Physikalischen Gesellschaft*, 16 (1935), 37.

the theory of electron motion in nonmetallic crystal lattices, Friedrich Hund did not fail to emphasize achievements by émigrés: "The analyses of electrons in a field of force adjusted to a crystal lattice led to important results at the hands of *Pauli, Sommerfeld, Bloch, Brillouin, Bethe, Peierls, Nordheim* et al."[14] Michael Eckert has already pointed out the importance of this convention.[15]

A third guiding theme, mechanical oscillations, including noise control (19 presentations), was dealt with on the second to last day with the collaboration of the Society for the Promotion of Radio Engineering. There were an additional 51 presentations on miscellaneous topics and, in conclusion, a study tour to Friedrichshafen with an excursion in a research vessel for cosmic-ray measurement on Lake Constance guided by Erich Regener. The range of topics covered is broad. One talk offered in the general part was about Dirac's spin theory and nonlinear field equations, and another offered by the Heinrich Hertz Institute and classified under the third guiding theme was on the development of the squeaking noise in brakes. Peter Paul Ewald continued to take part as session chair and speaker. Patrick Meynard Stuart Blackett from London delivered an introductory survey on the second guiding theme, "On the cosmic-ray problem." There were other contributions by foreign guest speakers. At the opening ceremony, Karl Mey welcomed the visitors from Austria, Hungary, England, Holland, and Czechoslovakia. The *Zeitschrift für technische Physik* printed only excerpts from his address.[16] He again quoted an encouraging word by the *Führer* and concluded, "the State will do everything within its power to promote these two fundamental sciences [mathematics and physics] for the good of the German people, the good of German industry, and our reputation in the world."

The convention in 1936 took place in Bad Salzbrunn under the chairmanship of Jonathan Zenneck.[17] After a historical talk by Walther Kossel about Otto von Guericke and a presentation by Karl Mey on new light sources, the opening session offered three reviews on astrophysics by prominent researchers. Although the text of the chairman's inaugural address was not published, it still exists among Jonathan Zenneck's

[14] Thus in the published text: Friedrich Hund, "Theorie der Elektronenbewegung in nichtmetallischen Kristallgittern," *Zeitschrift für technische Physik*, 16 (1935), 331; Sommerfeld and Brillouin were not émigrés.

[15] Michael Eckert, *Die Atomphysiker. Eine Geschichte der theoretischen Physik am Beispiel der Sommerfeldschule* (Braunschweig: Vieweg, 1993), 213ff.

[16] Opening address by Karl Mey in *Zeitschrift für technische Physik*, 16 (1935), 643.

[17] *Verhandlungen der Deutschen Physikalischen Gesellschaft*, 17 (1936), 35.

FIGURE 22. Physicists' conference in Bad Salzbrunn, 1936 (*Source*: DPGA)

papers.[18] The subsequent parallel sessions concern topics from electrical engineering (six presentations) and the physics of gas discharges (nine presentations). Under the first main theme, "Geometrical electron optics," there were 19 contributions, subdivided under "General and foundations," "Applied electron optics," "Elements of electron optics," and "Electron-optical emission research." An evening presentation with color movies offered an overview of the physics and technology of the Berthon–Siemens color-film process. Half of the 20 contributions to the second main theme, acoustics, notably concerned musical applications and room acoustics. Among the other 28 presentations on diverse topics, there was a talk by the Jewish physicist Alfred von Engel from the Siemens company on the physics of gas discharges. The subject of Erwin Wilhelm Müller's talk from Research Laboratory II of the Siemens Works, the theory of electron emission under the influence of high field strengths, evidently represented one developmental stage of his field emission microscope.

[18] Archives of the Deutsches Museum, Munich (Deutsches Museum Archiv, henceforth DMA), papers 53 (Zenneck), no. 12.

The initiation of the next convention in 1937, in Bad Kreuznach, is reminiscent of the convention 3 years earlier in Bad Pyrmont. Again, there was a lengthy inaugural address, this time published in the proceedings.[19] It surely pleased the ears of the official guests in attendance – one was a representative sent by the Reich Ministry for Science, Education and Culture (*Reichsministerium für Wissenschaft, Erziehung und Volksbildung*, REM). This time, a telegram to the "Führer and Reich Chancellor" was sent out right at the outset: "a respectful greeting together with a solemn pledge to collaborate with all our might... on the great missions posed this day to our nation." After a few commemorative words about the deceased Karl Scheel, the speaker turned to the convention's guiding themes. On this occasion, the first was again tuned to technology: "Physical measurement and regulation processes in engineering." This theme gave the speaker the opportunity to point out that physical measurement processes permitted "careful operation control, which revealed all that was not completely economical, thus doing its part in the 'battle against waste.' What this means in our current difficult situation, the expression of which has become the 'Four-Year Plan,' I do not need to tell you." (The *Kampf dem Verderb* was a well-known public campaign against wastefulness.) The importance of technical physics for "national defense" was also mentioned.

Regarding the second guiding theme, nuclear physics, there was the sober declaration: "Nuclear physics cannot as yet raise claim to having produced anything of technical utility.... But it would nevertheless be wrong to assume that nuclear physics was far away from the great missions of the present day." At this point in time, this could only be a general assumption applicable to all kinds of basic research. Nineteen presentations concerned themselves with the first main theme. An additional session under the rubric "Technical physics/Individual presentations of general content" assembled 17 contributions with a varied palette of topics ranging from transit-time influences in electron tubes to cruising compasses, newer applications of the piezoelectric measurement procedure in ballistics, and tone transitions for pipe organs (with performances).

The session on the second guiding theme, nuclear physics, was led by Walther Bothe and Werner Heisenberg. At the beginning of the session, the DPG chairman, Zenneck, announced the decision to award the Max Planck Medal to Erwin Schrödinger. The 13 presentations, five among

[19] *Verhandlungen der Deutschen Physikalischen Gesellschaft*, 18 (1938), 81; also published in *Zeitschrift für technische Physik*, 18 (1937), 346.

them reviews, can be regarded as a cross-section of the field, including cosmic radiation. The speakers were all well-known persons. From the published versions[20] one can gather that the intent was to maintain contact with the international literature. The convention closed with a session of 18 presentations under the heading "Pure physics/Individual talks of general content."

Peter Debye was in charge of the next annual convention in 1938, in Baden-Baden.[21] He had been elected chairman of the DPG in 1937 but left the welcoming address in the hands of Karl Mey in his capacity as president of the DGtP. Something about its content can be gathered from a general report about the convention.[22] One learns that Karl Mey greeted "especially the numerous guests in attendance from abroad as well as the members of the *Ostmark* [Austria], now finally counted as numbering among their German colleagues." He recalled "the discovery of electrical waves 50 years ago by Heinrich Hertz" in neighboring Karlsruhe and also welcomed "our fellow Germans from Czechoslovakia." In the meeting that followed, Abraham Esau expressed a few "words in memory of Max Wien," who had died at the beginning of the year. It was a matter-of-factual commemoration devoid of objectionable political references.

The first professional session was listed under the guiding theme "Dispersion and relaxation" and was cochaired by Peter Debye and Abraham Esau. The 10 presentations mainly concerned processes in fluids. Debye had made essential contributions to this area. One presentation also mentions an experimental contribution by Esau. This was followed by a large block of 28 presentations lasting 2 days under the heading "pure physics": contributions on solid-state physics (among other things, ferromagnetism and photocathodes), nuclear physics (nine contributions including a detailed report by Walther Bothe about research at the institute in Heidelberg), gas discharges, and other topics, such as the talk by Otto von Schmidt from the Ballistics Institute of the Air-War Academy in Gatow on shock-wave propagation in fluids and solid bodies.

On the third day, the participants had the opportunity to watch a performance presented by an "image reporter" of the German Post Office (*Reichspost*). The fourth day turned attention to technical problems under the theme light: optical apparatus, light generators (gas discharges), as well as the biological effect of optical radiation. Only the last day was

[20] *Zeitschrift für technische Physik*, 18 (1937), 497; "second Bad Kreuznach issue."
[21] *Verhandlungen der Deutschen Physikalischen Gesellschaft*, 19 (1938), 117.
[22] *Zeitschrift für technische Physik*, 19 (1938), 614.

devoted to technical physics. The first four presentations already indicated the variety of topics treated: "On optical methods for the examination of tillable soil," "On the question of the maximum velocities attainable with modern powdered fuels," "Analyses of nonlinear crosstalk by means of speech reconstruction," and "The acoustical signs of perfectly pitched violins." This last contribution, which came from the Institute of Oscillation Research at the polytechnic in Berlin, is remarkable in that even this so entirely nontechnical and not "useful" application of physics was being supported by the DFG, as can be gathered from the published version of the talk.[23] The 12 other contributions at this session concerned electron microscopy and high-frequency engineering. Two of the presentations about television technology originated from the research institute of the Post Office.

The program of the cancelled convention[24] – scheduled for September 1939 – began, as in the conventions for 1934 and 1937, with a focus on technical physics. The first guiding theme, "Metallic materials in technical physics," included 15 presentations with a strong showing from industry. The focus of the following 28 presentations, under the general theme technical physics, was improvements in measurement methods and apparatus. Three contributions from the PTR on musical acoustics are worth mentioning. One of them was a joint effort with the laboratory at the Siemens Works: "On the tonal effectiveness of the clavichord, harpsichord and grand piano." (This contribution was later, on 10 January 1940, the topic of a colloquium of the regional association for Berlin.)

The last 3 days of the convention were reserved for fundamental physics. The main theme of the first session was nuclear physics. It would have been the first convention since the discovery of nuclear fission. In accordance with the importance of this event, two introductory reviews had been planned. Otto Hahn was to discuss the bursting apart of the uranium and thorium nucleus into lighter atoms and Siegfried Flügge the physical and potential technical consequences of the discovery of uranium fission. Surveys were supposed to follow by Carl Friedrich von Weizsäcker (on astrophysical applications of nuclear physics), Helmuth Kulenkampff (on the mesotron in cosmic radiation), and Hideki Yukawa from Kyoto

[23] "Die akustischen Kennzeichen klanglich hervorragender Geigen," *Zeitschrift für technische Physik*, 19 (1938), 421. One also learns that these analyses formed part of the "Reich professional competition 1938 of the German Labor Front and string-instrument testing for the Reichsmusikkammer."

[24] *Verhandlungen der Deutschen Physikalischen Gesellschaft*, 20 (1939), 129.

(on the current state of the theory of the mesotron). From among the 11 remaining presentations, three treated cosmic radiation. Another 30 scheduled presentations were again assembled under the heading "pure physics": contributions on the physics of molecules, fluids, solid bodies, and gas discharges.

A final annual convention, limited to 2 days (Sunday and Monday), took place in September 1940 in Berlin.[25] As the chairman Peter Debye had left Germany, his deputy Jonathan Zenneck took his place, "sorely regretting" his absence, as he put it during the opening meeting. What he said in glorification of the *Führer* could hardly have been out-trumped. The end of the relevant section reads:

> We are today, more than ever, most profoundly thankful to our Führer; we all are instilled with the trust that he will carry the work he has started to a fortunate close for us all.

Unlike earlier, the importance of physics for the Armed Forces was set forth in detail. At the end, the speaker pointed out how many foreigners were members of the societies and saw in this

> a sign of the willingness of physicists of all countries to work together in the service of science and therefore for the good of all nations. We hope that after the end of this war this good harmony will soon be reinstated.

This latter remark does not, however, suit the general propaganda quite so well.

The first day was devoted to "pure physics." A presentation by Albrecht Unsöld on the frequency in the cosmos of the light elements is noteworthy. The announcement starts with the sentence: "According to Hans Bethe and Carl Friedrich von Weizsäcker, the generation of energy takes place in the interior of the stars." Thus both names, including the émigré Bethe, are mentioned. At the end there is mention of an "examination of a large amount of material about stellar spectra that Otto Struve and the author have taken up at the McDonald Observatory in Texas." In the published text of the talk,[26] Albrecht Unsöld "cordially thanks Prof. Otto Struve" for "his generous hospitality at the Yerkes and

[25] *Verhandlungen der Deutschen Physikalischen Gesellschaft*, 21 (1940), 31.

[26] Albrecht Unsöld, "Die kosmische Häufigkeit der leichten Elemente," *Zeitschrift für technische Physik*, 21 (1940), 301; Gustav Hertz, "Schallstrahlungsdruck in Flüssigkeiten und Gasen im Zusammenhang mit der Zustandsgleichung," *Zeitschrift für technische Physik*, 21 (1940), 301.

McDonald Observatory." Gustav Hertz spoke about sound radiation pressure in fluids and gases in connection with the equation of state. (A talk on this topic had been planned for the cancelled convention of 1939.) The topics of the remaining 18 presentations were distributed among nuclear physics, solid-state physics, and cosmic radiation.

The second day with 19 presentations fell under the heading technical physics. Three contributions on the topic of hard metals were devoted exclusively to technical applications, but this cannot be seen in all contributions. An investigation of the influence of various natural voltages on the coercive force and critical field strength of Barkhausen jumps by Martin Kersten and his co-workers also discussed the theoretical, microphysical background (whereby mention of the term *Bloch walls* was permissible, despite the fact that Bloch had been forced to leave Germany). A paper by Erwin Meyer and collaborators on a new sound-swallowing arrangement and the design of a sound-absorbing room is a surprise, however, especially when one reads the acknowledgments in the talk's published text[27]: "The German Research Foundation generously made possible the construction of the room. We would like to express our cordial thanks to Minister Director Prof. Mentzel and State Councillor Prof. Esau." The project was carried out at the Institute for Oscillation Research at the Berlin Polytechnic. Those attending the convention had the opportunity to view a sound-absorbing hall.

THE EVENTS ORGANIZED BY THE REGIONAL ASSOCIATIONS

Other events organized by the regional associations took place on the fringes of the annual conventions. The term used for these local organizations even prior to 1933 was *Gauverein*; it just happened to suit the new political era as well. Berlin assumed a special position. Its regional association, the Physical Society of Berlin (*Physikalische Gesellschaft zu Berlin*, PGzB), has – even now – the form of an independent society. The geographic concentration of its membership gave it the ability to offer individual talks throughout the year, usually during events jointly organized with the DGtP. The other regional associations organized smaller-scale conferences of 1 or 2 days' duration at a central location. Other societies also occasionally participated, such as the Society for Light Technology

[27] E. Meyer et al., "Eine neue Schallschluckanordnung und der Bau eines schallgedämpften Raumes," *Zeitschrift für technische Physik*, 21 (1940), 372.

(*Lichttechnische Gesellschaft*). In 1942, the PGzB exceptionally organized a 2-day event on 10–11 October (Saturday–Sunday!) instead of a major annual convention.

The spectrum of research areas treated does not differ substantially from that of the annual conventions. Purely technical topics, which define the individual sessions of the annual conventions, are rare. A few things stand out here, too, and will serve as examples. As late as 1937, the future Nobel laureate Hans Jensen explicitly referred to his scientific ties with the émigré James Franck in a talk before the regional association of Hessen.[28] In 1939, nuclear fission was still the subject of open discussion. At the Hessian regional conference on 8 July, Siegfried Flügge spoke about the splitting of the uranium nucleus by neutrons.[29] The announcement states: "In closing, the talk will address the issue of the chain reaction of fission neutrons and discuss more closely the option of perhaps taking this route to make the energies of atomic nuclei technically utilizable in the not too distant future."

Reichskrystallnacht, the pogrom of the evening of 9 November 1938, did not disrupt the scientific program. At the Berlin Polytechnic, a few hundred meters away from the nearest sacked synagogue, Carl Friedrich von Weizsäcker reported about nuclear transmutations as a source of stellar energy on 23 November.[30] On 14 December, Walther Bothe treated the problems and state of research on the atomic nucleus.[31] In between, on 30 November, there were two talks by staff members from Research Laboratory II of the Siemens Works (the "Hertz Laboratory") on field electron autoemission.[32] "Good" physics continued on – complete with citations of Jewish authors. One aberration – with regard to the organizers – was a series of four meetings of the Austrian *Gauverein* between 26 January and 17 March 1939.[33] These meetings had been "invited by the NSBDT" (the National Socialist League of German Engineers, *Nationalsozialistischer Bund Deutscher Technik*), which the DPG refrained from joining.

[28] *Verhandlungen der Deutschen Physikalischen Gesellschaft*, 18 (1937), 74.

[29] Siegfried Flügge, "Die Aufspaltung des Urankerns durch Neutronen," *Verhandlungen der Deutschen Physikalischen Gesellschaft*, 20 (1939), 123.

[30] Carl F. von Weizsäcker, "Kernumwandlungen als Quelle der Sternenergie," *Verhandlungen der Deutschen Physikalischen Gesellschaft*, 20 (1939), 2.

[31] Walther Bothe, "Stand und Probleme der Atomkernforschung," *Verhandlungen der Deutschen Physikalischen Gesellschaft*, 20 (1939), 10.

[32] *Verhandlungen der Deutschen Physikalischen Gesellschaft*, 20 (1939), 4.

[33] *Verhandlungen der Deutschen Physikalischen Gesellschaft*, 20 (1939), 57.

Research at the PTR's low-temperature laboratory continued under Eduard Justi's direction after the departure of Walther Meißner in 1934. In 1941, Justi was able to report a breakthrough on superconducting semiconductors with extremely high transition temperatures, including a demonstration of continuous current in superconductors.[34] It involved niobium compounds with transition temperatures of up to 20 Kelvin. In the book mentioned earlier, Per F. Dahl calls this achievement "a rather impressive milestone in the context of war-time Germany."[35] Throughout all these years, Max von Laue was constantly busy with the theory of superconductivity. His talk in Berlin on 10 July 1942 gave a thorough treatment of this theory,[36] as do his publications on the topic, linking it in multiple references with the name of the émigré Fritz London.

During a meeting of the regional association for Lower Saxony in Hamburg in 1942, Hans Jensen reviewed four papers by Niels Bohr, John Archibald Wheeler, and others that had appeared in the *Physical Review*.[37] The subject was nuclear fission. One paper by Bohr and Wheeler is a fundamental contribution about the droplet model. Another notification about research under way in America was the contribution in December 1942 by Georg Mierdel from the Siemens Works on the development of experimental physics in the United States in recent years.[38]

There was a curious prelude to the Berlin conference in October 1942.[39] It centered around a smoldering priority dispute between the laboratories at the German General Electricity Company (*Allgemeine Elektrizitäts-Gesellschaft*, AEG) and at Siemens over the electron microscope. Ernst Brüche from the AEG's laboratory wrote a letter to the PGzB's secretary, Hermann Ebert, to suggest that "everyone who registers a talk in this area be obligated to exclude all historical and polemical commentary" about its development.[40] The debate continued to rage on in the years that followed. One of the last papers in the *Physikalische Zeitschrift* of 1944 was a strongly polemical piece by the Siemens employees Ernst

[34] Eduard Justi, "Supraleitende Halbleiter mit extrem hohen Sprungtemperaturen. Demonstration des Dauerstroms in Supraleitern," *Verhandlungen der Deutschen Physikalischen Gesellschaft*, 22 (1941), 38. The paper on this topic was published by G. Aschermann et al. in *Physikalische Zeitschrift*, 42 (1941), 349.

[35] Dahl, *Superconductivity*, 257.

[36] *Verhandlungen der Deutschen Physikalischen Gesellschaft*, 23 (1942), 62.

[37] *Verhandlungen der Deutschen Physikalischen Gesellschaft*, 23 (1942), 53.

[38] G. Mierdel, "Die Entwicklung der Experimentalphysik in USA während der letzten Jahre," *Verhandlungen der Deutschen Physikalischen Gesellschaft*, 23 (1942), 195.

[39] *Verhandlungen der Deutschen Physikalischen Gesellschaft*, 23 (1942), 65.

[40] Brüche to Ebert, 7 July 1942, DPGA, no. 10024.

Ruska and Bodo von Borries about the historical development of the electron microscope.[41] There was evidently still time for such quarrels as this during "total" war!

The conference had three talks on electron microscopy on the program; the Siemens group was not represented. Nuclear physics took up much space with a total of 14 presentations. They include the "hot" topics of neutron-capture cross-sections and the products of uranium fission. Eduard Justi performed new demonstration experiments on superconductivity; one of them included a magnet with an excitation coil made of a superconductive ring that itself was made from the recently discovered material of high transition temperature – this was a first step toward modern technology in this area. Among the remaining 20 talks on various topics, the unassuming contribution by Helmut Scheffers (PTR) on a new kind of calculation method for Fraunhofer diffraction is remarkable. The Fourier method introduced there has since assumed a permanent place in modern optics; "important for the war" it certainly was not.

Not all the sessions were devoted to purely scientific subjects. On 9 February 1934, 11 days after Fritz Haber's death, the PGzB honored the deceased with a eulogy delivered by Max von Laue before the society's chairman Richard Becker started the day's agenda.[42] Haber's achievements as a researcher and inventor were acknowledged along with his decision to appoint "first James Franck, later Rudolf Ladenburg" to head a physics department at his institute: "Just from mentioning these names, you know what work on atomic physics came out of Haber's institute," von Laue added, also mentioning Haber's active engagement for the DPG: "Haber was for many decades a loyal member of the German Physical Society."

On 25 January 1935, the PGzB celebrated its 90th anniversary with a festive session.[43] The opening address was delivered by the chairman, Richard Becker. Karl Scheel delivered the celebratory speech on the history of the society, supplemented by Max Planck's personal reminiscences. Not a single sentence in the speeches expressed a connection to the political setting. The celebration proceeded as it would

41 Bodo von Borries and Ernst Ruska, "Neuere Beiträge zur Entwicklungsgeschichte des Elektronenmikroskops und der Übermikroskopie," *Physikalische Zeitschrift*, 45 (1944), 314–326.

42 *Verhandlungen der Deutschen Physikalischen Gesellschaft*, 15 (1934), 7.

43 *Verhandlungen der Deutschen Physikalischen Gesellschaft*, 16 (1935), 1; Karl Scheel, "Aus der Geschichte der Gesellschaft," *Verhandlungen der Deutschen Physikalischen Gesellschaft*, 16 (1935), 1; Max Planck, "Persönliche Erinnerungen," *Verhandlungen der Deutschen Physikalischen Gesellschaft*, 16 (1935), 1.

have under other conditions. As a matter of course, Karl Scheel included Emil Warburg, Heinrich Rubens, Haber, and Einstein in his list of former chairmen, adding in a footnote Peter Pringsheim and Rudolf Ladenburg. Walter Kaufmann gained mention as among those whose memberships spanned over 40 years – all of these were "incriminated" names.

One meeting on 9 June 1937 honored the discovery of X-ray diffraction by Wilhelm Friedrich, Paul Knipping, and von Laue 25 years before.[44] In a joint meeting by the PGzB and DGtP on 10 November 1937, Hans Geiger spoke a few words in memory of Lord Rutherford, who had died 1 month before.[45] The DPG and the DGtP participated in a scientific lecture evening on 16 March 1939 on the occasion of the 150th anniversary of the birth of Georg Simon Ohm.[46]

The PGzB celebrated Max Planck's 80th birthday with a special ceremony on 23 April 1938. A festive meeting and a banquet took place in the Kaiser Wilhelm Society's *Harnack-Haus*.[47] The celebration culminated in the award of the Max Planck Medal to Louis de Broglie. The French ambassador, who received the medal on behalf of the ailing awardee, had a few warm words for the DPG. He particularly expressed his appreciation of Max Planck, who in a foregoing address had invoked close ties between the French and German nations with the words: "May a kindly fate grant that Germany and France join hands, before it is too late for Europe." If you disregard the obligatory "Heil Hitler" salute by the chairman Carl Ramsauer at the opening, the political realities were ignored on this occasion as well, albeit not while the preparations were still under way. The original intention had been to award the medal to Enrico Fermi,[48] and he was even supposed to deliver the scientific address. However, consultations with the ministry revealed "reservations against it of a racial nature" – as a report about the PGzB's board meeting on 30 March 1938 specifies.[49] Fermi's wife was Jewish. In the end, Max von Laue delivered the scientific address. This proceeding in particular and the celebration in general exemplify the contradictory course taken by the DPG's board. Confrontation with the authorities that would probably have been futile, if not dangerous, was avoided, and attempts were made instead to make the best of the remaining freedom of action.

44 *Verhandlungen der Deutschen Physikalischen Gesellschaft*, 18 (1937), 77.

45 Hans Geiger, "Gedenkworte für Lord Rutherford," *Verhandlungen der Deutschen Physikalischen Gesellschaft*, 18 (1937), 114.

46 *Verhandlungen der Deutschen Physikalischen Gesellschaft*, 20 (1939), 55.

47 *Verhandlungen der Deutschen Physikalischen Gesellschaft*, 19 (1938), 57.

48 See the contribution by Beyler, Eckert, and Hoffmann to this volume.

49 DPGA, no. 20954.

Even during the war, memorial ceremonies did not cease. One took place on 24 April for Walther Nernst, who had died on 18 November 1941,[50] and on 4 December 1942 there was a "memorial hour for Julius Robert Mayer."[51] These were events conducted in conjunction with the Prussian Academy of Sciences and other societies. Time was also spent on more general topics. For instance, Carl Friedrich von Weizsäcker spoke at a joint meeting of the PGzB and the DGtP on 12 February 1941 in Berlin about the relationship between quantum mechanics and Kantian philosophy[52] and on 27 April 1942 in Vienna about the development of the atom concept.[53] Around this time, a paper appeared in the *Zeitschrift für Physik* by the same author on the interpretation of quantum mechanics, which is partly a discussion of the ideas of the Munich philosopher Hugo Dingler.[54] It suits the occasions during the final years of the war (discussed further later) that on 18 January 1945, the centennial of the DPG was celebrated seemingly worry-free (despite the unheated hall) in Berlin.[55] It even received press coverage.[56]

Physical Reports (Physikalische Berichte)

The first periodical covered by the current survey is the review journal *Physikalische Berichte*. It was issued by the two societies DPG and DGtP

[50] *Verhandlungen der Deutschen Physikalischen Gesellschaft*, 23 (1942), 1.

[51] *Verhandlungen der Deutschen Physikalischen Gesellschaft*, 23 (1942), supplement to issue no. 2, 17.

[52] Carl Friedrich von Weizsäcker, "Das Verhältnis der Quantenmechanik zur Kantschen Philosophie," *Verhandlungen der Deutschen Physikalischen Gesellschaft*, 22 (1941), 25.

[53] Carl Friedrich von Weizsäcker, "Die Entwicklung des Atombegriffs," *Verhandlungen der Deutschen Physikalischen Gesellschaft*, 23 (1942), 58.

[54] Carl Friedrich von Weizsäcker, "Zur Deutung der Quantenmechanik," *Zeitschrift für Physik*, 118 (1941/42), 488–509.

[55] See the contribution by Hoffmann to this volume as well as Armin Hermann, "Die deutsche Physikalische Gesellschaft 1899–1945," in Theo Mayer-Kuckuk (ed.) *Festschrift 150 Jahre Deutsche Physikalische Gesellschaft*, special issue of the *Physikalische Blätter*, 51 (1995).

[56] Carl Ramsauer was able to write an article on a century of physics (entirely in keeping with his campaign of the final years of the war): "Hundert Jahre Physik," *Deutsche Allgemeine Zeitung*, Reichsausgabe, 18 January 1945. See the reprint in Dieter Hoffmann and Mark Walker (eds.), *Physiker zwischen Autonomie und Anpassung – Die DPG im Dritten Reich* (Weinheim: Wiley-VCH, 2007), 632–635. There were other articles in the *Völkischer Beobachter* of 23 January 1945 and in the *Deutsche Allgemeine Zeitung* of 20 January 1945, the latter mainly concerned the honorary speech by the historian of physics Hans Schimank on the intellectual significance of physics in the 19th century. In these contributions, "objectionable" names were avoided, and one can assume that it was likewise so during the celebration itself. The Reich Youth Leader, Artur Axmann, was one of the guests present.

until 1937. After 1938, the DGtP was its official publisher with the DPG only acting in cooperation. It is a notable achievement that the periodical was able to maintain its straightforward review style without subjecting the topics or authors covered to any preselection criteria. A large staff of reviewers was used. The works under review came from as far away as the United States, the Soviet Union, and East Asia. "Relativity theory" was one of the headings in constant use in the systematic catalogue (this compared against the American publication *Science Abstracts*, which continued to use the heading "Relativity and Ether" up to 1940). The readers of the *Berichte* encountered the names of the many émigré authors – some of them occasionally as reviewers – apparently without any negative connotation being attached to their mention. Only examples can demonstrate this here.

In 1935, the even now still very much debated "EPR paper" by Albert Einstein, Boris Podolsky, and Nathan Rosen about the completeness of the quantum mechanical description was discussed.[57] In the same volume, one finds a detailed review by Prof. Karl Bechert of the paper written by Einstein and Rosen in English on "The Particle Problem in the General Theory of Relativity."[58] A review of a paper on the release of neutrons from beryllium, in which the émigré Leo Szilard was also involved, reveals that some scientific collaboration still existed between Berlin and London.[59]

Four years later, 25 names are mentioned that should have appeared objectionable to Aryan physics adherents for "racial" reasons, Max Born, James Franck, Victor Weisskopf, and Eugene Wigner among them.[60] A paper on atomic physics by the Jewish émigré Peter Pringsheim is reviewed in detail, surprisingly enough by a staff scientist at the PTR, Rudolf Ritschl, who was even a close collaborator on Johannes Stark's experimental attempts to verify an unusual quasi-classical model of the atom![61] There is a reference to an earlier paper by Einstein in the review of a book by Leopold Infeld.[62] Usually, there were only brief commemorative notices; an exception was made for an article on Michail W. Lomonossow as a physicist that had appeared in the *Mémoires de Physique Ukrainiens*. The reviewer spoke of an "interesting contribution to the history of science" and emphasized that this recipient of the first Russian professorship

57 *Physikalische Berichte*, 16 (1935), 1791.
58 *Physikalische Berichte*, 16 (1935), 2150.
59 *Physikalische Berichte*, 16 (1935), 550.
60 *Physikalische Berichte*, 20 (1939).
61 *Physikalische Berichte*, 20 (1939), 238.
62 *Physikalische Berichte*, 20 (1939), 5.

for chemistry in 1745 had formulated "clearly for the first time the law of the conservation of 'force.'" He continued: "At that time already, he regarded the energies of motion and heat and electrical force merely as different forms of energy."[63] Where the authors conducted their research was also always noted. The research by Ernst Bergmann and co-workers originated from the Daniel Sieff Research Institute in Rehovot.[64] There is also an acknowledgment by Robert A. Millikan (in English): "Albert Abraham Michelson. The first American Nobel Laureate."[65]

In 1935, the *Zeitschrift für Physik* exceptionally gave an Aryan physics follower, Karl Vogtherr, an opportunity to present his objections to relativity theory. Perhaps as a result of pressure by the editors, the argumentation was unusually unpolemical and did not venture beyond the scientific issues. The brief reviews by Karl Bechert of these three articles leave no room for further doubt. They are summed up in the statements: "A series of old objections against relativity theory, which have all been refuted in the literature"; "The modern experimental confirmations of the specific [sic] theory of relativity do not appear in the paper"; and "No mathematical formulations of the proposed statements are offered." As an exception, the restraint otherwise generally preserved was abandoned here, even to the point of outright rejection. (By contrast, the summarizing review of the three parts published by the *Science Abstracts* was entirely neutral.)[66]

One of the abstracted periodicals was a mainly German-language "Physics Journal for the Soviet Union." Besides reviews of its scientific articles, there is a reference to a welcoming address on the 90th anniversary of the DPG.[67] Another article about the society honoring the same occasion is referenced in the British journal *Nature*.[68] In volume 19 (1938), one encounters the name of the founder of *Die*

[63] *Physikalische Berichte*, 20 (1939), 125. The review appeared in the second issue of the year, on 15 January 1939; it therefore cannot be correlated with the changed political climate after the Hitler–Stalin pact at the close of August 1939.

[64] *Physikalische Berichte*, 20 (1939), 164, 798, 1556.

[65] *Physikalische Berichte*, 20 (1939), 721.

[66] *Physikalische Berichte*, 16 (1935), 1190, 1686, 2150; *Science Abstracts*, section A, physics, vol. 38 (1935), no. 3319.

[67] *Physikalische Berichte*, 20 (1939), 1189; A. F. Joffé, "Zum 90. Jahrestag der Deutschen Physikalischen Gesellschaft," *Physikalische Zeitschrift der Sowjetunion*, 7 (1935), 128.

[68] *Physikalische Berichte*, 20 (1939), 721; E. N. Andrade, "The Deutsche Physikalische Gesellschaft," *Nature*, 135 (1935), 55. The doctoral supervisor and teacher from 1910–1911 of the author of this well-intentioned article was Lenard (this information by courtesy of Andreas Kleinert).

Naturwissenschaften, Arnold Berliner. Because of his Jewish descent, Berliner had been forced to relinquish his editorship in 1935. He received mention here for a historical article that appeared in the Indian journal *Current Physics*.[69]

The volumes got slimmer during the war years, but there were otherwise no changes to the style. The familiar names of émigré scientists continued to appear. In 1942, for example, Max Born was mentioned with six papers.[70] Two papers by James Franck and others on photosynthesis were given a particularly thorough review.[71] On 22 March 1943, Arnold Berliner committed suicide in anticipation of being evicted from his apartment and deported. German periodicals took no notice. But there was an obituary by Peter Paul Ewald and Max Born in the journal *Nature*, which was cited in the *Physikalische Berichte*. The notorious book *Jewish and German Physics* by Johannes Stark and Wilhelm Müller was also reviewed.[72] This review by the coeditor Michael Schön does not exhibit any enthusiasm for the subject. A similar impression is gained from the review of Pascual Jordan's book on *Physics and the Secret of Organic Life* when it alludes to the "philosophical considerations that are loosely related to the topic."[73]

The last volumes before the war ended (volumes 24 and 25 published in 1943 and 1944) differ from the earlier ones by their leanness, lack of an alphabetical index, and poor-quality paper. Otherwise, there are no noticeable differences. The broad range of fields of research is still impressive, even among the German contributions; also impressive is the reference to an obituary on the deceased émigré chemist Richard Willstätter in the journal *Nature* (volume 24, p. 361).[74]

Journal for Physics (Zeitschrift für Physik)

The *Zeitschrift für Physik*, founded in 1920, was the leading German journal for basic research. It was published "with the collaboration of the DPG," with Karl Scheel serving as its editor until his death in 1936.

[69] *Physikalische Berichte*, 19 (1938), 1772.

[70] *Physikalische Berichte*, 23 (1942).

[71] *Physikalische Berichte*, 23 (1942), 656, 1452.

[72] Johannes Stark and Wilhelm Müller, *Jüdische und deutsche Physik* (Leipzig: Heling, 1941).

[73] *Physikalische Berichte*, 24 (1943), 2127.

[74] *Physikalische Berichte*, 24 (1943), 361; Pascual Jordan, *Physik und das Geheimnis des organischen Lebens* (Braunschweig: Vieweg, 1941).

Scheel's multifaceted engagement earned him honorary membership and a reputation as the society's true representative. He was also editor of the *Physikalische Berichte* until the end of his life. His successor at the *Zeitschrift für Physik* was Hans Geiger who had co-edited with him a famous 24-volume handbook on physics. Its final volume, covering modern physics and published in 1933, was one reason why the names Hans Bethe, Max Born, Karl F. Herzfeld, Wolfgang Pauli, Peter Paul Ewald, Rudolf Ladenburg, Peter Pringsheim, Otto R. Frisch, Otto Stern, and Lise Meitner continued to remain in the public eye. All had authored individual essays in this work.

No substantial change in the journal's scope is ascertainable. Twenty-one volumes were published between 1930 and 1932, and 32 between 1933 and 1938. Afterwards, another 12 volumes appeared before the war ended. No turning point marking the political revolution of 1933 emerges from an examination of the content. The journal continued to preserve its scientific character and did not welcome divergent topics. No discontinuities are evident in the coverage of any particular field of research. The journal was clearly conscious of its traditional responsibility. It would certainly have been otherwise if Johannes Stark, the champion of the Aryan physics movement, had succeeded in materializing his ambitious plans. As early as 1933, he had announced his intention to bring all professional periodicals under his personal control.[75]

Taking authorship into account in this assessment, the gradual disappearance of a few bylines that had appeared under important articles before 1933 becomes obvious. But these names continued to appear in citations, quotations, and acknowledgments, without any noticeable concessions made to the *Zeitgeist* ("spirit of the times") propagated by the Aryan physics movement.

The transition was not abrupt. By 1935, 36 articles had appeared that had been submitted after January 1933 by authors who were shunned elsewhere. There was even a paper written by Max von Laue together with the émigrés Fritz and Heinz London (University of Oxford) on the theory of superconductivity.[76] Fritz London was featured again in 1938

[75] Based on his speech at the physicists' convention in Würzburg in 1933: Johannes Stark, "Organization der physikalischen Forschung," *Zeitschrift für technische Physik*, 14 (1933), 433; translated in Hentschel, *Physics*, doc. 28. Max von Laue later courageously responded to the brief hints there in an address before the Prussian Academy on 14 December 1933; it was published much later in *Physikalische Blätter*, 3 (1947), 272.

[76] Max von Laue, Fritz London, and Heinz London, *Zeitschrift für Physik*, 96 (1935), 359.

with a rebuttal to a paper by Fritz Bopp[77] (the title of which explicitly mentions London's equations!).[78] Otto Stern – already in Pittsburgh – published a comment in 1938 on a paper by the spectroscopist Hermann Schüler.[79] The subject is the magnetic moment of the deuteron. The difference of opinion is settled in a placating tone, and we learn in passing from Schüler's reply that Otto Stern had telephoned him "in mid-August 1933 during his sojourn in Berlin while in the company of Dr. Berliner."[80] The controversy between Franz Simon – already at Oxford – and Eduard Justi from the low-temperature laboratory of the PTR was a much more heated battle.[81]

Hans Kopfermann reported in 1934 about the hyperfine-structure measurements at Copenhagen and thanked "Prof. Niels Bohr for the opportunity to be able to work in his institute again."[82] The analyses (on determining nuclear moments) were continued as a joint enterprise in the Berlin polytechnic and Bohr's Institute for Theoretical Physics in Copenhagen. A second publication from the following year is in the same volume,[83] which contains another historically important analysis on hyperfine structure as well. It was the first paper by Hermann Schüler and Theodor Schmidt from the Potsdam-Babelsberg astrophysics observatory on deviations of the atomic nucleus from spherical symmetry.[84] The subject was the discovery of the nuclear quadrupole moment and its interaction with the electron shell. This was atomic physics at its best.

Peter Pringsheim, Victor Weisskopf, Eugene Wigner, and "Privy Councilor Haber" are named in the acknowledgments. There are papers by the first three during this period as well. One doctoral student from the University of Göttingen who had started writing his thesis under James Franck expressed "his respectful heartfelt thanks" to his "highly esteemed teacher, Prof. Dr. James Franck."[85] Lise Meitner's name continued to appear among the authors until 1938 (in papers written together with

77 Fritz Bopp, *Zeitschrift für Physik*, 107 (1937), 623.

78 Fritz London, *Zeitschrift für Physik*, 108 (1938), 542.

79 Otto Stern, *Zeitschrift für Physik*, 89 (1934), 665.

80 Hermann Schüler, *Zeitschrift für Physik*, 89 (1934), 666.

81 Franz Simon, "Über neuere Verfahren zur Erzeugung tiefer Temperaturen. Bemerkungen zu der gleichnamigen Arbeit von Herrn Justi," *Zeitschrift für Physik*, 87 (1934), 815.

82 Hans Kopfermann and Ebbe Rasmussen, *Zeitschrift für Physik*, 92 (1934), 82.

83 Hans Kopfermann and Ebbe Rasmussen, *Zeitschrift für Physik*, 94 (1935), 58.

84 Hermann Schüler and Theodor Schmidt, "Über Abweichungen des Atomkerns von der Kugel-Symmetrie," *Zeitschrift für Physik*, 94 (1935), 457.

85 B. Duhm, "Die Diffusion von Wasserstoff in Palladium," *Zeitschrift für Physik*, 94 (1935), 434.

Otto Hahn and Fritz Strassmann – as first author, that is, out of alphabetical order!). Gustav Hertz is also represented in some papers until 1939.

From the perspective of the *Zeitschrift für Physik* of this period, it was possible to forget (or repress) any consciousness of the conditions under which one was living. The 1935 article in *Nature*, cited earlier, presented remarks about the *Zeitschrift für Physik*: "This publication is so well known to physicists in Great Britain as not to need commendation."[86]

From 1936 on, contributions by emigrants stop appearing. But that did not doom entire areas of research to oblivion in the periodical. Nothing changed about the journal's style. Even in the last volume of 1944, a paper by Walther Bothe starts with the sentence: "The discovery by Frisch, Na^{22} ..."[87] There was evidently no hesitation to name and point out achievements by émigrés, even then. The last article in this volume, and thus the last before the war ended, is part III of a paper by Werner Heisenberg on observable quantities in the theory of elementary particles.[88] The head of the uranium project was busying himself with such an abstract topic! The first part, published in volume 120 from 1942, introduced the so-called S matrix. Issues 7 through 10 of this volume contain a remarkable series of articles dedicated to the editor Hans Geiger in honor of his 60th birthday. Two papers with data on meson measurements in cosmic radiation came from abroad: from Bologna (Gilberto Bernardini) and from Paris (Louis Leprince-Ringuet and collaborators).

The first volume to appear after the war, volume 124 from 1948, contains 19 papers submitted before the war had ended. We cannot exclude the possibility that other papers existed on nuclear physics that could not be published because of the restrictions imposed by the Allies. Max von Laue – one of the post-war editors of the *Zeitschrift für Physik*, "where I know the circumstances at first hand" – pointed out in an article that 60 unprocessed manuscripts were lying around at the offices of the *Zeitschrift für Physik* after the war had ended, specifying, "since then the editors could have accepted 86 more, which still deal, to a considerable extent, with work carried out during the war years."[89]

86 E. N. Andrade, "Deutsche," 55.

87 Walther Bothe, "Die in Magnesium durch Deuteronen erzeugten Aktivitäten und die Frage des K-Einfangs bei Na^{22}," *Zeitschrift für Physik*, 123 (1944), 1.

88 Werner Heisenberg, "Die beobachtbaren Größen in der Theorie der Elementarteilchen," *Zeitschrift für Physik*, 123 (1944), 93.

89 Max von Laue, "Die Kriegstätigkeit der deutschen Physiker," *Physikalische Blätter*, 3 (1947), 424–425; reprinted in Hoffmann and Walker, *Physiker*, 640–643. His article

Journal for Astrophysics (Zeitschrift für Astrophysik)

The *Zeitschrift für Astrophysik* appeared at the Springer publishers in the same format as the *Zeitschrift für Physik*. It, too, was published with the collaboration of the DPG. Karl Scheel was on its board of trustees until his death. One of its two editors was Walter Grotrian, an active member of the DPG, who succeeded Scheel as managing editor.

This periodical was more slender than the *Zeitschrift für Physik*. The topical focus leads far afield from the problems of possible technical importance. Its editors remained characteristically concerned about upholding international cooperation within the field. The reader feels even less aware of the prevailing circumstances. Unlike other journals, the *Zeitschrift für Astrophysik* continued to publish many articles in English during the first few years after 1933; for example, 10 of 27 papers in volume 11 from 1936, plus one in French. As in other journals, it also includes contributions by foreign authors written in German. Book reviews within the specialty also fill some of the journal's pages but they are only few in number.

Different from other areas of physics, astrophysics had – with the sole exception of Erwin Finlay-Freundlich – no leading German researchers affected by the discriminatory measures. Thus, no problems arose for the publication in this respect. Erwin Freundlich's paper on the deflection of light in the sun's gravitational field, thus an interpretation of the general theory of relativity, still appeared; it was received by the journal in March 1933.[90] Experimental results of an expedition to Sumatra to observe the solar eclipse in May 1929 were its basis. Other authors in the journal treated the topic as well. Another paper from 1937 quoted Freundlich's papers at length.[91] A detailed report about nuclear transmutation as an energy source in stars by George Gamow from George Washington University in 1938 closed with thanks to the author's "friend Dr. Edward Teller."[92] The article by Karl Bechert commenting on the general theory of relativity alludes to the paper mentioned above that Einstein and Rosen

also appeared in English translation: "The wartime activities of German scientists," *Bulletin of the Atomic Scientists*, 4 (1948), 103, which is quoted here.

90 Erwin Freundlich et al., "Über die Lichtablenkung im Schwerefeld der Sonne," *Zeitschrift für Astrophysik*, 6 (1933), 218.

91 Albert von Brunn and Harald von Klüber, "Kritische Untersuchung zur Bestimmung der Lichtablenkung durch die Potsdamer Sonnenfinsternisexpedition von 1929," *Zeitschrift für Astrophysik*, 14 (1937), 242.

92 George Gamow, *Zeitschrift für Astrophysik*, 16 (1938), 113–160.

had just recently published.[93] In 1944, Albrecht Unsöld was still citing Hans Bethe's handbook article.[94]

Classical subjects in astrophysics were being pursued. For instance, the discovery of Nova Herculis in 1934 immediately produced three papers in volume 10 from 1935.[95] The topic reemerged in the journal in 1937, one of the three articles originating from the observatory in Lyon.[96] Not even the outbreak of war leaves any noticeable trace in this periodical. Volume 22 from 1943, the penultimate one to appear before the war ended, can serve as an example. It contains a paper by Ludwig Biermann (Potsdam-Babelsberg) on the oscillator strengths of some lines in the spectra of Na I, K I, and Mg II.[97] It forms a part of a larger project: "The present paper contains the first results of the quantum-mechanical computations started in Babelsberg of oscillator strengths of astrophysically important lines and limiting continua." At the end we read: "This opportunity is taken to thank the German Research Foundation [DFG] as well for the material support that has made the computations described here possible." There follow thanks to Dr. Friedrich Möglich, Prof. Georg Joos, Prof. Friedrich Hund, Prof. Albrecht Unsöld, and six other people for executing numerical computations. It must have been difficult to argue for the urgency of such a project. There are three other papers by Ludwig Biermann (the founding director of the Max Planck Institute for Astrophysics after the war) – about different topics – in the same volume, along with one paper each from Uppsala and Copenhagen, two from Helsinki, and three from Zurich.

This volume also includes six reviews of books in the specialty that had been published in 1941–1942. The book by Otto Heckmann, *Theorien der Kosmologie*, aroused much attention. The review by the coeditor of the *Zeitschrift für Astrophysik*, Emanuel von der Pahlen, fills six and one-half pages. One learns: "The third and last part of the work, finally,

93 Karl Bechert, "Eine Bemerkung zur allgemeinen Relativitätstheorie," *Zeitschrift für Astrophysik*, 12 (1936), 117.

94 Albrecht Unsöld, "Quantitative Analyse des B_6-Sterns τ Scorpii. Part 4: Druckverbreiterung der H- und He*-Linien," *Zeitschrift für Astrophysik*, 29 (1944), 75.

95 Walter Grotrian et al., "Über das Spektrum der Nova Herculis 1934 am 3. Januar 1935," *Zeitschrift für Astrophysik*, 10 (1935), 209. The other two articles appearing in this volume were on the same topic.

96 J. Dufay and M. Bloch (Observatoire de Lyon), "Spectre nébulaire de Nova Herculis 1934," *Zeitschrift für Astrophysik*, 13 (1937), 36.

97 Ludwig Biermann, "Die Oszillatorenstärken einiger Linien in den Spektren des Na I, K I und Mg II," *Zeitschrift für Astrophysik*, 22 (1943), 157.

describes Milne's second cosmology, as original as it is radical." The astrophysicist mentioned here, Edward Arthur Milne from Oxford, was a member of the board of the *Zeitschrift für Astrophysik* until 1944! His theory was a counterproposal to the general theory of relativity. He had reported about it in the *Zeitschrift für Astrophysik* besides contacting British publications.[98] The book is not confined to this topic, however. In the "corrected reprint" that appeared after the war in 1968, Heckmann declared in a footnote, not without an element of pride: "The second part of the book is perhaps the only positive description of Einstein's gravitational theory that appeared in our country in the period from 1933–1945."[99] There is a controversial discussion of this remark and of Heckmann's role in the Third Reich in the article by Klaus Hentschel and Monika Renneberg from 1995.[100]

The "cosmic-ice theory" propagated by the Ancestral Heritage Foundation (*Ahnenerbe*) of the SS could not be taken as a serious subject of controversy in sober-minded astrophysics. The topic appears once in a review of the popular-science book titled "Controversial Worldview" (*Umstrittenes Weltbild*).[101] The book's author himself subjected the doctrine to a "thorough critique." The reviewer (Emanuel von der Pahlen) added the rather ironical remark: "It can only be useful even to the expert, faced with such a bewildering number of outgrowths of the modern quest for an all-encompassing scientific worldview, to gain insight into the psychological motives underlying them and determining the usually harsh belligerency of their authors against 'school physics' that is so incomprehensible to him."

The review of Unsöld's 1938 book on stellar atmospheres, *Die Physik der Sternatmosphären*, is interesting because in the last part the reviewer, Paul ten Bruggencate, carefully weighed its strengths against another book that had recently appeared in England.[102] Max von Laue felt free to fill

[98] Edward A. Milne, *Zeitschrift für Astrophysik*, 6 (1933), 1–95, and *Zeitschrift für Astrophysik*, 15 (1938), 263–298.

[99] Otto Heckmann, *Theorien der Kosmologie*, corrected reprint (Berlin: Springer, 1968), 103.

[100] Klaus Hentschel and Monika Renneberg, "Eine akademische Karriere. Der Astronom Otto Heckmann im Dritten Reich," *Vierteljahrshefte für Zeitgeschichte*, 43 (1995), 581–610.

[101] Robert Henseling, *Umstrittenes Weltbild* (Leipzig: Reclam, 1939); review by Emanuel von der Pahlen in *Zeitschrift für Astrophysik*, 18 (1939), 366.

[102] Paul ten Bruggencate's review in *Zeitschrift für Astrophysik*, 17 (1939), 129; Svein Rosseland, *Theoretical Astrophysics* (Oxford: Oxford University Press, 1936).

two and one-half pages with a detailed discussion of a book by Richard Chace Tolman on *Relativity, Thermodynamics, and Cosmology* and give due acknowledgment to the author's achievements as a researcher.[103]

Annals of Physics (Annalen der Physik)

The DPG was also involved in issuing the well-established *Annalen der Physik*, founded in 1799. Its editors were the solid-state physicist Eduard Grüneisen and Max Planck, assisted by an editorial board of prominent figures. The journal fell somewhat under the shadow of the recently founded *Zeitschrift für Physik* when it rose to become the leading periodical of international stature and the main platform for quantum mechanics at the dawn of modern physics. The *Annalen*'s flagging position is also perceptible in the period considered here. The majority of the papers on nuclear physics, for instance, were published in the *Zeitschrift für Physik*. But indisputably, fundamental papers in other fields continued to appear in the *Annalen*.

What has been said earlier about the style and handling of ostracized figures for the *Zeitschrift für Physik* is also true of the *Annalen*. In 1937, there still appeared a paper each by Lise Meitner[104] and Rudolf Ladenburg,[105] in 1936 there were two papers by Peter Paul Ewald and collaborators,[106] and until 1936 there were also papers by Richard Gans.[107] A footnote in a paper by Werner Heisenberg about cosmic radiation from 1938 is interesting. It states: "The necessity for taking this retardation... into account was kindly pointed out to me by Mr. Heitler."[108] Not all bridges had yet been burned.

In special instances, the *Annalen* would print a portrait with a dedication. One was in celebration of its editor Eduard Grüneisen's 60th birthday in June 1937. The board and the publisher pointed out

[103] Max von Laue's review in *Zeitschrift für Astrophysik*, 8 (1934), 389.

[104] Lise Meitner, "Über die b- und g-Strahlen der Transurane," *Annalen der Physik*, 29 (1937), 246.

[105] Rudolf Ladenburg, "Die heutigen Werte der Atomkonstanten e und h," *Annalen der Physik*, 28 (1937), 458.

[106] Peter P. Ewald and H. Hönl, "Die Röntgeninterferenzen an Diamant als wellenmechanisches Problem," Parts I and II, *Annalen der Physik*, 25 (1936), 281 and 26 (1936), 673.

[107] Richard Gans, "Das magnetische Verhalten eines Nickeldrahtes unter starker Torsion," *Annalen der Physik*, 25 (1936), 77.

[108] Werner Heisenberg, "Die Absorption der durchdringenden Komponente der Höhenstrahlung," *Annalen der Physik*, 33 (1938), 594.

in their dedication that the focus of their honorable mention had "earned a position of trust that granted him the possibility of maintaining the level of achievement of this time-honored organ of physical literature not merely at its previous level but even gaining it new friends *during difficult periods of time.*"[109]

A double issue appeared in honor of Max Planck's 80th birthday. In this instance, too, some veiled criticism is perceptible in the dedication by the board and the publisher. It refers to the "teacher of a generation of physicists . . . who look up to him in devotion and respect and would have gladly praised their master in this journal in a larger number than circumstances allow."[110] The 19 contributions by German authors nonetheless exhibit a variety of research fields. To these are added contributions by Erwin Schrödinger (University of Graz) and Robert A. Millikan (Cal Tech). The introductory article on the topic of the quantum of action and the atomic nucleus came from Niels Bohr. His historical overview gives the achievements by Einstein, Born, Franck, Wigner, and others their due – highlighting their names in spaced type.

However, the signs of the times also left their mark in the *Annalen.* The special issue in honor of Arnold Sommerfeld's 70th birthday on 5 December 1938, is a clear example. During the initial preparations for its publication, Peter Debye and other pupils of Sommerfeld agreed to meet the publisher's request of only seeking contributions from "Aryans." This discriminatory proviso was unprecedented in physics publishing for the time and elicited strong protest by Wolfgang Pauli and other physicists:

> As concerns the unscientific supplementary conditions for authors set by the publishers, I hope that increasing numbers of authors will not want to use the journals of such publishers for their publications, irrespective of whether the authors count among the white or the black class of theoreticians. In the present case of Sommerfeld's 70th birthday this consequence has already occurred (entirely without my involvement) and various authors who are Sommerfeld's students will publish in the Physical Review of December 1st (somewhat similar to the form in which the Planck issue of Physica appeared, but not limited to those residing in America).[111]

This protest had no effect because the *Annalen* issue appearing at the end of the year contained articles exclusively by "not incriminated

[109] *Annalen der Physik*, 29 (1937), 109 (emphasis by the author).

[110] *Annalen der Physik*, 32 (1938), 1–224.

[111] Pauli to Heisenberg, Zurich, 15 August 1938, in Karl von Meyenn (ed.), *Wolfgang Pauli, Wissenschaftlicher Briefwechsel mit Bohr, Einstein, Heisenberg u.a., Volume* 2 (Heidelberg: Springer, 1985), 593.

authors" (Klaus Clusius, Peter Debye, Walther Gerlach, Werner Heisenberg, Helmut Hönl, Walter Kossel, Helmuth Kuhlenkampf, Wilhelm Lenz, Joseph Meixner, Heinrich Ott, Fritz Sauter, Otto Scherzer, and Albrecht Unsöld).[112] Besides the *Physical Review*,[113] the British journal *Nature*[114] also celebrated Sommerfeld's birthday.[115]

In 1937, four papers appeared by August Becker and his collaborators at the Philipp Lenard Institute of the University of Heidelberg.[116] This institute had received its name in a pompous inaugural ceremony in December 1935. Three other papers by this group appeared in the *Zeitschrift für Physik*. The topics are closely associated with Lenard's earlier research, as is conspicuously pointed out. They essentially constitute the transmission of experimental data without any attempt at theoretical profundity – entirely Lenard's style of doing things. These papers received normal reviews in the *Physikalische Berichte* and (with the exception of the last paper in 1939) were also included in *Science Abstracts*. The idiosyncratic reference in one paper to K^+-"carriers" is silently set right[117]; the reviews quite naturally use the more common term "ion," which had already long since come into general usage.

Another paper belonging in this section is the 46-page production published in 1941 by Ludwig Wesch, a fanatical advocate of the Aryan physics movement.[118] The institutional address given is the "Physico-Technical Department of the Philipp Lenard Institute." A footnote reveals it as the belated publication of a habilitation thesis presented at Heidelberg in 1935. Its veneration of Lenard is driven to the extreme. "High-frequency rays" are mentioned where Röntgen radiation[119] is meant (a note in the annotation at least clarifies this). Lenard's bitter resentment about

[112] *Annalen der Physik*, 33 (1938), 565.

[113] *Physical Review*, 54 (1938), December 1, 869–967. This issue contains 20 contributions with no direct allusions to the occasion; only seven explicitly refer to Sommerfeld's birthday.

[114] *Nature*, 142 (1938), 987.

[115] Dieter Hoffmann drew my attention to these proceedings.

[116] August Becker and E. Kipphan, "Die Streuung mittelschneller Kathodenstrahlen in Gasen," *Annalen der Physik*, 28 (1937), 465; W. Veith, "Elektronenanregung und Trägerreflexion beim Auftreffen von K^+-Trägern auf Metalle," *Annalen der Physik*, 29 (1937), 189; F. Frey, "Über die Geschwindigkeitsverteilung von der Kathodenstrahlen in Gasen ausgelösten sekundären Elektronen," *Annalen der Physik*, 30 (1937), 297; K. Kamm, "Über die Zinksulfid-Cadmiumsulfid-Phosphore," *Annalen der Physik*, 30 (1937), 333.

[117] Veith, "Elektronenanregung."

[118] Ludwig Wesch, "Über die optisch-elektrischen Eigenschaften der Lenardphosphore, 1. Der DK-Effekt," *Annalen der Physik*, 40 (1941), 249.

[119] Also called X-rays.

Röntgen having been granted priority in the discovery of the new radiation is the obvious background to this divergent terminology.

These few, professionally "domesticated" contributions by followers of Aryan physics are completely obscured by the abundance of other papers and alter nothing about the scientific impression of the journal as a whole. Glancing from the last page of the paper by Wesch to the first page of the following article, a dissertation written under the guidance of Fritz Sauter from the Institute of Theoretical Physics at the University of Königsberg,[120] one is no longer urgently referred to Lenard, rather to Bethe and the Bethe–Peierls theory and method, and the paper by August Becker and his collaborators[121] precedes the work mentioned above by Rudolf Ladenburg from Princeton University.[122]

One has to search hard for traces leading beyond the scope of scientific research. The paper by Wolfgang Finkelnburg on the theory of detonation processes contains a reference to "a detailed communication currently in press in the journal for gunnery and explosives," the *Zeitschrift für das gesamte Schieß- und Sprengstoffwesen.*[123] Johannes Picht notes in a paper on electron optics that his workplace is the "chair for theoretical optics at the Faculty for Defense Technology of the Berlin Polytechnic."[124] The first volume of the new, sixth series of the *Annalen* in 1947 published a dozen papers that had been received before the war's end.

Journal for all Science (Zeitschrift für die gesamte Naturwissenschaft)

For serious physics, the *Zeitschrift für die gesamte Naturwissenschaft* was entirely insignificant. It is mentioned here to point out how far away it was from the level represented by the other professional journals. This periodical was established in the form of interest here in 1937, when it became the organ of the physics section of the Reich student leadership, thereby becoming the mouthpiece of the Aryan physics movement.

Around this time, the influence of Aryan physics was already receding; its followers doggedly continued to hold the course against the changing

120 U. Firgau, "Zur Theorie des Ferromagnetismus und Antiferromagnetismus," *Annalen der Physik*, 40 (1941), 295.

121 Becker and Kipphan, "Streuung."

122 Ladenburg, "heutigen Werte."

123 Wolfgang Finkelnburg, "Zur Theorie der Detonationsvorgänge," *Annalen der Physik*, 26 (1936), 116.

124 Johannes Picht, "Bemerkungen zu einigen Fragen der Elektronenoptik," *Annalen der Physik*, 36 (1939), 249.

tide. The first issue (May/June 1937) pointed out in no uncertain terms the periodical's style and intentions. Philipp Lenard is celebrated in a number of flattering articles, which in and of itself was an obvious demonstration of the journal's opposition to the DPG. An article with the title "Physics and astronomy in Jewish hands" reads: "Indeed, the consequences of the elimination of the aether by *Einstein* and his Jewish team of helpers, *Max Born*, *Weyl*, et al., have been drawn with a frivolity and brutality as only an alien conqueror can do in the land of his enemy."[125] The author was one of the journal's three editors: Bruno Thüring, an astronomer at the State Observatory (*Sternwarte des Staates*) in Munich. For comparison, note that at that very time, volume 18 of the *Physikalische Berichte* (1937) was offering factual reviews of 59 papers under the rubric relativity theory. Ten of them had been drawn from German periodicals, and one of the others was by Einstein himself.[126]

Papers that directly addressed the theory of relativity were not the only concern of the proponents of Aryan physics. They objected to the way in which such results were being accepted and applied. The controversial successor to Sommerfeld's professorship, Wilhelm Müller, wrote in 1939 in his paper on the "Jewish spirit in physics":

> It seems as if theoretical physics were already so strongly laden with the forces of resistance and with foreign bodies that a lasting sense of the essential German kind simply cannot find any foothold anymore. The scales are still tipped far in favor of physicists inclined to acknowledge relativity theory as the norm of a universal physics.[127]

This verdict on manifestations of the Jewish spirit also extends to quantum mechanics, the second pillar of modern physics. The same article states:

> As regards the worldview and natural philosophy, *Heisenberg*'s quantum mechanics is a direct extension of *Einstein*'s doctrine, which *Heisenberg* himself explicitly recognizes as his basis.[128]

[125] Bruno Thüring, "Physik und Astronomie in jüdischen Händen," *Zeitschrift für die gesamte Naturwissenschaft*, 3 (1937/38), 55, quote on 62.

[126] *Physikalische Berichte*, 18 (1937), XXIX, systematic index entry no. 6. The above-mentioned paper is Albert Einstein and Nathan Rosen, "On gravitational waves," *Journal of the Franklin Institute*, 223 (1937), 43ff.

[127] Wilhelm Müller, "Jüdischer Geist in der Physik," *Zeitschrift für die gesamte Naturwissenschaft*, 5 (1939), 162–175, quote here on 170.

[128] Müller, "Geist," 173.

The elsewhere highly respected scientific journal *Nature* was, according to one article in the *Zeitschrift für gesamte Naturwissenschaft*, an "abominable Jewish rag."[129] This invective was in reaction to some critical articles that had appeared in the British journal about the political situation in Germany (which was here styled as "agitating articles against National Socialist Germany"). This "abominable rag" could not even hold back generous praise in a lengthy review of Albrecht Unsöld's book on the physics of stellar atmospheres: "The author is a distinguished astrophysicist.... His theoretical work shows a rare insight into the conditions of observational physics."[130] The reviewer praising this work would not have overlooked the acknowledgment of Einstein's contribution to the theory of radiation in the introductory chapter (his name even appears in a heading) or the naming of Rudolf Ladenburg, Hans Bethe, Victor Weisskopf, and Eugene Wigner in other contexts, and the recommendation of Max Born's textbook on optics.

In early 1939, an article appeared in the *Zeitschrift für gesamte Naturwissenschaft* under the title "Jews in physics: Jewish physics" by Ludwig Glaser. It closes with an explicit reference to the pogrom of November 1938. The "Jewish-minded influence" is blamed on Hermann von Helmholtz, Emperor Wilhelm II and his advisors, as well as Arnold Sommerfeld as the "patron" of a "crystallization point" for Jewish physicists. Finally, Glaser concludes:

> But then in 1938 a November storm raged throughout the land – it swept away the withered foliage. The remaining Jews, part-Jews and those related to Jews through marriage have disappeared from the academies and libraries, from the lecture halls . . . [131]

The final paragraph starts with the refrain: "We thank our Führer, Adolf Hitler, for having freed us from the plague of Jews." At this critical moment, the scourged Arnold Sommerfeld was celebrating his 70th birthday. At a festive session of the Bavarian regional physical society with prominent invited speakers, the chairman of the DPG, Peter Debye,

129 H. Rügemer, "Die 'Nature' eine Greuelzeitschrift," *Zeitschrift für die gesamte Naturwissenschaft*, 3 (1936/37), 475ff. The expression "jüdische[s] Greuelblatt" appears on 479. The author of this article was another employee at the *Sternwarte des Staates* in Munich.

130 Review of Albrecht Unsöld, *Die Physik der Sternatmosphären*, in *Nature*, 142 (1938), 975.

131 Ludwig Glaser, "Juden in der Physik, Jüdische Physik," *Zeitschrift für die gesamte Naturwissenschaft*, 5 (1939), 272–275; translated in Hentschel, *Physics*, doc. 77, quotes on 226, 229, 233f.

ceremoniously handed him a certificate of honorary membership.[132] One could hardly imagine a starker contrast.

The article discussed above represented a low point in the Aryan physics campaign. It does have the involuntary merit, though, of listing all the important Jewish physicists of the period. The fact that some unbiased articles of scientific content, the majority of them not from physics, are also included in the journal does not improve its overall image. In 1944, the journal was forced to cease publication – precisely when the DPG succeeded in pushing through its *Physikalische Blätter*.

Journal for Technical Physics (Zeitschrift für technische Physik)

Whereas the DPG was responsible for the *Zeitschrift für Physik*, the DGtP was the guiding association behind the *Zeitschrift für technische Physik*. This journal was founded in 1920, right on the heels of the founding of its mother association, hence in tandem with the *Zeitschrift für Physik*. Its editors were Carl Ramsauer and the high-frequency expert Hans Rukop. When Richard Swinne died, Ramsauer took on the chief editorship in 1939 as well, which had formerly been in the hands of Swinne's predecessor Wilhelm Hort until 1937.

Unlike the *Zeitschrift für Physik* and the *Annalen der Physik*, the *Zeitschrift für technische Physik* published not only original papers but also book reviews, notices by the society, and larger numbers of obituaries and anniversary addresses. The spectrum of subjects treated by the papers it published is broad. The talks held at the societies' conferences can serve as an orientation guide, because most of the written versions appeared in the *Zeitschrift für technische Physik*. Among the original papers, there is no noticeable tendency indicating any politically defined selection or polemics. From 1933 on, however, details in the general section reveal in which era the journal was being issued.

This does not apply in every case, however. One review in 1933 ranked Gustav Hertz's revision of Emil Warburg's textbook on experimental physics as "certainly suited for students of physics and neighboring fields."[133] Praise was also meted out to the textbook *Optik*, appearing in 1933 by "the famous theoretical physicist" Max Born[134]; likewise

132 *Verhandlungen der Deutschen Physikalischen Gesellschaft*, 20 (1939), 7.

133 Review of Emil Warburg, *Lehrbuch der Experimentalphysik in Zeitschrift für technische Physik*, 14 (1933), 523.

134 *Zeitschrift für technische Physik*, 14 (1933), 375.

the book *Moderne Optik* by the same author, containing "seven presentations about matter and radiation. Elaborated by Fritz Sauter."[135] The second volume of Max Abraham's *Theorie der Elektrizität* appeared in the same year 1933, newly revised by Richard Becker. The reviewer applauded its "consistent cross-references with the special theory of relativity," which had been lacking in Abraham's original version.[136] The "booklet written so charmingly close to research," on the structure of the atomic nucleus by Lise Meitner and Max Delbrück was introduced in 1935[137] and the famous book *The Quantum Theory of Radiation* by Walter Heitler in 1938.[138] A collection of essays by the authors Grete Hermann, Eduard May, and Thilo Vogel (1937) treated the importance of modern physics in epistemology. The reviewer complimented "the clear presentation of the main principles of quantum theory and relativity theory by Grete Hermann." The contribution by Eduard May, which argued for strict causality, Euclidean space, and absolute simultaneity as a priori givens, the reviewer glibly dismissed with the statement: "The reviewer was unable to follow Mr. May."[139] It is as if turning over a new leaf from theorist to technical physicist to read the review of Pascual Jordan's work from 1935 on physical thought in modern times:

> The reviewer finds the last section less fortunate which, far too swayed by the times, aims to derive the future value of science from the importance of the power of defense technology and, in our view, only musters lame emphasis on the fact that path-breaking research in new territory has been, and must always remain, the purpose-unbound pacesetter of task-bound application.[140]

The book by Ernst Zimmer on a revolution in the worldview of physics, which appeared in its fourth edition in 1938 with a foreword by Max Planck, received a positive review as "a book . . . that in a layman's hands, just as in a specialist's, can only do good and be a boon."[141] For the sake of comparison, consider the conception of modern physics depicted in

[135] *Zeitschrift für technische Physik*, 14 (1933), 215.
[136] *Zeitschrift für technische Physik*, 15 (1934), 166.
[137] Review of Lise Meitner and Max Delbrück, *Der Aufbau der Atomkerne* in *Zeitschrift für technische Physik*, 16 (1935), 316.
[138] *Zeitschrift für technische Physik*, 19 (1938), 20.
[139] Review of Grete Hermann, Eduard May, and Thilo Vogel, *Die Bedeutung der modernen Physik für die Theorie der Erkenntnis* in *Zeitschrift für technische Physik*, 18 (1937), 204.
[140] Review of Pascual Jordan, *Physikalisches Denken in der neuen Zeit* in *Zeitschrift für technische Physik*, 17 (1936), 142.
[141] Review of Ernst Zimmer, *Umsturz im Weltbild der Physik* in *Zeitschrift für technische Physik*, 19 (1938), 325.

the parallel review published in the *Zeitschrift für die gesamte Naturwissenschaft*: "It is almost superfluous to note in this journal that we cannot share this interpretation, which thus prohibits our finding any value in this book."[142] The trench between the two journals is just as deep in the reviews of volume one (on diatomic molecules) of Gerhard Herzberg's book on molecular spectra and molecular structure that appeared in 1939. The reviewer for the *Zeitschrift für technische Physik* pointed out the author's own "numerous fundamental researches," the "clear and concise language," and the comprehensiveness of the work.[143] Conversely, the reviewer for the *Zeitschrift für die gesamte Naturwissenschaft* could only find fault with a work authored by a "Jew by marriage," specifically criticizing it as a book: "in which the papers by Jews... are cited with rare completeness"; which was entirely devoid of references to Philipp Lenard; and in which the English literature was overvalued. The "confusion" in its handling of modern theory was one more weakness he found. His quintessence: "A highly unwelcome book that must be repudiated in the German physical literature."[144] Walter Grotrian's appreciation of the work in the *Zeitschrift für Astrophysik* reads: "Thus this book can be most highly recommended particularly for astrophysicists."[145]

In 1941, the first issues appeared of the comprehensive historical work on technology in modern times, edited by Friedrich Klemm, a librarian at the *Deutsches Museum*. The review in 1943 emphasized that the old rule, "When weapons speak, the Muses fall silent," did not apply in this case. "In any event, Clio, the Muse of history, is apparently thoroughly enjoying the favor of the god of war these days."[146]

Concessions are made to the Aryan physics camp in discussions of polemical writings. There is a lengthy report about the dedication ceremony of the Philipp Lenard Institute in Heidelberg on 13–14 December 1935, which the small circle of supporters of Aryan physics celebrated among themselves in the form of a 2-day symposium.[147] No judgment is passed on it, but the detailed quotations without commentary speak for

[142] Review of Zimmer, *Umsturz in Zeitschrift für die gesamte Naturwissenschaft*, 5 (1939), 154.

[143] Review of Gerhard Herzberg, *Molekülspektren und Molekülstruktur* in *Zeitschrift für technische Physik*, 20 (1939), 327.

[144] Review of Herzberg, *Molekülspektren* in *Zeitschrift für die gesamte Naturwissenschaft*, 6 (1940), 272.

[145] Review of Herzberg, *Molekülspektren* in *Zeitschrift für Astrophysik*, 19 (1940), 68.

[146] Review of Friedrich Klemm (ed.), *Die Technik der Neuzeit* in *Zeitschrift für technische Physik*, 24 (1943), 43.

[147] *Zeitschrift für technische Physik*, 17 (1936), 143.

themselves. The presentations were later assembled in a volume under the title *Scientific Research in Upheaval*, edited by the director of the institute August Becker. A brief review of it in the *Zeitschrift für technische Physik* is essentially a reference to an earlier report about the symposium, but it closes with the statement: "May this book, adorned with Lenard's portrait, help German scientists toward clear insight into the dependency of race on their science."[148]

The reviewer (Richard Swinne) was also benevolent toward the "Reich award-winning paper in the first Reich Achievement Contest among German students 1935/36." The book called *Philipp Lenard, the German Scientist: His Struggle for Nordic Research* was written by "10 comrades at the Philipp Lenard Institute."[149] The reviewer expected "this [will] contribute toward placing his [Lenard's] personality in the proper light."

Numerous editions of Ernst Grimsehl's physics textbook continued to appear until 1942, edited by Rudolf Tomaschek, one of Lenard's and Stark's followers who stands out for his professional qualifications. Reviews of his revised edition highlighted its thorough handling of the experimental material. Criticism was rather directed at the last part of the work under the subtitle "Matter and the aether," in which this advocate of Aryan physics naturally defended the theory of the aether against (other) "speculations." The writer of the last two reviews (Walter Weizel) became quite clear on this point: "A tendentious strain that expresses itself, e.g., in excessive quotings of *Lenard* and *Stark* is regrettable.... The fact that the aether... is gradually robbed of its palpable properties... in section X, many physicists will smirkingly note.... The addition of the so-called *Stark*ian atomic model is no gain for the new edition."[150] A subsequent brief review of a reissued edition noted: "In the 3rd volume one would have liked to see the known biases more mildly put."[151]

Although appreciating the survey of the experimental data in Johannes Stark's book on the *Physics of Atomic Surfaces* (1940), another reviewer found that "the overwhelming successes of quantum mechanics in all fields of atomic structure remain, however, completely unmentioned,"

[148] Review of August Becker (ed.), *Naturforschung im Aufbruch* in *Zeitschrift für technische Physik*, 17 (1936), 667.

[149] Review of *Philipp Lenard, der deutsche Naturforscher. Sein Kampf um nordische Forschung* in *Zeitschrift für technische Physik*, 18 (1937), 492.

[150] Review of Ernst Grimsehl, *Lehrbuch der Physik* in *Zeitschrift für technische Physik*, 22 (1941), 138.

[151] *Zeitschrift für technische Physik*, 23 (1942), 317.

and, for want of quantitative relations, an experimental verification of "*Stark*'s new conceptions is impossible."[152]

The substantive statements by "Aryan physics" are unequivocally rejected. One only needs to compare the examples quoted above against the overwhelming applause for Arnold Sommerfeld's *Atomic Structure and Spectral Lines*,[153] as well as reviews of other monographs of this period. In contrast, there are also visible attempts at reconciliation with the "other side," with the occasional compromise also in the vocabulary used. The book *Physical Optics* by Robert Williams Wood was, for the reviewer, "the well-known description of physical optics by this excellent North American experimental physicist." He added, however – borrowing the Aryan physics idiom – that the author was a "research personality, who . . . hardly values speculative notions."[154]

There is frequent use of the word *Führer*, the characteristic inference during that era to describe leadership in its various forms, whereas earlier (as now again) another word surely would have been chosen. The obituary of the founder of the *Deutsches Museum*, Oskar von Miller, for example, states that German technology had lost "with him a Führer and path-breaker."[155] From another obituary one gathers that Carl Friedrich Siemens had been "Führer of the German delegation to the world economic conference in Geneva."[156] The word recurs in embarrassing frequency in an article written by Carl Ramsauer in memory of Karl W. Haußer, a pupil of Lenard's who left a career in industry to found the physics department of the Kaiser Wilhelm Institute of Medical Research in Heidelberg. The reader learns that "a large part of his leadership [*Führertum*] constituted teaching." And further down: "As a technical physicist, Haußer was a pioneer and a Führer. . . . All of us gladly acknowledged the physicist and technician Haußer as a Führer. . . . There has rarely been a Führer so willingly followed as Haußer was by his followers." Ramsauer closed the article with a quote from Lenard: "My dear Haußer was the most Nordically typed among my students."[157]

[152] Review of Johannes Stark, *Physik der Atomoberfläche* in *Zeitschrift für technische Physik*, 21 (1940), 187.

[153] Review of Arnold Sommerfeld, *Atombau und Spektrallinien* in *Zeitschrift für technische Physik*, 22 (1941), 23.

[154] *Zeitschrift für technische Physik*, 17 (1936), 244.

[155] *Zeitschrift für technische Physik*, 15 (1934), 289.

[156] *Zeitschrift für technische Physik*, 22 (1941), 303.

[157] *Zeitschrift für technische Physik*, 15 (1934), 4.

Not all the articles in memoriam are as importune. One contribution after the death of the pioneer in glass technology, Otto Schott, described the departed not as a *Führer* but as an "honorable patriarch." Considering his willingness to sign his glassworks over to the Carl Zeiss Foundation, he is attributed "noble power and strength of mind which has nothing... in common with cheap salon Socialism." A second contribution draws the eye to the political milieu only toward the end, where at the funeral his "followers" reportedly took leave of the deceased "with the German salute."[158]

Birthday addresses honoring Ludwig Prandtl,[159] Karl Scheel,[160] and Karl Mey[161] are free of conscious linguistic adjustments to contemporary political correctness. In the cases of Prandtl and Mey, there is not even any reference to a close link with the modern mission of technology. In Carl Ramsauer's case also, an address in celebration of his 60th birthday only emphasizes the "influence of the great experimenter and teacher *Lenard*."[162] But it is astonishing to read in a contribution honoring Walther Nernst's 70th birthday: "otherwise as well, Nernst along with Lenard, Ernst Wiechert and Oliver Lodge, has proven to be one of the most zealous advocates of the hypothesis of the existence of a physical world aether."[163]

The statements made about Lenard's 80th birthday in 1942 reveal a steep gradient between the journals. In an address published in the *Zeitschrift für technische Physik*, Karl Mey praised the great physicist in the name of the DGtP as "the great German, who already early on had harnessed his entire personality to the Führer's mission."[164] The *Annalen* merely published a portrait with the short caption: "The venerable master of experimental physics – For his 80th birthday, 7th June 1942."[165] The position in between is occupied by the *Physikalische Zeitschrift*, where mention is made of "the master of experiment... champion of

158 *Zeitschrift für technische Physik*, 17 (1936), 1.
159 *Zeitschrift für technische Physik*, 16 (1935), 25.
160 *Zeitschrift für technische Physik*, 17 (1936), 65.
161 *Zeitschrift für technische Physik*, 20 (1939), 65.
162 *Zeitschrift für technische Physik*, 20 (1939), 33.
163 *Zeitschrift für technische Physik*, 15 (1934), 212. Nernst did work on cosmology as well and one model of a stationary universe did in fact introduce an aether. But its function as an energy reservoir was different from the aether in Lenard's meaning; and to my knowledge Nernst never drew any connection between them, either. For example, see Walther Nernst, *Zeitschrift für Physik*, 106 (1937), 633.
164 *Zeitschrift für technische Physik*, 23 (1942), 125.
165 *Annalen der Physik*, 41 (1942), 325.

the German essence."[166] A neutral article appeared in the journal *Die Naturwissenschaften* by Walther Kossel, wholly taken up by Lenard's experiments with electron beams.[167]

Walter Weizel's review of the published versions of talks given in 1941 during the inaugural session of the colloquium for theoretical physics at the University of Munich, a statement delivered in the name of the Faculty of the Sciences and Mathematics of the University of Bonn, reveals much about the struggle against the Aryan physics movement – including its double-edged aspects. "The authors attempt in this polemical work to use the National Socialist view of the world as a lead-in to their physical views, which have found little resonance in science." The remaining arguments are primarily directed against the "entirely unobjective method of suspecting quantum theory to be Jewish." "Writings like the present one, that attribute the merit of a great feat of German science to the Jews . . . will have the effect of pro-Jewish cultural propaganda, a success that the authors surely do not intend."[168]

There is the occasional surprise among the notices. One learns, for instance, that in 1937, the tenured professor of physics at the polytechnic in Breslau, Dr. Erich Waetzmann, had been elected a corresponding member of the Academy of Technical Sciences in Warsaw,[169] and that as late as 1939, Karl Mey proposed that the Institute of Experimental Physics of the University of Poznán (Poland) be welcomed into the DGtP.[170] One project on women as inventors, mentioned in a report from 1937, seems downright modern. An Institute for Social Work had started to investigate "the involvement of women in inventions of the past and present." The goal was "a statistical survey, such as long since exists in America."[171] In the same year we learn of a new campaign on aesthetically improving the workplace, the Beauty of Labor (*Schönheit der Arbeit*), by the agency aimed at improving air quality – all this running under the motto "Cheerful people in pretty firms of a happy Germany."[172]

[166] *Physikalische Zeitschrift*, 43 (1942), 137.
[167] *Die Naturwissenschaften*, 30 (1942), 317.
[168] Review of Johannes Stark and Wilhelm Müller, *Jüdische und deutsche Physik. Vorträge zur Eröffnung des Kolloquiums für theoretische Physik an der Universität München* in *Zeitschrift für technische Physik*, 23 (1942), 25.
[169] *Zeitschrift für technische Physik*, 18 (1937), 32.
[170] *Zeitschrift für technische Physik*, 20 (1939), 96.
[171] "Frauen als Erfinderinnen," *Zeitschrift für technische Physik*, 18 (1937), 112.
[172] "Frohe Menschen in schönen Betrieben eines glücklichen Deutschlands," *Zeitschrift für technische Physik*, 18 (1937), 237.

There are other kinds of news as well. A call for high-frequency specialists for the Air Force in 1943 was evidently connected with the belated attempt to rehabilitate physics as a technical and military factor of power in a drive closely associated with Ramsauer's name.[173] The following notice was posted in consultation with the Reich Aviation Ministry and the High Command of the Air Force: "The Air Force has a constant demand for physicists, certified engineers, engineers, technicians, wireless mechanics, and amateurs."[174]

Physics in Regular Reports (Physik in regelmäßigen Berichten)

In March 1933, the DGtP embarked on an ambitious project of publishing a series called *Physics in Regular Reports* (*Die Physik in regelmäßigen Berichten*). "An acknowledged specialist" was to report at intervals of roughly 3 years about advances made in his area of expertise. The plan was to divide the field into some 60 areas, not restricted to technical physics. The society preferred "here too, to place its powers at the service of physics as a whole."[175] The series continued until 1943. Most of the chosen areas became the subject of two or three contributions, the first of which appeared in 1933–1934 with the object of providing the basics on the latest research. These separate issues were delivered to the members of the DGtP as a free supplement to the *Zeitschrift für technische Physik*.

Carl Ramsauer took on editorship of this series as well. Richard Swinne was the managing editor until his death in 1939 whereupon Ramsauer stepped in, later to be assisted in these additional tasks by Rudolf Frerichs from the Osram Lighting Company. The manner of quotation used by the series was original. Besides referring to original papers, reviews appearing in the *Physikalische Berichte* were also cited – in the first few volumes only the latter appeared.

The articles on fundamental research were more frequent than those falling under what would count as technical physics. Among the authors are some famous names. Gregor Wentzel from the Physical Institute of University of Zurich reported in 1934 and 1939 on quantum theory and wave mechanics. It would be misplaced even to wonder whether the citations in these articles, introducing current problems of quantum

[173] See the contribution by Hoffmann to this volume.
[174] *Zeitschrift für technische Physik*, 24 (1943), 93.
[175] *Die Physik in regelmäßigen Berichten*, 1 (1933), preface.

electrodynamics and nuclear physics, followed some selection criterion distinguishing "desirable" from "undesirable" physicists. Friedrich Hund wrote two articles (1933 and 1937) on atoms and molecules; Wilhelm Hanle (1934 and 1933) on the excitation of gases; both Richard Becker (1935), and Walther Gerlach and his collaborators (1939) on magnetism, and so forth. Even the contributions by the "Aryan physicist" Rudolf Tomaschek (1934 and 1940) on optics and the electronics of solid and fluid bodies are unobtrusive. Nor does August Becker's essay on corpuscular rays (1934 and 1938) fall out of bounds, if one disregards the particular emphasis on Lenard's accomplishments in the first few pages of the first article. (But even this does not prevent the author, on page 5 already, from emphasizing as "significant" the "thorough development of *Born*'s collision theory by *Bethe*.")

The American *Science Abstracts* offered the same kind of neutral synopses of these essays as it did for articles in other journals. As a rule, chapters were listed and the amount of publications elaborated on was mentioned. For example, August Becker's second contribution received the following synopsis: "This report comprises a comprehensive summary with references to close on 400 original papers, of the progress in corpuscular-ray research since the issuance of part I, 4 years ago."[176]

This series was terminated along with the *Zeitschrift für technische Physik* at the war's end. Three articles from 1943–1944 were published retroactively in 1954 nonetheless. This last volume also contained an authors and subject index for the entire series – there was nothing needing to be retracted.

Physical Letters (Physikalische Blätter)

In 1943, the DPG proposed a project evidently connected with the initiatives by the society's chairman, Carl Ramsauer[177]: the founding of a journal designed for the broader public, the *Physikalische Blätter*. The fact that this was possible at a time of paper shortages and all the other economic problems in the midst of a "total war," is clear evidence of a change of direction by those in power. Twelve issues appeared between June 1944 and the end of the war – with some delays due to damage caused by the Allied bombing campaign. The journal was the organ of an "Information Office of German Physicists" that the DPG had founded

[176] *Science Abstracts*, section A, physics, XLI (1938), no. 2552.

[177] See the contribution by Hoffmann to this volume.

in 1943 with the mission of enlightening the "public about the work and importance of physical research as well as about the profession of a physicist, particularly in fundamental research" and of concerning itself with "attracting the next generation of scientists to a physicist's career."[178] Ernst Brüche, a department head at the AEG research laboratory and close contact of Carl Ramsauer's, became the journal's editor.

The underlying tone of the new periodical was to increase the prestige of physics as a "basic science" and to encourage its promotion, starting at school level. There was no shyness about getting to the point. The lack of support for physics and for the sciences in general was openly discussed, and the situations in the United States and elsewhere were presented as good models worth emulation. Aryan physics was no longer a topic worth speaking about; the *Zeitschrift für die gesamte Naturwissenschaft* had closed down. A report about a collection of images and film of German physicists[179] was now at liberty to name the "non-Aryans" James Franck, Gustav Magnus, Emil Warburg, Heinrich Rubens, and Fritz Haber.[180] Magnus even received special acknowledgment, complete with curriculum vitae and a full-page portrait, as one of the "German physicists, whose researches have an important share in the development of ballistics."

Each of the individual issues has its own guiding theme: "Research in distress!"; "Physics is basic science"; "Young talent is a life-and-death issue"; "Recruiting questions"; "Research also during the war!"; "Physics all over the world"; and "School reforms." The October issue served as a panel discussion about the journal itself, at which plans for the future were very candidly put forward. The November issue, as "History issue 1," was dedicated to Justus von Liebig and contains informative articles.

As can be gathered from the guiding themes, research played a central role alongside worries about the next generation of young professionals. Many articles pointed out important technical applications "for the economy and defense," based on results of physical research. But pure fundamental research, solely based on the urge for knowledge, was just as forcefully demarcated with the same persistence from "applied research." The editor evidently tried to encourage this research as well by the selection of articles as well as his own contributions. He gave the experiments

[178] *Physikalische Blätter*, 1 (1944), 1.

[179] Ernst Brüche, "Über ein Jahrzehnt Bild- und Filmsammlung Deutscher Physiker," *Physikalische Blätter*, 1 (1944), 44–46.

[180] Brüche, Jahrzehnt, 45.

on electromagnetic waves and electron emissions as examples of how research oriented purely toward scientific knowledge can yield the unanticipated application of radio communications using electron tubes.

In addition to writing his own essays, the editor, Ernst Brüche, also drafted lead articles and forewords to other contributions. Only a few of these authors were notable physicists. Jonathan Zenneck produced two articles,[181] one of which mentioned the Haber–Bosch process, and thus Haber's name![182] The article on the essence and importance of physics came from Wolfgang Finkelnburg's pen, with the closing remark: "It may, and indeed must be said without any exaggeration that the current state of physical research of a nation and a state is critical for the development of its economic and military strength."[183] This article appeared in volume nos. 3–4 ("Physics is basic science"), which also contain three articles by Carl Ramsauer. The only contribution in the entire volume essentially aimed to inform people about current branches of physical research originated from Wilhelm Westphal and was on problems of physics. But even there the journal's slant comes through at the beginning and end: "So we see the one otherworldly physicist busy with unsolved problems of pure scientific knowledge, while the other has both legs firmly planted in the reality of life and the struggle for survival.... May our accounts also serve to show to aspiring young physicists that, besides the now urgent tasks of practical applied physics, the most fascinating tasks of the future still also exist for a physicist focused only on scientific knowledge."[184] The accounts mentioned had been borrowed from other publications or were the topics of talks.

Right in the first issue an article appeared on "the position of science and research in the Third Reich"; it was based on a compilation of statements made by Reich Minister Goebbels.[185] Parts of this article read as if they were a direct response to the complaints Carl Ramsauer had previously lodged:[186] "The nation is still completely unaware of the military

[181] Jonathan Zenneck, "Die Bedeutung der Forschung," *Physikalische Blätter*, 1 (1944), 6–12; Jonathan Zenneck, "Forschung tut Not auch im Kriege," *Physikalische Blätter*, 1 (1944), 110–112. (See the facsimile of the title page in Hoffmann and Walker, *Physiker*, 519.)

[182] Zenneck, "Forschung," 110.

[183] Wolfgang Finkelnburg, "Wesen und Bedeutung der Physik," *Physikalische Blätter*, 1 (1944), 29.

[184] Wilhelm Westphal, "Probleme der Physik," *Physikalische Blätter*, 1 (1944), 33.

[185] "Wissenschaft und Forschung im Dritten Reich. Nach Ausführungen von Reichsminister Dr. Goebbels," *Physikalische Blätter*, 1 (1944), 20–23.

[186] See the contribution by Hoffmann to this volume.

successes it owes to the inventions and research efforts by our physicists, chemists..."[187] It continues: "Germany owes its world renown even more to its artists, scholars, scientists, researchers and inventors than to its statesmen, soldiers and economists"[188] and warns: "A nation without any respect for intellectual labor would sooner or later no longer offer to her intellectual workers any opportunities for their development and exercise. But it is a prerequisite for lasting national success."[189] A letter by the Reich Minister of Armament and War Production then gives the reader an idea of how he evaluated basic research: "I have been easily able to arrange in many cases that specifically named members of the Armed Forces be released for scientific research projects," adding: "It is extremely important to me that basic research be at work with full intensity and without suffering any frictional loss."[190]

Rudolf Mentzel, REM official and president of the DFG, identified in the editor's introduction as head of the Reich Research Council's managerial board, also had something to say about German wartime research.[191] His conclusion was: "Total war teaches us that we must constantly... continue to develop and improve our assault and defense weaponry.... At the same time, however, there is the realization that this scientific mobilization should not lead to a neglect of basic research and that there, too, work should be conducted along the broadest front, even such work as at first glance is not even necessarily directly connected with the demands of the war."[192]

Representatives of the Hitler Youth were contacted about how the next generation of professionals was to be encouraged. The result was the extensive article about youth and technology by a local official, Heinrich Hartmann.[193] This is the only place in the 12 issues that uses language reminiscent of the Aryan physics movement. But the author makes a concession that never would have occurred to its defenders. Under the subheading "A glance at the enemy" one reads: "Jewified, degenerate and rotten though the political and economic leadership of the Anglo-American coalition may be, it must not be overlooked that despite all

[187] *Physikalische Blätter*, 1 (1944), 23.
[188] *Physikalische Blätter*, 1 (1944), 23.
[189] *Physikalische Blätter*, 1 (1944), 23.
[190] "Einschätzung der Grundlagenforschung. Aus einem Schreiben von Reichsminister Speer," *Physikalische Blätter*, 1 (1944), 23.
[191] Rudolf Mentzel, "Deutsche Forschung im Kriege," *Physikalische Blätter*, 1 (1944), 103–106.
[192] Mentzel, "Forschung," 106.
[193] Hauptbannführer Heinrich Hartmann, "Jugend und Technik," *Physikalische Blätter*, 1 (1944), 75–81.

degenerative effects and mixing tendencies, its technical accomplishments in particular have been carried and still are being pushed through by the same Faustian urge as here at home.... We are being confronted by a formidable enemy, obligating us to deploy our utmost not just in the clash of arms but particularly in the clash of science, technology and inventions."[194] Here, one can detect something of the "blessed" influence of Aryan physics. For a long time, its attacks on modern physics prevented such a breakthrough to the unrestricted recognition and integration of modern science into the National Socialist state and war effort.

Issue no. 9, with the guiding theme "Physics all over the world," offered two informative articles about how science was being assessed in England[195] and America,[196] respectively. In his foreword to the second case, the editor pointed out: "This importance is a strong emphasis on research...an indication of the phase-shift between research and the progress achieved by it; something that – more or less clearly put – can give us pointers."[197] Even science and technology in the Soviet Republics received acknowledgment in this issue with an article reprinted from *Nature.*[198] A primitive-looking Russian galvanometer was positively assessed as a model of simplicity and practicality.[199]

In accordance with the guiding themes, much space in the journal was devoted to the problem of professional recruitment. Teachers were the main authors of its articles and letters to the editor. Brüche found it important to involve the Hitler Youth in this discussion, as "the second major educative factor."[200] This mission was evidently more important to him than just as editor of this journal. The outcome of these efforts was a campaign called "Youth and Technology." The Reich Youth Leader, Arthur Axmann, issued an appeal about this campaign that was published in the journal in July 1944,[201] along with the above-mentioned

194 Hartmann, "Jugend," 77.

195 William Bragg, "Englische Wissenschaft im Kriegsdienst," *Physikalische Blätter*, 1 (1944), 127–128.

196 Bruce Bliven, "Wie die amerikanische Wissenschaft unsere Welt sieht," *Physikalische Blätter*, 1 (1944), 129–133.

197 Bliven, "amerikanische," 129.

198 "Wissenschaft und Technik in den Sowjet-Republiken," *Physikalische Blätter*, 1 (1944), 135–138.

199 Walter Zapp, "Ein russisches Galvanometer," *Physikalische Blätter*, 1 (1944), 138–140.

200 Ernst Brüche, "Nachwuchsfragen," *Physikalische Blätter*, 1 (1944), 74.

201 "Zur Aktion 'Jugend und Technik.' Aufruf des Reichsjugendführers Artur Axmann vom Juli 1944," *Physikalische Blätter*, 1 (1944), 68–69.

commentary by Hartmann, a close co-worker of Axmann. The editor remarked in the October issue, in view of the less than satisfactory interest in the campaign by the press: "We note with gratification, however, that many articles in the Physikalische Blätter have been adopted in the H[itler] Y[outh]'s official guide for the campaign 'Youth and Technology.'"[202]

A problem discussed on more than one occasion was that of "Air Force helpers." Schoolboys from the higher grades were being deployed as helpers in anti-aircraft batteries; at the same time, they continued to attend some classes at their military positions. It was argued that they also be given the opportunity to perform experiments in their physics classes out there, despite the substantial obstacles involved. Ernst Brüche chose as his crown witness about the recruitment problem the lieutenant-colonel and headmaster Prof. Karl Hahn. Three articles were published on this topic (one of them a historical essay) by this "special commissioner for the Air Force helper deployment in one air district." His second contribution on aspiring scientists during the fifth year of the war[203] describes the difficulties posed by "the preponderance that the demands of the front rightfully have,"[204] by "the overweighting of philosophical subjects,"[205] and by the adverse external conditions for recognizing physics and conveying enthusiasm about the subject. But it was "no stab-in-the-back approach for us to secure, now already, what we need during and after the war: a favorable view of the importance and esteem for a scientist's activities and an influx of the really diligent and competent."[206] Ernst Brüche's foreword lends the author support: " Let us heed his warning of not mistaking technical school courses for the war youth with what Germany is in desperate need of: raising young people with intellectual, idealistic leanings in the field of science and therefore true researchers from whom alone true progress can come."[207] Elsewhere, Ernst Brüche argued for a "preferential right" for physics: "Owing to the importance of physics as a fundamental science for technology, physics must have first priority in the selection of suitable workers."[208] Further down

[202] "Die Physikalischen Blätter im Urteil der Leser," *Physikalische Blätter*, 1 (1944), 155 (footnote).

[203] Karl Hahn, "Wissenschaftlicher Nachwuchs im fünften Kriegsjahr," *Physikalische Blätter*, 1 (1944), 82–84.

[204] Hahn, "Nachwuchs," 82.

[205] Hahn, "Nachwuchs," 83.

[206] Hahn, "Nachwuchs," 82.

[207] Brüche's foreword to Hahn, "Nachwuchs."

[208] Ernst Brüche, "Aus der Arbeit der Informationsstelle," *Physikalische Blätter*, 1 (1944), 163.

he specified: "Physics only needs relatively few workers from out of the great reservoir of young labor, but it needs the really suitable people."[209]

The argument that such great utility can be gained from just a few scientists is used many times. One letter by one of the journal's readers, a secondary school teacher,[210] was even published anonymously "in order to allow the author to express himself frankly."[211] The teacher supported having carefully selected students of physics receive dispensation from military service. His letter culminates in the italicized declaration: "One shouldn't retort that every man is needed on the front-line! We shall win the war even without these couple of hundred men, but without them, we shall perhaps lose the peace."[212] It would have been fatal to question the first part of this statement publicly; so why not take the second part of it seriously, too?

A government building officer and certified engineer, Dr. Vogt, complained in a letter[213] about how little graduate pupils knew about science, blaming it with surprising directness on the distractions of other activities. He did so by means of a quote from the president of the Reich Chamber of Music. In his view, one only had to replace the words music, piano, and violin with science, mathematics, and physics:

> Since, however, the existence or nonexistence of our nation as a musical nation of the highest caliber is in fact at stake, no opportunity should be missed in recalling that this educational issue is more important than many other things that are happening for young people. It goes without saying that one does not have to leave it to the children to decide whether they would prefer to practice piano or violin or whether they would like to be a part of processions or other activities by the organization.[214]

There is embarrassing ideological conformity in the language Ernst Brüche used. In an introductory editorial to an essay on activating the potential of future academics, he demanded: "those suited to physical research must be sought out from among the hereditary pool of new arrivals in the labor force and guided early on into good training in

[209] Brüche, "Informationsstelle," 166.
[210] Studienrat Dr. N.N., "Ununterbrochener Studienurlaub," *Physikalische Blätter*, 1 (1944), 167–168.
[211] Dr. N.N., "Studienurlaub," 167 (footnote).
[212] Dr. N.N., "Studienurlaub," 168.
[213] Staatlicher Baurat Dr.-Ing. Voigt, "Auswahl der Geeigneten," *Physikalische Blätter*, 1 (1944), 166.
[214] Voigt, "Auswahl," 166.

physics."[215] Brüche's contribution on fundamental research in wartime instructs us[216]: "It is the same racial disposition that found expression in the invention of gunpowder, the discovery of roentgen rays or the development of quantum theory by our great German physicist Max Planck."[217] This fundamental research "corresponds to the specific characteristics of our race."[218] There are similar passages in Ramsauer's article on the key position of physics.[219] Ernst Brüche even cited *Führer* sayings[220] and quoted extensively from *Mein Kampf*.[221]

But one does have to grant to the editor how skillfully he prosecutes his case. Older articles were also accepted in the periodical. For example, there is an excerpt from an address by Max Planck from 1922 in which the successes of technology in Germany are attributed "very substantially to the circumstance that a science could develop here in isolation from economic interests."[222]

The editor Brüche's personal opinion shows through, for instance, in an opaque remark safely tucked away in his introductory editorial to an excerpt from a book by Bruce Bliven about how American science sees our world. Brüche justified an omission there with the words:

> albeit political considerations, such as, e.g., about whether American scientists prefer democracy or other state forms, such as liberalism, had an effect on scientific progress and whether it was right to subject scientists to racial laws, have been omitted, since we know the opinions represented in America on these points thoroughly enough.[223]

The historian of physics Hans Schimank wrote, on one hand, about the influence of physics on the military in changing times,[224] and on the other hand – entirely even-handedly – about Franco-German relations in

[215] Brüche's foreword to "Aktivierung des Akademikernachwuchspotentials," *Physikalische Blätter*, 1 (1944), 57.

[216] Ernst Brüche, "Grundlagenforschung im Kriege," *Physikalische Blätter*, 1 (1944), 112–115.

[217] Brüche, "Grundlagenforschung," 113.

[218] Brüche, "Grundlagenforschung," 115.

[219] Carl Ramsauer, "Die Schlüsselstellung der Physik," *Physikalische Blätter*, 1 (1944), 33.

[220] *Physikalische Blätter*, 1 (1944), 109.

[221] *Physikalische Blätter*, 1 (1944), 125, 162.

[222] Max Planck, "Aus der Eröffnungsansprache des Deutschen Naturforscher- und Ärztetages 1922," *Physikalische Blätter*, 1 (1944), 25.

[223] Ernst Brüche's introductory editorial to Bliven, "amerikanische," 129.

[224] Hans Schimank, "Der Einfluß der Physik auf das Wehrwesen im Wandel der Zeiten," *Physikalische Blätter*, 1 (1944), 119–124.

science.[225] The last issue lists factual information about schooling and school reforms throughout the past 100 years. Some of it is reminiscent of problems we are still familiar with today. A few anecdotes about physicists and the portraits already alluded to above, partly including biographical notes, served to lighten up the content. The personal anecdotes about Faraday, Millikan, and Bohr in issue no. 9 fit within the issue's guiding theme: "Physics all over the world."[226]

Thus, despite the obvious bias, a volume was created that did not impress obtrusive propaganda upon the reader in every line of its text. There is an astounding notice in the August issue of 1944. Some news originating from London is quoted from a Swedish paper under the heading "Yet another utopia": "In the United States scientific tests of a new bomb are being carried out. Uranium serves as the fuel and if the forces bound within this element were released, explosive effects of hitherto unimaginable potency could be generated."[227]

RESEARCH DURING THE WAR YEARS

Surveying the traces of scientific research in journals and conference proceedings necessarily excludes projects classified as secret. Entirely contrary to the astonishing openness about nuclear research in physics periodicals, we learn nothing about radar research, for example. There are clear references to secret work under way at the beginning of the war. In his speech opening the conference in 1940, Jonathan Zenneck mentioned that because of the war "some things had to be concealed that in other times would have been discussable and would perhaps be among the specially interesting subjects."[228]

But that did not mean an end to all "open" research, as the examples from this period in the foregoing sections have demonstrated. We gain a clear picture of the situation from the FIAT Review of German Science, a series covering scientific research and medicine conducted in Germany between 1939 and 1946.[229] Acknowledged specialists reported

[225] Hans Schimank, "Deutsch-französische Beziehungen in der Naturwissenschaft," *Physikalische Blätter*, 1 (1944), 141–143.

[226] "Physiker-Anekdoten," *Physikalische Blätter*, 1 (1944), 148.

[227] "Noch eine Utopie," *Physikalische Blätter*, 1 (1944), 118.

[228] *Verhandlungen der Deutschen Physikalischen Gesellschaft*, 21 (1940), 31.

[229] *Naturforschung und Medizin in Deutschland 1939–1946*, 84 Volumes (Wiesbaden: Dieterich'sche Verlagsbuchhandlung, 1947 ff.); published by the Field Information Agency. This is the edition intended for Germany of the *FIAT Review of German Science*.

about the wartime research conducted and published in their fields of expertise. Fifteen volumes concern physics, including astrophysics and biophysics. In a foreword to volumes 8/9 (*The Physics of Solids*, edited by Georg Joos), Arnold Sommerfeld wrote: "The abundance of experimental and theoretical research conducted during the war, even such as is not connected with war projects, will amaze the reader." At the end of this double volume, 33 "important books appearing in the reported period" are listed as pertinent to the volume's subject matter, more than half of which are assignable to physics proper. The scope and form of the reported research is similar in the other volumes. In the introduction to volume 12 (*Physics of the Electron Shells*), Max von Laue pointed out that "a large number of precise new methods [were applied] with the best of success." For the current investigation of the research actually performed, we may disregard whether or not these reports are incomplete owing to deliberate omissions.[230]

It must have remained possible to find maneuvering room for fundamental research and publish such results, even in cases where the research was officially geared toward technical military goals. In the chapter on electricity conduction and the photoelectric effect in semiconductors of volume 9 in the above series, the author, Wilhelm A. Meyeren, stated in his general overview: "The investigation of infrared-sensitive radiation receivers was particularly fostered during the last war. Only very little about these investigations has been published. The same is valid of the rapid development of barrier layer rectifiers." For his survey of this entire branch, at that time at the cutting edge, the author was nevertheless able to refer to many general publications on this subject from the war years, for example, by the team led by Walter Schottky and Eberhard Spenke in the Siemens electrical firm.

Robert Rompe from the Osram firm (a company specifically devoted to the study of electrical lighting) wrote to Hans Kopfermann on 6 May 1940, who was planning a conference that summer in Hannover for the regional association of Lower Saxony: "As concerns the talk topics, we would have one or two problems that could be presented. However, precisely the interesting things are of a nature that one would not be allowed to speak about them, and this probably is nowadays everywhere the case."[231] It sounds as if secret research was the main thing

[230] Mark Walker examined this issue in his account of the German "uranium project." See Mark Walker, *German National Socialism and the Quest for Nuclear Power 1939–1949* (Cambridge: Cambridge University Press, 1989), 210.

[231] Rompe to Kopfermann, DPGA, no. 40023.

going on there. In fact, there were investigations under way at Osram on high-pressure discharges, for instance, in which Walter Weizel from the University of Bonn was participating. One technical outcome was the high-pressure mercury lamp so important in many physical experiments as a light source of extremely high luminance. There was a report about its latest design in the *Zeitschrift für Physik* in 1944, including a reference to a talk about the same topic by Weizel at the Berlin convention of 1942.[232] In February 1942, Rompe and Friedrich Möglich took part in a conference by the regional association of Lower Saxony in Hamburg. Their topics were purely theoretical and biophysical.[233] In volume 10 of the above series (*Physics of Liquids and Gases*), Wolfgang Finkelnburg also described research being conducted by the Rompe–Weizel team in his essay on electric discharges in gases.[234] He listed a number of publications in the period 1939–1944, only one of which, cited as "German aviation research no. 1933 (1944)," was evidently a secret communication.

The apparent contradiction to Rompe's statement of spring 1940 may have had to do with the fact that secrecy regulations were loosened in the later war years, because they were deemed an obstacle to coordination and cooperation in research and development.[235] Scientists perhaps overestimated their ranking in secret projects, out of sheer self-importance, if not simply to stave off being called into active military service.

A similar case is known from Göttingen. In response to the inquiry mentioned above by Hans Kopfermann about a regional conference in 1940, Georg Joos wrote: "After asking around, there seems to be no possibility to hold any talks from Göttingen, since all the institutes are solely engaged in secret research."[236] This also seems to be an exaggeration. It certainly is not valid for the entire war period. Even the conspicuous figure Karl-Heinz Hellwege, a lecturer in the 2nd Physical Institute

[232] Robert Rompe, Wolfgang Thouret, and Walter Weizel, "Zur Frage der Stabilisierung frei brennender Lichtbögen," *Zeitschrift für Physik*, 122 (1944), 1. The same volume contains three other papers listing Robert Rompe and Walter Weizel as author or coauthor.

[233] Friedrich Möglich and Robert Rompe, "Strahlungseigenschaften dichtgelagerter, gleichartiger Atome," *Verhandlungen der Deutschen Physikalischen Gesellschaft*, 23 (1942), 46ff.; also published (with a slightly altered title) in *Zeitschrift für Physik*, 120 (1943), 741.

[234] On 239ff. of the German edition.

[235] Ruth Federspiel, "Mobilisierung der Rüstungsforschung? Werner Osenberg und das Planungsamt im Reichsforschungsrat 1943–1945," in Helmut Maier (ed.), *Rüstungsforschung im Nationalsozialismus. Organization, Mobilisierung und Entgrenzung der Technikwissenschaften* (Göttingen: Wallstein, 2002), 72–105, especially 82.

[236] Joos to Kopfermann, DPGA, no. 40023.

and professed Nazi, published a number of papers between 1941 and 1944 from his specialty, solid-state spectroscopy of the rare earths,[237] this despite having managed as acting director (since Georg Joos's departure in 1941) to get the institute recognized as a "special operation of the armaments industry."[238]

The situation at Göttingen changed entirely when Hans Kopfermann was appointed institute director in 1942. After the war he wrote in a letter of 11 December 1945 to the Allied Control officer: "Research work on the subjects mentioned has been continued up to the occupation of Göttingen by Allied troops.... It may be added that all research work the continuation of which is proposed has in the past being [sic] considered as fundamental research exclusively and was not underlying any restrictions concerning publishing of any details of apparatus and results."[239] This is remarkable when you consider that Hans Kopfermann's team was also linked to the uranium project, performing spectroscopic research on mass-spectroscopic isotope separation.[240] But it only achieved isotopic separations of small amounts[241] and was therefore just useful for atomic measurements in physics. A publication on the topic appeared as late as 1944 by Kopfermann and Wilhelm Walcher in the *Zeitschrift für Physik*.[242] The same volume contains two other related papers by Walcher.[243] Kopfermann was the author of an internationally acclaimed monograph about nuclear moments that appeared in 1940; this work was surely a model example of generous, unbiased

[237] Published in *Zeitschrift für Physik*, 117 (1941), 198 and 596; *Zeitschrift für Physik*, 119 (1942), 325; *Zeitschrift für Physik*, 121 (1943), 588 as well as (together with Georg Joos) in *Annalen der Physik*, 39 (1941), 25.

[238] Klaus Hentschel and Gerhard Rammer, "Nachkriegsphysik an der Leine: eine Göttinger Vogelperspektive," in Dieter Hoffmann (ed.), *Physik im Nachkriegsdeutschland* (Frankfurt am Main: Deutsch, 2003), 27–56, esp. 31. About Karl-Heinz Hellwege, see also Gerhard Rammer, "Göttinger Physiker nach 1945. Über die Wirkung kollegialer Netze," *Göttinger Jahrbuch*, 51 (2003), 83–104, esp. 91.

[239] Hentschel and Rammer, "Nachkriegsphysik," 36 (original English).

[240] Walker, *Quest*, 53 and 273 (report G-196).

[241] Wilhelm Walcher, "Isotopentrennung in kleinen Mengen," *Naturforschung und Medizin in Deutschland 1939–1946*, vol. 14: *Kernphysik und Kosmische Strahlung*, 94ff.

[242] Hans Kopfermann and Wilhelm Walcher, "Trennung der Thalliumisotope, II. Optische Untersuchung verschiedener Thalliumgemische," *Zeitschrift für Physik*, 122 (1944), 465.

[243] Wilhelm Walcher, "Über eine Ionenquelle für massenspektroskopische Isotopentrennung," *Zeitschrift für Physik*, 122 (1944), 62; and Wilhelm Walcher, "Trennung der Thalliumisotope, I. Massenspektroskopische Trennung," *Zeitschrift für Physik*, 122 (1944), 401.

acknowledgment of the contributions made by physicists in exile.[244] Even Samuel Goudsmit, the critical head of the *Alsos* mission, had "a very high opinion of Kopfermann,"[245] and the émigré Victor Weisskopf later wrote a respectful obituary about him for the journal *Nuclear Physics*.[246] Hans Kopfermann had participated in the Munich "synod" in November 1940 and was otherwise active in the DPG, serving during the war years as chairman of the regional association for Lower Saxony. His exemption from active military duty, with its welcome financial support, evidently served as legitimization for continued work on other topics "on the side" – perhaps in fact with even greater enthusiasm.

CONCLUDING REMARKS

Studies about the conduct of physicists during the Nazi era remain incomplete and can easily draw premature conclusions if they concentrate on individual spectacular topics and a few prominent persons. A full picture would require including wider circles and "lower" levels within the community. This chapter is one attempt to contribute toward this goal.

Without a doubt, like all totalitarian regimes, the National Socialists' goal was to impose on all levels and in every corner a normative behavior dictated from above. Reality did not meet this goal. Quite evidently, not all research performed by physicists was directed toward specific applications. The struggle for international acknowledgment in basic research continued, which is scarcely understandable without a strong traditional interest in the pursuit of pure knowledge. In the final years of the war, this interest could even enjoy legitimation in the campaign led by the *Physikalische Blätter*. Where else – besides the other research mentioned – could one fit in experiments serving the exclusive purpose of clarifying general basic principles? To name a few examples: verification of the quadratic Doppler effect in the optical range[247]; an experiment on coherence of moving beams using a rotating mirror, following a suggestion made by Einstein[248]; a repetition of the famous Bothe–Geiger

244 Hans Kopfermann, *Kernmomente* (Leipzig: Akademie Verlag Gesellschaft, 1940).

245 Goudsmit to Weisskopf, 11 February 1948, quoted after Rammer, "Göttinger," 91.

246 Victor Weisskopf, *Nuclear Physics*, 52 (1964), 177–183.

247 Gerhard Otting, "Der quadratische Doppler-Effekt," *Physikalische Zeitschrift*, 40 (1939), 681 (also planned as a talk for the annual convention of 1939).

248 Heinz Billing, "Ein Interferenzversuch mit dem Lichte eines Kanalstrahls (Dissertation)," *Annalen der Physik*, 32 (1938), 577. The paper mentions that the experiment involved one "that had already been suggested by Einstein in 1926." This thesis, like

experiments on energy conservation with Compton scattering for γ-rays[249]; and verification of the Fresnel diffraction of electron beams along a macroscopic edge.[250]

Applied research was also always under way, particularly in industrial laboratories, which became increasingly but not exclusively oriented toward military applications. One must bear in mind that the genuine role of a physicist even in such practical research always involves grappling with fundamental science, so there is some basic research "fallout" in this case, too, although not according to any systematic grid defined by the disciplinary specialty. Yet the thesis reached by Moritz Epple in his paper on the wartime "technoscience" conducted at the Kaiser Wilhelm Institute of Fluid Dynamics between 1937 and 1945 is not universally valid; that is, that "research retrospectively described as fundamental research" in fact served as "preparation for newer and better . . . techniques that were used elsewhere in projects of technological development."[251]

In industry, a firm's business interests took precedence over the scientific interests of its staff. A firm may well have soberly calculated the post-war period into its strategy as well. The Siemens Works, for instance, started developing the betatron right in the middle of the war. It was conceived as a radiation source for medical applications and soon did reap good profits after the war ended. In 1941, an American company sought to obtain a license for the patent,[252] and in 1943 there was even a public priority dispute about it with an American author.[253] The freedom with which Gustav Hertz was able to operate in a laboratory set up especially for him at Siemens is also astonishing.

The very fact that there was a separate society for technical physics besides the DPG, with its own journal, is indicative of the value system conventional among renowned physicists in particular: For them, technical physics was "just" technical physics. They were helped along by the lack of organization in science. The National Socialist *Führer* state was

the preceding one by Gerhard Otting, had been performed under the supervision of Walther Gerlach.

249 Walther Bothe and Heinz Maier-Leibnitz, *Zeitschrift für Physik*, 102 (1936), 143.

250 Hans Boersch, "Fresnelsche Elektronenbeugung," *Die Naturwissenschaften*, 28 (1940), 709 (from the AEG's research institute).

251 Moritz Epple, "Rechnen, Messen, Führen, Kriegsforschung am Kaiser-Wilhelm-Institut für Strömungsforschung 1937–1945," in Maier, *Rüstungsforschung*, 320.

252 Max Steenbeck, *Impulse und Wirkungen* (Berlin: Verlag der Nation, 1978), 118.

253 Max Steenbeck, "Beschleunigung von Elektronen durch elektrische Wirbelfelder," *Die Naturwissenschaften*, 31 (1943), 234. This paper includes a footnote by Carl Ramsauer as "chairman of the German Physical Society."

not a monolithic block. Setting many actors at the same level in accordance with the *Führertum* standard motivated turf battles to extend one's own influence at another's expense.[254] In this wrangling, it was possible to safeguard a few niches for fundamental research. Not even in such a sensitive field as aviation research were such attempts at tightening the reins successful. According to Helmuth Trischler, "the history of aviation research in National Socialism . . . [can be] described as a series of futile attempts at meeting this challenge" of coordinating the interplay between science, the state, and the economy. "In the chaos of the final war years, the chances for individual scientists to operate their own projects as 'small science' were bigger than ever."[255] Kai Handel notes about radar research that "the all-in-all parsimonious research projects in this field . . . were run completely without any coordination, even after the war had begun, in the research institutes of the Reich Aviation Ministry, universities, academy and industry."[256] It was only in 1942–1943 that things changed. Notker Hammerstein argues that the DFG could achieve "astonishing results, thanks to generous research funding" and gives as one reason that the director Rudolf Mentzel had to work hard "in order to be able hold his own before the professional public," just to keep his own position.[257] I refer here to the examples quoted earlier from astrophysics and musical acoustics.

Physicists received unintentional support from the Aryan physics movement. A small group centered around Philipp Lenard and Johannes Stark attempted to counter the challenge of modern physics with their own Aryan physics. Jewish influence, they supposed, had degenerated modern physics, which was based on quantum mechanics and relativity theory. The "success" of this movement so intent on serving the Nazi state was paradoxical. It shifted the debate about the role of physics onto an unproductive ideological plane, thus distracting attention away from its potential technical utility and preventing prominent physicists from

[254] For example, see Michael Grüttner, "Wissenschaftspolitik im Nationalsozialismus," in Doris Kaufmann (ed.), *Geschichte der Kaiser-Wilhelm-Gesellschaft im Nationalsozialismus. Bestandsaufnahme und Perspektiven der Forschung*, Volume 1 (Göttingen: Wallstein, 2000), 557.

[255] Helmuth Trischler, "'Big Science' or 'Small Science'? Die Luftfahrtforschung im Nationalsozialismus," in Kaufmann, *Geschichte*, Volume 1, 361.

[256] Kai Handel, "Die Arbeitsgemeinschaft Rotterdam und die Entwicklung von Halbleiterdetektoren. Hochfrequenzforschung in der militärischen Krise 1943–1945," in Maier, *Rüstungsforschung*, 255.

[257] Notker Hammerstein, "Die Geschichte der Deutschen Forschungsgemeinschaft," in Kaufmann, *Geschichte*, Volume 1, 608.

obtaining responsible political positions earlier. Imagine if the inaugural ceremony at the Philipp Lenard Institute in Heidelberg in December 1935 had culminated in a clarion call to mobilize physics for the technical military projects of the new state – which one did feel so very attached to! Instead, passions were squandered on crusades against certain "un-Aryan" developments in physics. One philosopher contended that statistical description in microphysics was "the ontological victory of the democratic majority vote."[258] We also learn there that the focus of the attention, Lenard, "might perhaps not be able to count as a friend of industry."[259]

Lenard's primary concern undoubtedly was the pursuit of pure scientific knowledge. In this sense, he certainly was a fundamental scientist, just one with his own, misguided conceptions. The best way to refute them was to continue to do science on the basis of modern theories and carry it through to success in a variety of fields. This is what in fact happened, to a considerable degree – thanks not least to the inadequate organization of science.

Apparent insignificance did afford physicists a certain amount of protection during the first few years of the regime. Physics stood in the shadow of chemistry. Chemistry was (and still is) closely associated with the chemical industry, which traditionally fostered close ties with the research institutions – and the man on the street can at least roughly understand its accomplishments. When the Kaiser Wilhelm Society was founded, the institutes for chemistry and physical chemistry were immediately set up. Physics had to wait another 25 years for a comparable institute to be founded. A chemist, Carl Krauch, from the I.G. Farben dye concern, was appointed director of a Department for Research and Development included in the Four-Year Plan (the department was later transformed into a national agency for economic development, the *Reichsamt für Wirtschaftsausbau*). It would scarcely have been conceivable that a physicist be chosen for this office. The secret situation reports by the Security Service of the SS from 1940 quote "university-teacher circles" saying that "the vanishingly small number of physics students against the number of chemists... is also attributable to the circumstance that all

[258] August Becker (ed.), *Naturforschung im Aufbruch, Reden und Vorträge zur Einweihungsfeier des Philipp-Lenard-Institutes der Universität Heidelberg am 13. und 14. Dezember 1935* (Munich: Lehmann, 1936). Quote from the speech by Prof. Wolfgang Schultz, "Deutsche Physik und nordisches Ermessen."

[259] August Becker, *Naturforschung*. The remark about Lenard is in the contribution by the industrial physicist Hans Rukop, 69.

public addresses and publications in the press mostly only speak about a need for chemists."[260]

In the early 1930s, the division between basic research and applications in physics was still deep. By ceding some branches to technology, "pure" physics seemed thus to have done its duty. It never occurred to physicists to call a factory that was producing X-ray or radar instruments part of any "physical industry." Goudsmit suggested that, up to a certain point, the broad basic training of German engineers made involvement by scientists in such applications superfluous.[261] Eckert has demonstrated with examples taken from nuclear energy and radar technology that the role of physicists is generally overrated against that of engineers.[262]

After such a rich harvest of technical applications from classical physics, one could not imagine that modern physics with its – for many, notorious – fundamentals could come up with any more results of great practical import. In 1933, quantum mechanics was just 7 years old. Atomic and molecular physics seemed to be purely of theoretical interest. Solid-state physics was making a slow start. A spectacular topic like nuclear fission was well over the horizon. When nuclear fission was discovered at the end of 1938 and possible applications became the subject of discussion, even the uranium project remained an enterprise occupying a mere handful of physicists,[263] granting many of them the opportunity to work even on problems of more general interest as well and to publish their results besides.

There was no uniform attitude among German physicists relentlessly steering all research toward technical purposes for the military, true though it is that armaments research was going on in Germany – no different than in the other warring nations. This should not be called resistance; but a certain resilience was necessary, in the sense of a buffer, to be able to pursue research on a substantial scale on current problems in science according to conventional standards, whereby – as suggested – external circumstances were conducive. Only thus is the relatively rapid revival of scientific research after the war comprehensible as well as the interest the victor powers on both sides took in physicists from Germany.

260 Boberach, *Meldungen*, annual situation report for 1940, 1051.

261 Samuel Goudsmit, "War Physics in Germany," *The Review of Scientific Instruments*, 17 (1946), 49–52, here 49.

262 Michael Eckert, "Theoretische Physiker in Kriegsprojekten. Zur Problematik einer internationalen vergleichenden Analyse," in Kaufmann, *Geschichte*, Volume 1, 296.

263 See the list in Walker, *Quest*, 53.

We cannot be completely content with this. Even in the pursuit of "pure" science, Germany's physicists served the Nazi system in that they helped conceal the actions and goals of the regime and erect a façade of normality. Such "cultural services" were not the work of physics and physicists alone, of course, and the question remains: What would have been the alternative?

8

The German Mathematical Association during the Third Reich

Professional Policy within the Web of National Socialist Ideology

Volker R. Remmert

At the onset of the Third Reich, mathematicians in Germany had recourse to three organizations devoted to the protection of their professional and particular interests. The eldest, founded in 1890, was the German Mathematical Association (*Deutsche Mathematiker-Vereinigung*, DMV), which primarily represented university mathematicians. The Reich Mathematical Federation (*Mathematische Reichsverband*, MR), founded in 1921, considered itself the mouthpiece of schoolteachers and was closely connected with the DMV and the German Association for the Promotion of Mathematical and Scientific Instruction (*Deutscher Verein zur Förderung des mathematischen und naturwissenschaftlichen Unterrichts*). The Society for Applied Mathematics and Mechanics (*Gesellschaft für angewandte Mathematik und Mechanik*, GAMM) had emerged out of the Association of German Engineers (*Verein Deutscher Ingenieure*) in 1922 as a forum for application-oriented mathematicians who had felt insufficiently represented by the DMV.[1]

[1] On the MR and GAMM, see above all: Herbert Mehrtens, "Angewandte Mathematik und Anwendungen der Mathematik im nationalsozialistischen Deutschland," *Geschichte und Gesellschaft*, 12 (1986), 317–347; Herbert Mehrtens, "Die 'Gleichschaltung' der mathematischen Gesellschaften im nationalsozialistischen Deutschland," *Jahrbuch Überblicke Mathematik*, 18 (1985), 83–103. About the GAMM, see Helmuth Gericke, *50 Jahre GAMM*, supplementary issue of *Ingenieurarchiv*, 41 (1972). On the MR, also see the commentary in Sanford L. Segal, *Mathematicians under the Nazis* (Princeton: Princeton University Press, 2003). Concerning the history of the DMV, see Helmuth Gericke, *Aus der Chronik der Deutschen Mathematiker-Vereinigung* (Stuttgart: Teubner, 1980), 28 (expanded revision of the article in *Jahresbericht der Deutschen Mathematiker-Vereinigung*, 68 [1966], 46–74); Martin Kneser and Norbert Schappacher, "Fachverband – Institut – Staat," in Gerd Fischer, Friedrich Hirzebruch, Winfried Scharlau, and

The real influence of these three organizations on state agencies was minimal. At the end of 1933, the DMV's chairman, Oskar Perron, complained about the impotence of his representative position for the profession with respect to science policy making. He explained: "very little is known about the DMV as an entirely private association of friends of mathematical science, and one does not need to know much, either."[2] This changed for the DMV by the end of the Second World War, but neither in the Weimar period nor during the first few years of the Third Reich did the DMV, GAMM, or MR make any major gains in professional or education policy.

The response by the MR to the seizure of power by the National Socialists was very different from that of the GAMM. Whereas the MR immediately declared its support for the "leadership principle" (*Führerprinzip*) in 1933, designating its former chairman Georg Hamel as its new leader, the GAMM reacted with extreme caution. Considerable tensions arose in the DMV (see the next section). The MR increasingly lost importance during the war years, and the DMV eventually took on its activities, in accordance with the declared goal of its chairman, Wilhelm Süss. In spring 1941 he wrote, "according to Mr. Hamel, [the MR regarded itself] rather as a branch institution of the DMV," and the DMV ought to adopt its responsibilities. This was in tune with the DMV's interests because, as Süss put it, the DMV had "the imperialistic goal of procuring for her [the DMV] alone all the rights and obligations of mathematics."[3] This statement would prove to be the DMV's agenda, but its relationship with the GAMM remained unaffected. The GAMM retained its independent position particularly because of the pivotal role of its chairman, Ludwig Prandtl, in aviation research.[4]

Little of novelty can be reported about the MR and the GAMM beyond the fundamental article by Herbert Mehrtens on the "realignment" (*Gleichschaltung*) of the mathematical societies in National Socialist Germany cited earlier, this because the availability of sources for either association

Willi Törnig (eds.), *Ein Jahrhundert Mathematik 1890–1990* (Braunschweig: Vieweg, 1990), 1–82.

2 Perron to Bieberbach, 31 December 1933, Archives of the University of Freiburg (UFA), E4/36.

3 Süss to Feigl, 3 April 1941, UFA, C89/51.

4 See Moritz Epple, "Rechnen, Messen, Führen: Kriegsforschung am Kaiser-Wilhelm-Institut für Strömungsforschung 1937–1945," in Helmut Maier (ed.), *Rüstungsforschung im Nationalsozialismus. Organisation, Mobilisierung und Entgrenzung der Technikwissenschaften* (Göttingen: Wallstein, 2002), 305–356.

is poor.[5] Researchers on the history of the DMV, by contrast, have gained access since the end of the 1990s to dense documentation.[6] Accessibility to sources is not the only justification for a focus on the DMV. The development of a formerly weak professional organization into an effective instrument of professional policy is worth some attention. This development was the fruit of the assiduous labors of Chairman Süss since 1938 and provides an illustration, reaching beyond the history of the discipline, of how professional policy could successfully be pursued within the ideological web of the period.

THE CRISIS YEARS OF THE DMV: 1933–1935

When the National Socialists seized power in Germany, the DMV counted more than 1,100 members inside and outside the country. The chairman's term lasted 1 year. Longer-term business was handled by the board, which then consisted of the secretary (Ludwig Bieberbach, since 1921), the treasurer (Helmut Hasse, since 1932), and the three editors of the association's journal, *Jahresbericht der Deutschen Mathematiker-Vereinigung* (Otto Blumenthal, since 1924; Ludwig Bieberbach, since 1921; and Helmut Hasse, since 1932).

The icy breath of the new era began to leave its traces in the files of the DMV in spring 1933 after the Law for the Restoration of the Professional Civil Service in April 1933 had created the legal conditions for the dismissals of politically undesirable and Jewish civil servants. On behalf of his Jewish colleague Otto Blumenthal from Aachen, who had been suspended from office, Otto Toeplitz inquired of Bieberbach in May whether Blumenthal would be taking "the interests of the DMV better into account by leaving or by staying" as coeditor of the *Jahresbericht*.[7] Bieberbach left no doubt that he thought a departure by Blumenthal was

[5] However, see the references to the GAMM in the Archives of the Max Planck Society (*Archiv der Max-Planck-Gesellschaft*, MPGA), Prandtl papers.

[6] The files in the University Archive of Freiburg, which had been kept at the Mathematisches Forschungsinstitut Oberwolfach until 1996, form the most important basis; namely, the papers of Wilhelm Süss (UFA, C89) and the files of the DMV (UFA, E4). They became accessible to historical research after their inventory in 1997–1998, through the support of the Volkswagen-Stiftung. See Volker R. Remmert, *Findbuch des Bestandes E 4 – Deutsche Mathematiker-Vereinigung (1889–1987)* (Freiburg: Uni-Archiv, 1999); Volker R. Remmert, *Findbuch des Bestandes C 89 – Nachlaß Wilhelm Süss: 1913–1961* (Freiburg: Uni-Archiv, 2000).

[7] Toeplitz to Bieberbach, 25 May 1933, UFA, E4/36.

the better solution and told him so outright as well.[8] Konrad Knopp succeeded Blumenthal as coeditor.

Ludwig Bieberbach, the Berlin mathematician who had given this confidential advice to the Jewish mathematicians Blumenthal and Toeplitz, had his share in shaping the events and policies in the DMV for years to come. Since the 1920s, he had a reputation as an influential and esteemed mathematician and served as secretary of the DMV and coeditor of the *Jahresbericht der Deutschen Mathematiker-Vereinigung*. Bieberbach was no stranger to German nationalism during the Weimar period, but his open advocacy of National Socialism in 1933 still astonished many.

Bieberbach sided with the National Socialists on two fronts. He worked hard – albeit with limited success – to nazify academia and the way it was organized, principally in the field of mathematics. At the same time, he developed a racist doctrine for mathematics under the label "Aryan mathematics" (*Deutsche Mathematik*) that he vehemently defended against all criticism.[9] He first presented this program in April 1934 in a public talk titled "Personality structure and mathematical creativity."[10] Bieberbach identified a "nurturing of the German type in science" as the first "essential duty of National Socialist science."[11] According to him, the "influences of blood and race" determine the choice of problems that a scientist takes up and thus they influence "the reservoir of secure results in science." Bieberbach relied on the research of the

[8] Bieberbach to Toeplitz, 14 June 1933, UFA, E4/36.

[9] On Bieberbach, see Herbert Mehrtens, "Ludwig Bieberbach and 'Deutsche Mathematik,'" in Esther R. Phillips (ed.), *Studies in the History of Mathematics* (Washington, DC: Mathematical Association of America, 1987), 195–241. See also Helmut Lindner, "'Deutsche' und 'gegentypische' Mathematik. Zur Begründung einer "arteigenen" Mathematik im Dritten Reich," in Herbert Mehrtens and Steffen Richter (eds.), *Naturwissenschaft, Technik und NS-Ideologie. Beiträge zur Wissenschaftsgeschichte des Dritten Reiches* (Frankfurt am Main: Suhrkamp, 1980), 88–115.

[10] See the press report: "Neue Mathematik. Ein Vortrag von Prof. Bieberbach," *Deutsche Zukunft* (8 April 1934), 15. The talk by Ludwig Bieberbach, "Persönlichkeitsstruktur und mathematisches Schaffen," was published in various versions in: *Unterrichtsblätter für Mathematik und Naturwissenschaften. Organ des Vereins zur Förderung des mathematischen und naturwissenschaftlichen Unterrichts*, 40 (1934), 236–243; *Forschungen und Fortschritte. Nachrichtenblatt der Deutschen Wissenschaft und Technik*, 10 (1934), 235–237, there significantly under the rubric: philosophy, psychology, and pedagogy. Later it appeared in full length under the title: "Die völkische Verwurzelung von Wissenschaft (Typen mathematischen Schaffens)," *Sitzungsberichte der Heidelberger Akademie der Wissenschaften, Mathematisch-Naturwissenschaftliche Klasse*, no. 5 (1940).

[11] Bieberbach, "Persönlichkeitsstruktur," *Forschungen und Fortschritte*, 235.

Marburg psychologist Erich Rudolf Jaensch and his school to ascertain the real influence of blood and race.

Bieberbach's renown at home and abroad was far-reaching. His political transition into active National Socialism posed serious problems for the DMV because he did not hesitate to use the DMV as a forum for these views.[12] Bieberbach's racial doctrine immediately encountered hefty criticism in a Danish newspaper article in May 1934 by the Danish mathematician and DMV member Harald Bohr.[13] Bieberbach responded with an open letter in which he accused Harald Bohr of being a "parasite to all international collaboration."[14] His open letter to Bohr appeared in the *Jahresbericht der Deutschen Mathematiker-Vereinigung* against the explicit wishes of his coeditors. They legitimately feared that a publication by the DMV's secretary in the association's official organ would raise the mistaken impression that Bieberbach's views represented those of the DMV. This was the beginning of Bieberbach's intense and lengthy quarrel with the DMV.[15]

In September 1934, Bieberbach's petition to introduce the Leadership Principle during the DMV's annual convention in Bad Pyrmont failed along with his bid to have the Göttingen mathematician Erhard Tornier assume this role as "leader" of the DMV. The obvious National Socialist comportment of the association's secretary completely isolated him from the DMV's other decision makers. The convention opted for a modified "leader" (*Führer*) model instead with a biannual chairmanship; the Hamburg mathematician Wilhelm Blaschke was elected to this post. The alterations to the statutes called for by the convention's resolutions were a source of renewed argument between Bieberbach and the DMV's board, composed of Helmut Hasse, Konrad Knopp, and the chairman, Blaschke. Bieberbach presented legal objections to the modifications to the statutes

[12] On the following, see Kneser and Schappacher, "Fachverband," 57–62; Mehrtens, "Gleichschaltung," 85–93; Segal, *Mathematicians*, 263–288.

[13] Harald Bohr, "'Ny Matematik' i Tyskland," *Berlingske Aften* (1 May 1934), 10f., quoted after UFA, E4/71 (translated by Egon Ullrich).

[14] Ludwig Bieberbach, "Die Kunst des Zitierens. Ein offener Brief an Herrn Harald Bohr," *Jahresbericht der Deutschen Mathematiker-Vereinigung*, 2nd div., 44 (1934), 1–3. On the complex context of the affair, see Kneser and Schappacher, "Fachverband," 59–62; Mehrtens, "Bieberbach," 221f.; Volker R. Remmert, "Mathematicians at war. Power struggles in Nazi Germany's mathematical community: Gustav Doetsch and Wilhelm Süss," *Revue d'histoire des mathématiques*, 5 (1999), 7–59, esp. 14–17.

[15] On the following, see Kneser and Schappacher, "Fachverband," 62–69; Remmert, "Mathematicians," 22–24.

and in his capacity as secretary stalled legal implementation of the ratified changes.

Moreover, the convention was unable to settle the Harald Bohr affair. On one hand, it lent support to Bieberbach by repudiating Bohr's public criticism; on the other hand, it merely regretted Bieberbach's open letter.[16] Because of this timid and ambivalent resolution, published with the convention's proceedings in the association's annual periodical, many prominent foreign members of the DMV withdrew their memberships in protest, with Bohr taking the lead.[17] In January and February, Hermann Weyl, John von Neumann, and Richard Courant followed suit, and the feeling that the DMV's reputation had been severely damaged spread.[18]

The crisis at the DMV reached its climax in January 1935 when Bieberbach and Blaschke finally resigned from all their offices in the association. Bieberbach largely lost his direct influence in the DMV as a result. But he remained a constant source of friction for his fellow colleagues and for the DMV all the same, because of his persistent agitation for the National Socialist cause and his close ties with the mathematician Theodor Vahlen, a veteran National Socialist who headed the Science Office at the Reich Ministry for Science, Education and Culture (*Reichsministerium für Wissenschaft, Erziehung und Volksbildung*, REM) from 1934 until the end of 1936 (see the later section on Wilhelm Süss). The "Leader" of the MR, Georg Hamel, assumed chairmanship of the DMV, following Bieberbach's proposal, which he had put forward in agreement with Vahlen.[19] Emanuel Sperner from Königsberg took over the office of secretary.

The DMV had managed to stave off Bieberbach's attempt to force the association to conform and become ideological, but it was forced to relinquish some of its independence nevertheless. The new chairman, Hamel, reached an oral agreement with Vahlen in 1935 that the DMV's board "regularly consult with the ministry about whether the person envisaged for the chairmanship of the DMV... would be acceptable there."[20] So the REM Science Office had to be consulted every year about the chairmanship.

[16] *Jahresbericht der Deutschen Mathematiker-Vereinigung*, 2nd div., 44 (1934), 87.

[17] Transcription of Bohr's letter to Blaschke, 1 January 1935, UFA, E4/72.

[18] Letters of withdrawal in: UFA, E4/38.

[19] Bieberbach to the committee members, 19 January 1935, UFA, E4/68.

[20] Müller to Wacker, 18 September 1937, UFA, E4/54.

CONFLICTS BETWEEN THE DMV AND LUDWIG BIEBERBACH AND HIS FOLLOWERS

When Bieberbach tendered his resignation at the DMV on 31 December 1935, it was merely a formal end to a shattered relationship.[21] But that did not mean a retreat from policy making for Bieberbach, by any means. It was not the end of the political wrangling over leadership of the German community of mathematicians, which Bieberbach claimed for himself and his Aryan mathematics. In 1936, the journal *Deutsche Mathematik* was founded as his mouthpiece, edited nominally by Vahlen but in fact by Bieberbach.

Already in early 1936, it became evident that Bieberbach and his allies had a variety of means at their disposal to continue the fight against the DMV. The first conflict arose out of an error in the DMV's membership list of December 1935. The statistician Emil Julius Gumbel was entered there as a member of the association and professor at the University of Heidelberg. Gumbel had earned his habilitation degree at Heidelberg in 1923 and obtained the title of professor in 1930. But as an active pacifist and social democrat, he had been the target of political attacks by nationalists since the beginning of the 1920s. Gumbel was considered *persona non grata* among party circles, and his name had appeared next to Albert Einstein's on the first expatriation list drawn up by the National Socialists soon after they had taken over power. Gumbel had emigrated to France in 1933.[22]

The appearance of his name on the DMV's membership rolls elicited written protests to then-chairman Erhard Schmidt in early January 1936 by the heads of the National Socialist University Lecturers Leagues at the University of Heidelberg and the Karlsruhe Polytechnic.[23] Schmidt promptly apologized for the mistake and informed them: "It goes without saying that Gumbel is unworthy of belonging to a German association. Gumbel has not been a member of the DMV for a long time now, in fact;

[21] Resignation statement, Bieberbach to Müller, 31 December 1935, UFA, E4/53.

[22] For example, see Wolfgang Benz, "Emil J. Gumbel: Die Karriere eines deutschen Pazifisten," in Ulrich Walberer (ed.), *10. Mai 1933. Bücherverbrennung in Deutschland und die Folgen* (Frankfurt am Main: Fischer Taschenbuch Verlag, 1983), 160–198; Christian Jansen, *Emil Julius Gumbel: Portrait eines Zivilisten* (Heidelberg: Wunderhorn, 1991); Sébastien Hertz, "Emil Julius Gumbel (1891–1966) et la statistique des extrêmes," unpublished dissertation (Lyon: University of Lyon, 1997).

[23] Leiter der Dozentenschaft der TH Karlsruhe to Schmidt, 8 January 1936, UFA, E4/73, and Leiter der Dozentenschaft der Universität Heidelberg to Schmidt, 10 January 1936, UFA, E4/73.

on the occasion of his withdrawal he received a letter of unsurpassable severity by the DMV's secretary at the time, Prof. Bieberbach."[24]

On the same day, Schmidt wrote in great concern to Hasse, Knopp, and the new secretary, Emanuel Sperner, that "the matter [could] carry with it grave consequences." The next day, he informed Knopp that he viewed "the situation pessimistically" and that it was "very possible that the matter [would lead] to the appointment of a commissioner."[25] This did not in fact happen, but no documentation has survived about the exact circumstances. Hasse later suspected that Bieberbach had held himself back because Chairman Schmidt was a colleague of his in Berlin.[26] The hastily printed errata page corrected both mistakes: "E. J. Gumbel, emigrated, neither member of the DMV nor professor at the University of Heidelberg."[27]

Beyond these problems, the overall political situation and the Harald Bohr affair also triggered a series of withdrawals from the DMV. Between 1933 and the outbreak of war, the membership dropped by more than 10 percent, from roughly 1,120 to 988.[28] The report covering the year 1935–1936, for instance, counted 23 deceased members, 29 withdrawals, and 21 whose memberships had "expired." These 73 departures were set against only 40 new enrollments.[29] The category "membership expired" mainly included individuals who had accumulated substantial sums of unpaid membership fees or had not been locatable for some time. Three years later, Jewish mathematicians and politically undesirable émigrés fell under this category as well (see the later section on the "Jewish question"). Thus, by 1935 it was already obvious that the DMV needed to attract new members if it wanted to counter this shrinkage.

These examples demonstrate that the DMV was in a politically precarious situation even after Bieberbach's exit because it was at all times vulnerable to any campaigns by student associations or the university lecturers, whether or not they arose spontaneously or were instigated and orchestrated by Bieberbach behind the scenes. As far as the chairmanship issue was concerned, which changed hands on a yearly basis, there was

24 Schmidt to Leiter der Dozentenschaft der TH Karlsruhe, 12 January 1936, UFA, E4/73.

25 Schmidt to Hasse, Knopp, and Sperner, 12 January 1936, UFA, E4/73, and Schmidt to Knopp, 13 January 1936, UFA, E4/73.

26 Hasse to Müller, 9 July 1936, UFA, E4/56.

27 UFA E4/91, 33.

28 Figures according to Gericke, *Chronik*, 28.

29 "Angelegenheiten der Deutschen Mathematiker-Vereinigung," *Jahresbericht der Deutschen Mathematiker-Vereinigung*, 2nd div., 46 (1936), 85–88, esp. 85f.

FIGURE 23. Wilhelm Süss (*Source:* University of Freiburg Archives)

the double problem that, on one hand, the REM Science Office had to approve the decision and, on the other hand, each chairman had to reach some understanding with Bieberbach.

WILHELM SÜSS: ELECTION AS CHAIRMAN AND ATTEMPT TO REINTEGRATE BIEBERBACH

The elections for the chairmanship were consequently a delicate subject. But when the DMV's board, composed of Helmut Hasse, Conrad Müller, and Emanuel Sperner, was looking for a new candidate in August 1937, they found strong arguments in favor of the University of Freiburg mathematician Wilhelm Süss, who had taken his doctorate at the University of Frankfurt am Main in 1920 under Bieberbach and had followed him to the University of Berlin in 1921 as his assistant until 1922. From 1923 to 1928, Süss served as instructor in German in Kagoshima, Japan. After returning to Germany, he obtained his habilitation degree at the University of Greifswald, where he obtained a salaried teaching appointment.

Until 1933, Süss was considered a second-rank mathematician with hardly any serious hope of obtaining a professorship. In the winter semester of 1934–1935, however, he became successor to the dismissed Jewish mathematician Alfred Loewy at Freiburg. There Süss entered the National Socialist German Workers Party (*Nationalsozialistische Deutsche Arbeiterpartei*, NSDAP) in 1937 and in 1938 became a member of the National Socialist German University Lecturers League (*Nationalsozialistische Deutsche Dozentenbund*, NSDDB). From 1938 to 1940, he was dean of the Faculties of Science and Mathematics. He became rector in 1940 and exercised this function to the great satisfaction of his colleagues at Freiburg and the REM until the end of the war.[30]

In August 1937, the DMV's board spoke in favor of Süss as the new chairman because it was known that he was interested in professional policy and in the DMV's affairs in particular. As Süss was a member of the NSDAP, they knew that the REM's approval could be anticipated. The board explicitly attached to Süss' candidacy the expectation that "as a pupil of Bieberbach . . . [he should] moreover not be subject to attacks from that quarter."[31] In particular, Hasse, Müller, and Sperner hoped that Süss could restrain Bieberbach to some form of DMV discipline or at least limit his separate and willful National Socialist policy making for the profession. The relations between Süss and Bieberbach were in fact good. Süss even served on the editorial board of the *Deutsche Mathematik* from 1936 to 1940. Tentative inquiries at the REM yielded no objections there to his candidacy, and Süss was elected the new chairman at the annual convention in Bad Kreuznach.[32]

Süss was perfectly well aware of the fact that the board expected him to take steps to reconcile with Bieberbach. In December 1937, he wrote Bieberbach accordingly, who had applauded his election as chairman.[33] Bieberbach's reaction was sobering, though. He stressed that the DMV

[30] See Volker R. Remmert, "Wilhelm Süss," in Bernd Ottnad and Fred Ludwig Sepainter (eds.), *Baden-Württembergische Biographien*, vol. III (Stuttgart: Kohlhammer, 2002), 418–421; Volker R. Remmert, "Vom Umgang mit der Macht. Das Freiburger Mathematische Institut im 'Dritten Reich,'" *Zeitschrift für Sozialgeschichte des 20. und 21. Jahrhunderts*, 14 (1999), 56–85; Volker R. Remmert, "Mathematicians"; Bernd Grün, "Das Rektorat des Mathematikers Wilhelm Süss in den Jahren 1940–1945 und seine Wiederwahl 1958/59," *Freiburger Universitätsblätter*, 145 (1999), 171–191.

[31] Sperner to Müller, 26 August 1937, UFA, E4/43.

[32] Hamel to Müller, 17 September 1937, UFA, E4/53; Müller to Wacker, 18 September 1937, UFA, E4/54; Dames (REM) to Müller, 2 October 1937, UFA.

[33] Süss to Bieberbach, 12 December 1937, UFA, C89/44.

could "with proper organization and proper guidance, develop a beneficial internal and external influence." But this, in his opinion, would require not just "close contact with government authorities" but "close contact with party authorities" as well. He specifically pointed out the future roles of the National Socialist Teachers League and the NSDDB, because the latter would "sooner or later bring forth a professional organization" of its own. These plans for the future drew into question the DMV's very survival. Bieberbach attached to his possible reentry into the DMV an additional demand: "that Mr. Hasse disappear from the board."[34]

Satisfying this condition was out of the question, but it led to temporary hard feelings between Süss and Hasse. In May 1938, Süss finally set straight in a letter to Müller that he had "no intention of making Mr. Hasse cede the way to Mr. Bieberbach"; nor did he wish "to propose as his [Hasse's] successor a man who is amenable to the group around Bie[berbach]." Although Süss assured him at the same time that he "strictly" rejected "any influence by this circle," it noticeably contradicted his subsequent assurances of intending "to find a way" to make possible "Bieberbach's readmittance."[35] Süss could count on the support of Müller and Sperner because at Müller's suggestion the three had already agreed in March that extending Süss' term of office for another year would be welcomed.[36] Müller and Sperner continued to view Süss as a good guarantor against Bieberbach's potential founding of a National Socialist League of Mathematicians and, at the same time, interpreted a re-electable chairmanship as a clear signal that the DMV was somewhat approximating the leadership principle.[37] After the necessary change had been made to the statutes, Süss was regularly re-elected and remained the DMV's chairman until the end of the war.

As Süss did not make any further advances to Bieberbach in spring and summer 1938, the tensions between him and Hasse were soon forgotten. They worked in concert with Müller and Sperner toward strengthening the DMV's position as a representative of the profession in government policy making, above all at the REM Science Office. At the beginning of 1937, Bieberbach's most important ally there, Theodor Vahlen, had been

34 Bieberbach to Süss, 27 December 1937, UFA. Bieberbach only re-registered membership in the DMV in 1975.

35 Süss to Müller, 22 May 1938, UFA., E4/46.

36 Müller to Hasse and Sperner, 15 March 1938, UFA, E4/54; Müller to Hasse and Sperner, 16 March 1938, UFA, E6/43; Sperner to Müller, 26 March 1938, UFA.

37 See Kneser and Schappacher, "Fachverband," 69.

replaced by the *Schutz-Staffel* (SS) member Otto Wacker. This change spelled the end of Bieberbach's strong influence at the Science Office.

But even without Vahlen's direct backing at the REM, Bieberbach continued to challenge the DMV's claim to sole representation of the profession through his *Deutsche Mathematik* movement, his own policies for the discipline, his connections with the Reich Student Leadership, and his ambitions within the National Socialist Teachers League to found a National Socialist League of Mathematicians. The DMV's board had no choice but to regard him as a threat. Bieberbach relied not only on the support of Vahlen but also on that of Erhard Tornier, who after embroiling himself in a feud with Hasse had left the University of Göttingen in 1936 for the University of Berlin. The mathematically brilliant Oswald Teichmüller, a doctoral graduate of Hasse's in 1935 who had allied himself mathematically as well as politically with Bieberbach at Berlin in 1937, was another friend along with Fritz Kubach, a leading member of the Reich Student Leadership who had submitted a thesis on the history of mathematics at the University of Heidelberg in 1935.[38] Together they caused the DMV a variety of worries in the mid-1930s.

Although Süss and the DMV formally kept Bieberbach at arm's length, he influenced their actions in three ways: (1) directly by his personal influence on Süss, such as on the question of the reorganization of mathematical journals and review publications (see the next section); (2) indirectly through the friction that arose from the DMV's and Süss' positions being at odds with those of Bieberbach or by letting themselves be impressed by his threats, for example on the so-called Jewish question in the DMV (see the related section later); and (3) by Bieberbach's allies, such as Harald Geppert in Berlin, who gained substantial influence on the DMV's international policy. The most important field for Süss and the DMV to distinguish themselves from Bieberbach was in the area of mathematical

[38] See Thomas Hochkirchen, "Wahrscheinlichkeitsrechnungen im Spannungsfeld von Maß- und Häufigkeitstheorie – Leben und Werk des "Deutschen" Mathematikers Erhard Tornier (1894–1982)," *NTM. Internationale Zeitschrift für Geschichte und Ethik der Naturwissenschaften, Technik und Medizin*, new series, 6 (1998), 22–41; Norbert Schappacher and Erhard Scholz, "Oswald Teichmüller – Leben und Werk," *Jahresbericht der Deutschen Mathematiker-Vereinigung*, 94 (1992), 1–39. On Kubach, see Michael Grüttner, *Studenten im Dritten Reich* (Paderborn: Schöningh, 1995), 509; Volker R. Remmert, "In the Service of the Reich: Aspects of Copernicus and Galileo in Nazi Germany's Historiographical and Political Discourse," *Science in Context*, 14 (2001), 333–359, especially 341f.; Reinhard Siegmund-Schultze, *Mathematische Berichterstattung in Hitlerdeutschland. Der Niedergang des Jahrbuchs über die Fortschritte der Mathematik* (Göttingen: Vanderhoeck & Ruprecht, 1993), 117.

war research (see the later sections on mathematical war research and on the DMV and DPG).

REORGANIZING MATHEMATICAL JOURNALS AND REVIEW PUBLICATIONS

Soon after his election as chairman of the DMV, Süss tried to exert influence on mathematical publishing. The plans he pursued at the beginning of the war to reorganize mathematical periodicals are sketched below.[39] Immediately after the National Socialists had seized power, there were discussions about reducing the number of academic journals to counteract what the National Socialists considered a growing fragmentation. The physicist and Nobel laureate (1919) Johannes Stark, an early enroller in the NSDAP in the 1920s and a standard-bearer of "Aryan physics" (*Deutsche Physik*) together with his colleague and fellow Nobel laureate (1905) Philipp Lenard, demanded in autumn 1933 a "reorganization of physical publications" under a common editorial board. Nothing came of all these plans, however, in the years that followed.[40]

It was presumably in spring 1939 that Bieberbach sent out a copy of his "Proposals for the organization of the field of scientific journals," addressed "to the attention of P[arty] C[omrade] Süss."[41] Taking Stark's conceptions as a model, Bieberbach presented mathematical journals as an illustration to explain his ideas. In his opinion, one should "proceed accordingly in the other fields of science as well." He complained in particular about the previously mentioned fragmentation as causing articles that actually belong within a single field of research to be dispersed among various journals. This practice, Bieberbach continued, was efficient neither for scholars in the profession nor for editorial boards. This state of

[39] See also Volker R. Remmert, "Mathematical publishing in the Third Reich: Springer Verlag and the Deutsche Mathematiker-Vereinigung," *Mathematical Intelligencer*, 22/3 (2000), 22–30.

[40] Michael Knoche, "Scientific journals under National Socialism," *Libraries & Culture*, 26 (1991), 415–426, especially 418; Heinz Sarkowski, *Der Springer Verlag. Stationen seiner Geschichte*, part I: *1842–1945* (Heidelberg: Springer, 1992), 329–331, as well as the contribution by Paul Forman to the German version of this volume, "Die Naturforscherversammlung in Nauheim im September 1920," in Dieter Hoffmann and Mark Walker (eds.), *Physiker zwischen Autonomie und Anpassung – Die DPG im Dritten Reich* (Weinheim: Wiley-VCH, 2007), 29–58.

[41] Ludwig Bieberbach, "Vorschläge zu einer Planung auf dem Gebiete der wissenschaftlichen Zeitschriften (probably dated April 1939)," UFA, E4/78.

affairs had the additional negative economic effect of an extreme rarity of private subscriptions. Bieberbach's reorganization proposals concluded with the suggestion that the DMV assume central monitoring and coordination of professional journals.

Süss supported Bieberbach's initiative. When he met with ministerial official Kummer at the REM in November 1939 to discuss an imminent fusion of the review publications *Zentralblatt* and *Jahrbuch über die Fortschritte der Mathematik*, Süss likewise brought up the subject of a restructuring of the system of mathematical journals.[42] Unaware of a prohibition issued by the Foreign Office for prestige reasons against mergers between professional journals, Kummer and Süss agreed on a new organizational principle for mathematical journals: specialization. This would have doomed traditional journals, which offered a broad thematic spectrum of articles, such as the *Journal für die reine und angewandte Mathematik* (*Crelle's Journal*), the *Mathematische Annalen*, and the *Mathematische Zeitschrift*. Bieberbach's suggestion was that *Crelle's Journal* should specialize on algebra and number theory, the *Mathematische Annalen* on analysis, and *Mathematische Zeitschrift* on geometry. Süss accepted the idea right away and decided to conduct negotiations with the responsible editors and publishers.[43] Faced with unanimous reservations by the respective editors and open rejection by the Springer publishing house, Süss decided to put his plans aside for the time being.[44]

When he became rector in autumn 1940, however, Süss regarded it as a new chance to resurrect this professional policy. At the beginning of his term of office, he presented to Minister Rust a wish list of four proposals for the agenda of the German Rectors' Conference in Prague in December 1940, which pertained to both university policy and professional policy.[45] The fourth of these proposals exposed his interwoven interests as a professional policy maker. Under the neutral heading "Scientific societies and periodicals," Süss pleaded for a reorganization of German journals under

[42] Süss to Kummer, 28 May 1940, UFA, E4/45.

[43] Süss to Sperner, 14 December 1939, UFA, E4/76.

[44] Springer publishers had every reason to be skeptical about Süss; see Remmert, "Mathematical."

[45] On this and other amalgamations by Süss of the offices of rector and DMV chairman, see Volker R. Remmert, "Zwischen Universitäts- und Fachpolitik: Wilhelm Süss, Rektor der Albert-Ludwigs-Universität Freiburg (1940–1945) und Vorsitzender der Deutschen Mathematiker-Vereinigung (1937–1945)," in Karen Bayer, Frank Sparing, and Wolfgang Woelk (eds.), *Universitäten und Hochschulen im Nationalsozialismus und in der frühen Nachkriegszeit* (Stuttgart: Steiner, 2004), 147–165.

the baton of the German scientific societies. Completely adopting Bieberbach's original argumentation, he demanded that the REM avail itself of the national scientific societies as "the most suitable advisors" on a "necessary reorganization of the German system of scientific periodicals."[46] As the war dragged on, however, these proposals receded increasingly into oblivion despite the new weight that Süss had intended to give them by his office as rector.

Süss' high-flown plans to reorganize German scientific journals were not his only attempt at applying influence on the world of mathematical publishing. The Springer publishing house felt the greatest pressure. In 1931, it had advanced into the area of review publications when it founded the *Zentralblatt für Mathematik und ihre Grenzgebiete*. Its editors were Otto Neugebauer and Richard Courant. From its inception, it directly and consciously rivaled the long-standing *Jahrbuch über die Fortschritte der Mathematik* (founded 1869), which was issued under the auspices of the Prussian Academy of Sciences in Berlin since 1927–1928.[47]

The *Jahrbuch* was notorious for long delays in the appearance of its reviews. So the *Zentralblatt* soon made a name for itself for its efficiency. Until 1939, the *Jahrbuch* had to contend with increasing ideological interference by Bieberbach, who had designated himself as spokesman for the journal's board at the Prussian Academy. He specifically demanded that it dispense with Jewish reviewers. National Socialist ideology also posed problems for the *Zentralblatt*. Soon after its founding there were efforts, particularly on the part of the DMV, to arrange for a collaboration between it and the *Jahrbuch*. As direct market competitors, the two publishers de Gruyter and Springer were naturally opposed to such an arrangement. An ideological incommensurability existed between Bieberbach and Springer as well. All the same, discussions were started in the late 1930s about a merger or at least cooperation between the *Zentralblatt* and the *Jahrbuch*.

In late 1938, news spread that the American Mathematical Society was intending to found a new review periodical, the *Mathematical Reviews*, in the United States. This prospect troubled mathematicians and publishers in Germany, irrespective of their opinions about National Socialism. In January 1939, Bieberbach pressured de Gruyter and Springer to fuse the two journals, immediately submitting detailed proposals about how to go about it. The DMV's chairman, Süss, also tried to apply pressure on de

[46] Süss to Rust, 4 November 1940, UFA, B1/1439.

[47] For details, see Siegmund-Schultze, *Mathematische*.

Gruyter and Springer in early 1939 and indefatigably promoted a merging of the journals.[48] But when he tried to negotiate this union between the *Zentralblatt* and the *Jahrbuch* at the REM in early November, he found out that the Foreign Ministry had prohibited "any consolidations of scientific journals during the war." As Süss explained, it "had been portrayed as a propagandistic requirement, to continue if possible our entire scientific production and publications in the normal numbers, at best with restrictions on their breadth." So the union between the *Zentralblatt* and the *Jahrbuch* was not to be. Agreement was reached nonetheless on a liaison between the two under a general editorial board in Berlin headed by the convinced National Socialist Harald Geppert.[49] Thus, the unification that the DMV, Süss, and Bieberbach had been striving for was somewhat approximated, including at the same time the kind of management according to ideological precepts that Bieberbach in particular had been wanting, because Geppert was made responsible for the assignment of reviewers for both the *Zentralblatt* and the *Jahrbuch*.

As the developments described in the foregoing sections have documented, the DMV's board and its chairman, Süss, were willing to collaborate closely with the REM to further the DMV's own goals in tandem with those of the ministry. As a result, any disassociation from the ideological preconditions promoted by the regime, such as anti-Semitism and anti-internationalism, became negligible; likewise for the avenues it offered, such as denunciation.[50] The DMV consequently drew quite close to Bieberbach's positions, allowing Süss to calm the waves between the DMV and Bieberbach at least for a time. Of particular importance in the development of the relations between the professional association of mathematicians and Bieberbach just as with the REM was the so-called Jewish question in the DMV.

THE "JEWISH QUESTION" IN THE DMV

As did other professional associations, the DMV or its board also thought about how to handle its Jewish and emigrated membership. Süss's involvement in the "Jewish question" – *Judenfrage*, as Secretary Müller's

48 Nathan Reingold, "Refugee mathematicians in the United States of America, 1933–1941: Reception and reaction," *Annals of Science*, 38 (1981), 313–338; Remmert, "Mathematical"; Siegmund-Schultze, *Mathematische*, 167ff.

49 Siegmund-Schultze, *Mathematische*, 224–226, appendix 14.

50 On this see Remmert, "Mathematical."

pertinent file was labeled – can be traced back to a conversation in March 1938 at the REM. The DMV's handling of its Jewish and émigré members was clearly formulated by the DMV's Chairman Süss, supported by the DMV's board, and did not arise out of any procedure dictated by the REM. In 1938, Süss and his fellow board members, Hasse, Müller, and Sperner, were worried that the DMV might lose its lobbying influence for the profession to Bieberbach. After his visit to the REM in March 1938, Süss reported to the board:

> If we want to clarify the state of doubtful émigrés, the Ministry is ready, upon request, to give the serving chairman confidential information from which it may be gathered whether or not the émigré is acceptable as a member.... I would advise that notices about the whereabouts of Jews and émigrés, for example, also address changes, not be announced anymore in the *Jahresbericht*. Neither their membership nor their leaving should cause any stir – this I would like to advise be the guiding thread of our actions. We are, I assume, all in agreement that we use suitable occasions to get rid of our Jewish and undesirable members among the émigrés as soon as we can.[51]

It soon turned out that the discussions about Jewish and émigré members were but a prologue to further developments. Their numbers were, in fact, considerable. Substantially more than 100 mathematicians had left the German Reich by 1939.[52] During the DMV's annual convention in September 1938, its officially approved chairman, Süss, heard that the REM would soon be demanding that all scientific societies "implement the Aryan principle." He emphasized to Müller that this issue would become "current in the foreseeable future," and it therefore appeared reasonable that he "immediately carry this matter forward at the Ministry." He expected "that during the course of the winter we shall still have to discuss these things amongst ourselves, perhaps also at the responsible office in the Ministry."[53]

After the orchestrated anti-Jewish pogroms of 9 and 10 November 1938 (*Reichskrystallnacht*), matters began to evolve quickly. A few days later, in a letter to Hasse that he also forwarded in excerpt to Müller and Süss, Sperner reopened the discussion of the treatment of "non-Aryan

[51] Reference file Conrad Müller *Judenfrage*, letter to Hasse, Müller, and Sperner, 9 March 1938, UFA, E4/46. Süss' papers (UFA, C89) do not include his correspondence on the "Jew issue."

[52] See the list in Reinhard Siegmund-Schulze, *Mathematicians Fleeing from Nazi Germany. Individual Fates and Global Impact* (Princeton: Princeton University Press, 2009), 344–357.

[53] Süss to Müller, 5 October 1938, UFA, E4/45.

members and émigrés." He did not believe "that we [the DMV] can bear these kinds of members among our ranks any longer; rather we probably must attack the problem, finally."[54]

On 15 November 1938, the REM issued what came to be known as the "academy decree" (*Akademie-Erlaß*). It prescribed to scientific academies – not societies – statutory changes stipulating the exclusion of Reich Germans who were either Jewish, "related to Jews," or of "mixed blood." It is unclear to what extent this decree triggered the subsequent developments on the Jewish question within the DMV. Hasse informed his fellow board members about the decree by letter on 24 November,[55] but a hint about it already exists in a letter by Bieberbach to Süss of 16 November. Bieberbach used the recent events as an opportunity to reiterate his conditions for reentry into the DMV:

> I haven't heard anything from which I could draw the conclusion that the German Mathematical Association had undertaken anything to shape herself, of her own accord, into an Aryan [*Deutsche*] association and to turn her attention to tasks whose solution appear necessary. From what I hear, I can only conclude that the DMV is waiting for things to fall into her lap that, for instance, by force of a regulation liberated her from her Jewish members. But that would ultimately only be the start. Much certainly would be gained, though, if the DMV did have enough youthful sprightliness to take a decision on this on her own. You don't have to take as a model the fact that the ancient academies, governed by the Gerousia, first needed a push from the outside, in order to shake off the Jews. At the end of your letter you take up the question of my reentry again. I have always told you what the conditions are. It includes that the DMV take a decisive step toward solving its German obligations. Expecting things to descend from the skies just isn't an autonomous step.[56]

Süss did not remain idle in this situation. He solicited the advice and backing of the National Socialist Regional Head (*Kreisleiter*) at Freiburg, Willy Fritsch, who held a doctorate in mathematics, before writing to his fellow board member Sperner on 18 November: "very imminently we will have to delete all current and former *German* Jews as members."[57]

Süss thought that "silent exclusion" of the Jewish members was "surely diplomatically the best and indeed the easiest."[58] Hasse contradicted this

54 Sperner to Hasse, 14 November 1938, UFA, E4/46.

55 Hasse to Müller, Sperner, and Süss, UFA, E4/46.

56 Bieberbach to Süss, 16 November 1938, UFA, C89/44.

57 Süss to Sperner, 18 November 1938, UFA, E4/46 (original emphasis). About DMV members of Jewish extraction, see the condensed exposition in Kneser and Schappacher, "Fachverband," 69f.

58 Süss to Hasse and Müller, 18 November 1938, UFA, E4/46.

approach a few days later, preferring that the German "Jews inside and outside the country . . . be explicitly contacted about their exclusion or in any case be instructed in a suitable form, as the Minister put it to the academies, that they withdraw their memberships of their own volition." Hasse conceded, "it is very difficult to find the right words for such a message," and he was also willing "to fall back on mute expirations of the memberships in 1939. In any event, those "affected" could be "referred to in the newsletter under withdrawn or expired memberships." For the DMV as for the REM, this procedure had the advantage that the underlying reasons would avoid receiving a public airing.[59]

A few days later, Süss was able to report progress on this issue of the expulsion of Jewish members. He had meanwhile discussed the Jewish question in the DMV with the head of the REM Science Office, Wacker, and Freiburg regional head, Fritsch, who joined the DMV himself in 1940:

> I said [to Wacker] that we had been preparing the elimination of all German Jews from our association for some time already, because it seemed to us that after the new situation that has arisen in the meantime has overtaken the earlier agreements reached with the Ministry about a gradual departure of the Jewish members. As concerns the foreign Jews among our members, I would have to insist that a state or party authority take on the responsibility of verifying who among our foreign members should be regarded as Jews . . .
>
> With the Regional Head, Dr. Fritsch, I reached agreement that he undertake, along party channels, the identification of who among the cases that appeared doubtful to us among our domestic members were Jews. I gave out an address list of the members falling under consideration . . .
>
> If we want to forestall an order by the Ministry through steps of our own, we will have to act quickly now. In any case we cannot delay the negotiations until we have drawn up really reliable lists of every type of member whose continuance in the DMV will be the subject of discussion.
>
> I would like to suggest the following as the wording for the message to the German Jews inside and outside the country: "Your membership in the German Mathematical Association is no longer possible in future. We shall therefore cease to list your name on our membership rolls as of 1 Jan. 1939." Whether this wording outdoes that by Mr. Sperner in coarseness or whether it sounds more obliging, I dare not decide for myself. I ask you please to judge.[60]

This formulation of the message to the Jewish members passed muster with Hasse and Sperner but not with Müller. A bureaucratic coldness toward these former fellow members nevertheless emanated from the final

59 Hasse to Müller, Sperner, and Süss, 24 November 1938, UFA, E4/46.

60 Süss to Hasse, Müller, and Sperner, 10 December 1938, UFA, E4/46.

letter. The one to the Munich mathematician Friedrich Hartogs from July 1939 reads:

Dear Professor,

You cannot in future remain a member of the German Mathematical Association. That is why I suggest you announce your withdrawal from our association. Otherwise we shall publish the expiration of your membership at the next opportunity.

Most respectfully,
The Chairman[61]

The German Physical Society (*Deutsche Physikalische Gesellschaft*, DPG), under the chairmanship of Peter Debye, had solved this problem otherwise; the DMV's board had seen the DPG's letter from December 1938 as well:

To the members of the German Physical Society,

Under the compelling prevailing circumstances the continued membership in the German Physical Society by German-Reich Jews in the sense of the Nuremberg Laws can no longer be maintained.

I therefore request, in agreement with the Board, that all members who fall within this provision inform me of their withdrawal from the Society.
Heil Hitler![62]

Compared with this invitation to decide for oneself whether or not withdrawal of membership in the DPG was necessary, the DMV's board chose the more thorough method of identifying the members concerned and then acting accordingly. In executing this business, Süss had, as we have seen, first secured the backing of the REM and the NSDAP. The discussions about who was to be counted a Jew and who was not began in December and dragged on until April 1939. That January, the REM finally did decree that "German Jews" were, in principle, also ineligible for membership in professional associations.

At the beginning of April 1939, Süss summarized the situation for his fellow board members under the heading: "cashiers and racial

61 Süss to Hartogs, 22 July 1939, UFA, E4/62.

62 An die deutschen Mitglieder der Deutschen Physikalischen Gesellschaft, Berlin, 9 December 1938, UFA, E4/46; see the transcription in Hoffmann and Walker, *Physiker*, 511, 565; translated in Klaus Hentschel (ed.), *Physics and National Socialism: An Anthology of Primary Sources* (Basel: Birkhäuser, 1996), doc. 67. Stefan L. Wolff's contribution to this volume discusses the treatment of the DPG's Jewish members.

matters." He pointed out that both the National Socialist Regional Head at Freiburg and the REM had checked the lists he had submitted. He also emphasized that Heinz Hopf in Zurich, "because surely not a full-blooded Jew (≤50%) and owing to his prestige, besides probably soon to become Swiss," as well as Paul Bernays, "owing to his prestige, [were] not removable without consequences for the DMV, as I have now learned in Switzerland." Thus he narrowed down the issue to the problem of how to keep the matter as low-key as possible in order not to set off a wave of professionally and politically unwanted withdrawals in reaction to the expulsions of Jewish members.[63]

Jewish and émigré membership was a very sensitive topic, not just internally but externally as well. If their exclusion was politically well received at home, the impression it raised was not a favorable one abroad, either for the association or for the Reich. That was why the REM and the Foreign Office repeatedly emphasized the need to proceed cautiously. The outbreak of war changed matters. At the beginning of September 1939, Süss wrote to the board members that after the war, "the situation [will] anyway be such that we can clarify matters without hesitation again."[64] Leaving this aside, the DMV had two alternatives to counter its shrinking membership – at the turn of 1938 to 1939 it had lost a total of 47 members to withdrawals along with another 40 to expirations.[65] The first was increased advertisement, as Süss suggested around the middle of April 1939, in the Protectorate of Bohemia and Moravia as well as in personal letters to specific mathematicians.[66] The second alternative was establishing dual memberships, such as with the Italian Mathematical Union (*Unione Matematica Italiana*, UMI; see the later section on international ties).

The DMV's rapid settlement of their Jewish question was substantially based on Süss' engagement, supported by the same board that had ushered him into office in 1937 on considerations of professional policy: Helmut Hasse, Conrad Müller, and Emanuel Sperner. Yet Müller was hardly involved in the intense debate that ensued at the end of 1938. In the Jewish question, Süss proved that the DMV was willing and able to collaborate with the system and to help ensure a smoothly functioning ministerial bureaucracy and its National Socialist goals, which were beyond his and

63 Süss to Hasse, Müller, and Sperner, 4 April 1939, UFA, E4/46.

64 Süss to Hasse, Müller, and Sperner, 9 September 1939, UFA, E4/45.

65 See the statistics in *Jahresbericht der Deutschen Mathematiker Vereinigung*, div. 2, 49 (1939), 76.

66 For instance, Süss to Hasse, 14 April 1939, UFA, E4/45, and Süss to Weiss, 6 April 1939, UFA, C89/85.

the DMV's influence or control. By doing so, he qualified himself and the DMV as reliable partners for the REM. As this collaboration progressed, the DMV adopted important National Socialist components as its own policy, especially anti-Semitism and anti-internationalism. On this foundation, Süss developed into one of the most influential representatives of German mathematicians during the Third Reich and, as chairman of their professional association and later also as rector at the University of Freiburg, into a close and loyal collaborator of the REM. This collaboration was a precondition for the DMV's success in implementing its active professional policies during the Second World War (see the later sections on mathematical war research and on the DMV and DPG).

The DMV's board regarded the developments in 1938 and 1939 as strictly confidential. In the midst of preparing his annual report as chairman in December 1939, Süss wrote Müller that he had not been idle, "but nothing can be said in public about most of the things, for example about the opinions on the professorship appointments in Vienna and Prague, concerning the international congress, the Jewish question, the review publications." So he had just "jotted down a few lines of generalities."[67] The issues he mentioned certainly were too delicate to be published in the *Jahresbericht*. The "Report by the chairman" contained, besides the comment that "the chairman's duties in the last few years have experienced considerable expansion against former times," just a few lofty words about the DMV, which "like any scientific society in Germany in its labors always keeps the general public in mind."[68]

The silent assumption was that Jews were no longer a part of it. In the correspondence between these party comrades or aspirants Hasse, Sperner, and Süss, Jews were transformed into bureaucratic items. The terminology used in these letters on the Jewish question reveals the central and ominous role of bureaucracy in the Third Reich and its tendency to dehumanize bureaucratic processes, its fatal intrinsic dynamics of bureaucratic zeal, which from the initial stage of "purging" the German Reich of Jews – making it "Jew free" (*judenrein*) – steered the nation down the path to the Holocaust.[69] Whereas Müller chose in his letter not to close his exposition of Alfred Pringsheim's and Blumenthal's professional

67 Süss to Müller, 18 December 1939, UFA, E4/45.

68 Wilhelm Süss, "Bericht des Vorsitzenden," *Jahresbericht der Deutschen Mathematiker-Vereinigung*, div. 2, 49 (1939), 77f.

69 Zygmunt Baumann, *Modernity and the Holocaust* (Ithaca: Cornell University Press, 1989), especially 102–106.

merits with the board's standard "Heil Hitler!" his fellow board members peppered their letters with such distancing vocabulary as "remove," "elimination," "gradual departure," "lists of the members of every type," "cashiers and racial matters," and so forth.

INTERNATIONAL TIES IN THE IMAGE OF POLITICS

Multiple repercussions arose from the DMV's dependence on foreign policies and the political context in general. The Harald Bohr affair, the Jewish question, and the reorganization of review publications are, without exception, inconceivable outside the context of National Socialism and its isolationist policy for the German Reich.[70] The "orientation of relations abroad" counted, as Bieberbach reminded Süss in November 1938, among the DMV's most essential future tasks.[71] But it was a difficult exercise for the chairman to model this orientation according to the conditions and demands of the day, in other words, taking National Socialist policy as a guideline, especially since the political requirements were not always transparent enough to serve as a basis for specific action. From 1939 on, the implications of the Jewish question and the war encumbered any effort to maintain international ties not only beyond the country's borders but even within them.

The Reich's complex and volatile foreign relations obliged the DMV to consult with the REM and the Foreign Office frequently about international exchanges it wanted to encourage. From 1938 on, Süss and the association's board followed a two-pronged strategy: They redoubled their efforts to invite foreign mathematicians to the DMV's annual conventions, which owing to the obligatory currency exchanges required the approval of the REM, and they tried to intensify their relations with specific countries and the professional mathematical associations of those countries. Friendly nations, such as Italy, Japan, and Spain, played a prominent role here. Both these approaches were permeated by a strong

[70] The international ties of mathematicians in Germany under National Socialism has thus far only been treated by Reinhard Siegmund-Schultze. See his publications: "*Mathematische Berichterstattung*; Faschistische Pläne zur 'Neuordnung' der europäischen Wissenschaft. Das Beispiel Mathematik," *NTM. Internationale Zeitschrift für Geschichte und Ethik der Naturwissenschaften, Technik und Medizin*, 23 (1986), 1–17; "The Effects of Nazi Rule on the International Participation of German Mathematicians: An Overview and Two Case Studies," in Karen Hunger Parshall and Adrian C. Rice (eds.), *Mathematics Unbound: The Evolution of an International Mathematical Research Community, 1800–1945* (Providence: American Mathematical Society, 2002), 335–351.

[71] Bieberbach to Süss, 16 November 1938, UFA, C89/44.

emissary sense of carrying out a cultural mission that during the heady early years of the war took on an almost imperialistic character.

In January 1938, the DMV's board discussed the possibility of issuing invitations to select foreign speakers for the annual convention that September in Baden-Baden. It quickly became apparent that English, French, Italian, and Swiss mathematicians would make desirable guests.[72] Their mathematical attractiveness, that is, their professional reputations, were not the only criteria used in the selection of these candidates. There also had to be some predisposition to accepting such an invitation. Süss also pointed out that only those "persons came into consideration as speakers and members of the DMV from abroad" who were "positively disposed toward us." He advocated inviting "alongside the whizzes," "one or two younger foreigners" on whom "the future of our relations with foreign countries" relied. He explicitly saw good prospects in a campaign "to advertise membership among pro-German foreigners."[73] Eventually, in February Müller and Süss applied for the REM's permission for the association to issue invitations, as the justification reads: "in order to be able to make the attractive pull of her convention available to German interests." The desirable candidates listed were Élie Cartan and Claude Chevalley (both from Paris), Andreas Speiser and Ernst Stiefel (both from Zurich), Georges de Rham (Lausanne), Francesco Severi (Rome), Enrico Bompiani (Bologna), and Marston Morse (Princeton).[74] The original invitees that showed up were Chevalley, Speiser, Bompiani, Severi, and de Rahm. Other foreigners also spoke at the annual convention besides the numerous mathematicians from the recent Austrian "appendage" to the German Reich: Otto Varga (Prague), Dan Barbilian (Bucharest), and Anton Huber (Fribourg) of Austrian origin.[75] The hope of gaining new foreign members was even fulfilled: in 1938 from among these speakers, Barbilian, de Rham, and Stiefel joined the DMV.

Thus from the board's point of view, this endeavor could be regarded as a double success, and plans were made to invite foreign mathematicians again for 1939. In March 1939, Süss submitted a list of candidates to the REM and again emphasized the DMV's intention "of making the attractive pull of her convention available to German interests in cultural

72 See the correspondence from January 1938, UFA, E4/58.

73 Süss to Hasse, Müller, and Sperner, 22 January 1938, UFA, E4/58.

74 Müller and Süss to the REM, 23 February 1938, UFA, E4/58.

75 See the program, UFA, E4/58, 388–396.

policy."[76] The convention had to be cancelled, however, because of the outbreak of war.

Specific details about the foreign invitation practices of other scientific societies during this period are lacking. But we certainly can say that the DMV's "cultural policy" sketched above differed at least from the DPG's. In August 1939, in reply to the DMV's inquiry about its foreign speakers in physics, the DPG stated that "just one Japanese until now" had been involved.[77] The war hampered the DMV's initiative, of course, and as a result the convention that took place in Jena in autumn 1941 was a little irregular.

Because the conditions of war either prohibited invitations to foreign mathematicians or made them practically infeasible, the DMV's focus shifted foremost to developing its relations with other mathematical societies in friendly states. Ties with Italian mathematicians and the UMI were especially fostered. The cultural treaty signed between Germany and Italy in November 1938 had flanked the Berlin–Rome axis proclaimed in November 1936. This formed the backdrop against which the DMV felt called upon to improve its official relations with that nation, as its invitations to Italian mathematicians to the annual conventions of 1938 and 1939 exemplified.

During the convention in Baden-Baden in September 1938, Süss discussed with Enrico Bompiani the possibility of arranging joint memberships in the DMV and the UMI. This proposal was welcomed by the UMI's presiding board, as Bompiani wrote Süss in December.[78] In July 1939, the plan had matured to the point that Süss was able to apply for the REM's approval.[79] As always, the obstacle posed by Jewish members first had to be cleared; Italy had issued its own anti-Semitic laws in 1938. After a brief debate, the board agreed to continue to list Italian "non-Aryans" as members "until the demand [for change] was submitted to us by the Italians."[80] Political opportunism was not the only driving force behind arranging such a joint membership. The board was hoping to reap new members, which were so urgently needed if the total of 1939 was not to drop even further. Whatever their extent may have been, after Italy's capitulation in September 1943 the plans for closer cooperation between

[76] Süss to REM, 12 March 1939, UFA, E4/60.
[77] Grotrian to Süss, 9 August 1939, UFA, E4/59.
[78] Bompiani to Süss, 31 December 1938, UFA, E4/76.
[79] Süss to REM, 13 July 1939, UFA, E4/76.
[80] Hasse to Müller, Sperner, and Süss, 3 June 1939, UFA, E4/44.

the DMV and the UMI were not worth the paper they were written on. Other cooperative efforts with Japanese and Spanish mathematical societies that Süss was supporting in summer 1940 "for reasons of foreign policy" never advanced beyond the preliminary stage.[81]

The stable conditions in the DMV's board and chairmanship from 1938 were beneficial to this directed policy of developing and encouraging international ties. It was a component of wider plans, which included the system of mathematical publishing, and fit seamlessly within the framework of the REM's cultural policy – it was followed in part by conviction, in part by practical considerations. After initial successes, such as the participation of foreign mathematicians at the DMV's annual convention in 1938 and more intense collaboration with the UMI, international activities became more and more limited as the war dragged on and the first rush for victory flagged. To the restrictions on international exchanges imposed by the war were added new demands on the DMV and its chairman set by the changed conditions, particularly in the organization of war research in mathematics.

MATHEMATICAL WAR RESEARCH AND THE FOUNDING OF THE REICH INSTITUTE OF MATHEMATICS IN OBERWOLFACH

During autumn 1941 at the latest, German mathematicians started to discuss the founding of a centralized institute for mathematics, resting its legitimization primarily on war research. The relevance of mathematical methods in war research made it seem desirable to have a national institute for mathematics that was centrally situated and furnished with the necessary resources to devote itself to mathematical problems encountered elsewhere. The rethinking among party and government circles in 1942 away from a concentration on short-term research projects in a lightning war to mid- to long-term projects in a war of longer duration made the materialization of a Reich institute for mathematics or even a larger institution (*Mathematisch-Technische Reichsanstalt*) seem to fall within grasp. It was unclear, however, who specifically should be able to found such an institute and under whose auspices. Personal animosities and ambitions played as large a role here as elsewhere in the polycratic thicket of competing governmental agencies. In summer 1942, three serious candidates existed: (1) the Freiburg mathematician Gustav Doetsch, a ranking major in the *Luftwaffe* and member of the research leadership in

[81] Süss to Hasse, Müller, and Sperner, 17 July 1940, UFA, E4/45.

the Reich Ministry of Aviation who was seeking to build up an institute under the Air Force; (2) his colleague at Freiburg, Wilhelm Süss, who as chairman of the DMV wanted to establish such an institute under the association's very own authority, with the backing of the Reich Research Council (*Reichsforschungsrat*, RFR) and REM; and (3) the outsider Alwin Walther, who directed the Institute for Practical Mathematics at Darmstadt, which was already increasingly being harnessed by various quarters for war research (e.g., the Army Testing Station in Peenemünde).[82]

Under the direction of Doetsch – an archenemy of his fellow Freiburger Süss – at the beginning of July 1942 Wolfgang Gröbner started to erect an institute for higher mathematical methods at Göring's Aviation Research Institution (*Luftfahrtforschungsanstalt Hermann Göring*) in Braunschweig under the sole authority of the Reich Ministry of Aviation, and it later acquired the name Working Group for Industrial Mathematics (*Arbeitsgruppe für Industriemathematik*). All in all, this highly ambitious enterprise was only able to operate on a quite modest scale.

Süss followed other, higher goals with his plans for a Reich Institute of Mathematics. The importance he attached to such an institute did not hinge on war research alone. It was to be a research establishment continuing work beyond the anticipated victory on a broad spectrum of mathematics. In summer 1942, there were determined struggles behind the scenes to bring about its materialization. After Doetsch had opened the working group for industrial mathematics, Süss began to promote his own project intensely, leaving no room for doubt that the Braunschweig institute was incapable of fulfilling the role he was envisioning. For Süss, interests of the mathematical profession had precedence over the demands of the war. But his efforts only began to bear fruit in early 1944. In February, he addressed a letter to a number of his colleagues for suggestions about the imminent founding of this center and withdrew from other obligations that March in order to draft an official project

[82] On this and the following, see Moritz Epple, Andreas Karachalios, and Volker R. Remmert, "Aerodynamics and mathematics in National Socialist Germany and fascist Italy: A comparative study of research institutes," in Carola Sachse and Mark Walker (eds.), *Politics and Science in Wartime: Comparative International Perspectives on the Kaiser Wilhelm Institutes* [Volume 20 of *Osiris*] (Chicago: University of Chicago Press, 2005), 131–158; Herbert Mehrtens, "Mathematics and war: Germany 1900–1945," in Paul Forman and José M. Sánchez-Ron (eds.), *National Military Establishments and the Advancement of Science and Technology: Studies in Twentieth Century History* (Dordrecht: Kluwer, 1996), 87–134, especially 115–118.

application.[83] At the beginning of August 1944, the official petition for the establishment of a Reich Institute of Mathematics was finally submitted by Walther Gerlach in his capacity as head of the physics section of the RFR. The necessity for such a Reich institute was based on the "lack of mathematical tools and techniques in research and industry" that had evidenced itself in past years in an "ever growing demand for tabulated mathematical functions, equation solving, mathematical machines and calculating apparatus of every sort."[84] The justificatory rhetoric explicitly made pure mathematics a part of the institute's structure. The "first survey of the tasks of the Reich Institute" comprehended three working areas but was not concrete: (1) "Promotion of the mathematical sciences and their applications in the broadest sense," (2) "Development of divisions into *Calculation Institutes* and mathematical *Production Institutes* especially equipped with mathematical instruments and devices," and finally (3) "General tasks," including a "central office for mathematical reporting" as well as the "establishment of a mathematical card catalogue for the best exploitation of human resources and staffing purposes" and the "establishment of a central information office and verification office for mathematical questions."[85] The tasks sketched in the petition primarily concerned not true mathematical activities but rather the organization and concentration of resources, for which Süss as chairman of the DMV certainly did appear to be the right person.

Of course, toward the end of the war, these major plans could no longer be turned into reality. The Reich Institute for Mathematics that began to be established in autumn 1944 under Süss' direction in Oberwolfach had neither a war research program nor the staff that had originally been foreseen. The institute was quickly recast as the location of pure mathematical research, which Süss developed into an international venue now known as the Mathematical Research Institute Oberwolfach (*Mathematisches Forschungsinstitut Oberwolfach*).

THE DMV AND THE DPG: JOINT CONCERN FOR WAR RESEARCH

A regular contact point between the DMV and the DPG since the 1920s was the jointly organized and held Convention of German Physicists

83 Serial letter by Süss without addressee, 1 February 1944, UFA, E6/1, and Süss to Sperner, 29 March 1944, UFA, C89/77.

84 Gerlach's application to the RFR, 2 August 1944, UFA, C89/4.

85 Gerlach's application to the RFR, 2 August 1944, UFA, C89/4.

and Mathematicians, which, as a rule, took place in September. The members of the DMV's board were consequently often guests at the DPG's board meetings in order to coordinate the scientific programs and social events. The files of the DMV reveal little more about official contact with the physics society. A difference of opinion may have arisen over the Jewish question, as the earlier-quoted letter by Debye to the DPG's Jewish members documents, a copy of which is among the DMV's files (see the earlier section on the Jewish question).

During the war, the DMV and the DPG found themselves confronting the same group of problems – the organization of war research.[86] It is known that the DPG's then-chairman, Carl Ramsauer, and its board had submitted a petition to the REM in autumn 1941 to draw attention to the deficits in science and research policy and especially to the promotion of coming generations of physicists.[87] The ministry never responded. In 1942 and 1943, Ramsauer continued to point out faults in the promotion of science and warned about the related threat to the course of the war. In April 1943, for instance, he addressed the German Academy of Aviation Research about the efficiency and organization of physics in Anglo-Saxon countries, adding perspectives on German physics. His admonition was that "in Germany the fundamental belief [must] penetrate that physics is a factor of power of the first order and that funding for the promotion of physics should be gauged against the scale of armaments projects and not the scale of an individual science."[88] Notwithstanding all the discipline-specific myth building and legitimizing rhetoric, the conclusion could hardly be refuted. Ramsauer also extended his feelers on these important issues in the direction of the Reich Aviation Ministry – where he had an important ally in Ludwig Prandtl – and the High Command of the Armed Forces. In June 1943, he met with Minister Albert Speer's personal adjutant at the Reich Ministry of Armament and War Production to discuss the state of German physics, and in July 1943 the topic was on the agenda at a meeting between Ramsauer and Rudolf Mentzel, head of the REM Science Office.[89]

86 An indispensable source on the concept and context of war research is the editor's introduction to Maier, *Rüstungsforschung*, 7–29.

87 See the contribution by Hoffmann to this volume.

88 Carl Ramsauer, "Über Leistung und Organisation der angelsächsischen Physik," *Jahrbuch der Deutschen Akademie für Luftfahrtforschung* (1943/44), 86–88, especially 87.

89 For more details, see Dieter Hoffmann, "Carl Ramsauer, die Deutsche Physikalische Gesellschaft und die Selbstmobilisierung der Physikerschaft im 'Dritten Reich,'" in

At this point, the chairmen of the DMV and the DPG began to cooperate. In mid-July 1943, Süss received the order from Paul Ritterbusch, who was heading the war effort in the humanities at the REM, to give a presentation at the rectors' conference in Salzburg in August. The other speaker at this event with the guiding theme "The effects of total war on science and universities" was the head of the REM Science Office, Rudolf Mentzel. Ritterbusch suggested Süss report about "The current state of German science and German universities" and pointed out the necessity "to discuss this thoroughly in person," asking Süss to come and see him at the ministry that following week.[90] Süss actually did give a report under the suggested title on 23 August 1943 at the rectors' conference.

The exact circumstances of Ritterbusch's invitation are unknown, but Süss was not completely unprepared for it because at a meeting in Freiburg in June, he already asked Ramsauer to send him his talk "On the efficiency and organization of Anglo-Saxon physics with perspectives on German physics," telling him that he intended to conduct a parallel examination of the position of mathematics conducted by English speakers compared with that of Germans. Ramsauer gladly complied and wrote:

> In my current campaign I place great store by your findings; for even though the situation of German mathematics does not make the same impression on the influential authorities as the state of German physics, a surpassing of German mathematics by the USA would be taken as a substantially stronger and alarming symptomatic indicator.[91]

Süss replied with thanks for the "strong support by the strong sister science" and inquired of Ramsauer whether he still had any "substantial complaints about a lack of coordination among the various military authorities interested in physics and other technological research stations." He knew nothing specific about the situations of other sciences, but for mathematics "the coordination issue has been known to be one of our worries for quite some time as well."[92]

The conclusions in mathematics that Süss arrived at in his report at the rectors' conference in September 1943 were similar to Ramsauer's. To illustrate the imbalance of the "scientific war potential of the two competing sides," he drew a comparison between German and Anglo-American

Maier, *Rüstungsforschung*, 273–304, especially 283ff., as well as his contribution to this volume.

90 Ritterbusch to Süss, 14 July 1943, UFA, C89/7.

91 Ramsauer to Süss, 16 June 1943 UFA, C89/11.

92 Süss to Ramsauer, 30 June 1943, UFA, C89/11.

mathematical productivity, following Ramsauer's example. His analysis of citations from periodicals in 1937 yielded the result that the importance of German mathematics compared with that of American mathematics had sunk rapidly since the late 19th century, and especially during the 1930s. He explicitly pointed to the similar finding Ramsauer had reached for physics.[93] From it he drew the transparent conclusion that "the scientific potential of Germany" had to be exploited "in an optimal way and fully and completely for the war effort." In particular he demanded, a little self-servingly as DMV chairman, "besides certain possible institute improvements for subjects decisive to the war, an immediate perceptible increase in staff employees."[94] Although Süss put on record that he was not thinking of the establishment of new institutes, his interest in the founding of a Reich Institute of Mathematics was on file at the REM. Süss' admission "that the longer the war lasts, the larger a weight scientific discoveries can have on its outcome" should also be regarded against this background of professional policy making.[95] From the abundance of problems resulting from enlisting the sciences in the war effort, Süss named two particularly important ones; namely, (1) the "official competency difficulties" (i.e., the polycratic jumble in which necessities frequently went under in jealous bickerings) and (2) the low "ideal and material" position of German university teachers, which seemed hardly suited "to attract the best abilities of the nation" to science.[96]

His review was full of concrete suggestions for reform applicable to the entire scientific enterprise. It particularly demanded that scientists urgently needed by the "scientific war effort" and who would "perhaps decide the war" be granted an exemption from direct military service.[97] So Süss was not shy with clearly criticizing the known conditions that Ramsauer had also pointed out before him. At the same time, the professional policy line of strengthening mathematics as an important science for war research runs throughout the review like a subtext. How serious Süss was about his argumentation is documented by a list of officials to whom he had sent copies of a private printing of his talk: Martin Bormann and his deputy Gerhard Klopfer; Karl Dönitz; the head of the Army Ordnance Office, Fritz Fromm; the head of the Reich Chancellery, Hans Lammers;

93 Wilhelm Süss, *Die gegenwärtige Lage der deutschen Wissenschaft und der deutschen Hochschulen* (Freiburg: private publication, 1943).

94 Süss, *gegenwärtige*, 7.

95 Süss, *gegenwärtige*, 8.

96 Süss, *gegenwärtige*, 8 and 15.

97 Süss, *gegenwärtige*, 8.

the state secretary at the Reich Aviation Ministry, Erhard Milch; the director of the Planning Office at the RFR, Werner Osenberg; and the Reich Youth Leader, Baldur von Schirach.[98] Even Heinrich Himmler had read the report, Süss contended in January 1945. At that time, an SS agent had informed him that "I should in future contact the Reichsführer directly, if I have any such important things on my mind again."[99]

One historian's evaluation of this talk as having been a "brave, fearless rejection of political puppet-strings" is at least minimized by these circumstances.[100] Rather, Süss' Salzburg report took up a topic that, after consultation with the responsible officials at the REM, Ramsauer had been trying to propagate among various authorities almost 2 years before. It is hard to assess what impression the Salzburg report left as a single piece in the mosaic. However, it demonstrates coordination between the interests of the DMV and the DPG in such important issues as war research. Ramsauer's and Süss' demand for a dispensation of scientists from active duty was even granted at the end of 1943. By the end of 1944, the "Osenberg Action" (*Aktion Osenberg*) had "recalled" more than 3,000 scientists and engineers from the battlefield.[101] At the same time, the legitimizing subtext for mathematics in the Salzburg report provides a smooth transition for Süss' activities for the founding of a Reich Institute of Mathematics. These efforts, characterized by a collaboration with physics, incidentally also bore fruit. Süss had managed to secure support from Gerlach, who directed the physics section at the RFR (see the earlier section on mathematical war research).

CONCLUDING REMARKS: THE EVOLVING DMV

When Wilhelm Süss took office as chairman in 1937, the DMV, which in any case since 1933 had not been a powerful instrument of professional policy, had gone strongly on the defensive against Bieberbach and the Aryan mathematics camp. Bieberbach's threats to found a National Socialist League of Mathematicians exposed serious limitations to the

98 See the enclosure letter in: UFA, C89/73, Süss papers.

99 Süss to Gerlach, 25 January 1945, UFA, C89/12.

100 According to Notker Hammerstein, *Die Deutsche Forschungsgemeinschaft in der Weimarer Republik und im Dritten Reich. Wissenschaftspolitik im Republik und Diktatur* (Munich: Beck, 1999), 462f.

101 Ruth Federspiel, "Mobilisierung der Rüstungsforschung? Werner Osenberg und das Planungsamt im Reichsforschungsrat 1943–1945," in Maier, *Rüstungsforschung*, 72–105, especially 89.

DMV's claim to sole representation of mathematicians at German universities. Since 1935, its elections for chairman depended on the REM's approval. The DMV's scope of action was clearly constrained, and at the same time its reputation and therefore its attractiveness abroad suffered strongly from the Harald Bohr affair, notwithstanding a widespread political distancing from the National Socialist regime. So in 1937, Süss assumed the chair of an association whose best days appeared to be numbered.

During his 8 years in office, Süss expressed his goals for the DMV on various occasions. He never concealed his ambition to lend it a voice at government agencies. As he stated in a report to his fellow board members in April 1938, it was a matter of "extending feelers in a timely fashion" to the relevant authorities. Thereby, the DMV had to "be able to act in such a form that it stood in every respect, not only purely professionally, but also politically, as the sole legitimate advisor on mathematics for universities."[102] He repeatedly spoke in no uncertain terms in this sense in the years that followed. As we have seen, Süss pursued the self-assigned "imperialistic goal of procuring for her alone [the DMV] all the rights and obligations of mathematics" consistently, intensely, and with success.[103]

It was not just on exceptional issues like the restructuring of professional periodicals, the Jewish question, or the organization of war research that the DMV collaborated, in the person of its Chairman Süss, since the late 1930s in a relationship of trust with the REM. It also had the ministry's ear on everyday and longer-term questions of no less importance. Süss was consulted by the REM on practically all professorship appointments in the discipline, and as a rule his recommendations, which he expressed on the basis of the opinions of his fellow colleagues in the field, were followed as well.[104] This could not be taken for granted, as the refilling of the theoretical physicist Arnold Sommerfeld's position at Munich exemplifies. The relevant minutes for the DPG's board meeting in May 1943 record the wish that the society be involved "in an influential way at professorship appointments, etc."[105]

102 Süss to Hasse, Müller, and Sperner, 23 April 1938, UFA, E4/58.

103 Süss to Feigl, 3 April 1941, UFA, C89/51.

104 For example, see Süss' expert opinion on Gottfried Köthe, 16 January 1941, UFA, C89/53. For the REM's consultations, Süss had prepared a list of mathematicians deemed appointable candidates along with possible references. See UFA, C89/88.

105 Report about the board meeting on 31 May 1943, Archives of the German Physical Society (*Archiv der Deutschen Physikalischen Gesellschaft*, DPGA) no. 10023.

The DMV's success was possible, in my view, because the association and Süss substantially proved their reliability to the REM, also in the ideological sense, when the Jewish question was current at the DMV. Additionally, the advantages of continuity at the top of the DMV were obvious. With Süss, the DMV gained a diplomatically versed chairman who knew how to move on the political platform and to work toward the goals of party and government authorities.

The extent of the collaboration between the DMV and National Socialist agencies was not generally known after the war, and only very few mathematicians in Germany saw reason to maintain critical distance from the DMV or its longtime chairman, Wilhelm Süss.[106] In Germany reference was made, as if it were a matter of course, to the "compelling circumstances."[107] The situation was very different abroad. In 1950, Harald Bohr asked the chairman of the DMV (refounded in 1948), Erich Kamke, whether Süss was not "an opportunist of the not entirely harmless sort."[108] For Max Dehn, who as a Jew had been dismissed in Frankfurt in 1935 and had emigrated to the United States in 1941, the events of 1935 were already reason enough to decline Chairman Kamke's invitation to rejoin the DMV, even while emphasizing his sense of solidarity with mathematicians in Germany:

> *I cannot rejoin the Deutsche Mathematiker Vereinigung.* I have lost the confidence that such an association would act differently in the future than in 1935. I fear it would, once again, not resist an unjust demand coming from outside. The D.M.V. did not have to take care of very important values. That it did not voluntarily dissolve itself in 1935, and not even a considerable number of members left the association, leads me to this negative attitude. I am not afraid that the new D.M.V. will again expel Jews, but maybe next time it will be so-called communists, anarchists or "colored people."[109]

106 See Remmert, "Mathematical," 28f.; Remmert, "Mathematicians," 49–51.

107 The words "Zwang der Verhältnisse" appear not once but twice in the preface to the first post-war volume of the DMV's annual report: *Jahresbericht der Deutschen Mathematiker-Vereinigung*, 54 (1951).

108 Bohr to Kamke, 27 February 1950, UFA, E4/532.

109 Dehn to Kamke, 13 August 1948; cited after Siegmund-Schultze, *Mathematicians*, 393 (original emphasis).

9

"To the Duce, the Tenno and Our Führer: A Threefold Sieg Heil"

The German Chemical Society and the Association of German Chemists during the Nazi Era

Ute Deichmann

The German Chemical Society (*Deutsche Chemische Gesellschaft*, DChG) and the Association of German Chemists (*Verein deutscher Chemiker*, VdCh) were the two most important organizations representing German chemists. They were founded when the discipline and the industry based on aromatic organic chemistry were rapidly taking shape just after the mid-19th century. By 1900, Germany had become the international leader in chemistry, with Jews figuring prominently in both academic and industrial developments. This was demonstrated also in their strong participation in the activities of the two chemical societies. Consequently, anti-Semitic measures, which were initiated in 1933, had a drastic effect on staffing in laboratories, management, and the editorial offices of scientific journals and handbooks. Because of the international standing of the chemical societies (of a total membership of more than 4,000 in the DChG in 1932, for instance, about 40 percent were non-Germans), the political conformance displayed by the societies' prominent members during the National Socialist era had far-reaching and lasting consequences on chemistry in Germany.

This contribution examines the influence of National Socialist policies on the DChG and the VdCh.[1] The way in which they responded to the political conditions during the Nazi period is analyzed, including measures these societies took against their Jewish members and employees and statutory alterations and the propagation of ideological and political

[1] An account of chemists and biochemists during the Nazi period and of their science can be found in: Ute Deichmann, *Flüchten, Mitmachen, Vergessen. Chemiker und Biochemiker in der NS-Zeit* (Weinheim: Wiley-VCH, 2001).

goals as reflected in their publications. Finally, a comparison is made between these two chemical societies and the German Physical Society within the context of certain crucial differences between the disciplines of chemistry and physics. As will be shown, the economic and military importance of the two disciplines as well as the quite different scientific, economic, and political roles played by Jewish scientists within these two disciplines were particularly significant.

I start with a short review of the histories of both chemical societies up to 1933.

A BRIEF HISTORY OF THE DCHG AND THE VDCH BEFORE 1933

The DChG was founded in Berlin in 1867, a few years before the founding of the German Empire in 1871. This association of academic and other chemists sought to encourage contacts between its members as a means of contributing toward the dissemination of new scientific findings. In 1868, the first issue of the society's journal, *Berichte der Deutschen Chemischen Gesellschaft*, appeared and quickly gained an international readership (today it is *Chemische Berichte*).

The rapid expansion of the chemical industry, foremost in the field of synthetic dyestuffs, resulted in the founding in 1877 of another professional association, the *Verein Analytischer Chemiker*, which represented the "practically employed chemists." Ten years later, it adopted the name Society for Applied Chemistry (*Gesellschaft für Angewandte Chemie*), and in 1896 it became the *Verein deutscher Chemiker* (VdCh; from around 1940 VDCH). Its members received the *Zeitschrift für Angewandte Chemie*. Before the First World War, the VdCh was the largest chemical association in the world with local branches in many countries. After the war, the difference between the two societies – representation of scientific chemists versus promotion of professional interests – became increasingly blurred.[2]

The DChG's founding president was the chemist August Wilhelm Hofmann (later von Hofmann). His isolation and identification of aniline in coal tar laid the foundations of the dye industry, initially in England in the 1850s, but after 1865 in Germany. Hofmann exerted a strong influence on the DChG for almost two decades with respect to not only its scientific standards but also its avowed liberal and internationalist

[2] On the history of the DChG, see Walter Ruske, *100 Jahre Deutsche Chemische Gesellschaft* (Weinheim: Verlag Chemie, 1967).

policy. After completing his studies under Justus von Liebig, he became a professor at the Royal College of Chemistry in London in 1845 and accepted a chair in chemistry at the University of Berlin in 1865. The two-decade stay in England contributed toward his international reputation and liberal leanings. Hofmann actively defended these principles. He spoke up, for instance, in defense of the rights of his Jewish fellow university teachers. Religious-based restrictions on an academic career at German universities had *de jure* been eliminated with the founding of the German Reich in 1871. However, the situation was very different in practice. A virulent form of anti-Semitism developed and became organized during the economic depression shortly after the birth of the empire. It soon infected universities. The Anti-Semitic League founded in Berlin in 1879 demanded that the legal equality granted Jews be largely repealed. Academics at the University of Berlin published books, distributed leaflets, and gave talks spreading this animosity toward Jews. A particularly prominent figure in this regard was the historian Heinrich von Treitschke, notorious for his slogan: "The Jews are our misfortune." He and his fellow anti-Semites clamored for revocation of the right of Jews to academic teaching posts, leading to what came to be known as the "Berlin anti-Semitism Controversy" at the university. In response, Hofmann joined a group of influential liberals censuring the attacks by von Treitschke and his confederates. Hofmann, rector of the university from 1878, was one of 73 prominent Berliners to sign a public "Declaration of notables." The historian Theodor Mommsen and the physician Rudolf Virchow also signed it.[3]

Several chemists criticized Hofmann's liberal policies and his active opposition to Jewish discrimination. The organic chemist and *Ordinarius* at the University of Leipzig Hermann Kolbe was one of the most prominent of these critics. Like Hofmann, Kolbe enjoyed the highest esteem in Germany, but his political attitude was governed by his extreme chauvinism. Around 1870, he began to direct his anti-French prejudices against Jews. His particular hatred of Berlin Jews became evident in a controversy with the DChG: a scientific issue was at the core of this dispute. Kolbe was among a minority of European chemists refusing to accept the unique constitution-structural, or type, theory of organic chemistry

[3] See Kurt Mendelssohn, *Walther Nernst und seine Zeit. Aufstieg und Niedergang der deutschen Naturwissenschaften* (Weinheim: Physik-Verlag, 1973), 40; Ruske, *100 Jahre*, 48; Alan J. Rocke, *The Quiet Revolution. Hermann Kolbe and the Science of Organic Chemistry* (Berkeley: University of California Press, 1993), 358.

that originated in France. Embroiled in polemics with French chemists, he expected backing by the DChG and was stunned when the *Berichte der Deutschen Chemischen Gesellschaft* published a letter from the Russian Chemical Society protesting against the nationalistic attitude of German chemists like Kolbe. Thus, he struck back not just against the Russian and German chemical societies but also against Hofmann and his Jewish "henchmen."[4] In Kolbe's opinion, too many of the DChG's members were Jews or of Jewish extraction, and in 1871 he complained to the president of the society, Adolf Baeyer (later von Baeyer): "The Chemical Society is after all already known as a hotbed of Jewry in chemistry."[5]

Before 1933, voices like Kolbe's did not make up a majority in the DChG. Ironically, this lambasted "Jewry in chemistry" had helped German chemistry and its societies earn its international reputation since the 1860s. By 1918, four of the six German Nobel laureates in chemistry were Jews or of Jewish origin (von Baeyer, Fritz Haber, Otto Wallach, Richard Willstätter), and other names point to the importance of Jewish chemists in the German dye industry: Heinrich Caro, who founded the first central research laboratory in industry (at BASF), Paul Mendelssohn-Bartholdy, and Arthur von Weinberg, to name but a few. The DChG and the VdCh had many Jewish members and, as is shown later, Jewish chemists played an active part in editorial posts with the societies' periodicals and handbooks. Among the DChG's 28 former presidents and vice-presidents alive in 1933, eight (i.e., 28 percent) were Jewish or of Jewish ethnic origins: Herbert Freundlich, Fritz Haber, Kurt H. Meyer, Carl Neuberg, Arthur Rosenheim, Arthur von Weinberg, Richard Willstätter, and Arthur Wohl.

After the First World War, Jewish chemists were conspicuously involved in reintegrating the German societies within the international community of chemists from which they had been excluded through the boycott against German scientists imposed by the Allies. Ten years after the founding of the *Union Internationale de Chimie* in 1918, which had barred membership of German scientists, the DChG, the VdCh, and the Bunsen Society joined together to form a confederation of German associations under the presidency of Fritz Haber. Finally, in 1929 it entered the *Union Internationale*. It is remarkable that this took place precisely during Haber's term, because Haber had originally been placed on the

4 Rocke, *The Quiet Revolution*, 354.

5 See Rocke, *The Quiet Revolution*, 355. Kolbe seems not to have known that Baeyer's mother was a – baptized – Jew.

FIGURE 24. Fritz Haber and Albert Einstein around 1914 (*Source*: MPGA)

Allies' list of war criminals for having initiated and organized the deployment of chemical weapons in the First World War. In particular, the physical chemist Ernst Cohen, from Utrecht, was instrumental in the resumption of contacts between Western European scientists and their fellow German chemists.[6] That is why volume 130 of the *Zeitschrift für Physikalische Chemie* (1927) was dedicated to him: "Honorary volume to Mr. Ernst Cohen, the successful researcher and indefatigable pioneer in the resumption of peaceful relations between scholars from the nations

[6] Levi Tansjö, "Die Wiederherstellung von freundschaftlichen Beziehungen zwischen Gelehrten nach dem 1. Weltkrieg. Bestrebungen von Svante Arrhenius und Ernst Cohen," in Gerhard Pohl (ed.), *Naturwissenschaften und Politik, Proceedings of papers presented at the University of Innsbruck, April 1996* (Vienna: Gesellschaft Österreicher Chemiker, 1997), 72–80.

divided by the war." Cohen's subsequent destiny highlights dramatically and tragically the break with tradition in German science during the Nazi era. Cohen was arrested in February 1943 for not wearing his "star of David" while at work in his laboratory in Utrecht. In March 1944, he was murdered in Auschwitz at the age of 74.

The following sections describe and analyze the reactions of the two chemical societies to the political redistribution of power and the subsequent developments until 1945.

THE ASSOCIATION OF GERMAN CHEMISTS, 1933–1945

The chronological table on the home page of the Society of German Chemists (*Gesellschaft Deutscher Chemiker*, GDCh) lists the following events for its precursors, the DChG and the VdCh, during the National Socialist era[7]:

1934 First award of the Joseph König Memorial Medal (to Adolf Beythien).

1936 New statutes of the VdCh with Nazi content on May 16. The new statutes of the DChG avoid such Nazi formulations. First award of the Carl Duisberg Memorial Prize (to Rudolf Tschesche). Founding of the Division of Plastics.

1938 Integration of the society into the umbrella organization National Socialist League of German Engineers.

1942 Relocation of the "Imperial Division Chemistry" to Frankfurt am Main, Bismarckallee 25 (today Theodor-Heuß-Allee, seat of the DECHEMA, Society for Chemical Technology and Biotechnology (Gesellschaft für Chemische Technik und Biotechnologie)).

1943 Move of the Imperial Division Chemistry to Bockenheimer Landstr. 10 (former Palais Rothschild), Rudolf Wolf becomes secretary general.

1944 The DChG offices in the Hofmannhaus and the VdCh offices destroyed in air raids on January 29 and March 18, respectively. The VdCh office is transferred to Grünberg/Hessen.

These few events suggest that the daily business of both societies was generally normal, and with minimal change, throughout the Nazi period

[7] For the history of the GDCh and its predecessor organizations, see www.gdch.de/gdch/historie.htm, dated October 2004. (The web page in English is shortened.)

until their offices were destroyed in the war. Only the allusion to the statutory changes and the integration into the National Socialist League of German Engineers (*Nationalsozialistischer Bund Deutscher Technik*, NSBDT), a party-controlled confederation of associations in the technical sciences, suggest some political realignment. The fact that the Imperial Division Chemistry was a part (*Reichsfachgruppe*) of the NSBDT is not mentioned. The assertion, "The new statutes of the DChG avoid such Nazi formulations" suggests that such formulations *had been* added to the VdCh's statutes. The discussion below examines details of the political realignment by the *Verein* and its promotion of ideological and political goals.

The VdCh, with a membership that brought together university teachers, students, and industrial chemists, instituted the autocratic *Führerprinzip* quite quickly in 1933 under its chairman, Prof. Paul Duden (a member of the board of the Hoechst dye company), incorporated political goals into its statutes, and openly proclaimed allegiance to the Nazi state. During the general assembly in Würzburg in June 1933, its board formally named Duden the *Verein*'s *Führer*.[8] The journal *Angewandte Chemie* printed excerpts from the speech delivered by the engineer and party ideologue Gottfried Feder.[9] The minutes of the session paraphrase Feder emphasizing:

> Especially chemistry, with its extraordinary potential for development, must regard service of the nation as its highest priority. Hitherto, German chemists, just as German technicians generally, had busied themselves far too much with their own pure specialties, without keeping in view the political development of their country.... [Feder] could note with satisfaction, however, that since the coming to power by our People's Chancellor *Adolf Hitler*, the Association of German Chemists had now also understood the signs of the times and through the voice of her chairman had expressed her readiness to collaborate consciously on building up a new German national state.... Feder's speech, often interrupted by enthusiastic applause, left a deep impression on the assembly and came to an impressive close with three cheers [*Sieg Heil*!] to the German Fatherland and its chancellor *Adolf Hitler* along with the national anthem and the Horst Wessel song.[10]

The modifications to the statutes extended the association's goals. Henceforth, its furtherance of chemistry involved "scientific and technical

[8] *Angewandte Chemie*, 46 (1933), 789.
[9] *Angewandte Chemie*, 46 (1933), 369.
[10] *Angewandte Chemie*, 46 (1933), 369 (original emphasis).

inspiration and promotion as well as education of her members for the National Socialist people's community."[11] In 1934, the association joined the Reich Association for the Technical and Scientific Work (*Reichsgemeinschaft der technisch-wissenschaftlichen Arbeit*, RTA; transformed into the NSBDT in 1937), an amalgamation of Nazi-led associations in the technical sciences headed by Fritz Todt.[12] At that time, Todt was director of the Technical Department (*Hauptamt Technik*) of the NSDAP, becoming Reich Minister for Armament and Munitions in 1940. The majority of the new members of the *Verein*'s board were committed Nazis, many of whom had joined the NSDAP even before 1933.

The members of the board in 1935 were Prof. Paul Duden, not an NSDAP member, as chairman (he was replaced a few years later by Dr. Karl Merck, director of the Chemistry Division of the NSBDT); the engineer Dr. Kurt Stantien, a party member since 1925, as vice-chairman; and Dr. Walter Schieber, a party member since 1931 in addition to being a member of the *Schutz-Staffel of the NSDAP* (SS) where he became *Oberführer*, as treasurer.

The members of the smaller council in 1935 were Dr. Gustav Baum (joined the NSDAP in 1933, also the *Sturmabteilung* of the NSDAP [SA]), Prof. Burckhardt Helferich (not a party member), Dr. Hermann Kretschmar (joined the NSDAP in 1932), Dr. Karl Merck (joined the NSDAP in 1933), Prof. Rudolf Pummerer (not a party member), Prof. Otto Ruff (party affiliation not known), Dr. Hans Wolf (joined the NSDAP first in 1921, then in 1930), and Dr. Fritz Scharf (joined the NSDAP in 1933).

The association's main board apparently asked the boards of its local branches to supply information about the party affiliations of their members in 1933. Adolf Windaus wrote to one of his former students, Adolf Butenandt, in August 1933: "A request was forwarded to me from the Association of German Chemists that I declare which members of our local branch belong to the NSDAP. I am not going to respond, of course, and am withdrawing membership in the local branch."[13] Such brave and uncompromising decisiveness was characteristic of Windaus, who, from

[11] *Angewandte Chemie*, 46 (1933), 789.

[12] Records by Dr. H. Ramstetter and other materials about the VdCh he had compiled. They are deposited at the offices of the GDCh, Frankfurt am Main. See also Ruske, *100 Jahre*, 152.

[13] Windaus to Butenandt, 20 Aug. 1933, Archives of the Max Planck Society (*Archiv der Max-Planck-Gesellschaft*, hereafter MPGA), div. III, 84/1, Butenandt papers.

1933, did not flinch from expressing his repudiation of Nazism. Comparable reactions by other chemists are not known.[14]

The journal *Angewandte Chemie* and, in particular, its news supplement, *Der deutsche Chemiker. Mitteilungen aus Stand/Beruf und Wissenschaft*, first issued in 1935, published, apart from scientific articles, essays about the importance of chemistry in furthering the goals of the Nazi's Four-Year Plan, along with public appeals for its economic and military goals. Ideological and political approval by the association also came in the form of repeated declarations of loyalty to Hitler and the Nazi regime, as the following examples clearly illustrate. Under the heading "Tasks of chemistry in the new Germany," the editors published an appeal to the nation by the Nazi party in 1935 to cultivate and collect medicinal herbs; the lecturer Ilse Esdorn pointed out the role of chemistry in research on these kinds of plants.[15] Following the trend for a holistic conception of science and society, Alwin Mittasch published an article in 1936 on the topic "On wholeness in chemistry."[16] Karl Merck, chairman of the VdCh in 1940, emphasized the importance of chemistry for the war effort and called on German chemists to attend the national convention in Breslau with the words: "The professional research of the Reich convention should help whet the sword forged of German chemistry, further the efficiency of our professional comrades, and bear testimony before the whole world to the unbroken might of German chemical expertise and its deployment for the Führer and the Reich."[17] A month later, the journal published appeals to the German people, German technology, and German business by Hermann Göring, Todt, and Albert Pietzsch (president of the Reich Chamber of Commerce) to make donations of metals.[18] In 1941, the editors of the journals *Angewandte Chemie* and *Die Chemische Fabrik* explained to their readers that it was "one of our duties, . . . to report about the major goals staked out for German research during its final battle and in peacetime."[19] In the same year, a full-page notice appeared under the heading "Leadership in German technology" in celebration of Todt's 50th birthday.[20] An article by Ferdinand Sauerbruch "On the current state of all the sciences and medicine" closed with

[14] Windaus's standing during the Nazi period is discussed in Deichmann, *Flüchten*, 83f.

[15] "Aufgaben der Chemie im neuen Deutschland," *Angewandte Chemie*, 48 (1935), 255.

[16] "Über Ganzheit in der Chemie," *Angewandte Chemie*, 49 (1936), 417–420.

[17] *Der Deutsche Chemiker*, 6/1 (1940), 3.

[18] *Der Deutsche Chemiker*, 6/4 (1940), 11

[19] *Der Deutsche Chemiker*, 7/1 (1941), 1.

[20] "Führung der deutschen Technik," *Der Deutsche Chemiker*, 7/3 (1941), 11.

the words: "May this convention be an avowal of these goals. Long live Germany! Long live our Führer!"[21]

The attitude of the association's board during the debates about the planned memorial celebration in honor of Fritz Haber in 1935 is examined in a later section.

THE GERMAN CHEMICAL SOCIETY, 1933–1945

1933–1936: The Removal of Prominent Jewish Members and the Dismissal of Jewish Employees

Unlike the VdCh, the DChG did not immediately decide to join the RTA (later NSBDT) in 1933. No Nazi ideologues made their appearances at the assembly of its members on 6 May 1933 and no "Sieg Heil!" cheers resounded in applause – at least until that point. Some Gentile members of the board thought the most pressing problem was how best to remove their fellow Jewish members on the board without tarnishing the society's international image. A subtle method was contrived to immediately vacate Jewish members from their leadership posts in the society, including from the presidency itself. Alfred Wohl had been elected to the presidency 1 year before, in 1932; his deputy was Arthur Rosenheim. They, along with the long-time editor of the renowned *Beilstein* handbook, Bernhard Prager, were asked to tender their resignations. The *Berichte der Deutschen Chemischen Gesellschaft* commented on these measures of anticipatory obedience – a small fraction of the "Aryanization" under way – in its report on the members' assembly in 1933 as follows:

> The current president, Privy Councilor A. Wohl (Danzig) and the vice-president Prof. A. Rosenheim (Berlin) have, in consideration of the domestic political situation, made their positions available at the end of the first year of their terms; likewise the science editor of the *Berichte*, Prof. M. Bergmann (Dresden). The honorary committee member Prof. O. Warburg (B[erlin]-Dahlem) has resigned his post owing to other obligations.
>
> The board ratified this at a meeting held on the day of the plenary session, regretfully but in full appreciation of the situation.[22]

The Berlin inorganic chemist Karl Andreas Hofmann became the society's new president in 1933; Paul Duden became vice-president.

[21] Über die gegenwärtige Lage der gesamten Naturwissenschaften und der Medizin," *Der Deutsche Chemiker*, 7/4 (1941), 13–16. Sauerbruch's talk was presented in 1936.

[22] *Berichte der Deutschen Chemischen Gesellschaft*, 66 (1933), 57–58.

TABLE 3. *The DChG's Changing Membership from 1933*

Year	Members	Withdrawals	Cancellations Due to Unpaid Fees
1932	4,157		
1933	3,944	238	246
1934	3,723	139	233
1940	3,369		
1941	3,465		

Heinrich Hörlein, on the board of the chemical trust *I.G.-Farbenindustrie*, was the main impetus behind the scenes for the exclusion of Jews from leadership positions in the DChG. As Dr. Hermann Kretschmar (the head of the Chemistry Group within the *Kampfbund für deutsche Kultur*) saw it, Hörlein was "the only member of the [DChG's] board to put a stop to the estranging Jewish imbalance."[23] When Kretschmar complained on 23 April 1933 that "The last [board] meeting nevertheless took place under the chairmanship of Rosenheim,"[24] Hörlein responded 2 days later: "The plenary session will be led neither by Privy Councilor Wohl nor by Professor Rosenheim, but by Prof. Binz. The two aforementioned gentlemen will be making their mandates as members of the board available so that the changed circumstances of the times can be taken into account without damaging the Society's reputation abroad."[25]

The society's concern for its reputation abroad, behind which were tangible business interests due to the high number of the society's publications dispatched beyond the country's borders, influenced the society's procedure for removing its Jewish members in the following years. There are no exact figures on these expulsions. Based on information in the relevant issues of the *Berichte*, the DChG's changing membership statistics after 1933 proceeded as shown in Table 3. Presumably, most of these withdrawals and cancellations involved Jewish members. Between 1932 and 1940, the total membership shrank by 788. The rise in 1941 seems to have come in the wake of the German military victories.

[23] Kretschmar to Hörlein, 23 April 1933, DChG materials at the offices of the GDCh, Frankfurt am Main.

[24] Kretschmar to Hörlein, 23 April 1933, DChG materials at the offices of the GDCh, Frankfurt am Main.

[25] Kretschmar to Hörlein, 23 April 1933, DChG materials at the offices of the GDCh, Frankfurt am Main.

TABLE 4. *Jewish Science Editors and Editorial Staff of Periodicals and Handbooks of the DChG*

Publication	Individuals Employed
Berichte	Max Bergmann, Fritz Haber, Carl Neuberg, Richard Willstätter
Beilstein	Bernhard Prager, Dora Stern
Editorial office	
Zeitschrift für Physikalische Chemie	Emil Abel, Max Born, Georg Bredig, Ernst Cohen, Herbert Freundlich, Kasimir Fajans, James Franck, Viktor Goldschmidt, Fritz Haber, Gustav Hertz, Georg von Hevesy, Hartmut Kallmann, Rudolf Ladenburg, Hermann Mark, Lise Meitner, Fritz London, Kurt H. Meyer, Friedrich Paneth, Michael Polanyi, Ernst Riesenfeld, Otto Stern, Fritz Weigert
Zeitschrift für Elektrochemie	Georg Bredig, Kasimir Fajans, Fritz Haber

The society's worries about its image abroad did not, however, prevent it from promptly "purging" Jewish scientists and employees from the editorial offices for its periodicals and handbook series.

Jewish Science Editors of Journals. The *Berichte* was the first periodical to have its editorial staff "purged" of its Jewish employees – as early as May 1933. Its science editors up to that time had been Max Bergmann, Fritz Haber, Karl A. Hofmann, Wilhelm Marckwald, Carl Neuberg, Max Volmer, and Richard Willstätter. Afterward, the remaining editors, Hofmann and Volmer, were joined by Hermann Leuchs. Haber, Marckwald, Neuberg, and Willstätter were Jewish.

The title pages of two journals shown in Figure 25 exemplify the drastic effect the exclusion of Jews from editorial offices had on periodicals and other publications in chemistry, such as the *Beilstein-Handbuch für organische Chemie.*

The editors or other members of the DChG editorial staff listed in Table 4 were Jewish. ("Jewish" here is used in the general sense comprehending converts to Christianity of Jewish origin and individuals for whom only one parent was Jewish.) This list is not complete as in some cases biographical information is missing.

Six employees were immediately dismissed from the *Beilstein* editorial office in 1933. The editor Bernhard Prager and one other member of his

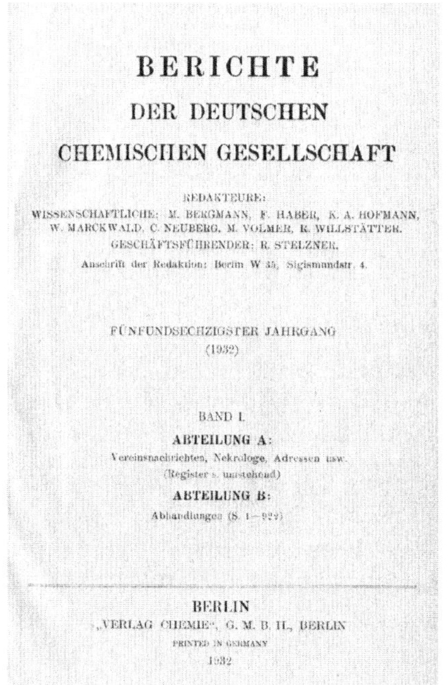

BERICHTE
DER DEUTSCHEN
CHEMISCHEN GESELLSCHAFT

REDAKTEURE:
WISSENSCHAFTLICHE: M. BERGMANN, F. HABER, K. A. HOFMANN,
W. MARCKWALD, C. NEUBERG, M. VOLMER, R. WILLSTÄTTER.
GESCHÄFTSFÜHRENDER: R. STELZNER.
Anschrift der Redaktion: Berlin W 35, Sigismundstr. 4.

FÜNFUNDSECHZIGSTER JAHRGANG
(1932)

BAND I.
ABTEILUNG A:
Vereinsnachrichten, Nekrologe, Adressen usw.
(Register s. umstehend)
ABTEILUNG B:
Abhandlungen (S. 1—922)

BERLIN
„VERLAG CHEMIE", G. M. B. H., BERLIN
PRINTED IN GERMANY
1932

BERICHTE
DER DEUTSCHEN
CHEMISCHEN GESELLSCHAFT

REDAKTEURE:
WISSENSCHAFTLICHE: K. A. HOFMANN, H. LEUCHS,
M. VOLMER.
GESCHÄFTSFÜHRENDER: R. STELZNER.
Anschrift der Redaktion: Berlin W 35, Sigismundstr. 4.

SIEBENUNDSECHZIGSTER JAHRGANG
(1934)

BAND I.
ABTEILUNG A:
Vereinsnachrichten, Nachrufe, Adressen usw.
(Register s. umstehend)
ABTEILUNG B:
Abhandlungen (S. 1—1133)

BERLIN
„VERLAG CHEMIE", G. M. B. H., BERLIN
PRINTED IN GERMANY
1934

ZEITSCHRIFT
FÜR
PHYSIKALISCHE CHEMIE
BEGRÜNDET VON
WILH. OSTWALD UND J. H. VAN 'T HOFF

UNTER MITWIRKUNG VON

ABEL-WIEN, BAUR-ZÜRICH, BENEDICKS-STOCKHOLM, BENNEWITZ-JENA, BILTZ-HANNOVER, BJERRUM-KOPENHAGEN, BONHÖFFER-FRANKFURT A. M., BORN-GÖTTINGEN, BRAUNE-HANNOVER, BREDIG-KARLSRUHE, BRÖNSTED-KOPENHAGEN, CENTNERSZWER-WARSCHAU, CHRISTIANSEN-KOPENHAGEN, COEHN-GÖTTINGEN, COHEN-UTRECHT, DEBYE-LEIPZIG, EBERT-WÜRZBURG, EGGERT-LEIPZIG, EUCKEN-GÖTTINGEN, v. EULER-STOCKHOLM, FAJANS-MÜNCHEN, FOERSTER-DRESDEN, FRANCK-GÖTTINGEN, FREUNDLICH-BERLIN, FRUMKIN-MOSKAU, FÜRTH-PRAG, GERLACH-MÜNCHEN, H. GOLDSCHMIDT-GÖTTINGEN, V. M. GOLDSCHMIDT-GÖTTINGEN, GRIMM-LUDWIGSHAFEN, HABER-BERLIN, HAHN-BERLIN, v. HALBAN-ZÜRICH, HANTZSCH-DRESDEN, HENRI-MARSEILLE, HERTZ-BERLIN, HERZFELD-BALTIMORE, v. HEVESY-FREIBURG I. BR., HINSHELWOOD-OXFORD, HUND-LEIPZIG, HÜTTIG-PRAG, JOFFÉ-LENINGRAD, KALLMANN-BERLIN, KOSSEL-KIEL, KRÜGER-GREIFSWALD, LADENBURG-BERLIN, LANDÉ-TÜBINGEN, LE BLANC-LEIPZIG, LE CHATELIER-PARIS, LONDON-BERLIN, LUTHER-DRESDEN, MARK-LUDWIGSHAFEN, MECKE-HEIDELBERG, MEITNER-BERLIN, MEYER-LUDWIGSHAFEN, MITTASCH-OPPAU, MOLES-MADRID, NERNST-BERLIN, J. UND W. NODDACK-BERLIN, PANETH-KÖNIGSBERG, POLANYI-BERLIN, RIESENFELD-BERLIN, ROTH-BRAUNSCHWEIG, SCHMIDT-MÜNSTER, SCHOTTKY-BERLIN, SEMENOFF-LENINGRAD, SIEGBAHN-UPSALA, SMEKAL-HALLE, SVEDBERG-UPSALA, STERN-HAMBURG, TAYLOR-PRINCETON, THIEL-MARBURG, TUBANDT-HALLE, VOLMER-BERLIN, WALDEN-ROSTOCK, v. WARTENBERG-DANZIG, WEGSCHEIDER-WIEN, WEIGERT-LEIPZIG, WINTHER-KOPENHAGEN, WOLF-KIEL UND ANDEREN FACHGENOSSEN

HERAUSGEGEBEN VON
M. BODENSTEIN · C. DRUCKER · G. JOOS · F. SIMON

ABTEILUNG A
CHEMISCHE THERMODYNAMIK · KINETIK
ELEKTROCHEMIE · EIGENSCHAFTSLEHRE

SCHRIFTLEITUNG:
M. BODENSTEIN · C. DRUCKER · F. SIMON
BAND 158
MIT 90 FIGUREN IM TEXT UND 1 TAFEL

LEIPZIG 1932 • AKADEMISCHE VERLAGSGESELLSCHAFT M. B. H.
PRINTED IN GERMANY

ZEITSCHRIFT
FÜR
PHYSIKALISCHE CHEMIE
BEGRÜNDET VON
WILH. OSTWALD UND J. H. VAN 'T HOFF

HERAUSGEGEBEN VON
M. BODENSTEIN · K. F. BONHOEFFER · G. JOOS · K. L. WOLF

ABTEILUNG A
CHEMISCHE THERMODYNAMIK · KINETIK
ELEKTROCHEMIE · EIGENSCHAFTSLEHRE

BAND 174
MIT 174 FIGUREN IM TEXT

LEIPZIG 1935 • AKADEMISCHE VERLAGSGESELLSCHAFT M. B. H.
PRINTED IN GERMANY

FIGURE 25. Title pages of two journals before and after the editorial "purge" (*Source*: *Berichte der Deutschen Chemischen Gesellschaft*, 1932 and 1934; *Zeitschrift für Physikalische Chemie*, 1932 and 1935)

staff resigned. Prager, who had started working there in 1899 under the editorship of Paul Jacobson, died a year later from the consequences of a heart condition. In 1934, the *Berichte* published an obituary in his honor authored by his successor, Friedrich Richter.

These large numbers of dismissals caused such a massive production backlog that deadlines previously agreed upon with the handbook's publisher could not be met, and the DChG feared compensation claims from the publisher, Springer.[26] As a result, the board decided to postpone further dismissals from the *Beilstein* editorial staff until projects up to 1937 had been completed.[27] The next wave of anti-Semitic dismissals by the society came in 1936 and 1937. They affected Dr. Dora Stern and Dr. K. Loria from the *Beilstein* editorial staff, Mrs. Heymann from the *Gmelin* editorial office, and Käthe Fiegel from the *Zentralblatt*.[28] Other Jewish employees, whose exact duties are not known, were Dr. Gregor Brilliant, Dr. Gustav Haas, Dr. Edith Josephy, Dr. Hedwig Kuh, and Dr. Fritz Radt.[29]

1936–1938: Alfred Stock's Presidency

The inorganic chemist Alfred Stock succeeded Hofmann as president of the society in 1936. Stock had been a member of the Nazi party since 1933. The vice-presidents were Arthur Schleede and Burckhardt Helferich, the former likewise a member of the party, though the latter was not. The autocratic *Führerprinzip* was instituted under Stock's leadership, and members were required to fill out a questionnaire about their ancestry. The minutes of the special assembly of the members on 8 February 1936 record that the board "resolved to propose at the next plenary session or at a special general assembly, for the sake of administrative

[26] Hörlein to Binz, 21 March 1933, DChG materials at the offices of the GDCh, Frankfurt am Main, folder: Chemische Erinnerung.

[27] Hörlein to Binz, 21 March 1933, DChG materials at the offices of the GDCh, Frankfurt am Main, folder: Chemische Erinnerung.

[28] There seems to have been comparatively few Jewish chemists among the editors and science staff for the *Gmelin Handbuch für Anorganische Chemie*, as indicated by a perusal of a few volumes of the *Gmelin* predating 1933 (F, Cl, Br, J, B, Zn, Cd, Fe, Pd, Sr). Biographical data on many of the chemists are, however, lacking, such as on Sybille Cohn-Tolksdorf (Br) and Hans Ehrenberg (Fe). Maximilian Pflücke remained the editor of the *Chemisches Zentralblatt* before and after 1933.

[29] These relate to a list of "non-Aryan" employees formerly insured under an Allianz policy. See Dörfel to Arndt, 29 July 1957. DChG materials at the offices of the GDCh, Frankfurt am Main, folder: (1956).

simplicity, adoption of the Führerprinzip and corresponding changes to the statutes."[30] After the war, this section was conspicuously truncated to: The board "resolved to propose at the next plenary session a corresponding changes [sic] to the statutes."[31]

At a December 1937 consultation concerning the issue of the DChG's "non-Aryan" members, Stock pointed out why it was not advantageous for the society to act too hastily in excluding Jews. The society had

> over 40 % (about 1,500) foreign members (a large, not precisely ascertainable number of them non-Aryans), who as subscribers to the large, costly publications of the Society (particularly the *Chemische Zentralblatt* and the handbooks for inorganic [*Gmelin*] and organic chemistry [*Beilstein*], are of the greatest cultural importance (advertisement for German science and language) as well as of considerable commercial value (annually for 3/4 million RM foreign currency). Around 100 non-Aryan German members still belong to the Society, some internationally very well-known ones within the world of science among them (e.g., the Nobel laureate Willstätter).[32]

To avoid a concerted wave of protests and boycotts of its publications abroad, the DChG decided, like other scientific societies with strong foreign interests, not to completely exclude "non-Aryan" members:

> 1. For foreign members of the DChG, the race issue should not be raised. 2. As regards the still registered non-Aryan German members, a reduction in their numbers in the Society, if possible to the point of vanishing, should, of course, be sought.[33]

Soon after the *Anschluss* of Austria into the German Reich, Stock opened the meeting of the DChG on 14 March 1938 with the words:

> In general, the waves of political events stop at the doors of this temple of science. The majesty of the events of the last few days, arousing every German, blows up the bonds of custom. Our first thought today is for the union of the two German

30 *Berichte der Deutschen Chemischen Gesellschaft*, 64 (1936A), 50.

31 The "denazified" version in: DChG materials at the offices of the GDCh Frankfurt am Main, folder: (DChG). I am unaware of how widely this reprinting was distributed.

32 Besprechung betr. nichtarische Mitglieder der Deutschen Chemischen Gesellschaft, 14 December 1937. DChG materials at the offices of the GDCh, Frankfurt am Main.

33 Besprechung betr. nichtarische Mitglieder der Deutschen Chemischen Gesellschaft, 14 December 1937. DChG materials at the offices of the GDCh, Frankfurt am Main.

lands, our deepest gratitude is to the Führer for his new feat in the history of the world.[34]

It was while still under Stock's presidency that the DChG became subsumed in the NSBDT as an affiliated professional group – quite contrary to the German Physical Society, which never entered this formation of the Nazi party.

1938–1945: Richard Kuhn's Presidency

On 7 May 1938, the board of the DChG decided – in agreement with the head of the NSBDT's chemistry division, Karl Merck – to choose Richard Kuhn for the society's presidency.[35] Merck was also a member of the DChG's board. The currently serving vice-presidents were Arthur Schleede (since 1936) and Burckhard Helferich (since 1937).

With this decision, the DChG was placed under the guidance of a chemist of world renown. Kuhn thus had a direct influence on the character of the society and its international image for what became a period of 7 years. No other president could match this. That is why a brief biographical portrait is given before our examination of the society's further developments.

Kuhn, born on 3 December 1900 in Vienna, was one of Germany's most successful chemists in the field of natural substances. Appointed professor at the Zurich Polytechnic in 1926, his career continued to follow a rapid ascent. In 1929, at the age of 29, he became the director of the chemistry department of the Kaiser Wilhelm Institute (*Kaiser Wilhelm-Institut*, KWI) of Medical Research in Heidelberg. Together with Otto Meyerhof, who had become the director of the institute's physiology department in the same year, he managed to turn this KWI into an international center for biochemistry in the early 1930s. It was at the beginning of the decade that Kuhn turned his focus onto the chemistry of natural substances. His research on the biologically important carotenoids and flavonoids contributed toward the discovery that vitamins are preliminary stages of coenzymes. In 1939, Kuhn was chosen for the 1938 Nobel Prize in Chemistry for his work on carotenoids and vitamins. But, like Gerhard Domagk and Adolf Butenandt, he refused the award because

34 *Berichte der Deutschen Chemischen Gesellschaft*, 71 (1938A), 115.

35 *Berichte der Deutschen Chemischen Gesellschaft*, 71 (1938A), 149.

Richard Kuhn

FIGURE 26. Richard Kuhn (*Source*: MPGA)

of the National Socialist government's prohibition of Germans accepting the prize after 1936.[36]

Kuhn very quickly gained great influence on chemistry in Germany. In 1937, he was promoted to director of the entire KWI of Medical Research. As president of the DChG he was, in addition, head of the newly founded Special Section (*Fachsparte*) for Organic Chemistry of the Reich Research Council (*Reichsforschungsrat*, RFR) from 1940 to 1945. He was directly responsible for the distribution of research funding and also for war-related projects.

[36] Elizabeth Crawford, "German scientists and Hitler's vendetta against the Nobel prizes," *Historical Studies in the Physical Sciences*, 31 (2000), 37–53.

In contrast to his colleague Peter Adolf Thiessen, head of the Special Section for General and Inorganic Chemistry of the RFR, Kuhn did not give preferential treatment to party members.[37] As president of the DChG, besides heading the special section at the RFR and sitting on many other science policy panels, Kuhn was one of the most powerful representatives of chemistry in wartime Germany. He retained the directorship of the KWI of Medical Research in 1945 (renamed a Max Planck Institute [*Max-Planck-Institut*] in 1948), was appointed a full professor at Heidelberg in 1950, and became vice-president of the Max Planck Society in 1959. He died in Heidelberg in 1967.

Kuhn was not a member of the Nazi party; there are no indications that he had any particular anti-Semitic convictions. Yet opportunism and anticipatory obedience toward the political authorities characterized his treatment of his Jewish fellow colleagues. He was on good terms with the 1922 Nobel laureate Meyerhof, for instance, at least until 1933. But in 1936, Kuhn went out of his way to inform the general management of the Kaiser Wilhelm Society for the Advancement of the Sciences (*Kaiser Wilhelm-Gesellschaft zur Förderung der Wissenschaften*, KWG) that Meyerhof was continuing to engage Jewish employees.[38] Kuhn had dismissed his own Jewish assistants in 1933 without the slightest effort to assist them. (Efforts of support and assistance for Jewish students and collaborators are documented in a number of comparable cases: Haber, Meyerhof, Windaus, and Max Bodenstein.[39])

Several examples show that Kuhn conformed with the Nazi regime in many respects and supported its policies. In several cases, his actions,

[37] Deichmann, *Flüchten*, sec. 5.4.

[38] Kuhn wrote to the general secretary of the KWG, Friedrich Glum, on 27 April 1936: "An inquiry by the State Police gives me occasion to request that you have the questionnaires of those employed at our Institute of Physiology carefully checked. . . . There are supposedly 3 more persons of non-Aryan descent presently at work in the institute under Prof. Meyerhof (Mr. Lehmann, Miss Hirsch and another lady with whom I am not yet acquainted), a situation that places the Kaiser Wilhelm Society as a whole and the Heidelberg institute in particular in question. I would like to suggest that after looking over the questionnaires you give Prof. Meyerhof precise guidelines he should adhere to in selecting his staff." MPGA, div. 1, rep. 1A 540/2. Kuhn's letter shows that he assumed that these three people were employed by Meyerhof. However, he received the answer that Glum already knew about them: The three people did not work as employees but as guests or doctoral students in the department, and their names thus did not have to be reported to the general management of the KWG. Kuhn's denunciatory letter was therefore unnecessary, even according to Nazi law.

[39] Deichmann, *Flüchten*, sec. 2.3.4.

whether motivated by opportunism or conviction, far exceeded strict observance of Nazi laws and prescriptions.[40]

Unlike many of his fellow scientists, at the outbreak of war Kuhn did in fact redirect virtually all the research performed at his institute toward military and war-relevant applications. The research he conducted on poison gas was both for defensive and offensive purposes. He examined the effectiveness of new nerve agents of the tabun/sarin series for the Army Ordnance Office and also made available to it a similar agent he had developed called soman.[41]

[40] Examples are:

- Virtually all his letters close with "Heil Hitler," even ones to colleagues known not to be National Socialists, who (initially) omitted this ending (e.g., see Max Hartmann [MPGA, Hartmann papers, III/47]).
- His refusal of the Nobel Prize in Chemistry for 1938 was particularly strongly worded. Bending to pressure from political officials, Kuhn (like Domagk and Butenandt, who were also forced to reject their awards of the Nobel Prize) wrote to the Royal Swedish Academy of Science in 1939 not only declining the award but also complaining that the award to a German was essentially aimed at enticing the prize-winner to disobey the "Führer's" decree. Unlike Butenandt and Domagk, Kuhn added at the bottom of his letter the handwritten comment: "The Führer's will is our faith." (Alfred Neubauer, *Bittere Nobelpreise* (Norderstedt: Books on Demand, 2005), 44.)
- On 21 July 1942, Kuhn drew the rector's attention to Anthony Eden's critical reaction to the University of Heidelberg's alteration of its inscription from: "Dem lebendigen Geist" to: "Dem deutschen Geist." The tone clearly suggests that Kuhn approved of the change:

> "Esteemed Rector,
>
> On the assumption of amusing you a little, I let you know that, according to *Nature 148*, 403, Mr. Anthony Eden said the following, among other things, during a lunch address on the 25th of Sept. 1941 before the delegates of the conference of the British Council: 'No one action can more clearly reveal the present German spirit than the replacement at the University of Heidelberg of the inscription 'to the living spirit' by 'to the German spirit.' This German spirit has made German scientists slaves of the regime, and opposed to all that science represents. That spirit must be overcome.'
>
> Heil Hitler!
Respectfully yours,
Richard Kuhn"

(University archive, Heidelberg, Personalakte Richard Kuhn). The Eden quote is in the original English.

[41] In evaluating this research, note that the line between offensive and defensive poison and nerve gas research is very fine. A definite step was nevertheless taken here in the direction of offensive gas warfare. Kuhn began with "defensive" research on antidote development. When he realized that this – quite unexpectedly – led to new kinds of poison gas, he did not put a stop to it. His synthesis of soman created a new weapon of mass destruction. The nature of the Nazi regime puts this fact in a very different light from

The Development of the DChG during the Kuhn Era. As an affiliated professional group of the NSBDT, the DChG had to accept the Nazi party's claim to leadership under Kuhn's presidency. A circular by the NSBDT informed the society in 1938: "In accordance with the Führer's wish, the NSDAP has a claim to leadership in all areas of life inside Germany. For all the technical fields, this leadership has been consigned to the NSDAP's Main Office for Technology, to which the NS League of German Engineers, as an affiliated confederation, is subordinated, for implementation of these responsibilities."[42] As had already become customary under Stock, avowal of loyalty to Hitler became part of the routine at meetings, and ostracizing of Jewish members presumably intensified. The above statistics indicate a sharp drop in the membership by 354 between 1934 and 1940.

In October 1938, after occupation of the German-speaking Sudeten regions of Czechoslovakia, condoned by the treaty of Munich, the DChG applauded Hitler with a "Sieg Heil!" Vice-President Helferich (not a party member) opened the session on October 10 with the words: "For the second time this year Germany celebrates the home-coming of millions of Germans into the Reich! This great success, too, was made without war! A threefold Sieg Heil to the Führer, in expression of our thanks."[43]

The last Jewish names were struck from the title pages of the chemistry periodicals during the Kuhn era. Arthur Schleede, one of DChG's vice presidents, protested to Kuhn that in 1938 it was still possible for Jewish chemists to appear as coeditors of a German scientific journal:

In designing our German journals we must – irrespective of any acknowledgment of the genuine achievements of other nations – maintain a German stance. Not without reason do foreign journalists regard German science as a haven of reactionaries since it is still possible, 5 years after the awakening of the German people, for the names of Jews to appear in full, some even replete with current residence, on the title pages of German journals. No matter whether a Jew with acknowledged achievements, such as Willstätter, or a Jew without any, such as Paneth . . . , in the eyes of those abroad this stance of "objective science" creates

Fritz Haber's research. See Ute Deichmann, "Kriegsbezogene biologische, biochemische und chemische Forschung an den Kaiser Wilhelm-Instituten für Züchtungsforschung, für Physikalische Chemie und Elektrochemie und für Medizinische Forschung," in Doris Kaufmann (ed.), *Geschichte der Kaiser-Wilhelm-Gesellschaft im Nationalsozialismus. Bestandsaufnahme und Perspektiven der Forschung*, Volume 1 (Göttingen: Wallstein, 2000), 231–257.

42 Circular by the NSBDT, 20 December 1938. DChG materials at the offices of the GDCh, Frankfurt am Main.

43 *Berichte der Deutschen Chemischen Gesellschaft*, 71 (1938A), 188.

an image of criticism by the German intelligentsia of the principles of National Socialism, whereby the fact that German chemists are the ones to have contributed a major share in the German nation's world acclaim is of particular importance in this aspect.[44]

By then Friedrich Paneth had, incidentally, already become internationally famous for his research in the field of radiochemistry (together with Georg von Hevesy) and for demonstrating the existence of free alkyl radicals using radioactive tracers (together with W. Hofeditz). Kuhn's reply to Schleede's letter is not documented. However, Paneth was informed a short time later, then in exile in London, by the editors of the *Zeitschrift für Anorganische und Allgemeine Chemie*, Wilhelm Biltz and Gustav Tammann, that in the future the list of coeditors would be omitted from the title page, as "this index no longer agrees with the current situation of our journal."[45]

This example shows that precipitous obedience and denunciation played a major role in the exclusion of Jews also from the scientific editorial offices. Schleede's letter and the prompt reaction to it show, moreover, that Kuhn and the DChG had accepted the redefined "German stance," according to which German Jews were no longer Germans. Their recent roles in the advances that had brought about "the German nation's world acclaim" in chemistry were erased by German chemists and their associations within just a few years.[46]

Denunciation was also the cue for the DChG in the case of Georg-Maria Schwab. To prevent the appearance of further articles by Schwab in the *Berichte*, the inorganic chemist Wilhelm Klemm sent the editors an official notification about Georg-Maria Schwab's "non-Aryan" descent. The managing editor, A. Ellmer, reacted, with the approval of Kuhn, by shelving one of Schwab's newly submitted papers.[47]

44 Schleede to Kuhn, 18 March 1938, DChG materials at the offices of the GDCh, Frankfurt am Main.

45 Biltz and Tammann to Paneth, 17 November 1938, MPGA, div. III/45/121.

46 To name just a few examples: Patents by Jewish chemists in Germany had to appear anonymously or under other names after 1933. When Fritz Haber died in 1934, no obituary appeared in the *Angewandte Chemie*. Some other chemistry journals, for example, the *Berichte* and the *Zeitschrift für Elektrochemie*, did publish obituaries but ignored other deceased Jewish chemists. The DChG even resorted to refraining from mentioning Baeyer and Wallach "in order not to expose itself sooner or later to unpleasant accusations," as the managing editor of the *Berichte* wrote to President Kuhn on 14 July 1941. DChG materials at the offices of the GDCh, Frankfurt am Main.

47 Ellmer to Kuhn, 28 May 1943, DChG materials at the offices of the GDCh, Frankfurt am Main, regarding the publication of articles by Jews. Klemm's motives are unknown.

Good political conduct became the precondition for the honor of an obituary, as the example of George Barger demonstrates. This professor of chemistry at the University of Edinburgh had died on 5 January 1939. The minutes of the meeting of 13 March 1939 contained the information that an obituary on Barger would "be appearing soon in the *Berichte*."[48] This obituary did not appear until 1946. Barger had not only helped many German refugee Jews by procuring employment for them at his institute in Scotland but also joined other foreign chemists in sharply criticizing the dismissals.[49]

In 1940, another world-renowned chemist, Adolf Butenandt, was selected for the vice-presidency of the DChG. Thus, two Nobel laureates headed the society during the Second World War. The remainder of the board was unchanged (except for the addition of Walter Schieber in 1942): Besides Kuhn and Butenandt, it included Eduard Zintl (as the other vice-president), Rudolf Weidenhagen, Heinrich Hörlein (treasurer), Karl Merck, Rudolf Schenck, Ernst Späth, Kurt Stantien, Peter Adolf Thiessen, and Erich Tiede. Seven of these 12 – Butenandt, Hörlein, Merck, Stantien, Thiessen, Tiede, and Schieber – were party members, and Merck, Stantien, Schieber, and Thiessen were party activists.

In 1941, the NSBDT informed its member associations about a more stringent version of the Aryan article: "The regular members and their wives must be of German blood and must confirm this in writing; for extraordinary members the prerequisites for the right to Reich citizenship from 15 Sep. 1935 ... must apply; for foreign members the same is valid, respectively."[50] When Todt instructed Schieber to reorganize the Reich Chemistry Division in the NSBDT in 1942,[51] the qualification "affiliated professional group [*Arbeitskreis*] within the NSBDT" appeared behind the society's name on all subsequent title pages of the *Berichte*.

He had become a fully-tenured professor in Danzig in 1933 and became a member of the NSDAP in 1938. In 1951, he became professor in Münster. Throughout the Nazi period, the editors of purely scientific journals were not compelled to check whether their authors were Jewish. So prior to this denunciation by Klemm, Schwab was able to publish articles in some German periodicals (*Berichte*, *Zeitschrift für Physikalische Chemie*, and *Kolloid-Zeitschrift*) as well as in the Viennese *Chemiker Zeitung*. Schwab, himself a Catholic, had been dismissed because his father was Jewish and emigrated to Greece.

48 *Berichte der Deutschen Chemischen Gesellschaft*, 72 (1939A), 67.

49 For example, see Deichmann, *Flüchten*, sec. 2.3.2.

50 Circular letter by the NSBDT, 1 March 1941. DChG materials at the offices of the GDCh, Frankfurt am Main.

51 *Berichte der Deutschen Chemischen Gesellschaft*, 75 (1942A), 114.

Kuhn's public address on the occasion of the DChG's 75th anniversary in 1942 underscored his nationalistic attitude as much as his support for Hitler. As the following excerpt demonstrates, he drew a remarkably contorted link connecting Hitler's politics with August Wilhelm Hofmann's – and by association the DChG's – political conceptions:

The men who founded the "German" Chemical Society on 11 November of the year 1867 under the leadership of August Wilhelm Hofmann . . . sought and implemented with the new Society – in the field of chemistry – those ideas of a unification of all Germans that Otto von Bismarck was able to carry through to a certain extent more than 3 years afterwards, on 18 January 1871, which, however, only Adolf Hitler has brought comprehensively to victory in our day. . . . We commemorate the dead. They gave their lives in the duel between the winds, faraway from the homeland, in the glowing sands, on the open seas, in frigid winters, in the Russian lands. – Now earth covers them. Yet their spirits remained immortal, it lives on at the battle front. And we honor the front line: All clans of Germans, from the Elbe and Oder, from the Danube and Rhine, and the men of the Duce and the sons of the Tenno, at their sides.[52]

Kuhn closed his speech with the words:

If, in closing, we glance back once more at the 75 years lying behind us, we can see the important extent to which the history of chemistry of this period is reflected in the evolution and fate of the German Chemical Society. We see how over the course of time chemistry has risen to a factor of power on our Earth. Yet we also see what an overwhelming portion of the foundations of modern chemistry is due to those nations of the Occident who have given mankind a Scheele and a Berzelius, a Lavoisier and a Pasteur, an Avogadro and a Cannizzaro, a Liebig and a Wöhler. For the survival of this blood, for the further development of this culture of theirs, the peoples of Europe stand today at arms just as do those of the ancient East Asian cultural sphere for their own. We commemorate the men in whose hands our shared destiny lies: Three cheers to the Duce, the Tenno and our Führer: A threefold Sieg Heil![53]

CRITICISM AFTER THE WAR

Kuhn's actions as the *Führer* of German chemists were observed attentively outside the country. After the war his political compromises were strongly criticized. Because the *Berichte* had printed his 1942 speech,

[52] "Mussolini and the Japanese emperor," Kuhn's address on 5 December 1942, in *Berichte der Deutschen Chemischen Gesellschaft*, 75 (1942A), 147.

[53] *Berichte der Deutschen Chemischen Gesellschaft*, 75 (1942A), 200.

it reached a wide international audience.[54] His long-time colleague at Heidelberg, Otto Meyerhof, who after his dismissal in 1938 had first emigrated to France, then via Spain and Portugal to the United States (Philadelphia), summarized his views, which he shared with many colleagues, in an expert opinion for the American Military Government in Heidelberg, as follows:

> Professor Kuhn is non-politically minded. He has had a liberal education, has held democratic views during the German republic and was the faithful and devoted pupil of the famous German Jewish chemist R. Willstätter, Munich. In spite of this he has sided with the Nazi regime in some important matters.
>
> Apparently after I lost my restraining influence on him (we had been in close cooperation during eight years) and after he felt that the regime was irrevocably entrenched in power, he was ready to compromise his great scientific reputation without scruples. I am convinced that he did this from expediency and from weakness of character and that he never held any Nazi convictions. Probably he was no member of the party. But he was the leader of the "Deutsche Chemische Gesellschaft" during many years of the Nazi regime and the head of German chemical delegations to the International Congress in Rome (1939) and on other occasions.
>
> . . . I am convinced that now after this complete turn of fate he is sincere in his endeavor to cooperate with the American authorities and ready to help to repair the atrocious misdeeds and injustice which the Nazi regime committed. Probably he still justifies his former activities by the excuse that he has in this way saved some scientific values and prevented worse more crimes from being committed. But I don't share this view, which is now accepted by many German scholars.
>
> The scientific work of Richard Kuhn is outstanding and of great importance. I am strongly in favor of the proposal that his scientific work remains unhampered and that he can continue research with his staff for the benefit of science and industry. On the other hand he should not be allowed to represent German chemistry in a leading role or be entrusted with the education of university students.
>
> I think that my view is shared in this country by many colleagues who know the work and the personality of Professor Kuhn.[55]

One of these colleagues was Erwin Chargaff, a prominent biochemist living in exile in New York. This was his pithy synopsis of Kuhn's character:

[54] Émigrés particularly noticed the speech. This is indicated by the note Hans Krebs wrote about discussions with Otto Westphal in Hamburg on 29 July 1976. University of Sheffield, Krebs papers.

[55] Meyerhof to the American Military Government in Heidelberg, 29 January 1947, University of Pennsylvania, Philadelphia, archive, Meyerhof papers. Original English.

"a Karajan of chemistry. Essentially he was very good, but he degenerated, politicized in the same way as Heisenberg."[56]

KUHN'S POSITION ON THE DCHG'S POLITICAL REALIGNMENT

Kuhn's own statement with regard to his political compromises as president of the DChG was as follows:

> The German Chemical Society did not seek a connection with Nazi organizations in the years after 1933, in particular not with the NSBDT. Rather it was trying to retain all its members, especially the foreign ones. It successfully refused to incorporate the Aryan Clause into its statutes, and numerous Jews counted among its members and honorary members till the end (1945). The German Chemical Society was forcibly affiliated with the NS League of German Engineers only in 1942 and only from that year onward did the words "affiliated professional group within the NSBDT" appear in small print on the title page of the "Berichte der Deutschen Chemischen Gesellschaft". It was known that associations balking at integration were liquidated.

Kuhn continued to explain that, based on the fact that he was not a member of the NSDAP, the DChG had only a "very loose tie to the NSBDT."[57]

It is worth pointing out here that the German Physical Society was not dissolved, even though it did *not* join the NSBDT. The history of events presented earlier in this chapter indicates that, contrary to Kuhn's assertions, political realignment by the society started *before* it was incorporated into the NSBDT (which took place in 1938, not 1942). The DChG had already pressured its Jewish president and vice-president to resign in 1933; dismissed the majority of the editorial employees for the *Beilstein*, the *Berichte*, the *Gmelin*, and the *Zentralblatt* by 1937; and removed Jews from the periodical board and lost a large part of its German Jewish membership through withdrawals or exclusions. The initial ideological restraint and stepwise approach in the dismissals or exclusions of Jewish employees and members took place in consultation with NSDAP officials. The motivations behind this policy were opportunistic; they did not corroborate any distancing from the Nazi party.

The preparations for the Fritz Haber commemoration in 1935 caused a major stir in the chemical societies, and the same was presumably

[56] Erwin Chargaff in conversation with the author, New York, 28 January 1997; the conductor Herbert von Karajan made substantial compromises with the Nazi regime.

[57] Richard Kuhn, Deutsche Chemische Gesellschaft, 5 December 1950, MPGA, div. III; Butenandt papers, 84/1/529.

true for the German Physical Society. The political significance of these discussions justifies an analysis of the events within the two chemical associations.

UNEXPECTED OBJECTIONS: THE CHEMICAL ASSOCIATIONS AND THE FRITZ HABER MEMORIAL IN 1935

Fritz Haber was one of the most important German chemists of the first half of the 20th century. His synthesis of ammonia from the elements nitrogen and hydrogen earned him the Nobel Prize in Chemistry for 1918. Haber, a Jew who converted to Protestantism in 1892, also gained fame as a German patriot and the main organizer of German gas warfare during the First World War. In April 1933, he resigned from his post as director of the KWI of Physical Chemistry and Electrochemistry that he had held since 1911, this in protest against the National Socialist dismissals policy, which affected many of his employees. Haber emigrated to England shortly after and died of a heart attack while on a trip to Basel on 29 January 1934.

The DChG remembered Haber at its annual convention in 1934 with a memorial address by Wilhelm Schlenk, which appeared in the *Berichte*.[58] The *Angewandte Chemie*, on the contrary, did not even publish a brief obituary about its honorary member.[59]

The memorial ceremony organized by Max Planck in 1935 on the first anniversary of Haber's death was the only public gathering of German scientists in Nazi Germany that, though it was not forbidden, took place with the explicit disapproval of the political officials. It is often described as the sole demonstration of protest by German scientists under the Nazi regime.[60]

The event, organized by the KWG in conjunction with the DChG and the German Physical Society, took place on 29 January 1935 in the

[58] "Gedenkansprache Wilhelm Schlenks auf der Jahrestagung der DChG," *Berichte der Deutschen Chemischen Gesellschaft*, 67 (1934A), 20.

[59] Just the following regular death notice appeared under the "deceased" rubric: "Priv. Gov. Coun. Prof. Dr., Dr. of agriculture e.h., Dr. med. h.c., Dr.-Ing. e.h., Dr. of techn. sciences e.h. F. Haber, former director of the K.W.I. of Physical Chemistry and Electrochemistry, Berlin-Dahlem, honorary member of the Association of German Chemists, on the 29th of January at 66 years of age while on a trip to Basel." *Berichte der Deutschen Chemischen Gesellschaft*, 67 (1934A), 93.

[60] See also the contribution by Wolff to this volume.

Harnack-Haus of the KWG.[61] Max Planck personally made the arrangements. On the original program were memorial speeches by Otto Hahn and Karl-Friedrich Bonhoeffer, after opening words by Planck. Planck had also invited Haber's son, Hermann, and other members of the family. Hermann Haber declined to attend, however. In a letter to Prof. Coates in England, he and his wife, Margarethe, justified this decision: "We find – something we could not put in writing [in the official refusal] – that one has no right to celebrate a person *dead*, whom one would not tolerate *alive* today" (emphasis in the original).

On 15 January, shortly after the invitations had been sent out, the Reich and Prussian Minister of Culture Bernhard Rust prohibited all civil servants and employees under his purview from attending. The justification was that by his resignation, Haber had expressed his opposition to the state. The comment that followed was extraordinary in its cynicism: The plan to conduct a Haber commemoration would have to be interpreted as an affront against the National Socialist state because customarily only "the greatest Germans" were celebrated in events of the kind.

Planck responded with a remarkable letter to Rust in an effort to persuade him to lift his ban. It reads:

> The Kaiser Wilhelm Society has attested to its positive attitude toward the present State and its profession of loyalty to the Führer often enough by word and deed to be able to lay claim to receive the confidence that during its planned event any tendency giving rise to misconceptions be most strenuously avoided.[62]

Rust appreciated Planck's attitude, though he did not lift the ban to attend the celebration completely. He allowed it to take place as an event of the KWG: "Taking into account that the attention of the press at

[61] For a more detailed account of the background to this celebration, see Ute Deichmann, "Dem Vaterlande – solange es dies wünscht. Fritz Habers Rücktritt 1933, Tod 1934 und die Fritz Haber-Gedächtnisfeier 1935," *Chemie in unserer Zeit*, 30 (1996), 141–149.

[62] Planck to Rust, 18 January 1935, MPGA, div. V, rep. 13, 1850; transcribed together with Rust's decree from 15 January 1935 in the appendix to Dieter Hoffmann and Mark Walker (eds.), *Physiker zwischen Autonomie und Anpassung – Die DPG im Dritten Reich* (Weinheim: Wiley-VCH, 2007), 557–559. It appears that no publication on Haber has ever quoted this letter. Otto Hahn indicates in his reminiscences that it was "unfortunately no longer accessible to me." Otto Hahn, "Zur Erinnerung an die Haber-Gedächtnisfeier vor 25 Jahren am 29. Januar 1935 im Harnack-Haus in Berlin Dahlem," *Mitteilungen aus der Max-Planck-Gesellschaft*, 1 (1960), 3–13, here 8.

home and abroad has already been aroused about the matter, that foreign participants are expected at the ceremony and, finally, that private individuals number among the Kaiser Wilhelm Society's ranks," he suggested to Planck that "the event take place as a purely internal and private celebration of the Kaiser Wilhelm Society." The daily press should not report on it. Rust asked Planck to submit a list

> of those professors who in their capacities as members of the Kaiser Wilhelm Society or the German Chemical Society or the German Physical Society may have announced their attendance. I reserve for myself to grant dispensation from the prohibition to attend to the named professors, if they considered it important.[63]

Planck had copies of this letter sent by special delivery to the relevant societies, the directors of all the Kaiser Wilhelm Institutes, the presiding secretary of the Prussian Academy of Sciences, and to the rector of the University of Berlin.

As much as the conduct of the two chemical associations differed in this situation, they both revealed a distancing from their honorary member Fritz Haber and political conformity. The then-president of the DChG, Karl Andreas Hofmann, had the Berlin board members informed about the possibility of a dispensation. He thought it was already too late, however, to inform all the society's members about the minister's letter.[64] A subsequent inquiry showed that no dispensations were made. The existing documents allow the assumption that none of the board members applied for one.[65] As far as can be established, Adolf Windaus was the only member of the DChG who tried to obtain a dispensation.

[63] Rust to Planck, 24 January 1935, DChG materials at the offices of the GDCh, Frankfurt am Main, folder: Chemische Erinnerungen.

[64] Hofmann to Binz, 26 January 1935, DChG materials at the offices of the GDCh, Frankfurt am Main, folder: Chemische Erinnerungen.

[65] Prof. Erich Tiede wrote to Hofmann that, in his opinion, "under the changed circumstances participation by the German Chemical Society in the celebration, if it really does take place, does not appear right. An abstention, which would spare board members in the civil service, especially members on the smaller board, from anxious deliberations, is in my view arguable also because we have already remembered Haber in a most honorable manner in our Society at a solemn session on the 12th of February 1934.... Of course, the decision about whether the Chemical Society should decide now lies in your hands. For me personally, attendance upon petition for a dispensation is out of the question now." Tiede added that Prof. Leuchs, who "through his accident is prevented from going out and from writing," also asked "that he in any case not be placed on the list of those board members falling under consideration for a dispensation petition." DChG materials at the offices of the GDCh, Frankfurt am Main, folder: Chemische Erinnerungen.

His petition was declined. The DChG, which Haber had presided over from 1922 to 1924, appeared on the program of the festivities, albeit as co-host.

The VdCh made a contribution of a special kind to the Fritz Haber memorial: It forbade its members to participate. Postcards were sent out to all its members on 25 January 1935 with the following wording: "Pursuant to the instruction by the president of the RTA, Dr. Ing. Todt, participation in the commemoration of Fritz Haber on the 29th of January 1935 in the Harnack-Haus is denied to all members of the Association of German Chemists, reg. assoc."[66]

Nevertheless, objections were unexpectedly raised. They were not directed against the National Socialist government, though, but against the association's board. Its prohibition from attending the commemorative event elicited a flurry of protest letters and even some withdrawals, forcing the association to take a stand and to engage in correspondence for a period of more than 2 months. Members of the *Verein* felt unfairly and illegitimately imposed upon by such a prohibition. But they also did not see why a commemorative celebration for a German scholar and patriot of such stature should be banned. Among these letters, addressed directly to the board in Berlin, were the following: "If I had heard about the celebration in time and had had the intention of participating, I would not have let myself be stopped by the postcard from attending" (Privy Government Councilor Dr. Karl Süvern, 31 January 1935). Only his 40-year-long membership in the association prevented him from withdrawing, "despite the disrespect for a great deceased figure expressed by the prohibition and the lack of gratitude toward a man whom Germany still has much to thank for today." The reaction from Dr. Löhmann on 7 February was similar: "I kind[ly] request notice about the reason behind this prohibition being issued. It otherwise seems entirely incomprehensible, as Privy Councilor Haber was not just one of Germany's most important chemists but was recognized by the whole world for his unique achievements." Dr. Karl Bittner from Vienna announced his withdrawal from the association on 12 February 1935.[67]

[66] See this and the following quotes concerning the Haber celebration, DChG materials at the offices of the GDCh, Frankfurt am Main, folder: Chemische Erinnerungen. Most of the quoted letters did not provide the first names. Some have been added on the basis of information from the Bayer-Archiv, Leverkusen.

[67] DChG materials at the offices of the GDCh, Frankfurt am Main, folder: Chemische Erinnerungen.

At least in some cases the board felt obliged to justify its decision.[68] The ban on attending the Haber memorial event caused considerable concern in the VdCh's regional branch for the Upper Rhine, as First Chairman Dr. Hans Wolf informed the association on 13 February 1935. The complainants were primarily chemists "who consciously or unconsciously were setting themselves in opposition to the present government, for no rational reason." The events in the regional branch for the Rhineland are particularly well documented: More than 10 protests of this kind had been received by mid-February, all of them from Ph.D. chemists, presumably from industry. None came from university teachers.

The following are excerpts from some of these letters: "Sent back because I must refuse to accept such orders" (Dr. Hans Niedeggen). "Acknowledged but with consternation, reasons missing for comprehension! Return to the board" (Dr. Hess). "On the basis of which statutory by-laws does the Association consider itself authorized to issue such decrees impinging on the private lives of its members?" (Dr. Erich Mayer). "For members of an independent organization such a heavy-handed ban is not explicable, especially when it involves a case such as this one, honoring a man whom Germany owes almost 2 years continuation of the war. I would be grateful if you can indicate to me on the basis of which statutory article the association is granted the possibility to intervene against individual members in the form of a ban" (Dr. Kurt Zimmermann). "In deep indignation, I return to the Association of German Chemists the notice re the Fritz Haber commemoration. Content and form are so grotesque that they are beyond words" (Dr. Heinz Clingestein).

Dr. Albert Gundlach and Dr. Leonhardt pointed to the reluctance by many members in the Rhineland regional branch. Specifically, "it is incomprehensible that [the Association's chairman] Prof. Duden did not find the ways and means to make clear to Dr. Ing. Todt in time how highly regarded Haber was for his immense accomplishments and what an unfavorable impact a ban composed in this form would have."

[68] Süvern's charges were "most resolutely rejected" in a lengthy letter from 15 February 1935: The board justified its action as a duty toward the discipline to pass on orders by the head of the RTA, Dr. Todt, to members without criticism. The board otherwise regretted the unfortunate course of the matter. An inquiry to Berlin by the Austrian regional branch of the VdCh at the instigation of Bittner received the reply, in a letter written on the same day, that "the prohibition was *not* imposed because Haber was a Jew but because he had resigned his posts in opposition to the new Reich." DChG materials at the offices of the GDCh, Frankfurt am Main, folder: Chemische Erinnerungen.

Dr. Otto Müller quoted from Richard Wagner's *Meistersinger*, "Scorn not my masters and honor their art," and, referring to the importance of Paul Ehrlich and Richard Willstätter alongside Haber, professed his adherence to the universalist principle of achievement: "We must not ask: Is the man a Jew and Christian? Rather, we ask: What has he achieved?"[69]

The reactions by the board members diverged. Duden and Stantien wanted to let things stand as they were. The regional chairman for the Rhineland, Walter Schieber, on the contrary, was of the opinion that the nature of the association's leadership had been exposed by the whole chain of events and declared he would feel obliged to tender his resignation if the main association did nothing. In reply, Stantien wrote Schieber to remind him about the *Führerprinzip*:

> Professor Duden, too, is of the opinion that you should make clear to your members, who conceive of the Association as a purely private establishment, that the Association of German Chemists as a member association of the RTA certainly has an official character and should be regarded as a professional representation of German chemists. In the Third Reich our members simply have to comply with the orders issued by organs designated by the Führer.[70]

Schieber wrote a circular letter to all the members on 24 March 1935 "to settle the large number of inquiries received":

> We consider it necessary, in the inherent interest of chemists, that this affair not continue to be handled in public, as the political demonstration that has taken place thus far has burdened the status of German chemists already enough. We might, however, have expected that the members of the Association would not have raised their charges of neglect toward a great deceased figure and of lack of appreciation without checking the reasons behind the prohibition. . . . As concerns the Association of German Chemists, it was not its duty to assume any kind of critical stance on this [ban proclaimed by the RTA's president Todt]; rather it is an obligation toward the discipline to pass this order on to its members.

Schieber then announced that he would write personal letters to members "who had succumbed to scolding the leaders of German chemists and German technicians in general in such an impossible form," instead of addressing them in a purely factual manner, "to demand an unconditional

[69] DChG materials at the offices of the GDCh, Frankfurt am Main, folder: Chemische Erinnerungen.

[70] DChG materials at the offices of the GDCh, Frankfurt am Main, folder: Chemische Erinnerungen.

retraction of their accusations by a specified date." He hoped this circular would finally put an end to this "unpleasant affair."[71]

Neither the personal letters Schieber had promised nor any replies are documented. But there was at least one more noteworthy reaction. Dr. Kurt Zimmermann wrote from Wuppertal on 5 April 1935:

> Unfortunately I cannot draw the line under this very "unpleasant" affair – referring not to the arrangement of the celebration but to the imposition within our Association of a command-like prohibition on personally independent members – as long as full clarity has not been reached. I therefore request the following questions be answered:
>
> 1. Do the officiating organs of the VdCh have the right to interfere with the personal decisions of individual members?
> 2. If yes: On which § of the statutes is this right anchored?
> 3. If no: How does one propose to prevent such breaches of the statutes in future?

Then follows a commentary on individual passages of Schieber's letter:

> Why is the status of German chemists already "burdened enough" by a memorial celebration that nobody regarded as political or anticipated any political effect from it? . . . Furthermore, what is the meaning of: it lies in the "inherent interest," etc.? As a front-line soldier, I, for example, have no reason not to speak about our great professional colleague Priv. Coun. Prof. Dr. Haber, a ranking major of the r[eserves] during the World War, organizer of the first gas assault, which I had the honor of participating in, etc.; you yourself describe him as a "great deceased figure." I even advocate – I would like to interject the remark here that I have been an opponent of Jews since the time I learned to think – the view that this man should be respected not just for his science but also as a person, because, in my opinion, he did not resign from his posts and leave Germany "in opposition" to the National Socialist state but as a man who drew the consequences that a person with a sense of honor must draw in such a situation. And I still have "respect" for "decency" and "honorableness," even if the person concerned is not a fellow countryman (*Volksgenosse*).[72]

Zimmermann was the only protesting member of the VdCh to go far beyond praising Haber's patriotism and to support his resignation. The remaining question is why this professedly anti-Semitic chemist and a few other colleagues in industry protested so resolutely against their association's board, completely unlike chemists at universities and the Kaiser

[71] DChG materials at the offices of the GDCh, Frankfurt am Main, folder: Chemische Erinnerungen.

[72] DChG materials at the offices of the GDCh, Frankfurt am Main, folder: Chemische Erinnerungen.

Wilhelm Institutes. Protest was presumably not reconcilable with the tradition of obedience within the German civil service. Assistants and private lecturers also had to fear being passed over in promotions.

There is no exact documentation on who exactly attended the commemoration. From Otto Hahn's reminiscences, we gather that the large hall in the *Harnack-Haus* was filled to capacity with the wives of Berlin professors, members of the KWG, and Fritz Haber's personal friends.[73] Max Planck opened the event with a Hitler salute and a few words of introduction.[74] Two memorial addresses were held by Otto Hahn and the retired colonel Joseph Koeth. During the First World War, Koeth had headed the department for raw materials of the Prussian Ministry of War, and his acquaintance with Haber, who headed the department for chemical warfare, dates back to this period. According to Hahn, the director of the KWI of Anthropology, Eugen Fischer, then serving as rector of the University of Berlin, called him up a few days before the ceremony to forbid him from delivering his address. Since Hahn had already left the Faculty of Philosophy, and therefore was no longer a member of the university, Fischer's authority over him had expired.

Max von Laue had written a remarkable obituary in which he compared Haber with Themistocles, a historical figure remembered "not as the pariah at the court of the Persian king, but as the victor of Salamis."[75] But even von Laue did not dare attend the event.[76] Neither did Karl Friedrich Bonhoeffer participate. He was prohibited from delivering the originally planned memorial address; Hahn read it out instead. The organic chemist Richard Willstätter, who had resigned from his

[73] Hahn, Zur Erinnerung.

[74] Richard Willstätter, *Aus meinem Leben. Von Arbeit, Muße und Freunden* (Weinheim: Verlag Chemie, 1949), 277.

[75] "Themistocles went down in history not as the pariah at the court of the Persian king, but as the victor of Salamis. Haber will go down in history as the genius inventor of the procedure of binding nitrogen with hydrogen, which underlies the synthetic extraction of nitrogen from the atmosphere. He will be remembered as the man who, in the words used at the award of his Nobel Prize, had created in this way 'an exceedingly important means towards promoting agriculture and human prosperity.' He will be remembered as the man who had made bread out of thin air and who triumphed 'in the service of his country and of the whole of humanity.'" Max von Laue's obituary on Fritz Haber in *Die Naturwissenschaften*, 22 (1934), 97; translated in Klaus Hentschel (ed.), *Physics and National Socialism: An Anthology of Primary Sources* (Basel–Boston: Birkhäuser, 1996), doc. 29, especially 77–78.

[76] Ruth Sime, *Lise Meitner: A Life in Physics* (Berkeley: University of California, 1996), 156.

professorship at the University of Munich in 1924, came to the celebration along with Privy Councillor Carl Bosch, Arthur von Weinberg, Director Kühne, and a few other managers from the *I.G.-Farbenindustrie*, whom Bosch had urged to attend. His Excellency Schmidt-Ott, Dr. A. Petersen, and Dr. Johannes Jaenicke from the *Metallgesellschaft* in Frankfurt, Wolfgang Heubner and Elisabeth Schiemann from the University of Berlin, as well as Lise Meitner, Fritz Straßmann, and Max Delbrück from the KWI of Chemistry were also there. Georg Melchers, assistant at the KWI of Biology, likewise came without first requesting permission.[77] Heubner, full professor of pharmacology, and Schiemann, extraordinary professor of botany, were known for their anti-Nazi positions; their attendance at the event caused them no direct professional disadvantages.

The above-cited correspondence with the VdCh indicates that the full professor of physical chemistry at the University of Vienna, Hermann Mark, was also in attendance. (He would be dismissed in 1938 because his father was Jewish.) Haber's former co-worker Hans Eisner, who had already emigrated to Spain, and the non-Jewish enzyme chemist Friedrich Franz Nord from Berlin also appeared at the celebration.[78]

All in all, based on the available information, certainly as pertains to the chemical societies, the Haber memorial cannot be seen as a demonstration against the Nazi regime. The DChG and its university members did not take part in the ceremony; the VdCh even forbade its members to attend. Only a few of the German Physical Society's members were present, for instance Meitner and Planck. University teachers generally, a few exceptions notwithstanding, did not try to use the available legal option of petitioning for a special dispensation in order to be able to attend the event (not to mention simply ignoring the ban altogether to appear). As far as is known, the only university teacher in chemistry to try (in vain) to get a dispensation was Adolf Windaus, who, incidentally, had declined to collaborate on poison-gas research in 1914.

By the absence of the majority of the elite in science and industry, the Haber memorial event symbolized the changeover from the old nationalism, which allowed Jewish scientists like Haber to develop and deploy poison gas for the German nation, to a new nationalism, in which there was no place for Jews irrespective of their political views, not even in memorials of battlegrounds of the past.

77 Prof. Georg Melchers, personal information to Prof. Benno Müller-Hill.

78 On Mark, Eisner, and Nord, see Deichmann, *Flüchten*, chap. 2 and chap. 4.

SUMMARY: COMPARISON WITH THE GERMAN PHYSICAL SOCIETY

Realignment (*Gleichschaltung*)

In contrast to the German Physical Society (*Deutsche Physikalische Gesellschaft*, DPG), the DChG as well as the VdCh incorporated themselves into the NSBDT as affiliated professional groups. The two chemical associations differed in the rapidity with which they realigned themselves politically. The VdCh introduced the *Führerprinzip* already in 1933, adopted ideological goals in its statutes in 1934, and was subsumed by the RTA (the later NSBDT) in the same year. The DChG, initially very timid about making political or ideological statements; instituted the *Führerprinzip* in 1936 and was incorporated into the NSBDT in 1938. All the same, it, too, was overly servile in ousting its Jewish members from leadership positions and editorial posts in 1933.

Mobilization for Economic and Military Goals

The political leaders of the Nazi state were undivided about the national and economic importance of chemistry. The head of the Special Section for General Chemistry at the RFR, the physical chemist Peter Adolf Thiessen, was able to cite Hitler in emphasizing the importance of chemistry in a speech. The topic was "Science and the four-year plan," delivered during a demonstration by the Berlin chapter of the National Socialist German University Lecturers League (*Nationalsozialistischer Deutscher Dozentenbund*, NSDDB): "At the end of his speech [in 1936], the Führer declared: We shall prevail, even though we are a poor nation suffering from shortages; we shall prevail, because we have the fanatic will to help ourselves, and because we have chemists and inventors in Germany who will have our needs under control."[79]

This economic importance of chemistry was one of the reasons why, in contrast to physics, no apparent effort was made to influence the subject matter and content of academic chemistry ideologically. Unlike in physics, there was no massive ideological movement for a *Deutsche* or "Aryan science" in chemistry. The ideologized *Deutsche Chemie* by the

[79] *Wissenschaft und Vierjahresplan. Reden anläßlich einer Kundgebung des NSD-Dozentenbundes, Universität Berlin* (Berlin: NSD-Dozentenbund, Gau Groß-Berlin, 1937), 18 January 1937, 4–17, Federal German Archives, Berlin (*Bundesarchiv*, BA), R73/15159.

likes of Karl Lothar Wolf remained marginal. There were no controversies and power struggles between representatives of an "Aryan chemistry" and its opponents in either of the chemical societies. Again in contrast, the mobilization of members for the goals of the Four-Year Plan, economic autarky and war preparedness, had high priority, particularly in the VdCh.

Strife with Prominent Jewish Members and Employees

Both the chemical associations and the DPG had, with Haber and Albert Einstein, Jewish members who were not only renowned scientists but who also played prominent political roles. These roles were, however, very different. Haber's political prominence came from his patriotism, his success in synthesizing ammonia from the elements, which provided nitrogen compounds for explosives during the war, and the organization of gas warfare for Germany. Einstein, in contrast, excited the hatred of conservative circles by his critical attitude toward the nationalistic German policies during the First World War. It was therefore much easier to reach consensus on excluding Einstein from the DPG in 1933 than on excluding Haber from the chemical associations. After Haber's death, the issue surrounding attendance of the anniversary function triggered debates at the DChG and the VdCh and even drew protests against the latter's board because of the imposed ban. There is no information about any comparable debates within the DPG, even though Haber had likewise served there as chairman (1914–1915).

The DChG had a large number of prominent Jewish members inside as well as outside the country and also counted many Jewish scientists and editors in its periodical and handbook editorial offices. The sales of its periodicals and handbooks were an important source of foreign income. This complicated cancellations or dismissals of Jewish members or employees.

The Role of the Presidency

With Richard Kuhn, the DChG was presided over from 1938 to 1945 by one of the most important German chemists of the time. Like Carl Ramsauer and the other DPG chairmen, he was not a member of the Nazi party. The available information provides no indication that Kuhn had shielded the society from even greater Nazi party influence, as he later alleged. On the one hand, Nazi politicians appreciated Kuhn as a scientist and science policy maker, as is shown, for example, by the fact that he became president of the DChG and special section head at the RFR with the involvement of party officials. On the other hand, Kuhn's actions are characterized by conformism and anticipatory obedience to the Nazi regime. He bolstered the regime with his reputation and his scientific and organizational abilities out of nationalistic conviction and opportunistic reasons. Without becoming a member of the NSDAP, as president of the society he publicly advocated unconditional support for Hitler during the war. The combination of his high achievements in science and complete political conformism harmed the society and the reputation of chemistry in Germany far beyond the era of National Socialism.

The DChG's founding president, August Wilhelm Hofmann, had influenced the society with his internationally oriented liberalism and his antidiscriminatory stance on behalf of his Jewish colleagues, whose influence in chemistry was far greater than that in other sciences during the 19th century. Hofmann's successors during the Nazi era abandoned these universalistic principles – in part under political pressure and in part motivated by nationalism, anti-Semitism, and opportunism. As a guide to good science and its representation, Hofmann's maxims still hold today.

10

Distrust, Bitterness, and Sentimentality

On the Mentality of German Physicists in the Immediate Post-War Period

Klaus Hentschel

Mentality, according to Jacques Le Goff, is that which a person shares with other individuals, that which, as it were, subliminally allows or inhibits thoughts, words, deeds, and feelings.[1] The history of mentality frequently focuses on extended structures *de longue durée* that sometimes persist for centuries. So the period covered here from the collapse of the Nazi regime in 1945 until the founding of the two German republics in 1949 may come as a surprise. Mentality does not change overnight, of course. But during this brief period it evolved relatively quickly as a result of altered defining conditions in German politics. For this reason, these five post-war years are an extremely interesting transition phase containing *relics* of a waning mentality along with *germs* of its successor. There were signs of a National Socialist mentality even after the demise of the Third Reich. Contemporary stories already reveal it. Hartmut Paul

[1] For historical and methodological surveys of the history of mentality, see the references in my more detailed book, *Zur Mentalität deutscher Physiker in der frühen Nachkriegszeit* (Heidelberg: Spektrum Verlag, 2005). An expanded English translation appeared under the title *The Mental Aftermath: The Mentality of German Physicists 1945–1949* (Oxford: Oxford University Press, 2007).

The German Research Foundation, the *Deutsche Forschungsgemeinschaft*, generously granted me a stipend for conference travel and archival research. It afforded me the means to peruse the papers of James Franck, Eugene Rabinowitch, and Michael Polányi at the Regenstein Library of the University of Chicago. I thank the archivists there as well as Herr Ralf Hahn at the Archive of the German Physical Society (*Deutsche Physikalische Gesellschaft*, DPG) and Frau Kazemi at the Archive of the Kaiser Wilhelm/Max Planck Society for their advice and for granting me permission to reprint excerpts from documents among these holdings. Ann Hentschel, Dieter Hoffmann, Jost Lemmerich, Gerhard Rammer, and an anonymous referee offered helpful criticism of earlier drafts of this chapter.

Kallmann, for instance, had been dismissed from his position as department head at Haber's Institute of Physical Chemistry in 1933 and deprived of his permission to teach at university because three of his grandparents were of Jewish origin. He decided not to leave Germany and worked under the protection of I.G. Farben's president on a private fellowship. Although he was rehired in 1945, becoming director of the Kaiser Wilhelm Institute of Physical Chemistry, he said that he ultimately decided to leave Germany in 1949 because a Nazi mentality continued to dominate.[2]

I do not intend to pass self-righteous moral judgment from the safe vantage point of retrospection. Nor is this intended as an apology for commissions and omissions of the period. The aim here is a neutral description of the prevailing mentality.[3] If we can better understand *how* people of that time thought and felt, we can better understand *why* they acted and wrote the way they did. Let there be no doubt about my own position. It was one of the most depressing experiences I ever had as a historian to see reflected in the documents how very soon *after* 1945 the chance of coming to grips with the National Socialist regime was allowed to slip away, thus missing the opportunity to make a frank assessment of the facilitating conditions the regime had set.

The focus here is not on biographical descriptions of individual perceptions. Our view will reach beyond the personal level to aspects valid for many individuals at once. It will circumscribe a "space or horizon typical of the period, the 'expressive frame,' within which comportment is formed, while allowing a certain leeway for individual deviations."[4] It is not possible to go into the details about the many orientation points lending this landscape its topology: the ordinances and laws imposed by the Allies, the wretched working conditions in many places after the war, and so forth. But it is obvious that a mutual relationship existed between the prevailing mentality and the immediate living conditions. Squalid living quarters in overcrowded, make-shift housing, a dire scarcity of food

[2] See Ute Deichmann, *Flüchten, Mitmachen, Vergessen: Chemiker und Biochemiker in der NS-Zeit* (Weinheim: Wiley-VCH, 2001), 484 (letter by M. Polányi, 9 December 1947).

[3] On the problem of neutrality, see, for example, Michael Balfour and John Mair, *Four-Power Control in Germany and Austria 1945–1946* (Oxford: Oxford University Press, 1956), 5: "if he succeeds in preventing himself from expecting pigs to fly, [he] is apt to sound as if he pitied the animals for their inability to do so."

[4] Peter Dinzelbacher, "Zu Theorie und Praxis der Mentalitätsgeschichte," in Peter Dinzelbacher (ed.), *Europäische Mentalitätsgeschichte* (Stuttgart: Kröner, 1993), XV–XXXVII, here XXII.

and clothing, freezing cold in the wintertime, and a constant influx of more dispossessed refugees: these typical elements have to enhance the mental picture I draw.[5]

According to George Duby, mentality comes in layers, with one very broad layer encompassing nearly all of the population, a next one characterizing particular strata, such as the *Bildungsbürgerliche* elite, and others limited to even more specific subsets of it. German physicists, as the specific group chosen for this study, thus share many elements of their mentality with the population at large but also deviate in certain respects, related to their specific status and function in post-war Germany. My focus is on the mentality of physicists who had stayed in Germany during the National Socialist period. But the perspective of their discriminated colleagues, who had been expelled and in many cases remained unwelcome even after 1945, is useful, too. Their distance as émigrés endows them with the function of a remote mirror – albeit not a plane one – of what was said and thought in Germany after 1945. A few unknown figures, both young and old, will have their say besides more famous physicists. I shall leave open to what extent the mentality of German physicists differs from that of other scientists, members of other disciplines in the humanities and social sciences, indeed the population as a whole. Nor can I go into possible differences in attitudes between male and female physicists, though it is indeed quite striking that the few female voices in this study (Martius and Meitner) are strikingly different from those of their male colleagues. In addition, there might also be subtle generational differences, so prominent in other periods (First World War or the 1968 rebellion), though I found my reference group surprisingly homogenous in mentality across all ages in this respect. All these fine-tuning questions could ultimately only be answered by comparison against similar studies

[5] Living conditions during that time are described, for example, by Balfour and Mair, *Four-Power*, 113ff. Heidelberg, Göttingen, and a few other places were spared from the bombing. About the working conditions of physicists at universities and research institutes elsewhere, see the articles in *Neue Physikalische Blätter*, 2/1 (1946), 23f.; 2/2, 14f.; 2/3, 65–70; 2/4, 85–89; 2/5, 116–121, and so forth. The sufferings by the many refugees and homeless from famine and the cold are described, for example, in the supplements by F. H. Rein from 1946 in the Archives of the Max Planck Society (*Archiv der Max-Planck-Gesellschaft*, hereafter MPGA), div. III, rep. 14A, no. 5730. The *Physikalische Blätter*, 4 (1948), 42f, published a declaration by the Technical University of Munich on the nutritional situation signed, among others, by Ludwig Föppl, F. von Angerer, and G. Hettner. For intellectual workers it demanded "extra ration coupons at least at the level of semi-hard laborers." M. von Laue wrote L. Meitner on 27 November 1947 about the destruction, clothing shortage, and consequences of malnutrition: Jost Lemmerich (ed.), *Lise Meitner–Max von Laue. Briefwechsel 1938–1948* (Berlin: ERS Verlag, 1998), 309.

on the mentality of other scientific communities, which unfortunately do not seem to exist yet (at least for the post-war period).[6] A cursory glance at Ute Deichmann's major study on chemists under National Socialism reveals many points of agreement with the mentality profile of physicists examined here. Be this as it may, the documentation assembled here stems almost entirely from physicists, categorized here by membership, short-lived though it may have been, in the German Physical Society (*Deutsche Physikalische Gesellschaft*, DPG). Hence, those rejected by the society in 1938 who explicitly decided not to rejoin after 1945 are also among them.

ABOUT THE SOURCES USED

A main source of this study is the journal *Physikalische Blätter* (founded in 1944).[7] The hundreds of complete articles, shorter notices, and obituaries in its first few volumes constitute a serial source as used in the French tradition of writing the history of mentality.[8] Its broad scope beyond that of a professional journal makes it particularly appropriate for our purposes. Originally conceived as a "physico-political journal" by its founding editor, the specialist on electron microscopy Ernst Brüche, this "informal noticeboard" accepted brief announcements by all the universities and colleges, published letters to the editor covering topics of the

[6] Fritz Ringer describes in *Die Gelehrten. Der Niedergang der deutschen Mandarine 1890–1930* (Stuttgart: Klett-Cotta, 1983) the mentality of academic "mandarins" based primarily on examples taken from the humanities. Jonathan Harwood, "'Mandarine' oder 'Außenseiter'? Selbstverständnis deutscher Naturwissenschaftler (1900–1933)," in Jürgen Schriewer et al. (eds.), *Sozialer Raum und akademische Kulturen* (Frankfurt am Main: Lang, 1993), 183–212, as well as the articles by Jonathan Harwood, Jeffrey Johnson, and others in Rüdiger vom Bruch and Brigitte Kaderas (eds.), *Wissenschaften und Wissenschaftspolitik. Wissenschaften und Wissenschaftspolitik: Bestandsaufnahmen zu Formationen, Brüchen und Kontinuitäten im Deutschland des 20. Jahrhunderts* (Stuttgart: Steiner, 2002), juxtapose the Weber ideal of the specialist or expert, more typical of scientists or technologists of the post-war period. Hannah Arendt, *Besuch in Deutschland* (Berlin: Rotbuch-Verlag, 1993), provides a description of the mentality of the German population in general that is as caustic as it is applicable.

[7] Translated article titles from this professional newsletter appear here within quotation marks, with more specific referencing in the annotation. See the relevant subsection in Gerhard Simonsohn's contribution to this volume.

[8] For examples of this research approach, see Jean-Michel Thiriet, "Methoden der Mentalitätsforschung in der französischen Sozialgeschichte," *Ethnologica Europea*, 11 (1978/80), 208–225. For some criticism of it, see Ralf Reichardt, "Für eine Konzeptualisierung der Mentalitätstheorie," *Ethnologica Europea*, 11 (1978/80), 234–241. I subscribe to Le Goff's assertion: "tout est source pour l'historien des mentalités," in Jacques Le Goff, "Les mentalités. Une histoire ambigüe," in J. Le Goff and P. Nora (eds.), *Faire de l'histoire* (Paris: Gallimard, 1974), 76–94.

day, and also reprinted articles from other publications (some in excerpt or translated by Brüche). According to Brüche's foreword in its first issue from May 1946, the *Neue Physikalische Blätter* (as it was then called) was also intended as a forum for chemists and mathematicians, teachers, and engineers. One of them, Arnold Sommerfeld, was incidentally not impressed by it. He wrote to the publisher, Vieweg & Son, in early 1946: "May the 'Physikalische Blätter' expire, they are not quite worthy of your publishing house."[9] Notwithstanding this scathing verdict, the *Physikalische Blätter* attracted a large readership, becoming "a physicist's homily [*Hauspostille*]," thanks to its up-to-date and multifaceted reporting. Thus, it is a useful contemporary source for the mentality of physicists. We must bear in mind, however, that these articles as well as those from the *Göttinger Universitäts-Zeitung* or the *Neue Zeitung*, for instance, could only appear after receiving the stamp of approval of the military government. So the image it reflects is of the official side of the coin. That is one reason why excerpts from contemporary correspondence would have to supplement it. For lack of space here, we will have to make do with a few samplings.

SCIENTISTS IN GERMANY SEEN FROM THE OUTSIDE

The primary purpose of Richard Courant's trip in 1947 for the Office for Naval Research was to evaluate German progress on the development of calculating machines at a number of universities and technical colleges. At the same time, he was on the lookout for promising young researchers. His reconnoitering of the general situation among university teachers and their students serves us well as a kind of mood barometer. After landing at Frankfurt, he and his company were amazed at how difficult it was to find one's way around in the labyrinthine ruins of the city and "at the demoralization of the population. They were shaken by the sight of great

[9] Sommerfeld to Vieweg from 24 January 1946, Archives of the Deutsches Museum, Munich (*Deutsches Museum, Archiv*, hereafter, DMA), collection 89, 014. The history of the *Physikalische Blätter* is outlined by Ernst Dreisigacker and Helmut Rechenberg, "50 Jahre Physikalische Blätter," in *Physikalische Blätter*, 50 (1994), 21–23; Ernst Dreisigacker and Helmut Rechenberg, "Karl Scheel, Ernst Brüche und die Publikationsorgane," in Theo Mayer-Kuckuk (ed.), *Festschrift 150 Jahre Deutsche Physikalische Gesellschaft*, special issue 51 (1995), F135–F142, here F139ff. Armin Hermann, "Die Deutsche Physikalische Gesellschaft 1899–1945," in Mayer-Kuckuk, F61–F105, there F102f. discusses Brüche's original intention. For a history of the *Deutsche Physikalische Gesellschaft* after 1945, see Wilhelm Walcher, "Physikalische Gesellschaften im Umbruch (1945–1963)," in Mayer-Kuckuk, F107–F133.

crowds of ragged, hungry Germans, many of them begging."[10] At the polytechnic in Darmstadt, Courant observed that the registered students, numbering more than 2,000, all put in a half day every fortnight toward clearing and reconstruction work in the totally bombed-out campus. This required labor was common elsewhere as well. Its president, Richard Vieweg, had complaints about the young helpers, not only because of their underfed condition but also because of their poor academic quality and lack of any sense of ethics: "for some, it seemed that the only thing the Nazis had done wrong was to lose the war." At Heidelberg, Courant perceived "a slight lack of resonance" in a conversation with the publisher Ferdinand Springer. We shall encounter this peculiar incommensurability between the emotional and linguistic worlds of resident versus exiled German. At Marburg, the mathematician Kurt Reidemeister told Courant that the age group between 20 and 25 was too poorly educated to come under consideration for his purposes. Those between 30 and 35, however, had another drawback: political *Belastung*, that is, the burden of a Nazi past:

> thus, carrying the stigma of having been National Socialist Party members – but he is sure that many of the belastet have not mentally been Nazis. [They] had to join the Party in order to keep the positions.... [Still] he said that 90 per cent of the population, including academic people, are dangerously but not hopelessly nationalistic. In natural science people in general, much less, however.[11]

The impression Courant's fellow Germans at Göttingen left in July 1947 was not favorable: "absolutely bitter, negative, accusing, discouraged, aggressive."[12] An interesting dichotomy between Werner Heisenberg's superficial and inner mental profiles was revealed in conversation. During his first few interviews with Courant, Heisenberg was prepossessing and basically positive, as he had been before the war. But that was evidently a

[10] Constance Reid, *Courant in Göttingen and New York: The Story of an Improbable Mathematician* (New York: Springer-Verlag, 1976), 259. For example, see Balfour and Mair, *Four-Power*, 11ff. on the extent of the destruction in Germany.

[11] Reid, *Courant*, 260f. (original English). A delegation of the Association of Scientific Workers that visited various university and industrial laboratories in 1948 came away with a more extreme impression after seeing German scientists and speaking with Allied Control officers. "The Control officers were of the opinion that all Germans they had come into contact with were the same nationalists as before. As one of them put it: 'If they could only invent a weapon that would destroy the whole occupying force without lifting a hair of any Germans, they would be jubilant.'" Roger C. Murray, "Social Action: Die Diskussion um Wissenschaft und Wissenschaftler im heutigen Deutschland," *Göttinger Universitäts-Zeitung*, 4th series, no. 17 (2 September 1949), 171 and 15.

[12] Reid, *Courant*, 261 (the following passage: 265).

façade reserved for outsiders. Heisenberg was completely different once he had reassured himself that it was safe to speak frankly: "But another day, discussing politics, he found that Heisenberg came out finally with the same stories and aggressiveness against Allied policy of starvation [evidently meaning the dismantling of German factories] as the less cool and more emotional people."[13]

As this example shows, communication failed both at the level of superficial friendliness between colleagues as well as at the emotional level. Their basic moods and values were much too far apart. Courant wrote a friend: "I found very few people in Germany with whom an immediate natural contact was possible. They all hide something before themselves and even more so from others."[14] All the same, Courant's final verdict at the end of his visit in 1947 was that the Germans definitely had to be helped. He wrote to Warren Weaver of the *Rockefeller Foundation*:

> In spite of many objections and misgivings, we feel strongly that saving science in Germany from complete disintegration is a necessity first because of human obligations to the minority of unimpeachable German scientists who have kept faith with scientific and moral values.... It is equally necessary because the world cannot afford the scientific potential in German territory to be wasted.[15]

Sources such as these are very instructive. From an aggregation of numerous individual observations and conversations, some witnesses of the time managed to reach a higher plane and give a summary description of the general mood among their contacts.

TENSIONS WITH THE ALLIES

Superficial Admiration and Opportunistic Friendliness

Michael Balfour, a historian working for the British branch of the Control Council, interpreted the surprising "submissiveness in defeat" of the Germans, so contradictory to the image of the enemy, as the extreme psychological opposite to their arrogance and aggressiveness in a position of strength.[16]

[13] Reid, *Courant*, 261f.
[14] Courant to Winthrop Bell, cited Reid, *Courant*, 263.
[15] Courant to Weaver, cited Reid, *Courant*, 265.
[16] See Balfour and Mair, *Four-Power*, 53, on the "submissiveness in defeat" and 58f. about alternative attempts to play the Allies off against each other by feeding bits of information to one occupying force about misconduct by the other.

Numerous notices in the *Physikalische Blätter* acknowledge the competency of their new rulers outright, as far as technological accomplishments and scientific know-how are concerned. Examples include advances in color television in the United States (in its second volume of 1946, p. 21, based on a report in *Die Neue Zeitung*, 7 January 1946) or the detection of hydrodynamic rolling waves of the earth's surface during tests of the atomic bomb (on p. 49, taken from an article in *Time*, 28 January 1946), or the radio-controlled igniter design deployed against the V-1 rocket, attaining a downing rate of 70 percent by the end of the war (on p. 20, from *Yank*, 25 November 1945). The awe-inspiring total of 3,800 collaborators on American radar development is quoted, roughly 700 of them being physicists: "twice as much as had worked on the atomic bomb" (p. 18), and so forth.[17]

This last example illustrates the ambiguity of many of these compliments for Allied science. A reviewer of the book *Science at War* praised the apparatus in Rotterdam similarly in 1949 as the high point of British radar development: "The authors point out emphatically that this astonishing development was primarily the result of uninhibited and intimate collaboration among all the participants. The problems were pursued, ignoring rank and distinction, in discussion and critique between Air Force marshals, admirals, scientists, pilots, laboratory assistants, development and production engineers."[18] To every German reading between the lines, it was clear that the author was decrying the lack of such "uninhibited and intimate collaboration" between German researchers and the military. The almost dictatorial power of institute patriarchs like Robert Pohl in Göttingen and gray eminences like Sommerfeld in Munich hardly changed after the war until as late as the 1960s. The structure of physics departments at American universities, in contrast, led to a much stronger democratization, at least within the faculties.[19]

Brüche's notice in the *Physikalische Blätter* entitled "Flying eyes" indicates how strongly the war continued to govern the visions of the future:

> The battlefields of World War III will know no secrets. Flying televisors will watch how cities disintegrate and they will project the fighting onto the generals' viewing screens in deep bunkers. Even bombs will observe their own fall to the target with unperturbable electronic eyes, before report and bomb vanish simultaneously.[20]

[17] *Physikalische Blätter*, 2 (1946).
[18] H. Jetter, "Science at war," *Physikalische Blätter*, 5/1 (1949), 43.
[19] Theo Mayer-Kuckuck once said to me, quoting Maier-Leibnitz: "Departments are an extension of the *Ordinarius* principle to the level of assistants."
[20] Fliegende Augen, *Physikalische Blätter*, 2 (1946), 71, taken from *Time* (1 April 1946).

The Allies were in the difficult position of deciding whether to heed the call for immediate severe punishment of the guilty among the defeated enemy or to look to the future and foster their reeducation into good democrats.[21] Even clearly definable goals like dismissing the most seriously incriminated persons from university faculties conflicted with the equally clear goal of reopening important academic institutes as soon as possible. These included university clinics and hence physics institutes, where medical students received an important element of their training. As the period of occupation dragged on, the Allies, particularly the British and the Americans, were increasingly tempted to identify themselves with their charges, against the official prescription of maintaining "restrained arrogance" and at odds with the prohibition to fraternize.[22]

This same dilemma between the drive to purge and reform and practical considerations troubled the reopening of other scientific institutions and professional publications as well. The pressure to fall back on existing structures and staffing continued to mount. The case of the Kaiser Wilhelm Society (*Kaiser-Wilhelm-Gesellschaft*, KWG) exemplifies this. This national research society was dissolved in 1945 only to be reopened under a new name: Max Planck Society.[23] The Allies took their time approving a few of the traditional journals, such as the *Annalen der Physik* and the *Zeitschrift für Physik*. This left a vacuum during the first two postwar years that new publications were happy to fill: the *Zeitschrift für*

[21] The Allied perspective on these conflicts is provided, for example, by Balfour and Mair, *Four-Power*, 14–50, 162–183, Anonymous, "The fate of German science. Impressions of a BIOS officer," *Discovery*, 8 (1947), 239–243, here 243; Henry Kellermann, *Cultural Relations as an Instrument of U.S. Foreign Policy: The Educational Exchange Program between the United States and Germany, 1945–1954* (Washington, DC: U.S. Government Printing Office, 1978). James F. Tent, *Mission on the Rhine. Reeducation and Denazification in American-Occupied Germany* (Chicago: University of Chicago Press, 1982), goes into the guiding idea of "reeducation" as the Americans understood it.

[22] Officer Edward Y. Hartshorne describes this tense situation in detail in his diaries and correspondence, published in James F. Tent (ed.), *Academic Proconsul. Harvard Sociologist Edward Y. Hartshorne and the Reopening of German Universities 1945–46* (Trier: Wissenschaftlicher Verlag Trier, 1998). Hartshorne was responsible for the press and the reopening of the universities in the American sector. See 5f. on the prescribed attitude of "restrained arrogance" and the problems regarding the prohibition to fraternize.

[23] See, for example, Manfred Heinemann, "Der Wiederaufbau der Kaiser-Wilhelm-Gesellschaft und die Neugründung der Max-Planck-Gesellschaft (1945–1949)," in Rudolf Vierhaus and Bernhard vom Brocke (eds.), *Forschung im Spannungsfeld von Politik und Gesellschaft* (Stuttgart: DVA, 1990), 407–470; Armin Hermann, "Science under Foreign Rule; Policy of the Allies in Germany 1945–1949," in Fritz Krafft and Christoph J. Scriba (eds.), *XVIIIth Int. Congress of History of Science Hamburg–Munich. Final Report* (Stuttgart: Steiner, 1993), 75–86, here 83ff. See *Neue Physikalische Blätter*, 2 (1946), 124, *Physikalische Blätter*, 3 (1947), 136. See also Lemmerich, *Meitner–Laue*, 468, for further context about the Allies' wishes regarding the Kaiser Wilhelm Society.

Naturforschung and the *Physikalische Blätter*. The latter had originally been authorized by the chairman of the German Physical Society, Carl Ramsauer, in 1944 as the official publication of the "Information Office of German Physicists."[24] Its first volume in 1944–1945 had unabashedly emphasized the crucial military importance of physical research. Its editor Brüche nevertheless managed to obtain permission, first from the American and British sectors and soon afterward from the French sector, to publish an "emergency journal," which appeared in obvious continuation of its former title as the *Neue Physikalische Blätter*.

As we can gather from a carbon-copy circular letter of March 1946 by the founding editor to his colleagues, Brüche envisioned "an undemanding journal to reestablish contacts within physics and to discuss issues of the day." So it quite specifically also set its sights on science policy: "Particularly our neighbor to the west seems to show an interest in the journal and in collaborating. I consider it beneficial, because I believe that in future, scientific relations with France, in particular, should be fostered on an equal basis."[25]

With such headlines as "Research in distress!" the first issues of the *Physikalische Blätter* in 1944 certainly could have been interpreted as a political mobilizer of the country's last reserves. Isolated criticism was voiced about the journal's staffing continuity after the war, but it never became public:

> Concerning Mr. Brüche, I do find it unfortunate that such a man as Mr. Brüche be the editor of the Physikalische Blätter. Judging from the content of the Physikalische Blätter, Mr. Brüche was, during the Nazi period, certainly moving in the Nazi wake. So I do wonder that this journal was given permission to appear again. But nothing can apparently be done about it practically, at the moment, but the Physikalische Blätter must remain the publication of the Physical Society.[26]

This publication evidently also was subject to the scrutiny of Allied Control. *Censorship* is perhaps too strong a word, but it was clear that the

[24] The background to the founding of the *Physikalische Blätter* and its publishing outlet is discussed, for example, by Ernst Brüche, "Die Arbeit der Informationsstelle Deutscher Physiker," *Physikalische Blätter*, 3 (1947), 224–226 and Dreisigacker and Rechenberg, 50 Jahre, in the commemorative issue. The 150-year-old *Annalen der Physik* was refused its bid for paper consignments: see the introductory article in: *Neue Physikalische Blätter*, 2 (1946), 2. Other scientific journals are discussed in *Neue Physikalische Blätter*, 2 (1946), 111–114; see also the contribution by Dieter Hoffmann to this volume.

[25] Brüche's circular, March 1946.

[26] Meißner to von Laue, 15 December 1947, MPGA, div. III, rep. 50, no. 1315. Brüche elaborates on his motives in a letter to Planck, 21 October 1945, MPGA, div. III, rep. 50, no. 2390.

good will at the outset could easily be spent. Self-imposed control and careful testing of the limits of free speech were the rules of engagement. Paper consignments played as important a role for the journal as its licensing. The first issues in 1946 were printed at a run of 5,000. Even that "did not even remotely suffice to cover the demand," but the run for early 1947 had to be reduced to 2,500 copies because only about 500 kg of paper could be made available to the Physikalische Blätter per quarter instead of more than twice that amount. At the end of the second issue for 1947, Brüche wrote: "this had been interpreted by some readers as a sign of the Control authorities' purported discomfort with the journal. Pessimists were already saying that the journal would now go to its demise for economic reasons. The editor can declare today that the Military Government has restored the earlier printing and has moreover basically granted its approval for an increase in the printing by 50 %."[27]

Brüche took this upgrading to 7,500 copies as confirmation of his opinion that:

> honest discussion is regarded as detrimental only to those who are not able to distinguish between candid expressions of opinion and mean-spirited criticism; nor does the forwarding of contributions from America through the *Information Control Division* – the first among these to appear "exclusive to your magazine" in one of the coming issues – speak against it.[28]

This change of policy to include contributions by the Information Control Division without comment approached the tolerance limit of the *Physikalische Blätter* in conforming with the new government. Tinctured stories, such as on "50 years of American physics," came dangerously close to the boundary between reporting and propaganda. Brüche laid particular store by independence.[29] The following issue (no. 5) already included a printed note on red paper, justifying a reduction from 40 to 32 pages owing to an acute paper shortage. It also urged its readers to send

[27] Ernst Brüche, "Neue Auflage der Phys. Blätter," *Physikalische Blätter*, 3/2 (1947), 64.

[28] Brüche, Neue, Balfour, and Mair, *Four-Power*, 211–228 discuss the supervision of journals and other media in post-war Germany. On the Information Control Division, see also Tent, *Mission*, 77f., 86ff.

[29] Gordon F. Hull, "50 Jahre amerikanische Physik," *Physikalische Blätter*, 3/4 (1947), 107–110, among other notices of the same tenor. Brüche remembered his effort to maintain this independence in Ernst Brüche, "Nec temere nec timide," *Physikalische Blätter*, 6 (1950), 25–28 (draft manuscript MPGA, div. III, rep. 50, no. 370). See Dreisigacker and Rechenberg, "Karl Scheel," F140f., Dieter Hoffmann and Thomas Stange, "East-German physics and physicists in the light of the 'Physikalische Blätter,'" in Dieter Hoffmann (ed.), *The Emergence of Modern Physics* (Pavia: Univ. degli Studi di Pavia, 1997), 521–529.

any available recyclable paper directly to the publishing house, which at that time was "Volk und Zeit," run by Wilhelm Beisel in Karlsruhe.

COVERT RESERVE AND DISTRUST

Despite the official rhetoric, only a few physicists in Germany seemed to have viewed the arrival of Allied troops as a liberation from the National Socialist dictatorship. Max Steenbeck, interned by Soviet soldiers more by chance than owing to his position as foreman at a Siemens factory, wrote retrospectively:

> While we men sat in captivity in camps, our wives and children had to get by alone in a topsy-turvy world of violence, hate, and murder – and that was through our own making. No, on the eighth of May nineteen forty-five, I, in any case, absolutely did not feel liberated; perhaps liberated from the constant thought: Will you be alive tomorrow, at all? But that was insignificant against the life we then saw ahead of us.[30]

A directive by the Joint Chiefs of Staff (JCS 1067), initially only released to a few people in leadership positions on 21 May 1945, later to be published on 17 October 1945, determined that only "after removal of the characteristic traces of Nazism and of Nazi staff" could a reopening of educational institutions be permitted.[31] The dangling Damocles sword of dismissal for political involvement was the more threatening because for a long time no one knew how exactly such "traces of Nazism" were to be defined: on the basis of formal criteria like membership in National Socialist organizations or merely from eyewitness reports and affidavits?[32] A new wave of accusations and anonymous denunciations similar to the

[30] Max Steenbeck, *Impulse und Wirkungen* (Berlin: Verlag der Nation, 1977), 151. Nor did the Allies have the impression that their presence among the Germans was seen as a liberation: for example, see Edward Y. Hartshorne in Tent, *Academic*, 19.

[31] For example, see Balfour and Mair, *Four-Power*, 23ff., 66f., Conrad F. Latour and Thilo Vogelsang, *Okkupation und Wiederaufbau. Die Tätigkeit der Militärregierung in der amerikanischen Besatzungszone Deutschlands 1944–1947* (Stuttgart: DVA, 1973), 17 and 177. Klaus Schlüpmann, *Vergangenheit im Blickfeld eines Physikers (eine Wissenschaftsstudie), Entwurf einer Biographie von Hans Kopfermann*, www.aleph99.org/etusci/ks/ (version 18 January 2002), 454, points out that this directive was replaced by JCS 1779 on 11 July 1947, emphasizing reconstruction and the changed course; also see David Cassidy, "Controlling German Science," *Historical Studies in the Physical Sciences*, 24 (1994), 197–235 and 26 (1996), 197–237. For the response by physicists to these directives, see, for example, Brüche's columns on a "revision of prejudice," *Physikalische Blätter*, 5 (1949), 48, and "I certainly do hear the message," *Physikalische Blätter*, 5 (1949), 392.

[32] On the problem: "who is/was a Nazi?" see, for instance, Balfour and Mair, *Four-Power*, 52. Some Allied Control officers tried not to rely on the mere formality of membership

FIGURE 27. Max Steenbeck (*Source:* Archiv DH)

early period of the Nazi regime swept over the land. Some of the charges targeted real culprits, whereas others, it turned out, were just a settling of old scores.

John Gimbel, himself an officer of the occupying force as a young man, wrote in his book *Science, Technology and Reparations*: "The Germans continued to be *fearful and suspicious*, and eventually their worst fears came to pass." He was referring specifically to the abuse of Allied Control laws for industrial espionage.[33] But Law No. 22 on the surveillance of materials, installations, and equipment in the area of atomic energy by the Atomic Energy Commission (AEC), which law came into force as late as April 1950, placed such cramping limitations on German nuclear research that Heisenberg as chair of the German Research Council (*Deutsche Forschungsrat*) spoke of a "strangulation" of this branch of research.[34]

in the National Socialist Party or its suborganizations, looking also at individual political affinity with such Nazi ideologies as Aryan superiority, the authoritarian *Führer* principle, or the exercise of unlimited dictatorial powers. Hartshorne's classification of political attitudes held by university professors are described by Tent, *Academic*, 103, who also reported on the frequent distinction made between party comrades and "only nominal Nazis" (*Parteigenossen* versus *Karteigenossen*), Tent, *Academic*, 75.

33 John Gimbel, *Science, Technology, and Reparations. Exploitation and Plunder in Postwar Germany* (Stanford: Stanford University Press, 1990), 180 (emphasis mine), particularly concerning the United States. Wolfgang Finkelnburg wrote to Arnold Sommerfeld on 28 May 1947 about "bitter feelings, that always need a long time to die out after a war," available online at www.lrz.de/~Sommerfeld/JahrDat/1947.html (July 2, 2010).

34 See the article: Control of Nuclear Research in Germany, intended for publication in the *Bulletin of the Atomic Scientists* in July 1950 together with a commentary by Werner Heisenberg in the form of a letter of protest to Mr. K. H. Lauder, Director of the Research Branch Göttingen. Although page proofs still exist among the *Bulletin*'s files, the article itself never actually appeared: Manuscript department of the Regenstein Library, University of Chicago (hereafter, RLUC), box 32, folder 7. Compare the contributions by

Profound distrust is evidently a general characteristic of the post-war mentality of scientists and engineers.

The following notice in the *Physikalische Blätter* exemplifies this "climate of suspicion" and the resulting misrepresentations and biased self-portrayals of physicists in Germany before the Allies during the early period:

> Whoever went to an American authority with some petition regarding the field of science during the first months of the occupation did well not to use the words "science" and "research" initially, confining the argument to aspects of science education. The American press and perhaps also the exaggerations of the National Socialist Ministry of Propaganda had fed suspicions that German laboratories were dangerous witches' kitchens needing to be forbidden, one and all, upon mere suspicion. Maybe the points of the Morgenthau Plan that, as an agrarian country and exporter of raw materials, Germany had no more use for research also had something to do with it.[35]

Interest in taking public office was also very low, because many potential candidates were afraid of being held responsible for the vexations with the occupying forces and being stigmatized as a collaborator.

Distrust on all sides remained the order of the day in subsequent years as well. At the end of 1947, Max von Laue complained to an English colleague about the difficulties they were having filling a director's position at the German National Bureau of Standards, the *Physikalisch-Technische Reichsanstalt*, and about the high turnover rates in the civil service:

> This bad situation cannot be done away with at the moment. For that, the whole situation would need to calm down considerably, which could only come from the highest political levels. Peace has to come, and not just peace on paper, but peace of mind, whereby the terrible distrust, everyone against everyone, which poisons all human relations nowadays, finally stops. But it surpasses the power of scientists to bring this about.[36]

Klaus Clusius, director of the Institute of Physical Chemistry and for a while also dean of the University of Munich, complained soon after its reopening that it had taken place "in due pomp and ceremony with the

Regierungsdirektor Friedrich Frowein, head of the German research supervisory office in Hessen, on research control in West Germany: *Physikalische Blätter*, 6 (1950), 222–225, as well as "Nochmals Gesetz 22," *Physikalische Blätter*, 6 (1950), 316–319, with critical commentary.

35 Ernst Brüche, "Förderung der Forschung," *Physikalische Blätter*, 3/12 (1947), 12, 432.

36 Max von Laue to Charles Darwin (National Physical Laboratory), 12 December 1947; Archives of the German Physical Society (*Archiv der Deutschen Physikalischen Gesellschaft*, hereafter DPGA), no. 40046.

dismissal of most of its professors and assistants." His other letters are full of cynicism, bitterness, and fear of worse to come. He complained, for example, that Heisenberg's appointment would never work out for purely political reasons, because "at the behest of men of this terrestrial world" he has to stay in the British sector for the time being:

> Measured against the enormities that we have the dubious pleasure of bearing witness to, this issue seems a triviality. But set yourself back in time to 100 years ago and imagine what would have happened if that which Europe is being blessed with today had taken place then, so . . . we better keep quiet![37]

A questionnaire from 1946 in the American sector reveals how direly denazification was needed among the population. Given the alternatives that Nazism was (i) a bad thing, (ii) a good thing, or (iii) a good thing badly carried out, 40 percent of the respondents chose the third. When the questionnaire was repeated in 1948, this figure actually rose to 55.5 percent.[38]

RESISTANCE TO DENAZIFICATION

We have just seen a surprising willingness by two physicists in responsible positions to bend history for the sake of an institute's future in the "new Germany." In the manner of social constructivism *avant la lettre*, denazification had become negotiable. "Decent impulses" (evidently meaning pangs of conscience about withholding incriminating information concerning a colleague's past) must be suppressed before an "immoral law" – out of a sense of duty. Allied Control directive nos. 24 (12 January 1946) and 38 (September 1946) specified the removal of all active members of the National Socialist German Workers Party (*Nationalsozialistische Deutsche Arbeiterpartei*, NSDAP) and other persons opposed to Allied objectives from public and semipublic offices and from responsible positions in major private companies.

Walther Gerlach issued "whitewash certificates" (*Persilscheine*) not just for his close friend Werner Köster, but also for a former leading science policy maker, Rudolf Mentzel. This *Schutz-Staffel* (SS) regiment leader had been president of the German Research Foundation (*Deutsche Forschungsgemeinschaft*, DFG) and head of the Science Office

37 Clusius to Bonhoeffer, 11 December 1945 and 21 January 1946, MPGA, div. III, rep. 23, no. 14, 1.

38 Balfour and Mair, *Four-Power*, 58; also see on denazification: Balfour and Mair, *Four-Power*, 169–183, 331–334, and Tent, *Mission*, 83–109.

in the Reich Ministry for Science, Education and Culture (*Reichsministerium für Wissenschaft, Erziehung und Volksbildung*, REM). Before 1945, Mentzel had certainly not always lent his support to Gerlach as head of the physics sections of the DFG and the Reich Research Council (*Reichsforschungsrat*, RFR), so this solidarity is surprising.[39] In the western zones, Arnold Sommerfeld contributed to the whitewashing with at least 12 documentable exonerating affidavits against only two refusals. He declined Hans Kneser because too much time had elapsed since their acquaintance. The former dean, Karl Beurlen, was the only applicant he refused outright for political reasons.[40] In a confidential letter from 1947, Max von Laue described the guiding principle behind such whitewash characterizations by him and his colleagues: "We are trying to implement a policy here in physics that is unfortunately not being adopted by the state, namely, carrying out one big amnesty for all Nazi fellow travelers, after harshly condemning the real culprits."[41] By this time, Otto Hahn also shared this attitude. "Drawing the line" (*Schlußstrich-Mentalität*) is how Norbert Frei described it, for the sake of the young Federal Republic of Germany. When a doctoral student confronted Hahn with incriminating facts about Pascual Jordan and Herbert A. Stuart, he replied to the young woman:

> In such cases I am often asked how I should respond. If it doesn't involve blatant cases, I answer that I certainly won't do anything for the gentlemen, on the other hand, I won't actively bring charges against them either. I am reluctant to continue to add to all these unpleasant things. We had enough trouble with all that snooping and telling off during the Third Reich, and I don't think that after these

[39] See DMA, Gerlach papers, file on the denazification process, in English translation in Klaus Hentschel (ed.), *Physics and National Socialism. An Anthology of Primary Sources* (Science Networks. Historical Studies, vol. 18) (Basel–Boston: Birkhäuser, 1996), 403–406 (also available online at Google Books). For the contextual reasons and other positive and negative statements about Mentzel, see Schlüpmann, *Vergangenheit*, 417f. Von Laue wrote Hahn, 23 August 1946, about Gerlach's political attitude, MPGA, div. III, rep. 14A, no. 2462. But von Laue's protest against Gerlach's involvement in the reopening of the physical society was ignored, and on 11 September 1946 Gerlach became one of the founding members of the Max Planck Society (*Max-Planck-Gesellschaft*, MPG) in the British zone: see Rudolf Heinrich and Hans-Reinhard Bachmann, *Walther Gerlach: Physiker – Lehrer – Organisator* (Munich: Deutsches Museum, 1989), 187.

[40] See the Web site of the Sommerfeld project under 1946 and 1947. Carola Sachse, "'*Persilscheinkultur.*' Zum Umgang mit der NS-Vergangenheit in der Kaiser-Wilhelm/Max-Planck-Gesellschaft," in Bernd Weisbrod (ed.), *Akademische Vergangenheitspolitik* (Göttingen: Wallstein, 2002), 217–245, here 231 speaks, with reference to Walker, of as many as some 60 *Persilscheine* issued just by Sommerfeld and Heisenberg.

[41] Von Laue to Pechel, 11 November 1947, DPGA, no. 40048. Von Laue wrote to his son Theo along similar lines on 16 July 1946, MPGA, div. III, rep. 50, suppl. 7/7, sheets 27f.

gentlemen have their tribunal proceedings behind them and are relieved about it, that they will suddenly come forward again as active or potential Nazis.[42]

Even among émigrés there was opposition to the Allied denazification scheme. Lise Meitner, for instance, generously issued certificates to people whom she knew had once actually denounced her politically.[43] But many other well-informed scientists were "openly annoyed at what has been going on recently [i.e., 1946–1947] at some German universities under the banner of 'denazification,'" as the radiochemist and co-discoverer of nuclear fission Otto Hahn put it in 1947 in articles to the *Göttinger Universitäts-Zeitung* and the *Physikalische Blätter*. The concrete circumstance was a new wave of official dismissals in the Universities of Munich and Erlangen following American orders. The military government had the impression that these two universities had been dragging their feet, making use of the leniency granted them "to denazify their faculty members gradually, so as to avoid interruptions in the courses."[44]

The chairman of the Max Planck Society in the British Zone, just recently elected on 1 September 1946, and the university president at Göttingen pointed out that the "new wave of such 'official' dismissals . . . triggered very serious debates about the sense and senselessness of 'denazification' and aroused vivid memories of the arbitrariness during the 'Third Reich.'" About some of the affected persons, "it would never have entered our minds to doubt their opposition to National Socialism."[45]

These two Göttingen scholars arrogantly dismissed the competency of the denazification authorities tackling this complicated problem of evaluating political involvement. Simple allusion to their personal acquaintance with those affected sufficed for them. There was, of course, a rhetorical function for the (skewed) comparison of denazification after 1945 with the Aryanization a dozen years before. They wanted to discredit the pressure applied by the Allies for such "intellectual and moral purging."

42 Hahn to Martius, 12 November 1947, MPGA, div. II, rep. 14A, no. 2726 (reference by courtesy of Gerhard Rammer).

43 Meitner to Franck, 10 July 1947, RLUC, Franck papers, box 5, folder 5. For excerpts from letters by Hermann Fahlenbrach and for replies by Droste and Meitner, see Ruth Lewin Sime, *Lise Meitner: A Life in Physics* (Berkeley: University of California Press, 1996), 350.

44 Anonymous, "76 Entlassungen an der Erlanger Universität," *Die Neue Zeitung*, 3/10 February 1947, 3rd series, 5. See also Tent, *Mission*, 92ff.

45 Otto Hahn and F. Hermann Rein, "Einladung nach USA," *Physikalische Blätter*, 3 (1947), 33–35, quote on 34.

We must add that Otto Hahn was another contributor to the inflation of whitewash certificates. He too was swept up in the boycotting spirit against serious denazification attempts that created such a peculiar sense of solidarity even among formerly frosty colleagues. This experience dismayed Allied officers in charge of denazification: "The proclivity of "irreproachable" (*einwandfreie*) Germans to rush to the support of their colleagues who were fools enough to compromise themselves with the Nazi cause is surely one of the most startling and depressing aspects of post-Nazi German academic society."[46] Yet exonerating witnesses often had little incentive to clarify the role of the accused properly. Although formally "irreproachable," having never entered the party or its affiliates themselves, some were still caught up in the intricate relations of command and collaboration.

Otto Hahn was initially positively disposed to the first "spontaneous denazification, which undoubtedly hit the mark."[47] By 1947, however, he had completely changed sides, fundamentally criticizing its practical implementation and denying benefits it might afford:

> We [scientists] are neither politicians nor lawyers but are accustomed to regarding matters perhaps a little more calmly and rationally than other professions. We profoundly regret how the "denazification" is flipping into its obverse through the many measures, pushing true peace further and further away. We do not understand how it can take so long to distinguish finally between "criminality" and "political mistakes"; the arbitrary muddle probably causes much of the denazification problems today including, for instance, the attacks against the science of our nation.[48]

The blatant lack of equal treatment resulting from regional variations and the many alterations to the guidelines of denazification were criticized, often legitimately. There was the initial tendency to pass draconian punishment on collaborators without regarding many of the major offenders, who "frequently under false names and with fake passports could hide among the great number of displaced Germans."[49]

[46] According to Edward Y. Hartshorne's diary entry dated 24 July 1945, in Tent, *Academic*, 82.

[47] Hahn and Rein, "Einladung."

[48] Hahn and Rein, *Göttinger Universitäts-Zeitung*, 2nd series, no. 6 (21 February 1947), 1–2.

[49] Latour and Vogelsang, *Okkupation*, 179f., describe denazification as "a genuine tragicomedy of errors." Particularly the actions in the American zone "have to be regarded as a moral debacle." Sommerfeld's whitewash certificate for Heinrich Ott from 13 February 1946 is another example of complaints about unequal treatment, www.lrz.de/~Sommerfeld/KurzFass/05160.html. Karl Bechert declined the appointment

A year later this critical view had not changed. In a survey of the elapsed year 1947, Ernst Brüche seconded Hahn's polemics against the "denazification evil" with the following characterization of the prevailing mentality: "A dangerous process of disenchantment is underway that does not stop at physicists. Employing the bad word 'renazification' would be incorrect, because that would imply that initially one could have spoken of the 'nazification' of physicists."[50]

If, with Mark Walker, we define "nazification" as "effective, significant, and conscious collaboration with portions of National Socialist policy"[51] (to be applied to groups, not individuals), then Brüche's impression of a profound "process of disenchantment" matches reports about a reversal of the general mood. What Lutz Niethammer has termed a "fellow-traveler factory," primarily on the basis of denazification trials in Bavaria, thus also applies to our smaller sampling of scientists.[52] So does Norbert Frei's finding of "an enormous social antipathy towards a thorough legal investigation of the crimes of the NS [National Socialist] period." Quantitatively speaking, it was possibly even more extreme. In search of members of the physics community who had been condemned

as Sommerfeld's successor at Munich and as Minister of Culture for Greater Hessen because "the university policy in the American zone ... is the opposite of reasonable" and he considered "the purging law in its present form a monstrous stupidity." See his letters to Sommerfeld, 4 February and 5 March 1947, www.lrz.de/~Sommerfeld/KurzFass/02465.html and www.lrz-muenchen.de/~Sommerfeld/KurzFass/02467.html.

50 Ernst Brüche, "Rückblick auf 1947," *Physikalische Blätter*, 4/2 (1948), 45f., quote on 46. For a definition of nazification, see Mark Walker, "The nazification and denazification of physics," in Walter Kertz (ed.), *Hochschule und Nationalsozialismus* (Braunschweig: TU Braunschweig, 1994), 79–89, here 81.

51 Walker, "Nazification," 81.

52 See Lutz Niethammer, *Entnazifizierung in Bayern. Säuberung und Rehabilitierung unter amerikanischer Besatzung* (Frankfurt am Main: Fischer, 1972). See Herbert Mehrtens, "Kollaborationsverhältnisse: Naturwissenschaft, Technik und Nationalsozialismus im NS-Staat und ihre Historie," in Christoph Meinel and Peter Voswinckel (eds.), *Medizin, Naturwissenschaft, Technik und Nationalsozialismus, Kontinuitäten und Diskontinuitäten* (Stuttgart: GNT-Verlag, 1994) on the collaborative relations during the Nazi period, and Schlüpmann, *Vergangenheit*, 396, on the awe-inspiring number of 2.5 million whitewash certificates in Bavaria alone; see 419–421 there on Frei and Bauer. Günter Schwarberg goes further in contending that German justice was a "murderer's washing machine": By 1986, from among the 90,921 suits filed against persons with records from the Nazi dictatorship, only 6,479 had been brought to conclusion. Roughly 84,000 had been aborted by decree without any public supervision. In the U.S. zone, more than 50 percent of approximately 950,000 cases ended with fellow-traveler decisions and a fine. Cornelia Rauh-Kühne, "Die Entnazifizierung und die deutsche Gesellschaft," *Archiv für Sozialgeschichte*, 35 (1995), 35–70, here 55, discusses these paltry tribunal results by August 1949. She also comments on the "out-of-hand investigative practices and abrupt changes" in British denazification, Rauh-Kühne, 61.

for anything above fellow-traveler status, I come up with a mere handful of examples. The astronomer and advisor at the REM, Wilhelm Führer, was condemned to 4 years of forced labor, and the physical chemist Mentzel, who was placed in category III: lesser offenders (*Belastete*), was given a prison sentence of 2½ years. Because Mentzel's internment in Nuremberg from the end of May 1945 to 23 January 1948 was deducted, he was immediately released after the court had announced its verdict. Göttingen-trained Wilhelm Führer, classed a lesser offender by the tribunal of North Württemberg in 1949, then fellow-traveler in 1950, also regained his liberty.[53]

One physicist who had migrated to the United States shortly before the Nazi rise to power, Rudolf Ladenburg, foretold this lukewarm outcome of the initially apparently so rigorous tribunals as early as mid-1947. He had left Germany in 1932 to accept an appointment at Princeton University. There is an element approaching pity for the defendant in his letter to von Laue:

> I still don't quite understand under what authority the so-called denazification courts operate. Over here it is said that they are German courts. But Meißner writes that the "military government," therefore the American one, decides.... I just read about the conviction of Joh[annes] Stark in a German émigré paper here... "4 years of forced labor" is too severe punishment for him, of course, but it probably won't come to that...[54]

Erwin Schrödinger also seems to have regretted the conviction of "Giovanni Fortissimo"; and of all people, Max von Laue even offered his support toward reducing his sentence. According to Sommerfeld, one of the witnesses at Stark's trial trying to plead for leniency, Paul Gottschalk, still considered the verdict as major offender justified.[55]

[53] Stark is discussed by Andreas Kleinert, "Das Spruchkammerverfahren gegen Johannes Stark," *Sudhoffs Archiv*, 67 (1983), 13–24. While the tribunal was still going on, Brüche tried to obtain information about it for the *Physikalische Blätter*, but the tribunal and the appeals court of Upper Bavaria refused his petition. On Mentzel, see Manfred Rasch's article in *Neue Deutsche Biographie* (Munich); on Wilhelm Führer and other astronomers, see Freddy Litten, *Astronomie in Bayern, 1914–1945* (Stuttgart: Steiner, 1992).

[54] Ladenburg to von Laue, 30 July 1947, MPGA, div. III, rep. 50, no. 1158. See the letters by Annemarie and Erwin Schrödinger to Sommerfeld, 16 August 1947, www.lrz.de/~Sommerfeld/KurzFass/02883.html; von Laue to Sommerfeld, 16 July 1947, www.lrz.de/~Sommerfeld/KurzFass/01183.html; and Gottschalk to Sommerfeld, 10 August 1947, www.lrz.de/~Sommerfeld/KurzFass/04709.html.

[55] See Sommerfeld's letter to von Laue, 24 July 1947, MPGA, div. III, rep. 50, no. 2394; von Laue to Hill, 10 May 1951, MPGA, div. III, rep. 50, no. 876, and various letters regarding Stark, MPGA, div. III, rep. 50, 1908. The engineer Paul Gottschalk, chairman

Ladenburg turned out to be right, incidentally. In an appeal in 1949, Stark's 4-year sentence was transformed into a fine of 1,000 deutschmarks. Freddy Litten has described another of the few cases of physicist major offenders in detail. Wilhelm Müller's initial verdict in May 1948 was guilty of category I. In October 1948, his charge was reduced to category II (incriminated activist, militant, or profiteer), and in May 1949 he was sentenced to a year of "special labor" and the confiscation of 20 percent of his private wealth. An appeal in September 1949, however, found him to be an "other-worldly, prejudiced dreamer" and he was classified as a minor offender in category III and sentenced to 2 years' probation and a fine of 1,000 deutschmarks plus legal costs.[56] I know of no other convictions of physicists that were actually carried out. Two seriously incriminated persons (Werner Straubel in Jena and Peter Paul Koch in Hamburg)[57] committed suicide right after the war. In the neighboring field of chemistry there were a few more well-publicized convictions. Major chemical trusts like I.G. Farben (*I.G.-Farbenindustrie*, founded in 1925) or the German Gold and Silver Refinery (*Deutsche Gold- und Silber-Scheideanstalt*, Degussa) were deeply implicated in the Nazis' criminal system and the business of mass murder.[58]

The verdict on the I.G. Farben case on 29 July 1948 followed 152 days of testimony by 189 witnesses, and hearing records filling 16,000 pages. Thirteen defendants were convicted to incarceration ranging from 18 months to 8 years; 10 others were acquitted. These relatively mild verdicts nevertheless raised an outcry in the *Physikalische Blätter*. O. Gerhardt's lengthy article excerpted a petition by the Society of German

of the denazification committee for the PTR in Göttingen, was one of the witnesses for the prosecution in the trial against Stark: see Kleinert, "Spruchkammerverfahren," 19ff.

[56] The legal grounds for the appeals verdict are reprinted in: Freddy Litten, *Mechanik und Antisemitismus. Wilhelm Müller (1880–1968)* (Munich: Institut für Geschichte der Naturwissenschaften, 2000), 219.

[57] On Straubel, see the letter by E. Buchwald to Sommerfeld, 25 January 1946, DMA, collection 89 (Sommerfeld), 006. Koch was accused of reporting about others to the Gestapo; see Monika Renneberg, "Die Physik und die physikalischen Institute der Hamburger Universität im 'Dritten Reich,'" in Eckhart Krause, Ludwig Huber, and Holger Fischer (eds.), *Hochschulalltag im Dritten Reich: Die Hamburger Universität 1933–1945*, vol. 3 (Hamburg/Berlin: Reimer, 1991), 1103, 1110ff. About cases of suicide, the "most contagious disease running rampant here among us," see von Laue to his son Theo, 25 Aug. 1946, MPGA, div. III, rep. 50, suppl. 7/7, sheets 43f.

[58] On this subject, see, for example, Joseph Borkin, *Die unheilige Allianz der I.G. Farben. Eine Interessengemeinschaft im Dritten Reich*, 3rd ed. (Frankfurt am Main: Campus Verlag, 1990), 136, 207; Deichmann, *Flüchten*, and references given therein, especially 433 (on the high rate of suicide among chemists after the end of the war), as well as 484ff. (on how chemists came to terms with Auschwitz and the I.G. Farben trial).

Chemists (*Gesellschaft Deutscher Chemiker*, GDCh) to General Lucius D. Clay that follows the same strategy used in the whitewash certificates:

From years of working together with the condemned, we know they are honorable men. We are of the opinion that the methods used by the prosecuting authority do not satisfy the methods prescribed either prior to the Hitler regime in Germany or currently in the United States of America. We are furthermore of the opinion that the judges did not take into account the circumstances of a total war within a dictatorial state governing by erroneous methods. We are unable to comprehend the severity of the sentences of imprisonment for men who are thus, in our opinion, being unfairly equated with common criminals.[59]

By reprinting this petition along with the official statement by the Society of German Chemists, the German Physical Society joined the ranks of defenders of I.G. Farben, who repeatedly insisted that the manufacturer of the crystallized form of the poison gas Zyklon B, the German Corporation for Pest Control (*Deutsche Gesellschaft für Schädlingsbekämpfung*, Degesch), was not a member of the dye trust but rather a subsidiary of Degussa. The defenders carefully omitted any mention of the other inhumane services I.G. Farben had provided for the Nazi government, including the exploitation of prisoners in concentration camps as laborers or as guinea pigs for their pharmaceutical experiments. The distinct impression remains, however, that this demonstration of unbroken self-assurance and unconscionableness was not primarily driven by concern about the fates of these "honorable men." It was more in indignation about the "discrimination against the German chemical [industry], particularly the exceptional law against the IG Farben Industry, Control Council Ordinance No. 9 from November 1945, that had been a guilty verdict without the taking of evidence and due process."[60]

RESTRICTIONS OF RESEARCH AND WAYS AROUND IT

After initial discussions between hardliners like U.S. Secretary of the Treasury Henry Morgenthau, Jr., and more liberal advisors like the German Science and Industry Committee in London, a compromise was reached. The resulting Control Law No. 25 imposed broad restrictions that led to a

[59] O. Gerhardt, "Das Nürnberger Urteil im Chemieprozeß," *Physikalische Blätter*, 4 (1948), 429–432, quote on 429. For background information, see: Deichmann, *Flüchten*, especially 484ff., and Schlüpmann, *Vergangenheit*, 415f.

[60] Gerhardt, "Nürnberger." Werner Heisenberg, "Wer weiß, was wichtig wird? Die Notwendigkeit wissenschaftlicher Forschung," *Die Welt*, no. 143 (9 December 1948), 3, also refers to the chemical industry in addition to the optical industry in his plea for "scientific research as a necessity."

consolidation of the opposition of German scientists against the Allies.[61] It came into effect on 29 April 1946, drawing a formal distinction between basic research and applied research. The former was restricted only insofar as it required apparatus that was also useful in military research and development. This formulation was rather ambiguous, and its severity depended on the interpretation of the individual Control Officers enforcing it. As a result, local implementation of the law varied considerably. Applied research was affected quite seriously overall by a long list of forbidden fields of research. Such modern topics as applied nuclear physics, applied aerodynamics, aeronautical optimization of airfoils, and so forth, rocketry, jet propulsion, and other aircraft development were declared off limits. It really ought to have been no secret to anyone that research conducted in these areas between 1933 and 1945 had been strongly oriented toward the wishes and goals of the National Socialist government and had benefited much from its potential military usefulness, even though some advances came too late to be applied (such as jet propulsion developed by Hans-Joachim Pabst von Ohain at Göttingen). Others (such as the uranium engine) were run on a quite low budget that had been scaled down in 1942–1943. Nevertheless, not many thought it made sense to forbid these areas of research categorically. On the contrary, it was generally viewed as a strategy calculated to hamstring Germany's research potential. This suspicion was particularly persuasive when seen in light of the success these areas were enjoying in the United States, sometimes with the help of imported German specialists. Some of these recruited scientists had been procured false new identities to cover up their Nazi pasts. An example is the rocketry team headed by Werner von Braun.[62] Quoting Otto Hahn (and F. H. Rein) again in 1947:

> Personal conversations with foreign scientists and individuals in charge of monitoring German science repeatedly fan the hope that the tiny remnant of science

[61] For the background on this law and the preliminary political decisions reached, see Thomas Stamm, *Zwischen Staat und Selbstverwaltung. Die deutsche Forschung im Wiederaufbau 1945–1965* (Cologne: Verlag Wissenschaft und Politik, 1981), 56, 230f., and Gimbel, *Science*, 175ff. The law was also reprinted without commentary in *Neue Physikalische Blätter*, 2 (1946), 49–52. On Morgenthau's motives, unbeknownst to his contemporaries, see Wolfgang Benz (ed.), *Deutschland unter alliierter Besatzung 1945–1949. Ein Handbuch* (Berlin: Akademie Verlag, 1999), 358ff.

[62] See Tom Bower, *The Paperclip Conspiracy: The Battle for the Spoils and Secrets of Nazi Germany* (London: Michael Joseph, 1987). Ulrich Albrecht, Andreas Heinemann-Grüder, and Arend Wellmann, *Die Spezialisten. Deutsche Naturwissenschaftler und Techniker in der Sovietunion nach 1945* (Berlin: Dietz, 1992), discuss the recruitment of specialists in the East. They refer to 2,370 cases by name and estimate a total of 3,500 such German specialists.

and research granted Germany will not be completely strangulated. Experts know how easy it is to prevent it from falling into the wrong hands through control measures. But what is happening here on the other front by politicians against science sounds desperate. Wouldn't it be possible to give even a glimmer of hope of a change for the better today, two years after the end of this evil, to the many who had viewed the zones by the occupation forces of England and America as the last and only hope for an end to the Hitler regime and a return of reason to its full rights? What is the purpose of evidently trying to push these people systematically into despair and apathy? The result cannot be peace for Europe.[63]

At the time, the overwhelming majority of physicists, including the ones who had distanced themselves from the Nazi ideology (or at least proclaimed as much), regarded the situation as merely a substitution of *one* hated system of restrictions, imposed by the Nazis (e.g., to stamp out relativity or quantum theory) with *another*. Many physicists long regarded the Allies as an impediment to research, without recognizing the potential of collaborating with them, as was being done at Göttingen since January 1946 in the German Scientific Advisory Council of eight members. Hahn, Heisenberg, von Laue, and Adolf Windaus negotiated with some success with Colonel Bertie Blount and from October 1946 with Ronald Fraser as representatives of the British military government. The issues included licensing of scientific journals, (re)certification of scientific societies, the continuation of the Imperial Physical-Technical Institute (*Physikalisch-Technische Reichsanstalt*, PTR), and coordinating the *FIAT Reviews of German Science*. The German Research Council formed in 1948 by this group of people under the leadership of Heisenberg soon merged with the refounded Emergency Society for German Science (*Notgemeinschaft*) to form the German Research Foundation.[64]

RUSSIAN PHOBIA

Richard Courant summarized the mood among the Germans he had spoken to during summer 1947 with the words: "absolutely bitter, negative,

[63] Hahn and Rein, "Einladung," 35. Max von Laue seems to have fully agreed with this article: see his letter to Meitner, 25 March 1947, cited in Lemmerich, *Meitner–Laue*, 484f.

[64] On these science policy and funding organizations, see, for example, Stamm, *Staat*; Michael Eckert, "Primacy doomed to failure: Heisenberg's role as scientific advisor for nuclear policy in the Federal Republic of Germany," *Historical Studies in the Physical Sciences*, 21 (1990), 29–58; Hermann, "Science," 81f.; Cathryn Carson, "Science advising and science policy in postwar West Germany: The example of the Deutscher Forschungsrat," *Minerva*, 40/2 (2002), 147–179.

accusing, discouraged, aggressive. Main point: Allies have substituted Stalin for Hitler, worse for bad. Russia looms as the inevitable danger." After numerous conversations in Southern Germany, Courant sketched the following overall impression:

> Fear of Russians. Bitterness against French. Rumors also of American mismanagement. General lack of understanding for what America actually does to help the Germans. Little contact between scientists in different towns. None with abroad, almost none with Austria... [Criticism] of German administration. Small-time politicians, no understanding for cultural issues. University has no support from them. Complaints about zone competition. French do not permit some scientists to travel to other zones. Americans and British likewise compete for scientists and allegedly, impose restrictions.... [Many scientists] do not dare travel through Russian zone for fear of kidnapping, which sounds unbelievable but is universally accepted as real danger.[65]

As we now know from various publications about the work of German specialists in the Soviet Union after 1945, the hiring methods that the Soviets used covered the full range from persuasion to outright force. The Operation *Osoaviakhim* (Union of Societies of Assistance to Defense and Aviation-Chemical Construction of the USSR) drive alone recruited more than 2,000 scientists and engineers with their families in October 1946. They were transported by train to the Soviet Union, where they lived and worked in closed compounds conducting research and development in nuclear physics, electronics and electrotechnology, optics, torpedo design, rocketry, and aircraft research, among other projects of military relevance. The Soviet Military Administration in Germany (SMAD) did its utmost to make these recruitments appear "completely voluntary," and some physicists were pressured into publishing open letters to that effect in newspapers and journals. At least until 1946 they appear to have been convincing. The propaganda even alleged that there were more applicants than the Soviet Union was able to accept.[66] These so-called volunteers had been surprised in the early morning by soldiers surrounding their homes and demanding they and their families immediately get ready for departure. Some nevertheless managed to escape (for instance, by jumping off

[65] The foregoing is quoted from Reid, *Courant*, 261. Tent, *Academic*, provides other examples from among the general population, some of them supported by testimonies about experiences under the Russian occupation, 15, 114, 123.

[66] See, for example, Steenbeck, *Impulse*, 169ff., Albrecht, Heinemann-Grüder, and Wellmann, *Spezialisten*. Lilli Peltzer, *Die Demontage deutscher naturwissenschaftlicher Intelligenz nach dem 2. Weltkrieg – Die Physikalisch-Technische Reichsanstalt 1945/1948* (Berlin: ERS-Verlag, 1995), 35ff. and 102ff. on the recruitment techniques used.

the train traveling eastward). As the quote from Courant documents, the rumors soon exposed such officious statements as propaganda.

Traces of this "Russian phobia" are also found in the *Physikalische Blätter*. The second issue of volume two carried a rather conciliatory article about the "responsibility of German scientists and engineers toward Russia." A hail of protest letters descended on Brüche after its appearance containing "individual details that leave no doubt that the descriptions of the events that have appeared in the press in the British zone are legitimate. The letters mention 'kidnapping' and 'armed assault.'" But Brüche still tried to uphold balanced reporting and published a few fresh "positive assessments that we have received" that "consider the Russian phobia as greatly exaggerated." But the side remark "that every once in a while, one of those who are collaborating with the Russians gets devoured" did not sound encouraging.[67] The censorship of correspondence from Russia could not be overlooked, and by mid-1948 Brüche's tone had become more resigned:

> It doesn't help to bury one's head in the sand and refuse to see the progressive separation of East and West Germany. The differing meanings Americans and Russians attach to the term democracy make further alienation inevitable right now.... We Germans are becoming different peoples on this side of the curtain as well as on that, and the differing educational reforms can only serve to turn the coming generations on both sides into strangers.[68]

This resignation about the situation in East Germany is also reflected in the correspondence of physicists. In March 1948, Wilhelm Hanle wrote to Max Born:

> At the moment food worries shift politics somewhat into the background. Don't you also have the feeling that the political situation is deteriorating rapidly? We were particularly shocked about the fate of Czechoslovakia.
>
> We very much welcome the imminent formation of a bloc of western nations, since we regard it as the only possibility of countering the Russian advance. But will it really be of any use? We constantly ask ourselves this question. It is widely

[67] Foregoing quote from: Ernst Brüche, "Nochmals die Ostverpflichtungen," *Physikalische Blätter* 3/1 (1947), 32; see Ernst Brüche, "Briefe. Deutsche Wissenschaftler in Rußland," *Physikalische Blätter*, 4 (1948), 271, 452–454. Balfour and Mair, *Four-Power*, 40–47, 76f., describe the economic and political situation in the Russian zone.

[68] Ernst Brüche, "Ost und West," *Physikalische Blätter*, 4/5 (1948), 221. On censorship, see *Physikalische Blätter*, 4/1 (1948), 39f. Brüche's conciliatory efforts between East and West continued: see Hoffmann and Stange, "East-German."

thought that an armed conflict between West and East cannot be maintained over the long term and that then the Russians will overrun us.

Many also ask themselves what they should do then. The example of some of our colleagues – like Schützen – in the eastern zone show us how little freedom there is under the Russians. I do not want to live in a totalitarian state again. I had enough from the 12 years of the Hitler regime. And then what would the future look like? Either Europe will remain Russian and western culture will decline in Europe, or the Russians will be thrown out again, just as we were first thrown out of France; then while they are retreating, the Russians will carry us off with them, of course, and we either go under then or at best end up in Russia. I found it bad enough that Hertz and other German scientists have to work in the Soviet Union, probably for the Russian war machine. I don't want to be on the Russian side in a battle between West and East.... Many wonder, if it does come to a – perhaps only temporary – evacuation of Germany (and perhaps of Europe) before the Russians, whether one would have to retreat with the Americans. But where else can one go? And what would happen to one's family? These are the thoughts preoccupying us now.[69]

In 1950, Hartmut Paul Kallmann pleaded with von Laue not to go to East Berlin again to attend the 250th centennial of the "Russian Academy." In his view, the Berlin Academy of Sciences was no longer an association of free scientists but "just a scientific fig leaf for the suppression by the Communists of free speech, free science, independent research, indeed independent thought in general." The presence of western scientists would only grant apparent legitimacy to the "slavery, repression and crimes," similar to the recognition of the Hitler government by other countries after 1933. It had "enlarged Hitler's power and reputation and above all made any domestic opposition impossible.... From the Nazi era I know how terrible it is to be repressed and to see other people, who are free, come voluntarily to sing along."[70]

Physicists and engineers who later migrated back to the West gave poignant descriptions of those years of anxious insecurity and fear, when they "learned the value of reserve and distrust as tools of survival." They were allowed to return when their expertise had become obsolete and only after it had been exploited to the fullest in ongoing Soviet projects.[71] These two supporting pillars of the mental profile of the first few post-war

[69] Hanle to Born, Prussian State Library, Berlin (*Staatsbibliothek zu Berlin Preußischer Kulturbesitz* SBPK), Born papers, no. 279 (reference to this letter by courtesy of Gerhard Rammer).

[70] Kallmann to von Laue, 6 June 1950, MPGA, div. III, rep. 50, no. 991.

[71] An example is the reminiscences of a doctoral candidate at the Institute for Applied Mechanics at the University of Göttingen: Kurt Magnus, *Raketensklaven. Deutsche Forscher hinter rotem Stacheldraht* (Stuttgart: DVA, 1993).

years – insecurity and fear of what the future might hold – existed not just in these isolated research compounds but throughout the whole of occupied Germany.

SENSE OF ISOLATION AND GRIEF OVER THE FRAGMENTATION OF GERMANY

A painfully strong sense of isolation shines through in the last quote from Courant's report. This is another motif prevalent in texts from the period. It was not a product of the post-war era but had its beginnings in the National Socialist policy to cease all German participation in international collaborative efforts. The university president at Göttingen, Hermann Rein, remembered in the foreword to the newly founded paper *Göttinger Universitäts-Zeitung*, edited by lecturers and students: "With bitterness do we feel that intellectual isolation which had descended on German science and academe over the course of the past 12 years by the closing of the borders."[72] Hopes of reopening such ties were high, but the Allies could not fulfill them for various reasons. For simple reasons of security, movements between the occupation zones were strictly controlled, and only few people obtained – after much effort – the much coveted interzonal passports. Anyone caught attempting to circumvent the border controls by crossing over to another zone in the dark of night without the necessary permit had to reckon with prying interrogations and hours of delay sometimes lasting days. Occasionally, they fell victim to pillaging by unscrupulous border officials.[73] This intentional sealing off of the interzonal borders was not completely hermetic, but it was perceived as yet another restriction on the freedom of movement, this time externally imposed by the victor powers. It was also seen as a muzzling of communication by cutting off external influences in a strictly cordoned-off system.

Physicists complained about a lack of free movement inside Germany and the barriers set against outside contact.[74] Brüche's initiative in 1946

[72] F. Hermann Rein, "Zum Geleit," *Göttinger Universitäts-Zeitung*, 1st series, no. 1 (11 December 1945), 1. The alleged closing of the borders for a dozen years is clearly not an accurate representation of the facts.

[73] One eyewitness account is given in Magnus, *Raketensklaven*, 32.

[74] See C. F. von Weizsäcker's letter to Sommerfeld, 11 February 1947, www.lrz.de/~Sommerfeld/KurzFass/04893.html. Weizsäcker deemed "a longer interim practical,"

to have Samuel Goudsmit's article on "Secrecy or science" reprinted[75] also arose indirectly from the fear of an increasing inclination among Germans and the Allies alike to close off laboratories and research institutes from the public eye on the justification that the research was of potential military importance. Goudsmit's call for scientific exchange among American institutions was cleverly refitted to promote the opening up of German science to the outside world.

The complaints about isolation thus went beyond the issue of the seclusion of Germany or at least of the three western zones from the rest of the world. Interzonal barriers were another cause for worry. The decision reached at Potsdam to decentralize Germany and divide it up into four zones, separately governed by one of the four Allied powers along completely different guidelines, had the inevitable consequence that former national organizations had to be cut up into separate entities as well. The German Physical Society was initially reopened in early 1946 as the "Deutsche Physikalische Gesellschaft in the British Zone," and the American equivalent soon followed suit. When Erich Regener began to reorganize a smaller subdivision within the French zone in Stuttgart as a regional association for Württemberg and Baden, Robert Wichard Pohl protested in the name of physicists in Göttingen against "this separatist founding, which jeopardizes German unity also in the cultural sphere."[76] But if only for legal reasons, there was no other solution. When the treaty between the Americans and the British merging their two zones came into force in January 1947, von Laue declared in a tone characteristic of the mentality of the time:

> The cultural crisis that has broken out throughout the world, and in particular throughout Germany, also interrupted at the beginning of 1945 the hundred years of work conducted by the German Physical Society. We cannot continue it as a

owing to "the problem of free movement" until the situation in Germany had been solved.

75 Samuel Goudsmit, "Wissenschaft oder Geheimhaltung," *Neue Physikalische Blätter*, 2 (1946), 203–207.

76 See Pohl's letter to Regener, 24 July 1946, DPGA, no. 40027; Pohl to Schön, 12 August 1946, DPGA, no. 40027; Ramsauer to Pohl, 13 August 1946, DPGA, no. 40027; as well as Pohl to Regener, 2 August 1946, MPGA, div. III, rep. 50, no. 1391. See also the minutes of the founding meeting of the DPG in the British zone on 5 October 1946, DPGA, no. 40028; Brüche's letter to Pohl, 6 Aug. 1946, DPGA, no. 40028, and Pohl to Ramsauer, 29 January 1946, DPGA, no. 40029. The statutes of the Southwest German Physical Society (cited here toward the end of the section on political apathy) contained a significant divergence.

nation-wide society right now. So regional societies are temporarily taking its place.[77]

This statement appeared in the *Physikalische Blätter*, which as of its second issue in 1947 became the publishing organ of the regional society of Württemberg and Baden. Its editor, Ernst Brüche, quite legitimately saw this as "the first publicly visible link to a revived interzonal 'German Physical Society.'"[78] In other instances as well, such as defining the official location of the largest review publication, *Physikalische Berichte*, the theme of maintaining Germany as a single entity reappears:

In view of the zonal divisioning of Germany, we Germans must emphasize all the more energetically that Berlin remains for us the capital city of the Reich. Relocating the Physikalische Berichte to the southwest of our fatherland would, under the prevailing circumstances, appear as just a piece of that retreat from the eastern territories that we unfortunately have to observe in so many other developments.[79]

In the British zone, Max von Laue and Walter Weizel chaired jointly over a membership of 485 of the regional association of Lower Saxony, Rudolf Mannkopff was managing director, and committee members included Erich Bagge, Hans Kopfermann, and Fritz Sauter. Hans Gerdien, Otto Hahn, Lise Meitner, Gustav Mie, Max Planck, Ludwig Prandtl, Hermann von Siemens, Arnold Sommerfeld, and Jonathan Zenneck were honorary members. In Württemberg-Baden, Walther Bothe presided over 160 members; in Hessen, Georg Madelung over 82; in Bavaria, Gerhard Hettner over 162; in Rhineland-Palatinate, W. Ewald in Mainz over a membership

[77] Max von Laue, in *Physikalische Blätter*, 3 (1947), 1. On the following page in a retrospective on the centennial celebration of the DPG on 18 January 1945, Ernst Brüche refers to the "modest celebration in the sixth year of the struggle between nations [*des Völkerringens*]," brashly playing down the world war by his choice of words. In *Physikalische Blätter*, 3 (1947), 2. Both citations are taken from Schlüpmann, *Vergangenheit*, 448.

[78] Cited from the notice: Ernst Brüche, "Der dritte Jahrgang," *Physikalische Blätter*, 3/1 (1947), 32. There we also find an answer to the inquiry about the delayed founding of a physical society in Hessen. According to a letter by Kurt Madelung from 22 February 1947, an application for a permit for the regional association had been submitted, but nothing more had been heard since from the responsible authority. There were reservations about having the *Physikalische Blätter* be the official publication for all physical societies in Germany. See Regener's letter to von Laue, 20 January 1947, MPGA, div. III, rep. 50, no. 2391: "there is not much support for it anymore, because we do not consider the level of the Phys. Blätter particularly high, especially concerning the foreign news reports, that make up the majority of the communications."

[79] Von Laue's letter as chairman of the DPG in the British zone to Ramsauer, 4 November 1946.

of more than 46; and in Berlin, Carl Ramsauer over some 100 members. The first commonly held meeting of all the regional societies took place from 11 to 15 October 1950 in Bad Nauheim, where it was resolved to consolidate into the Association of German Physical Societies (*Verband deutscher physikalischer Gesellschaften*). In 1963, the national association readopted its former name, German Physical Society.[80]

The German Geophysical Society (*Deutsche Geophysikalische Gesellschaft*) underwent a similar superficial regionalization after its dissolution in 1945. On 20 November 1947, a local geophysical society was founded in Hamburg "because the necessary permits for it are easier to obtain from the responsible military government,"[81] 2 years later to be reunited again with its old name. More was involved than a superficial reshuffling of names. It was a conscious hide-and-seek with the Allies, an expression of another element of the mentality of this period: A disingenuous urge to hide the true state of affairs from unwelcome monitors as soon as touchy points like topics of research or one's own past were involved. A reminiscence by a co-worker of Peter Adolf Thiessen is relevant here. Under Thiessen's directorship until 1945, the Kaiser Wilhelm Institute of Physical Chemistry had been chiseled into a model Nazi organization. After the war, Thiessen worked for 10 years in Sukhumi in a research program on methods of isotope separation, and from 1956 he directed the Institute of Physical Chemistry of the Academy of Science in East Berlin. During the German Democratic Republic period, Thiessen confessed to his former deputy: "My dear Linde, we pulled a fast one on the Nazis and we are pulling a fast one on the Communists in exactly the same way." I

80 The inauguration of the DPG in the British zone is covered in *Physikalische Blätter*, 2 (1946), 16. The first convention of physicists took place in Göttingen 1946, see *Physikalische Blätter*, 2 (1946), 178f., as well as Gerhard Rammer's contribution to this volume. On the rules of procedure, see *Physikalische Blätter*, 3 (1947), 29; on Bavaria and the Rhineland-Palatinate, *Physikalische Blätter*, 4 (1948), 124. Shortly after the founding of the Federal Republic of Germany, von Laue was no longer willing to apply to the military government for permission on matters concerning the DPG: "We Germans no longer need to ask the occupying forces about such matters of purely German concern, especially considering that the union of scientific societies certainly does not fall under the occupation statute." Von Laue to Bothe, 13 June 1949, MPGA, div. III, rep. 50, no. 330. The general history of the organization after 1945 and the first post-war conferences are treated by Wilhelm Walcher, "Physikalische Gesellschaften," as well as in Gerhard Rammer's contribution to this volume. Brüche's special concern for East German interests is discussed in Hoffmann and Stange, "East-German."

81 See M. Koenig, "Die Deutsche Geophysikalische Gesellschaft 1922-1974," in H. Birett, K. Helbig, W. Kertz, and U. Schmucker (eds.), *Zur Geschichte der Geophysik* (Berlin: Springer, 1974), 5.

would not agree with Hartmut Linde's emendation, "To him only science was important."[82] But a discussion of personal attitudes and convictions would lead us too far astray.

BITTERNESS ABOUT THE "EXPORT OF SCIENTISTS"

Besides the lack of new faces and ideas in barricaded Germany, there was the brain-drain problem: the departure of significant numbers of highly qualified researchers and teachers to other countries, whether voluntarily as émigrés or through forced recruitments. Specialists in the fields of weaponry or military technology such as aerodynamics, rocketry, or nuclear research were particularly sought after. On 21 February 1947, the recent Nobel laureate Otto Hahn drafted an appeal together with the president of Göttingen University, Hermann Rein, under the heading "Exportation of scientists to America" and published it in the *Göttinger Universitäts-Zeitung*. Brüche adopted it for his *Physikalische Blätter* under the less polemical title "Invitation to the USA."[83] As an introduction, a few American newspaper reports about the acquisition of German scientists and engineers are reviewed. The United States purportedly made a saving of about 1 billion dollars as a consequence.[84] The coauthors then lament:

> Science and scientists – without even mentioning German patents – are being described and treated as objects of "reparation." There is no question that things are going on that are probably unique in the history of science and are beginning to stimulate considerable bitterness among that relatively small group of those responsible for representing science in Germany.[85]

"Bitterness" is a word that recurs frequently in the sources and constitutes one of the components of the mentality profile we are assembling.[86] After

[82] Quoted from Guntolf Herzberg and Klaus Meier: *Karrieremuster* (Berlin: Aufbau-Taschenbuch-Verlag, 1992), 21. I owe this reference to Jens Jessen.

[83] *Physikalische Blätter*, 3/2 (1947), 33–35. See Hahn's considerably milder draft in MPGA, div. III, rep. 14A, no. 6194, and a list naming 17 physicists and chemists with invitations to the United States for a duration of at least 6 months, MPGA, div. III, rep. 14A, no. 5730, dated 23 January 1947.

[84] See *Neue Physikalische Blätter*, 2/2 (1946), 20, on 130 German scientists and engineers participating in military research in the United States. *Neue Physikalische Blätter*, 2/8 (1946), 260, Ernst Brüche notes in the editorial column a saving of 2 million dollars per invited specialist.

[85] Hahn and Rein, "Einladung," 33.

[86] For instance, Otto Hahn: "It is certainly understandable that the factory dismantlings still taking place four years after the capitulation are being greeted with bitterness,

complaining about a new wave of dismissals, cited above, Hahn and Rein continue:

> The official invitations to the USA mentioned at the beginning arrive in this new, unsettled, bitter situation, triggered by the threat to the employment of renowned researchers and to the livelihoods of their families. As much as we wish that these colleagues in particular, who in our opinion have been unfairly affected, may continue to work in a better atmosphere more favorable to science, we very much regret that they are condemned here and dismissed from their positions – while over there, they are sought after because of their expertise. They are unfairly blamed here by the public as deserters, over there they are viewed by leading scientists as unwelcome intruders.
>
> Most of the older professors leave Germany very unwillingly, because they feel that their place is here. Necessity compels them, because their livelihoods and working opportunities in their own country are taken away from them or else they are left in a constant state of fear of such an occurrence. All this, after our having experienced well enough what it means to replace competence with "politically irreproachable" dilettantes. But more depresses these men: the awareness that it is evidently not a matter of an honorable appointment to an independent American research institution or university of some rank but (at least according to the American press) forms a part of the "reparations." Centuries ago, princes sent their countrymen away as plantation workers or soldiers. Today, scientists are exported.[87]

After seeing repeated parallels drawn between Allied science policy and measures taken by the Nazi regime, this time we have a potent comparison with absolutist governmental practices. Ernst Brüche was fully aware that by reprinting the appeal in the *Physikalische Blätter* he was risking the loss of his publishing license from the military authorities. That is why he added a footnote to the article pointing out that he was just following the same procedure as had recently been done in the *Neue Zeitung*, a paper issued by the American military government. It had reprinted text from *Foreign Affairs* with the comment:

> We believe we are contributing better toward orienting public opinion by granting generous hospitality to a variety of ideas than if we identify ourselves with a particular direction. We do not assume responsibility for the views published

particularly among the academic youth." Otto Hahn, "Antwort an eine Delegation," *Göttinger Universitäts-Zeitung*, 4th series, no. 12, (1949), 3. Balfour and Mair, *Four-Power*, 152, mention "a great deal of bitterness" with reference to the bad food situation during the winter of 1946–1947.

87 Hahn and Rein, "Einladung," 34. Reparations are discussed in Balfour and Mair, *Four-Power*, 80ff., 131ff., 144ff., 162–168.

here, but we do assume responsibility for having given them an opportunity to be published.[88]

Formally, Brüche was in the clear, even though he was of one mind with the two authors. We gather this from his annual retrospective in the *Physikalische Blätter*. From the safe vantage point of a year's interval, he wrote that Hahn and Rein "had found clear and appropriate words about the professional obligations of German scientists in the United States, the essential points of which any physicist would agree with."[89]

Not everyone was pleased about this article, as Brüche had anticipated, especially scientists who happened to be away on such visits to the United States or England right then. They did not like to see this personal distinction suddenly reduced to a form of reparations payment. Georg Joos complained to Sommerfeld about the "presumptuous tone" of the *Göttinger Universitäts-Zeitung*, foretelling that it would be the source of the "next debacle."[90]

James Franck also carefully expressed reservations about it in a letter to his former close friend Hahn: "By the way, I am not quite sure that I completely agree with you. I don't love the 'expert-transfer' method for various reasons, that's true. On the other hand, under the present conditions that unfortunately happen to be, I am not sure whether at the moment Germany doesn't perhaps have more academics than it can feed; if this is the case, there is a good side to this export, noxious though it is to me in many respects."[91]

Rudolf Ladenburg's letter to Max von Laue, written right after he had read the article, adds another aspect:

Otto Hahn's article in the Göttinger Universitäts-Zeitung draws a link between dismissed German professors and American offers of professorships in the USA. Do you know of anyone from among our colleagues who was first dismissed & is now moved away to the USA? I only know that Westphal is in the USA & that

[88] Hahn and Rein, "Einladung," 34.

[89] Brüche, "Rückblick," 45. He concurred likewise with the contribution by Werner Kliefoth, "Physiker als Reparationen," *Physikalische Blätter* 2 (1946), 369f. Kliefoth had asked: "How should the 'dismantled physicists' be reckoned [against the reparations debt]?" ("Wie soll man die 'demontierten' Physiker anrechnen?") For example, see A. Eucken's letter to Sommerfeld, 28 August 1947, www.lrz.de/~Sommerfeld/KurzFass/04673.html. Eucken regarded Clusius's emigration as a "catastrophe" for German science; likewise Heisenberg, who feared "the big danger is that the very best emigrate." Heisenberg, "Wer weiß."

[90] Joos to Sommerfeld, 30 April 1948, in DMA, collection 89 (Sommerfeld) 009. I owe this reference to Oliver Lemuth.

[91] Franck to Hahn, 4 October 1947, RLUC, Franck papers, box 3, folder 10.

Joos left for here at the beginning of June.... Westphal was, as far as I know, never dismissed & Joos was re-engaged after a short period of time. In his case, the dismissal was based on an (unforgivable) confusion of names.[92]

Ladenburg sensed an ulterior motive in the article in correlating the dismissals with the importation of German scholars for exploitation in American weapons development. By insinuation, the article leaves an unpleasant aftertaste of duplicitous ethics and unscrupulous self-interest.[93] Wolfgang Finkelnburg is one example. He was vice-chairman of the German Physical Society until 1945. After the war, his year as acting head of the National Socialist German University Lecturers League (*Nationalsozialistischer Deutscher Dozentenbund*) in Darmstadt in 1940–1941 caused him trouble. His party membership (dating to 1937) was cited as the reason for his dismissal as vice-chairman of the German Physical Society. Finkelnburg portrayed the "fellow-traveler problem" from the point of view of someone affected, who saw no alternative to leaving for the United States. He generally agreed with the article by Hahn and Rein but offered another reason for why people felt obliged to accept "special job" offers from the United States, even though they would have much preferred to stay in Germany:

We youngsters have lost confidence in regular university appointments.... Calls like the one in Mainz inevitably give us the impression that merit as a researcher and university teacher has nothing to do with it, just a clean political record taken to the extreme and those notorious "connections."... It seems to me that people are far too timid again about just wanting to conduct science. That is why we have to pack up for good and go abroad![94]

The German Physical Society refused to assume the role of advocate of persons dismissed for political reasons. Max von Laue heard about Finkelnburg's letter from Otto Hahn and answered in May 1947: "This colleague of ours has always been a very good prosecutor of his own affairs. He is in this case, too. But I can only reply to him: If a university faculty turned to the German Physical Society about appointment matters, it would surely get a response. But none have done so thus far."[95] So it was

[92] Ladenburg to von Laue, 30 July 1947, MPGA, div. II, rep. 50, no. 1158, cited in Schlüpmann, *Vergangenheit*, 426f.

[93] M. Rubinstein points out another aspect of this transfer of know-how: "the importation of Nazi scientists is accompanied by the importation of Nazi ideas." M. Rubinstein, "Importation of German scientists into America and Britain," *New Times*, no. 10 (1947), 18.

[94] Finkelnburg to Hahn, 18 May 1947, MPGA, div. III, rep. 14A, no. 926.

[95] Von Laue to Hahn, 24 July 1947, MPGA, div. III, rep. 14A, no. 925.

not the German Physical Society as an association that decided but the network of influential referees and gray eminences behind the scenes, as Wolfgang Finkelnburg soon realized: "in Germany one only has prospects when one has been ordained at Göttingen."[96]

But the correspondence by physicists employed in the United States or Great Britain after 1945 reveals that a lack of opportunities in Germany was not the only element at work. Better economic and working conditions in science were a primary motivation in their decision to take such a step. A year later, in April 1948 Georg Joos justified his decision not to return to Germany for the time being with the following argument: "The population count has to be reduced or else we shall all just waste away like the Indians."[97] In an effort to slow the trend of emigration, von Laue even resorted to exaggerated polemics. In an interview with a reporter of the *Darmstädter Echo* in mid-1949, soon after returning from a trip to the United States at the end of July, von Laue recounted that he and his wife "were served particularly hospitably in the stores" when their identity as Germans was revealed by their questions and requests. However,

> many researchers employed in military research laboratories silently yearn for the peace of their former quest for truth, even though collaboration on the major projects does attract American scientists, of course, who are generally not very well situated financially. But some are compelled to find compensation in civilian industrial research for lack of state funding.... Some of those who had sold themselves to the USA in 1945 are doing quite poorly. They don't live much better than prisoners, receive 6 dollars pay – per day – and may only leave their assigned location with special permission.[98]

As had been the case with von Laue's earlier statements, this publication, too, triggered harsh criticism abroad, for instance from a former close friend of his. The crystallographer Peter Paul Ewald had resigned his office as university president of the polytechnic in Stuttgart in protest against the notorious racist Law for the Restoration of the Professional Civil

[96] Finkelnburg to Sommerfeld, 8 August 1947, www.lrz.de/~Sommerfeld/KurzFass/04691.html.

[97] Joos to Sommerfeld, 30 April 1948, www.lrz.de/~Sommerfeld/KurzFass/03073.html.

[98] Max von Laue, "Amerika und die deutsche Wissenschaft. Gespräch mit Prof. Max von Laue," *Darmstädter Echo*, 5th series, no. 173, 27 July 1949, 4; compare this interview with von Laue's letter to the editor of the *Darmstädter Echo*, 5 January 1950, MPGA, div. III, rep. 50, no. 2339. For the following, see Ewald's letters to von Laue, 23 November 1949 and 19 January 1950, MPGA, div. III, rep. 50, no. 562.

Service in 1933 and had emigrated 4 years later. From Brooklyn, New York, Ewald wrote his former colleague that the interview had caused "astonishment and repudiation" and drew a completely false picture of the situation. The subsequent exchange of letters could not change the fact that von Laue's friends in the United States were "very disappointed and disconcerted, particularly with respect to the effect it could have in official places, who are conducting the exchange of students and lecturers mentioned in the last sentence." Max von Laue's polemics had done some damage to the touchy business of normalizing international scientific exchange. Bitterness also characterizes the tone of various reports regarding "The situation in the area of patenting in Germany."[99] Just like after the First World War, all German patent rights were confiscated and released as property of the Allies "for mutual use" (according to an announcement in the *Tagesspiegel* of 10 August 1946, they numbered more than 100,000). In April 1946, the U.S. Department of Commerce published an appeal urging American companies to use the new possibilities to full advantage.[100] A decision on domestic patents and new patent applications was only reached later on, which led to a considerable backlog of pending claims. In August 1947, a law finally came into force in the United States making it possible (under certain conditions) for German citizens to file patent applications in Washington again. This had been feasible in France since April 1946, but it took 2 years longer (until April 1948) for it also to become possible in England. The protracted legal uncertainty and the impression of having been deprived of legitimate rights strengthened the widespread sense of victimization, even when this "plundering" took place with explicit reference to identical practices Germany had itself followed in the occupied territories until 1945. Reparations were a problem affecting the society as a whole, but the demoralizing effect had specific repercussions on universities and the mood of physicists.

[99] Thus the title of a longer article by the engineer and expert on patent law H. G. Heine, "Die Lage auf dem Patentgebiet in Deutschland," *Physikalische Blätter*, 3 (1947), 387–389. Also see: "Das Patentwesen nach dem Zusammenbruch," *Neue Physikalische Blätter*, 2 (1946), 169–174, Fr. Frowein, "Schutzrechte im Ausland," *Physikalische Blätter*, 5 (1959), 15–17, and Gimbel, *Science*, especially 63 (on the "Exploitation and plunder" by the Americans), and 173 (on patents).

[100] "Never before has American industry had such an opportunity to acquire information based on painstaking research so quickly and at such low cost. This is part of our reparations from Germany in which any American may share directly." Quoted from Hermann, Science, 79.

SELF-PITY, SENTIMENTALITY, AND SELFISHNESS

Michael Balfour and John Mair identified these three points as early as 1956: self-pity, sentimentalism, and selfishness are pinpointed as general characteristics of vanquished nations.[101] Their report on the Allied occupation of Germany for the Royal Institute of International Affairs is a balanced description of the attitudes on both sides, also outlining the Germans' expectations at the time. Another characteristic given there – a lack of objectivity – is omitted here because that probably generally applies to actors in any given historical situation. Hannah Arendt noted during her visit to Germany in 1950 the self-pity, listlessness, and inability to express emotions, the escapism, moral perplexity, and the refusal to grieve.[102] These observations were based on a cross-section of the general population. The same traits emerge among my sampling from the physics world, indicating that it did not differ much from other groups. Edward Hartshorne's diaries also reflect these characteristics – for instance, in his rejection of the idea of hiring Germans for evaluating filled-out questionnaires: "They were bubbling over with self-content. I must say I am against employing Germans at all. They are only seeking to improve their own lot and still despise the rest of the world."[103] Officer Hartshorne was in charge of reopening universities within the American occupied zone.

Max Steenbeck reported about his experiences in Soviet captivity right after the unconditional surrender in May 1945:

> For most of us the world had fallen apart or – for some, worse still – our personal bases for earning a living and therefore our meaning in life. Everyone came from positions of authority. Here no one wanted to listen, everyone was filled with his own, different thoughts; there was no common fate to be overcome. Everyone would have liked to talk about his own, if only he could find someone who was interested.– At first, people talked about what they had experienced before, often somewhat boastfully, often omitting some things for good reasons. But once you did start listening to someone and set demands on what was being reported, you very soon encountered the other person's worries about his future – as if you didn't have enough worries of your own. Not being able to listen anymore was the true crux of the matter.[104]

[101] See Balfour and Mair, *Four-Power*, 63 (section on: Attitude of the Germans): "the sentimentalism, the selfpity, the selfishness, and the lack of objectivity which tend to characterize a nation in defeat."

[102] See Arendt, *Besuch*, 24–26 and 46.

[103] Edward Y. Hartshorne on 30 April 1945, in Tent, *Academic*, 37f.; see Tent, *Academic*, 83.

[104] Steenbeck, *Impulse*, 157f.

The element of self-pity is also prominent in a notice reprinted in the *Physikalische Blätter* from the Swiss paper *Basler Nachrichten*. The article quotes from a conversation Otto Hahn had with a Swiss visitor and is originally dated 10 November 1946, shortly before Hahn's departure for Sweden to receive his Nobel Prize:

You see, I had hoped for years for the time when we would be rid of the heavy mental burden of National Socialism, and how much I looked forward to being able to work freely and without hindrance. But now I am sitting here, a head without a body; I am not allowed to return to my institute because it lies in the French zone, and I have little idea about the other institutes, and here come new people every day wanting a job or a political exonerating certificate or whatever else. I simply cannot help these people. Formerly, I really used to be a cheerful person and was actually never pessimistic, but if people just come with demands and one can hardly move for all the restrictions, I simply cannot go on. And imagine, ludicrous though it may sound, at the moment I don't even have a sound pair of shoes to put on. So, what use is it to me if the Nobel Prize is waiting for me in Sweden, which I am not allowed to pick up because I don't get a travel permit and meanwhile, I submit one application after the next for months on end in vain for a pair of shoe soles. If they would at least send me a pair of shoe soles against the Nobel Prize account, then I wouldn't have to walk around with wet feet all the time.[105]

Hahn's circumstances had been rather privileged, starting at least with his internment at the end of the war in the English manor house Farm Hall. At the time the article was written, he had just been appointed president of the Kaiser Wilhelm/Max Planck Society and was living in Göttingen, a town that was virtually unscathed by the war. If something negative had to come out, why did he not talk about the much worse lot of his less prominent colleagues? Hahn preferred to complain about his leaky shoes.

The following passage from the physical chemist Hartmut Paul Kallmann's letter in the same year to his émigré colleague Michael Polányi confirms this impression of general egocentricity:

It still depresses me when I see that in this country recognition of what is of real importance in life has still not penetrated even after these terrible years. The tough momentary situation is deplored much more than the evil of the past 10 years. It ought to be the other way round. The masses still don't know what a salvation the destruction of the Nazis was to the whole world and to Germany as well.[106]

105 Otto Hahn, in *Physikalische Blätter*, 2/8 (1946), 240.
106 Kallmann to Polányi, 22 May 1946, RLUC, Polányi papers, box 5, folder 2.

We gather from a report by the nuclear physicist Hans Jensen to the Minister of Culture of Lower Saxony about his trip to Copenhagen and Oslo in 1948 that the Danes and Norwegians he had spoken to repeatedly complained about the one-sided and exaggerated self-pity they noticed among the Germans they knew:

Many expressed their astonishment that letters from Germany and conversations with German guests always place the entire focus on the special desperate conditions in Germany; and very few noticeable attempts are made to regard the problems at least within a European context and as consequences of the past 15 years; and that there is rarely even any perceptible shame that the Nazis had wreaked so much havoc and misfortune in Germany's name. As much as I tried to explain such a closed-minded view by the truly dire need these last few years inside Germany, I do have to say that this astonishment, which does appear to me somewhat justified, did make me feel quite awkward.[107]

Lise Meitner's letters confirm this impression of saturated self-pity in Germany. In 1947, for instance, she wrote James Franck about her recent trip to England:

I recently had a chance to speak with many English persons who had been in Germany to attend German physics conferences – hence certainly good-willed about helping Germany retrieve decent living conditions. They all said to me that the Germans were full of "self-pity," saw nothing but their own predicament, always told the English first of all that they had not been Nazis. One of the English physicists... told me that Hahn and Laue and many others, he also mentioned Kopfermann, were absolutely not able to think beyond their own professional group. He only excepted Heisenberg and Weizsäcker and thought that even though they perhaps had been fascinated by the successes of Nazism, they did now have much broader points of view.... Heisenberg is certainly much more contemplative than Hahn and probably also smarter than Laue, he is also much younger – but is he sincere?[108]

Margrethe Bohr also met many significant Germans visiting her husband (before and after 1945) including Heisenberg and von Weizsäcker. She wrote to Meitner in 1948: "It is a difficult problem with the Germans, very difficult to come to a deep understanding with them, as they are always first of all sorry for themselves."[109]

[107] Hans Jensen's report to the Minister of Culture of Lower Saxony, end of 1948 (Universitätsarchiv Heidelberg, carbon copy in the personnel file Jensen), quoted from Schlüpmann, *Vergangenheit*, 440.

[108] Meitner to Franck, RLUC, Franck papers, box 5, folder 5.

[109] M. Bohr to Meitner, 10 June 1948, quoted from Sime, *Meitner*, 358.

The continuation of an open letter by an anonymous scientist is a good example of how easily self-pity and self-justification could slip into melodrama:

All this was reason enough for us not to grab hold of the flywheel with our bare hands. The mark of the true martyr, who voluntarily and consciously takes suffering upon himself, is a firm bond with a belief system. Scientists are almost never of this very rare type of person. Researchers in Germany as elsewhere see as their goal a life of activity and not mortal self-sacrifice.[110]

At this point, the author apparently noticed how dangerously close he was to slipping into self-pity and tried to rationalize it away:

With these explanations I must trust that you [the reader] will not suspect that I am making an appeal to sentimentality. On one hand, this fine word from French Romanticism, which means "understanding everything" as well as "forgiving everything," is not appropriate for our world – human understanding is important, not for the sake of forgiving but for doing better. On the other hand, such a suspicion could create very painful circumstantial conditions for us ...[111]

"PROPAGANDA-FREE DAY-TO-DAY" AND POLITICAL APATHY

The ubiquity in the print media of *Lingua Tertii Imperii* (LTI), as Victor Klemperer termed the idiom of the Third Reich, created a yearning for language thoroughly purified of such phraseology, an "ideology-free discourse" (using today's phraseology!). "We must part with untruths and slogans," one student demanded in the first issue of *Göttinger Universitäts-Zeitung*. A "respectable measure of clarity and honesty under the watchful eye of the academic public" ought to "safeguard the newspaper from the dangers of the free press."[112]

Until 1933, the word "propaganda" still held a positive connotation and was even selected as part of the official name of a new ministry. One variant of the post-war distaste for propaganda and phraseology expressed itself in a preference for a purposefully austere choice of words and simple sentence structure. This new factuality (*neue Sachlichkeit*) characterizes the styles of Friedrich Hund or Hans Kopfermann, for instance. There were, however, also smooth transitions into insensitive

[110] Anonymous, "Ein Brief nach Frankreich," *Neue Physikalische Blätter*, 2 (1946), 8–11, quote on 10.

[111] *Neue Physikalische Blätter*, 2 (1946), 10.

[112] All quotes, from this section's heading onward, come from Wolfgang Zippel (law student), "Zum Geleit," *Göttinger Universitäts-Zeitung*, 1st series, no. 1, 11 December 1945, 1. The dangers evidently refer to a relapse into the old jargon.

and euphemistic down-playing of past insufficiencies. The president of the Göttingen Academy of Sciences after 1945 took this course in his letter to Jews who had been pressured to withdraw from among the academy's illustrious ranks in 1933. He was not more specific than to allude to "deplorable circumstances." In response to protests against this evasive formulation, he offered the following justification:

> Confronted with the unspeakable things that have happened elsewhere, we did not want to emphasize this matter unduly. Added to this, however, is a profound abhorrence among us for fancy words, after the abominable inflated verbosity during the Third Reich. There was furthermore a distaste for strong words about the Third Reich, which nowadays are cheaply made and with which the fellow travelers of the Third Reich are... trying to shout each other down as fellow travelers of the present day, now that it is free of risk and advantageous to give the dead monster a few more kicks after the fact.[113]

In extreme cases, this aversion led to a total refusal to read the paper anymore or to inform oneself politically. When a delegation of the British Association of Scientific Workers criticized the "reactionary mentality at German universities," Otto Hahn attempted to defend his fellow countrymen against what he considered an unjustified accusation: "Students want to work and only want to work.... And there is less danger of a revival of Nazism or a rejection of any idealistic Socialism than of a certain cynicism against any state authority."

The response by the head of this delegation, R. C. Murray, was: "I would like to agree fully with Professor Hahn that *the basic characteristic of German students and likewise of German scientists is political apathy*. An ideological vacuum exists. Nazism is completely discredited but there is nothing of the same persuasive force to replace it yet. It may be understandable but it is nevertheless unfortunate. If this vacuum persists for too long, it could prepare the ground for a new form of fascism."[114]

The reactions by F. Hermann Rein and Max von Laue to this article reveal how thoroughly they misunderstood it. Von Laue countered Murray's charge of "political apathy" with:

[113] Smend to Franck, 1947, RLUC, Franck papers, box 10, folder 5. This incident is described in greater detail in Anikó Szabó, *Vertreibung, Rückkehr, Wiedergutmachung. Göttinger Hochschullehrer im Schatten des Nationalsozialismus* (Göttingen: Wallstein, 2000), 511f., and Jost Lemmerich, *Aufrecht im Sturm der Zeit: Der Physiker James Franck 1882–1964* (Stuttgart: Verlag für Geschichte der Naturwissenschaften und der Technik 2007), 265f.; see the forthcoming English translation: *Science and Conscience* (Stanford: Stanford University Press, 2011).

[114] Hahn, "Antwort," 3 (emphasis by the author). See also Murray, "Social," 13.

Of course, this is just what Murray does not like. He wants more "public relations." We would like to point out, on the contrary, that "public relations," albeit of an undesirable sort, were precisely what Hitler forced upon German scientists with varying degrees of success. One should therefore let caution reign in this regard and exercise restraint.[115]

The theoretical physicist Richard Becker also thought that only extraordinary events of "elementary impact," such as the development of the atomic bomb, could lure physicists out of a "zone of silence," in which, according to him, they evidently normally seemed to stay:

We have heard this admonition to consider the religious foundations of our culture often enough out of the mouths of priests and philosophers. Nowadays their voices are joined by sober scientists, who in view of the consequences of their research see themselves compelled to leave that zone of silence about things "of which they know nothing." They step out of the auditorium into the public in order to shake people into awareness and educate them about a humane use of their power.[116]

Becker also received cover from a physics student, who wrote an article in a later issue of the *Göttinger Universitäts-Zeitung* about why scientists did not concern themselves about current-day issues. He also found it strange "to see so few scientists express their opinions in the *GUZ* about issues treated in it. Scientists are often accused of caring too little about burning current issues, above all political ones." This candidate of physics did not intend to discount this but tried to explain the attitude of scientists. He limited his scope to physicists, as "typical representatives of modern science," and painted the following mental profile for this paradigm:

The mentality of a physicist differs very much from that of a scholar in the humanities, a politician, a businessman or even a technician. By comparison it seems to me that he probably more closely resembles an artist. Add to that a shot of philosophy, rubbed off from the humanities. But that, I would think, is where the comparability ends.[117]

After a column about profound and elementary curiosity being the main motivation behind physical research, the article takes a more apologetic turn:

[115] Max von Laue, "Public Relations?" *Göttinger Universitäts-Zeitung*, 4th series, no. 18, (23 September 1949), 12.

[116] Richard Becker, "Gefahren der Naturforschung," *Göttinger Universitäts-Zeitung*, 2nd series, no. 24 (21 November 1947), 1f.

[117] Rolf Hagedorn, "Verhinderte Naturwissenschaftler. Warum kümmern sie sich nicht um Gegenwartsfragen?" *Göttinger Universitäts-Zeitung*, 3rd series, no. 20 (24 September 1947), 13.

Pardon the scientist for his curiosity, his obsessiveness, his "playing," and that he forgets the normal routine in the process.... His research, the issues that move him, have virtually no relation to public problems.... Of course he does remain bound to the events of the day but it is rarely possible for him to grasp more than their broad outlines. And this alone is not enough to join in the debate. So his restraint is not just attributable to a lack of interest but, to a good part, also to purely practical reasons.[118]

Far be it from me to ridicule this quote, naïvely comical though it is. Some of the observations in it may well be right. What is clear, though, is that this text – the only explicit analysis of the mentality of physicists that I am aware of from this period – had an entirely different, apologetic function. Political aloofness was a traditional part of a scientist's self-image from as early as the 19th century. After 1945, political aloofness received an additional positive connotation: being apolitical signaled that one was also "politically unimplicated" (in the sense of denazification) – whether or not it agreed with the facts.

Political disenchantment also welled over into the *Physikalische Blätter*, but its force was partly broken against Ernst Brüche's unquenchable optimism. His editorial quoted below, entitled "Physicists at the crossroads," starts with a pithy review of the apolitical attitude of scientists into which he incorporated sociobiological and historical comparisons characteristic of the mentality of the day:

Scientists are individualists and their verdicts about politics are rarely kind. In Germany today, just as 14 years ago, this relationship is not improved by talk about their having historically important obligations with repercussions for the future. With these arguments come calls for a politicization of universities, which now as before do not fail to include a menacing undercurrent of thorough dissatisfaction with past performance at universities.... The special distaste for politics in Germany stems perhaps from the position of its intelligentsia who had always been safely encased within the state civil service, where they feel much more strongly beholden to authority than to their own judgments. The strong tendency toward specialization which emerges within this sociological framework (similar to a termite society) is often valuable and sometimes indispensable; but if it is carried too far, then whole professional classes lose their ability to judge and assess their own interests against those of others. This may explain why state aggression encounters only very little active opposition among the intelligentsia in Germany in particular, especially when it leaves the cells of the scientific beehive alone.[119]

118 Hagedorn, "Verhinderte." The tenor is, incidentally, quite similar in *Göttinger Universitäts-Zeitung*, 4th series, no. 118 (1949), 9.

119 Ernst Brüche, "Die Physiker am Scheideweg," *Physikalische Blätter*, 3/6 (1947), 207f. Richard H. Beyler, "The concept of specialization in debates on the role of physics in

This notice ends, however, with an about-face: Brüche pleads for involvement in one of the many newly founded organizations fostering the internationalization of science and a greater awareness among scientists of their responsibilities:

The lessons of the past few years are so obvious today that they are even reluctantly conceded by those who a short while ago let total encapsulation be their guiding motto. Acknowledgment of a scientist's share of responsibility in the common weal as well as in his own interest necessarily had to arise earlier in intellectually freer countries abroad and produce politically active scholars.

Then follows a discussion of the organizational consequences for the German Physical Society:

The administrative session – a novelty in the history of the Physical Society – has issued carefully considered formulations that point out that the society must fulfill not only scientific duties but also the obligation of keeping alive a sense of common responsibility in the forming of people's lives and of advocating the freedom, truthfulness, and dignity of science.... [The] unanimity on resolutions passed at the administrative session in Stuttgart concerning these issues permits the supposition that the majority of German physicists are no longer standing at the crossroads but have already set off down the right path into the future. Just as at the founding of the Physical Society 100 years ago, the younger members will hurry ahead and the circumspect will follow.[120]

Obtuse historical comparisons between post-1945 and post-1933 and quaint sociobiological metaphors of termite and bee communities are a strange context indeed for first genuine attempts at learning from history. The way the German Physical Society developed and the appearance of the Mainau and Göttingen manifestos against nuclear armament during

post-war Germany," in Hoffmann et al., *Emergence*, 389–401, discusses the objection to extreme specialization, so typical of the time.

120 Beyler, "Concept." Erich Regener introduced to Brüche's journal this change in the by-laws for the regional physical society of Württemberg and Baden in an article entitled: "The common responsibility of scientific workers": Erich Regener, "Mitverantwortlichkeit der wissenschaftlich Tätigen," *Physikalische Blätter*, 3/6 (1947), 169f. Article 2 of this by-law reads: "From the fact that the knowledge reaped from physics exerts a growing influence on the mental attitudes of people, that furthermore practical physical findings are having an ever stronger effect on all areas of human activity, the Physical Society accepts the obligation to keep alive a sense of shared responsibility among workers in science in the shaping of people's lives. It will always support the freedom, truthfulness, and dignity of science." Compare Regener's draft from 1946, DPGA, no. 40199, esp. Article 3: "The DPG obligates itself and each of its members to support the freedom, truthfulness, and dignity of science and furthermore to stay constantly aware that workers in science have an especially high degree of responsibility for the shape of public life."

the 1950s demonstrate that the path actually chosen ultimately did lead in the right direction. Although these manifestos have frequently been faulted for their apologetic purpose, they did have the positive attribute of making a clean break with the 19th-century apolitical self-image of scientists.[121] Their role models were physicists in the United States, who as builders of the first atomic bomb suddenly became only too well aware of the burden of social responsibility.[122] A letter to James Franck by one of the younger signers of the Göttingen manifesto shows how strong a model Franck's politically responsible actions continued to be after the war. The experimental nuclear physicist Heinz Maier-Leibnitz wrote him:

> Surely you have heard about our step concerning atomic bombs for Germany. In doing so I often thought of you. The responsibility we still bear today is, of course, small against what you took upon yourself in 1945. But we wanted at least to make a contribution within our area; maybe developments will prove us right.[123]

Responsibility discourse appeared after 1945 in declarations of intent, prefaces, and, as we have just seen, even in the by-laws of institutions and organizations like the German Physical Society. The special responsibility of a scientist is still a live issue today. The necessity "of substituting the former scholar type, entirely wrapped up in his special area of expertise, with a more generally communicative educator of the young and of raising awareness of his own responsibility with respect to cultural issues" is nevertheless a pertinent insight.[124]

[121] Another example of a constructive route toward acknowledgment of responsibility is the article: "Der Eid des Homo Sapiens," *Physikalische Blätter*, 2/2 (1946), 1. Brüche published this article at the suggestion of the American anthropologist Gene Weltfish, placing it at the beginning of the issue.

[122] Think of the Franck Report and Niels Bohr's memoranda.

[123] Maier-Leibnitz to Franck, 21 December 1957, RLUC, Franck papers, box 5, folder 3. See von Weizsäcker's correspondence with Eugene Rabinowitch in 1957, among his papers, RLUC, Franck papers, box 9, folder 8.

[124] Hahn, "Antwort," 4. He quotes there from article 2 of the new by-laws of the regional Physical Society of Württemberg-Baden. The editor of *Göttinger Universitäts-Zeitung* also attributes to Otto Hahn "a true political sense of responsibility" and accepts the interpretation of science as a "factor of political power." He regarded the founding of a German research council as an "important sign of an active sense of responsibility." *Göttinger Universitäts-Zeitung*, 4/18 (1949), 9 and 4. Other examples of such responsibility discourse include the article *Physikalische Blätter*, 2/4 (1946), 73f., and Max von Laue, "Public Relations?" 12 (last column).

"IF WE WANT TO LIVE, WE MUST REBUILD"

Everyone probably immediately associates the term *post-war Germany* with the reconstruction and the resulting economic boom, the *Wirtschaftswunder*. The rebuilding mentality only gained full force in the Federal Republic of the 1950s and 1960s. Many texts nevertheless traced early on the immense significance of this reconstruction motif, in metaphorical form.

In the very first post-war issue of the *Physikalische Blätter*, its editor and manager already prophesied:

> Active hope is possible, [however,] ... against which it is no small consolation that the intellectual foundations have been liberated again as building ground by the downfall of despotism. Those not sharing this hope will notice with skeptical wonder the energetic interventions even now perceptible at many places, efforts to mend torn threads and lay new foundations.[125]

What had happened could not simply be denied and repressed. Something else had to take its place, a contrasting image to give people new perspectives, hopes, and visions for the future. One could hardly imagine a more suitable theme for this than reconstruction. Without any clearing efforts or rebuilding, continuation of research or teaching was anyway unthinkable. In many places, students with political records were often set the prerequisite of "committing themselves to doing reconstruction work" for one or two semesters without compensation.[126] Elsewhere, this even became a general precondition for enrollment. The Mitscherlichs observed a "dogged determination with which the removal of the ruins were immediately begun." They diagnosed its "manic streak" as a defense mechanism against melancholy, grief, and feelings of guilt.[127] Even victims of National Socialism were subject to this psychological compulsion to concentrate on working for the future in order to come to terms with the current hardship. Hartmut Paul Kallmann, who had been dismissed from his position as department head at Haber's institute in 1933, received an appointment as extraordinary professor at the Technical University in Berlin after the war along with his former position at the Kaiser Wilhelm Institute of Physical Chemistry. Reporting to his former

[125] Ernst Brüche, "Zur Einführung," *Neue Physikalische Blätter*, 2 (1946), 2.

[126] The University of Munich is an example, according to *Neue Physikalische Blätter*, 2/2, 67f.

[127] Alexander Mitscherlich and Margarete Mitscherlich, *Die Unfähigkeit zu trauern. Grundlagen kollektiven Verhaltens* (Munich: Piper, 1967), 40.

colleague at the institute about the state of affairs in mid-1946, Kallmann also touched on the sense of scientific isolation:

We have now officially applied for a research permit and hope that we will receive it soon. You cannot imagine how ravenous we are to be able to be scientifically employed again; and I think this is the only thing that is left to us in this desolate situation, really working on something useful again and I think we can, in fact. I would very much welcome being able to visit you soon again in more active relations, because we feel very much abandoned in that regard. Those terrible years of horror did not go by me, in particular, without leaving a trace, of course, and I know precisely what is the only thing of importance, namely, really doing some proper work in peace.[128]

In 1947, the former chairman of the German Physical Society, Carl Ramsauer, wrote: "The time of need in National Socialism also had a good side to it for the German Physical Society. We collected ourselves and discovered the foundation upon which we can and will rebuild German physics."[129]

This reconstruction motif reappeared in a different light in an engineer's talk before the international conference on engineering training in Darmstadt that same year. The abbreviated transcript, published in the *Physikalische Blätter*, shifts the perspective beyond national borders. Research is regarded as "the most important champion and advocate of a union of European states."[130] This was not far from the truth, either, as such international joint efforts as the European Council for Nuclear Research (*Conseil Européen pour la Recherche Nucléaire*, CERN) and German Electron Synchrotron (*Deutsche Elektronen Synchrotron*, DESY) exemplify in historical retrospect.

The road to successful long-term international research was still long and tortuous. The first priority had to be local efforts. Comments from the period indicate how very close some people were to giving up:

The difficulties in rebuilding German science are immense. I do believe, though, that it is worthwhile joining in the efforts to overcome them at least somewhat over time. One is occasionally tempted to lose all hope; but I think this resignation would be misguided. We have to get through a minimum and then things will eventually get better again.[131]

[128] Kallmann to Polányi, 22 May 1946, RLUC, Polányi papers, box 5, folder 2.

[129] Carl Ramsauer, "Zur Geschichte der Deutschen Physikalischen Gesellschaft in der Hitlerzeit," *Physikalische Blätter*, 3 (1947), 110–114, here 114.

[130] See H. Klumb, "Naturwissenschaft, Technik und europäischer Aufbau," *Physikalische Blätter*, 3/7 (1947), 209–211.

[131] Hahn to Martius, 18 February 1947, MPGA, div. II, rep. 14A, no. 2726.

This incantation to rebuild on new foundations, letting everything else lie buried in the past and radically resorting to a *tabula rasa*, culminates in the following hymn-like refrain in a column of the *Physikalische Blätter*: "If we want to live, we must rebuild, after every last remnant has been torn down. We must start again from the very beginning. A scientist's research is the first precondition for the success of everything. Holy is his mission, and conscious departure on this noble quest makes him artist and herald of what shall be."[132]

Social psychologists have analyzed this connection between a rebuilding fixation and forgetting. It is barely veiled in passages like Otto Hahn's statement in his newspaper article from 1949: "the western zones are overcrowded with refugees, the worry about building an existence in human dignity overshadows all former prejudices."[133]

CONCLUSION

To recapitulate, the most prominent features of the mental profile of physicists and other scientists, technicians, and their contemporaries between 1945 and 1949 are insecurity and fear (of the Allies), bitterness (also in the face of extremely harsh living conditions right after 1945), insensitivity toward the sufferings of others, self-pity, and sentimentality, coupled with a cramping inability to imagine the situation of others and feel compassion. It explains the difficulties Germans encountered in communicating with émigrés. It also explains the widespread feeling inside Germany of living in a topsy-turvy world in which nothing was permanent or reliable anymore and in which they were being judged and sentenced according to incomprehensible norms. An official of the Allied occupying forces wrote in his diary:

> One had the painful realization that even well-bred and supposedly well-educated Germans regarded themselves as essentially right and misunderstood by the world. I found it a fascinating but depressing experience to struggle towards some kind of mutual understanding. If "re-education" is [im]possible even under such favorable circumstances, what can we hope to achieve with the masses?[134]

The sociopsychological antagonism against the Allies also explains the conspicuous and – for modernity – unique homogeneity of the mentality

[132] Dr. Hüttner (Werdau i. Sa.), "Das wissenschaftliche Buch," *Physikalische Blätter*, 4 (1948), 267.

[133] Hahn, "Antwort," 3.

[134] Edward Y. Hartshorne on 30 April 1945, in Tent, *Academic*, 37; see also 17.

characterizing these 5 years.[135] But compared with the Weimar period, with its famous trench battles between various physicist camps, these differences disappeared for a short while in the immediate post-war period, as it were, as terms of higher order against the backdrop of greater basic conflicts with the "occupiers." Belligerence against the Allies permitted the bridging of internal differences and strengthened the sense of solidarity among physicists as one very active and influential special interest group in science policy. Sincere and effective "denazification" was condemned to failure against this landscape of escapism and repression, a settling of accounts on the backs of others, and a conscious or unconscious inability to feel remorse.

[135] For example, see Philippe Ariès, "L'histoire des mentalités," in Roger Chartier and Jacques Revel (eds.), *La nouvelle histoire* (Paris: Retz, 1978), 422, on the tendency of a mentality to be pulverized into many micro-mentalities of small and smallest social groups. On conflicts within the physics community during the Weimar Republic, see, for example, Paul Forman, "The Financial Support and Political Alignment of Physicists in Weimar Germany," *Minerva*, 12 (1974), 39–66, and Hentschel, *Physics* (note 40), lxx ff.

11

"Cleanliness among Our Circle of Colleagues"

The German Physical Society's Policy toward Its Past

Gerhard Rammer

> Even to-day I see the 1947 Göttingen meeting of the Physical Society as a litmus test of the intellectual attitude of the German academic community and this was clear to me – and others – at the time.[1]

This retrospective assessment in 2002 by the former doctoral student of physics Ursula Franklin (née Martius) addresses central issues of this contribution. What intellectual and political positions did the academic community take regarding its involvement with National Socialism? What role did the German Physical Society (*Deutsche Physikalische Gesellschaft*, DPG) and the physics conferences it organized play in manifesting this community's stance? In other words, the topic of inquiry is the policy of physicists inside Germany and of their professional association toward the past. Norbert Frei saw this policy by Germans as

[1] Franklin to Rammer, 17–27 November 2002 (original English). My correspondence with Ursula Franklin in fall and winter 2002 was mostly conducted in English.

Besides the aforementioned correspondence and interviews with contemporaries, the primary sources of this chapter are mainly from the holdings of the Archives of the German Physical Society (*Archiv der Deutschen Physikalischen Gesellschaft*, hereafter DPGA). The papers of a number of physicists were consulted in addition. Those of Max von Laue and Otto Hahn at the Archives of the Max Planck Society (*Archiv der Max-Planck-Gesellschaft*, hereafter MPGA) in Berlin and of Samuel Goudsmit at the Niels Bohr Library & Archives in College Park, Maryland, proved most useful.

I thank Ursula Franklin for valuable background on her critical article and for biographical information. I am furthermore grateful to the participants of the workshop in Berlin and Washington for their critical commentary. Wolfgang Böker, Henning Trüper, and the editors of this volume subjected earlier versions of this chapter to their scrutiny. The latter also granted me access to important sources.

composed of three elements: amnesty, integration, and demarcation.[2] We shall encounter these again within the physics profession. In the following, the primary interest will be the processes of integration and exclusion within the scientific community, which took shape particularly during the first post-war physics meetings held in Göttingen. Established colleagues made various attempts, just as did outsiders, to draw lines within their profession to exclude individuals deemed politically or collegially unacceptable. Essentially different boundary lines were the result, depending on whether the yardstick used was political or collegial. Research on denazification has hitherto taken only inadequately into account collegial ties in academia. This chapter will devote special attention to them, considering their eminent importance in the processes at issue. It will also examine why the newly founded German Physical Society in the British Zone (*Deutsche Physikalische Gesellschaft in der britischen Zone*), and why Göttingen, in particular, occupied a prominent place in policy-making regarding the past. One of the DPG's most important activities was the organization of physics conferences; their special function during the first post-war years will be analyzed from the professional as well as political points of view.

The broad-scale rehabilitation and reintegration of politically incriminated individuals in West Germany during the first decade after the war has already been examined historically for many fields.[3] A whole bundle of amnesty and reintegration measures followed on the heels of the dismissals and punishments meted out by the Allies. In a broadly conceived "pardoning," the overwhelming majority of fellow travelers and even convicted war criminals were set "in their social, professional

[2] Norbert Frei, *Vergangenheitspolitik. Die Anfänge der Bundesrepublik und die NS-Vergangenheit* (Munich: Beck, 1996), 14; English translation: Norbert Frei, *Adenauer's Germany and the Nazi Past. The Politics of Amnesty and Integration* (New York: Columbia University Press, 2002), XII.

[3] Besides Frei, *Vergangenheitspolitik*, see also Norbert Frei (ed.), *Karrieren im Zwielicht. Hitlers Eliten nach 1945* (Frankfurt am Main: Campus Verlag, 2001); Helmut König, Wolfgang Kuhlmann, and Klaus Schwabe (eds.), *Vertuschte Vergangenheit. Der Fall Schwerte und die NS-Vergangenheit der deutschen Hochschulen* (Munich: Beck, 1997); Wilfried Loth and Bernd-A. Rusinek (eds.), *Verwandlungspolitik. NS-Eliten in der westdeutschen Nachkriegsgesellschaft* (Frankfurt am Main: Campus Verlag, 1998). Stronger emphasis is placed on universities in Bernd Weisbrod (ed.), *Akademische Vergangenheitspolitik. Beiträge zur Wissenschaftskultur der Nachkriegszeit* (Göttingen: Wallstein, 2002), as well as, in overview, in the essay review by Kay Schiller, "Review Article The Presence of the Nazi Past in the Early Decades of the Bonn Republic," *Journal of Contemporary History*, 39 (2004), 286–294.

and national... status quo ante."[4] A glance at the university landscape reveals, however, that the rehabilitations of professors did not operate as consistently as the legal situation would have permitted. Political and legal criteria were not the only determinants of academic policy concerning the past. The scientific merit of those affected played as considerable a role in their rehabilitation as whether they had formerly abided by the expected rules of collegial conduct. As a consequence, support was withheld from individuals with insufficient professional qualifications just as it was from those who had earlier proved disloyal by betraying their fellows in the profession or, as Oliver Schael put it, by breaking unwritten rules of professional conduct.[5]

The processes of inclusion and exclusion outlined here took place on two different planes: on a formal plane influenced by the legal provisions, such as the denazification stipulations and the regulations for Article 131 of the Basic Law of the Federal Republic; and on a social plane that was defined by the rules of conduct governing the scientific community. The effects of the processes operating on these two levels could be synchronous or run in opposite directions. A dismissed individual who had lost his professorship during denazification and thus formally became an outcast could continue to remain a component of the collegial community of university teachers. Other colleagues who had withstood the political scrutiny unscathed were not necessarily welcomed back in their professional community, however. Formally, the course of these processes is relatively easily followed. Socially, it is more hidden and less obvious. The professional ties between physicists particularly influenced the exclusions that did – or in fact did not – happen right after the war had ended.

Understanding these ties is less a question of political or moral attitude than of how strongly someone was bound within the collegial network.[6]

[4] Frei, *Vergangenheitspolitik*, 13f.

[5] Oliver Schael, "Die Grenzen der akademischen Vergangenheitspolitik: Der Verband der nicht-amtierenden (amtsverdrängten) Hochschullehrer und die Göttinger Universität," in Weisbrod, *Akademische*, 53–72, especially 60.

[6] The effectiveness of collegial networks in physics is discussed in Gerhard Rammer, "Göttinger Physiker nach 1945. Über die Wirkung kollegialer Netze," *Göttinger Jahrbuch*, 51 (2003), 83–104, and Gerhard Rammer, "*Die Nazifizierung und Entnazifizierung der Physik an der Universität Göttingen*," (Göttingen: University of Göttingen Ph.D. dissertation, 2004). Peter Mattes pointed out the low importance of political or moral criteria in the redefinition of the collegial community in his analysis of German psychology after 1945: "If political or moral reservations ever did arise, they had to be sacrificed for the preservation of orderly working conditions." Peter Mattes, "Die Charakterologen. Westdeutsche Psychologie nach 1945," in Walter H. Pehle and Peter

It seems to have been a characteristic of the academic world of that period that the preservation of collegiality held a higher status in the distinction between tolerable and disqualifying conduct than political acts and views. Further "fruitful collegial collaboration" – to borrow the contemporary usage – was apparently still deemed conceivable and welcome from whoever had behaved "decently" in this sense.[7] One consequence of this characteristic was that hardly any questions were raised about political missteps by colleagues subject to criticism. It was only in the beginning of the 1950s that this issue was pursued more seriously.[8]

To be able to evaluate the significance of the DPG in the way the past was addressed, one must first clarify what purposes the society was seeking and how it was willing and able to act. Immediately after the war, the Allies shut down the DPG along with all other scientific associations in Germany. Thus, each regional branch had to conduct separate negotiations with its governing occupier before it could reopen its doors within a particular zone as a separate physical society. The DPG in the British zone set down in writing the society's purposes in new statutes that differed only slightly from the former version dating back to the Weimar period: "The German Physical Society in the British Zone should serve the promotion and dissemination of the physical sciences, bring physicists closer together, and represent them as a whole to the rest of society. It seeks to attain this purpose through assemblies and scientific meetings."[9] The most important such convocations were the annual physics meetings, when the DPG sometimes passed policy decisions concerning the profession. Physicists attending these conferences also exchanged the latest news about the future of their academic discipline and their own career prospects within the profession. Such an organized meeting place was

Sillem (eds.), *Wissenschaft im geteilten Deutschland. Restauration oder Neubeginn nach 1945?* (Frankfurt am Main: Fischer Taschenbuch Verlag, 1992), 125–135, especially 135.

7 The quote is from Arnold Eucken as dean of the Faculty of Mathematics and the Sciences of the University of Göttingen in summer 1945, pointing out the impossibility of fruitful collegial collaboration with Kurt Hohenemser, who wanted to regain the assistantship he had been dismissed from in 1933. Hohenemser had not held his tongue about the political support physicists of his former institute had received in their careers; see Rammer, *Nazifizierung*.

8 The special mentality predominant among professors during the period of occupation is one explanation for this. See the contribution by Klaus Hentschel to this volume as well as his book: *The Mental Aftermath: The Mentality of German Physicists 1945–1949* (Oxford: Oxford University Press, 2007).

9 The statutes of the German Physical Society in the British Zone were approved on 14 October 1946 by Ronald Fraser, Research Branch. DPGA, no. 40040. They are published in *Physikalische Blätter*, 3 (1947), 29–31 and 163.

urgently needed in occupied Germany. In 1946, Göttingen was the first to achieve this goal.

REGIONAL REORGANIZATIONS OF THE DPG

When the DPG was suspended, its former leadership made attempts from its previous headquarters in Berlin to secure its continued existence. But political circumstances did not allow for a nationwide solution. Carl Ramsauer's hope as its last chairman was first to set up the western regional branches again in order to be able to do the same afterward in the Soviet zone and in Berlin.[10] He sought a competent member of the profession to pursue the matter in the west. Ernst Brüche took up the cause of his own accord, but Ramsauer discouraged his efforts. He preferred to empower the experimental physicist Robert Pohl from Göttingen with reorganizing the DPG in West Germany in December 1945.[11]

Pohl could easily survey the situation of physics at German universities and was able to exert some influence on hiring issues. His colleagues attributed to him an "almost innate human authority." Maier-Leibnitz called him "a patriarch of physics."[12] Another quality besides character favoring his selection was where he resided. The small university town of Göttingen had survived the war almost completely undamaged. Research and teaching in physics there was intact and productive, and the Göttingen Academy of Sciences was also situated there. These favorable circumstances caused Göttingen to become a "collecting basin of 'homeless' physicists" after 1945.[13] But science policy decisions reached in England had a greater impact, turning Göttingen into a "crystallization point"[14] of German post-war science. This constructive research policy

[10] See Ramsauer to Pohl, 13 August 1946, DPGA, no. 40027.

[11] Ramsauer to Pohl, 18 December 1945, DPGA, no. 40025. Sheer chance dictated that this decision survive on paper, thus coming within reach of historians. This letter informs us that Wilhelm Westphal had actually been planning to come to Göttingen in order to discuss this DPG affair with Pohl. Only because Westphal was prevented from traveling was this invitation made and discussed in writing.

[12] Georg Joos, "Robert Pohl 70 Jahre," *Zeitschrift für angewandte Physik*, 6 (1954), 339; Heinz Maier-Leibnitz, "Die große Zeit in Göttingen. Robert W. Pohl, ein Patriarch der Physik, wird neunzig," *Frankfurter Allgemeine Zeitung* (10 August 1974).

[13] Thus the description by a private lecturer in Göttingen at that time: Wilhelm Walcher, "Physikalische Gesellschaften im Umbruch (1945–1963)," in Theo Mayer-Kuckuk (ed.), *Festschrift 150 Jahre Deutsche Physikalische Gesellschaft*, special issue of the *Physikalische Blätter*, 51 (1995), F107–F133, especially F108.

[14] According to Hahn's diary entry following a conversation at the Royal Institution on 2 October 1945 together with Heisenberg and von Laue, on the German side, and

by the British ensured that the former national scientific society, the Kaiser Wilhelm Society (*Kaiser-Wilhelm-Gesellschaft*, KWG), survived the transition in 1945 by being authorized, first just in the British zone, later also in the other western zones, under the new name, Max Planck Society (*Max-Planck-Gesellschaft*, MPG), and at new headquarters: Göttingen.[15] The Max Planck Institute for Physics (*Max-Planck-Institut für Physik*, MPIfP) also profited by this policy, being likewise established in Göttingen in 1946 with most of the physicists who had just been released from Farm Hall.[16] That is how the important figures in post-war science policy, Otto Hahn, Werner Heisenberg, and Max von Laue, came to Göttingen.[17]

Commissioned with reorganizing the DPG, Pohl approached the official expert on scientific issues in the British zone, Colonel Bertie Blount, in January 1946. Pohl concluded from this meeting that "Göttingen was probably the right choice of location" to revive the DPG.[18] One advantage Göttingen offered was the collegial relations between local scientists and influential persons among the British Control officers responsible for

Patrick Blackett, George P. Thomson, William Lawrence Bragg, Henry H. Dale, and Archibald V. Hill, on the British side. See Otto G. Oexle, "Wie in Göttingen die Max-Planck-Gesellschaft entstand," *Max-Planck-Gesellschaft Jahrbuch* (1994), 43–60, especially 54f.

15 Mark Walker, *German National Socialism and the Quest for Nuclear Power* (Cambridge: Cambridge University Press, 1989), 188f.; Oexle, "Göttingen"; Otto G. Oexle, "Hahn, Heisenberg und die anderen. Anmerkungen zu 'Kopenhagen,' 'Farm Hall' und 'Göttingen,'" *Ergebnisse. Vorabdrucke aus dem Forschungsprogramm "Geschichte der Kaiser-Wilhelm-Gesellschaft im Nationalsozialismus*," No. 9 (Berlin: Forschungsprogramm, 2003; cited hereafter as *Ergebnisse*, No., and year). Hans Joachim Dahms, "Die Universität Göttingen 1918 bis 1989: Vom "Goldenen Zeitalter" der zwanziger Jahre bis zur "Verwaltung des Mangels" in der Gegenwart," in Rudolf von Thadden and Günter J. Trittel (eds.), *Göttingen. Geschichte einer Universitätsstadt*, vol. 3 (Göttingen: Vandenhoeck & Ruprecht, 1999), 395–456, especially 430.

16 About the internment of these atomic physicists in the English manor house Farm Hall, see Dieter Hoffmann (ed.), *Operation Epsilon. Die Farm-Hall-Protokolle oder Die Angst der Allierten vor der deutschen Atombombe* (Berlin: Rowohlt, 1993), especially 49f.

17 In early 1946, Hahn, Heisenberg, von Laue, Weizsäcker, Wirtz, Korsching, and Bagge came to Göttingen from Farm Hall. The political importance of Otto Hahn and Werner Heisenberg is discussed in Mark Walker, "Von Kopenhagen bis Göttingen und zurück. Verdeckte Vergangenheitspolitik in den Naturwissenschaften," in Weisbrod, *Akademische*, 247–259; "Otto Hahn: Responsibility and repression," *Physics in Perspective*, 8/2 (2006), 116–163. For specifics on Hahn's role as president of the MPG, see: Carola Sachse, "'Persilscheinkultur.' Zum Umgang mit der NS-Vergangenheit in der Kaiser-Wilhelm/Max-Planck-Gesellschaft," in Weisbrod, *Akademische*, 217–246.

18 Pohl to Ramsauer, 29 January 1946, DPGA, no. 40029.

research. As a consequence, the effect of the formal imbalance between occupiers and the defeated was much weakened.[19]

Table 5 presents a chronological overview of the individual regional societies until they were reunited as the Union of German Physical Societies (*Verband Deutscher Physikalischer Gesellschaften*) in 1950.

THE SEARCH FOR A SUITABLE CHAIRMAN

Pohl did not consider himself the most suitable of chairmen but regarded his role merely as a start-up helper or "custodian"[20] entrusted with ensuring the society's continued existence. He agreed with Ramsauer not to demand a resurrection of the German Society for Technical Physics (*Deutsche Gesellschaft für technische Physik*). Their reasons were political, primarily because technical physics could so easily be linked with military applications. They thought a single society sufficed.[21] In negotiations with the military government in June 1946, Pohl succeeded in having the old regional associations of Rhineland-Westphalia and Lower Saxony resume operations as the "German Physical Society in the British Zone" under the "statutes valid before 1933."[22] To have the American zone follow suit, Pohl asked his colleague in Stuttgart, Erich Regener, to initiate the necessary measures. He wanted to avoid raising the slightest appearance that the Göttingen physicists were trying to establish a "monopolizing position."[23]

It was particularly important to him that the other regional branches follow the Göttingen model closely. So he asked Regener to found a DPG

[19] Rammer, *Nazifizierung*. In the case of the chemist Bertie Blount, for instance, there were closer ties with German science by virtue of his having earned his doctorate at the University of Frankfurt in 1931.

[20] Pohl to Ramsauer, 29 January 1946, DPGA, no. 40029.

[21] For example, see Pohl to Gerlach, 22 June 1946, DPGA, no. 40026.

[22] Pohl described the old statutes valid since 1925 as those from "before 1933" even though statutory alterations as those excluding Jewish members were only made in 1940. See Pohl's circular letter of 21 June 1946 as well as the approval by Ronald Fraser, Research Branch, 14 June 1946, DPGA, no. 40026. The changes to the statutes in 1940 are discussed in Alan D. Beyerchen, *Scientists under Hitler. Politics and the Physics Community in the Third Reich* (New Haven: Yale University Press, 1977), 76; Dieter Hoffmann, "Zwischen Autonomie und Anpassung: Die Deutsche Physikalische Gesellschaft im Dritten Reich," *Max Planck Institute for History of Science Preprint*, no. 192, (Berlin: Max Planck Institute for History of Science, 2001), 12–14, as well as the contributions by Stefan L. Wolff and Dieter Hoffmann to this volume.

[23] Pohl to Ramsauer, 25 June 1946, DPGA, no. 40026; Pohl to Regener, 22 June 1946, DPGA, no. 40027; Ramsauer to Pohl, 2 July 1946, DPGA, no. 40027.

TABLE 5. *Post-War Regional Physical Societies*

Society	Founding Date	First Chairman (Deputy)	First Meetings	Remarks
German Physical Society in the British Zone	5 Oct. 1946	Max von Laue (Clemens Schaefer)	4–6 Oct. 1946, Göttingen 12–13 Apr. 1947, Göttingen (spring meeting of the Regional Association of Lower Saxony) 5–7 Sep. 1947, Göttingen 9–11 Sep. 1948, Clausthal 22–24 Apr. 1949, Hamburg 21–25 Sep. 1949, Bonn 15–17 Apr. 1950, Münster	Includes the two earlier Regional Associations of Lower Saxony and Rhineland-Westphalia; in 1950 renamed Northwest German Physical Society
Physical Society of Württemberg-Baden	15 Aug. 1946 (retroactively approved)	Erich Regener (Walther Bothe, Ulrich Dehlinger)	5–6 July 1947, Stuttgart 15–16 Nov. 1947, Heidenheim 5–6 June 1948, Stuttgart 30–31 Jan. 1949, Heidelberg 10–11 Dec. 1949, Freiburg 10–11 June 1950, Karlsruhe	Merger on 16 Jan. 1950 with parts of the Society of Rhineland-Phalia into the Society of Württemberg-Baden-Phalia
Physical Society of Hesse	19 July 1947	Erwin Madelung (Richard Vieweg)	24 Apr. 1948, Frankfurt 11 Jun. 1949, Frankfurt 22 Apr. 1950, Frankfurt	Merger on 16 Jan. 1950 with parts of the Society of Rhineland-Phalia into the Society of Hesse-Central Rhine
Physical Society in Bavaria	8 Dec. 1947	Walther Meißner (E. Rüchardt)	29 July to 2 Aug. 1949, Munich	
Physical Society of Rhineland-Phalia	29 Apr. 1948 (retroactively approved)	Hans Klumb (Karl Wolf)	14 May 1949, Mainz	Existed only until 16 Jan. 1950
Physical Society in Berlin	7 Dec. 1949	Carl Ramsauer (W. Schaaffs)		
Union of German Physical Societies	13 Oct. 1950	Jonathan Zenneck (Max von Laue)	11–15 Oct. 1950, Bad Nauheim (15th physics meeting) 19–23 Sep. 1951, Karlsruhe	Reunification of the five regional societies
Physical Society in the German Democratic Republic	26 Sep. 1952	No official chairman until 1970; officiated by Robert Rompe and from 1955 Gustav Hertz	Fall 1952, Berlin April 1953, Freiberg July 1953, Greifswald September 1953, Halle April 1954, Dresden	

"in the American zone," likewise on the basis of the earlier statutes to facilitate future consolidation into a nationwide society. When Regener did not follow this suggestion to the letter and founded a Physical Society of Württemberg-Baden (*Physikalische Gesellschaft Württemberg-Baden*), Pohl criticized it as separatist. This charge is very revealing because it shows how Pohl was placing the DPG within a larger political context in an effort to strengthen Germany under the four occupying powers:

> We all are aware that we shall just be forming a cultural unit, not a political one, in Germany for a long time. Shouldn't we try everything to attain uniform establishments at least in the cultural sphere? For us physicists this is only possible if we do not replace our old physical society by founding new separatist local societies.[24]

Heisenberg supported Pohl's criticism, arguing that the DPG should avoid any subdivisioning into smaller units that the zonal borders did not prescribe. On the one hand, the zonal borders were enforced; on the other hand, they were "geographically so unnatural that there's no risk of future fragmentation."[25] In retrospect, Pohl's fears turned out to be exaggerated. After the founding of the Federal Republic of Germany, the western regional societies reunited quite rapidly into a Union of German Physical Societies, despite their differing names and statutes.

The moment the regional society obtained its permit from the British, and thereby could resume operations, Heisenberg assumed the responsibilities of chairman and organized a physics meeting in Göttingen for the beginning of October 1946. Pohl's colleague Hans Kopfermann, who had chaired the Regional Association of Lower Saxony (*Gauverein Niedersachsen*) since 1938 and supplied important support to Pohl during the negotiations with the Allies, had originally suggested his nomination.[26] It would have been expected that Heisenberg be ratified as chairman at the first regular meeting;[27] but he was not, despite his international reputation as a Nobel laureate.

[24] Pohl to Ramsauer, 24 July 1946, DPGA, no. 40027. He was more emphatic in his letter to Schön from 12 August 1946: "In the opinion of us Göttingers, every separatist new founding that endangers German unity in the cultural sphere must be avoided." DPGA, no. 40028.

[25] Heisenberg to Regener, 26 July 1946, DPGA, no. 40029.

[26] See Pohl's circular letter, 21 June 1946, DPGA, no. 40026. Pohl and Kopfermann to Heisenberg, 21 May 1946, DPGA, no. 40029; Heisenberg to Blount, 23 May 1946, DPGA, no. 40029. See also the reports about Heisenberg's preliminary management of the affairs in *Neue Physikalische Blätter*, 2 (1946), 85 and 116.

[27] Pohl and Kopfermann wrote to Heisenberg on 21 May 1946 that at the first meeting they would "nominate [him] as chairman of our regional association and propose at the

On the first day of the physics meeting, Pohl sent a handwritten letter to his fellow Göttingen resident Heisenberg suggesting he submit von Laue's name for the chairmanship during the administrative session scheduled for the following day. This unusual procedure needs some explanation. The reason Pohl gave to Heisenberg was the failure of the original plan to reopen the regional associations in the British zone because the "mother society," the DPG, had been eliminated. That was why refounding a society with new statutes and a new chairman and deputy had become necessary. Max von Laue had just realized this predicament "in the nick of time," he explained, adding: "In the few remaining days available, Mr. von Laue had settled everything and therefore please be so kind as to accept the consequences and propose Mr. von Laue as first chairman."[28] That von Laue had been the one to take the initiative in acting to forestall anticipated problems in the founding of the society may well have been important in Pohl's decision. Whether Pohl was alluding in this letter to the deciding factor leading to this last-minute nomination of von Laue is doubtful, however. Von Laue's political capital made him clearly the better candidate, and Pohl knew it.

The reputation von Laue had as an opponent of National Socialism placed him in a much stronger position to counter the political obstacles physics was facing after 1945.[29] One obstacle was the international isolation of German science that National Socialism had caused. It was still in effect after 1945 because the involvement of German physicists during the Nazi regime was not so easily forgotten abroad. In some cases, the resentment was quite considerable. The alienation demonstrated by Philip Morrison's judgment, "they worked for the cause of Himmler and Auschwitz, for the burners of books and the takers of hostages. The community of science will be long delayed in welcoming the armorers of the Nazis,"[30] was not representative. But it does show that such reservations existed and stood in the way of closer international exchanges. In fact,

same time that you be regarded as the focal authority for both regional associations." DPGA, no. 40029.

[28] Pohl to Heisenberg, 4 October 1946, DPGA, no. 40029.

[29] Max von Laue's political acts during the period of National Socialism are discussed in Beyerchen, *Scientists*, 34, 40, 64–66, 113, 115f., 170, 200, 207f., 209; Friedrich Herneck, *Max von Laue* (Leipzig: Teubner, 1979), especially 60–68; as well as Katharina Zeitz, *Max von Laue (1879–1960). Seine Bedeutung für den Wiederaufbau der deutschen Wissenschaft nach dem Zweiten Weltkrieg* (Stuttgart: Steiner, 2006).

[30] Philip Morrison from Cornell University in a review of Samuel Goudsmit's book *Alsos* in *Bulletin of the Atomic Scientists*, 3 (1947), 365. See MPGA, Atomforschung, V, 13, Atomforschung und Folgen, 1.

no Germans attended the first major physics conferences held in the immediate post-war period. But this was less a result of the mentioned political reservations than of the restrictive travel regulations imposed by the occupiers. It applied to the conference on elementary particles in Cambridge of 1947, the Shelter Island conference of the same year, which was described by Richard Feynman as the most important global conference of theorists of all time, and to the Solvay conference of 1948.[31]

Max von Laue was the first German physicist able to break through this isolation. As early as July 1946, he was granted permission to travel to London to attend an international conference of crystallographers; two other invited Germans were refused travel permits.[32] By issuing this invitation to von Laue, the British evidently wanted to send a political signal. Opponents of the Nazi regime ought to be lent support and appointed to important positions even in science. This was also an essential reason behind von Laue having been taken prisoner together with nine other scientists of the Uranium Club (*Uranverein*) at the end of the war and deported to England. The British had selected him out as a suitable leader for the rebuilding of academic life. This retired full professor willingly accepted such a role as well. In his autobiographical report, von Laue mentioned as the main reason why he had not emigrated during the National Socialist period that he wished "to be directly on location when, after the ever anticipated and hoped-for collapse of the 'Third Reich,' the opportunity presented itself for a cultural rebuilding on the ruins created by this Reich. A large part of my activities after 1945 were subsequently governed by this aspect."[33]

31 Gabriele Metzler, *Internationale Wissenschaft und nationale Kultur. Deutsche Physiker in der internationalen Community 1900–1960* (Göttingen: Vandenhoeck & Ruprecht, 2000), 218f. On the Shelter Island conference and Feynman's assessment, see Michael Eckert, *Die Atomphysiker. Eine Geschichte der theoretischen Physik am Beispiel der Sommerfeldschule* (Braunschweig: Vieweg, 1993), 254. Applied physics was not spared exclusion from international conferences. The 6th and 7th international congress for applied mechanics took place in Paris in 1947 and London in 1948 without German attendance. See *Physikalische Blätter*, 3 (1947), 93; *Physikalische Blätter*, 4 (1948), 148.

32 Max von Laue, "Mein physikalischer Werdegang. Eine Selbstdarstellung," in Hans Hartmann, *Schöpfer des neuen Weltbildes. Große Physiker unserer Zeit* (Bonn: Athenäum-Verlag, 1952), 178–210, especially 205. See also Max von Laue, "Royal Society. Deutsche Physiker auf britischen Tagungen," *Göttinger Universitätszeitung*, no. 18 (1945/46), 12.

33 Von Laue, "Werdegang," 203. One early confirmation of this cultural mission of his is found in a letter to Herbert Mataré from 9 January 1945: "I wish you survive the rigors of the war well. Afterwards men will be needed for the reconstruction of some culture." American Institute of Physics, Niels Bohr Library, College Park, MD (hereafter AIP),

The reorganization of physics periodicals should be understood as part of this cultural revival. "The 'German Physical Society in the British Zone' regards as its first task helping work toward the reissuance of the *Zeitschrift für Physik*."[34] Pohl joined von Laue in editing this most important of physics journals besides the *Annalen der Physik*. Its reappearance was obstructed less by political problems than by material ones. Despite good relations with the Control Officers and notwithstanding their support, it was impossible to obtain the amounts of paper required for it to be printed.[35] One task von Laue successfully settled during his stay in England in summer 1946 was this paper issue.[36] His colleagues at home did not take his speedy success for granted. "It is very astonishing that he has been treated there so very nicely," the Munich professor Walther Meißner wrote to Pohl, for example.[37] This friendly reception von Laue was receiving was quite a contrast against the attacks Heisenberg was being subjected to right at that time. In the United States, the Dutch émigré physicist Samuel Goudsmit was publicly criticizing Heisenberg's behavior as a scientist, stylizing him as the mastermind of German uranium research.[38] Rumors about his visit with Niels Bohr in 1941 caused many physicists abroad to consider Heisenberg politically suspicious as well. This situation in fall 1946 made Heisenberg seem even less suitable as future chairman of the DPG against von Laue's candidacy. So we may presume that Pohl's instructions to Heisenberg to decline the chairmanship were based on political calculations.

Although Heisenberg had conducted the society's business as a substitute and sent out the invitations to the meeting in Göttingen, he was denied the opportunity to make a prominent appearance at the meeting

Mataré, Herbert Franz, MP 2002–523. Von Laue's opening words in the first issue of the *Physikalische Blätter* of 1947 are also revealing: "The cultural crisis that has broken out throughout the world, and in particular throughout Germany, also interrupted at the beginning of 1945 the hundred years of work conducted by the Deutsche Physikalische Gesellschaft." (English translation in Hentschel, *Mental*, 77.)

34 Pohl, 21 June 1946, Archives of the German Physical Society (*Archiv der Deutschen Physikalischen Gesellschaft*, hereafter DPGA), no. 40029.

35 The appearance of the *Zeitschrift für Physik* in Göttingen, hence in a western zone, was an important political decision in the midst of growing tensions between East and West. It served as a counterweight to the *Annalen der Physik*, which was appearing in Leipzig.

36 Pohl to Rüchardt, 10 July 1946, DPGA, no. 40028.

37 Meißner to Pohl, 28 September 1946, DPGA, no 40028.

38 Samuel A. Goudsmit, "Secrecy or science?" *Science Illustrated*, 1 (1946), 97–99 (in German translation in *Neue Physikalische Blätter*, 2 [1946], 203–207). On the uranium research in Germany, see Walker, *Quest*.

itself. Von Laue held the inaugural speech as "the most senior of the officiating physicists here."[39] The founding meeting of the DPG in the British Zone on the following day lay firmly in Pohl's hands. He opened the session, initially ceding the podium to Heisenberg, who briefly reviewed the prehistory of the society's founding. Then, at Pohl's behest, he proposed that von Laue be appointed chairman. Pohl then allowed Ramsauer to speak. Before the 76 physicists present, Ramsauer declared that under his chairmanship, the society had been "directed in an anti–National Socialist vein." His phraseology concerning the history of the DPG also purported in effect that the board had exercised resistance against National Socialism. Pohl thanked him for arranging "that within the confines of what could possibly happen, everything possible did happen." Finally, Pohl asked von Laue to read out the draft statutes but continued to keep a tight rein on the proceedings, even after, as expected, von Laue was elected chairman, with Clemens Schaefer as his deputy – they were, incidentally, the sole candidates for the posts.[40]

As we have seen, Pohl exerted considerable influence on this first refounding of a regional society. His intervention at the eleventh hour resulted in von Laue being nominated and elected as chairman.[41] With its most suitable chairman, from the political point of view, the society could carry on its business with renewed energy. One of its most important tasks was hosting conferences.

FUNCTIONS OF THE PHYSICS MEETINGS

This section will only briefly discuss the apparent purpose of the conferences, namely, the presentation of research results and their debate among

39 Manuscript of the opening speech from 4 October 1946, MPGA, div. III, rep. 50, von Laue papers, no. 124.

40 See the three different records of the minutes of this founding meeting on 5 October 1946. Reinhold Mannkopff, serving as secretary, was also appointed the new managing director. DPGA, no. 40038. Auwers, Heisenberg, Jensen, Justi, and Unsöld from the regional association of Lower Saxony were elected as board members along with Försterling, Gerlach, Kratzer, Meixner, and Rogowski from the regional association of Rhineland-Westphalia. The shortened terms of Rogowski and Försterling were later filled by Schlechtweg and Fucks. MPGA, div. III, rep. 50, von Laue papers, no. 2390. The entire board was reelected unchanged in 1947 at Pohl's proposal. See the minutes of the administrative meeting of 6 September 1947, DPGA, no. 40043.

41 The society's official history written on the occasion of its 150th anniversary does not acknowledge Pohl's influence. His name is not even mentioned. See Walcher, "Physikalische," F108.

experts.[42] Instead, our focus will be on their functions arising out of the political situation in which the first few post-war conferences were set. These functions can be subsumed under two key concepts: internationalism and the policy toward the past. Both already influenced the course of the first meeting in 1946. The influx of refugees in Göttingen dictated that the attendance be limited to at most five participants per university,[43] so many debates were postponed to the following, far larger meeting of 1947. These two meetings are at the center of the following analysis. It will attempt to throw light on the internationalism practiced by the DPG along with its policy toward the past, not just within the context of the meetings but in view of their overall importance.[44]

Internationalism

In science, especially physics, internationalism was of special importance. It is understood as the endeavor to foster scientific ties among physicists from the various nations. The underlying ideal was that physical knowledge knows no national boundaries. In addition to its professional meaning of the international exchange of ideas or even real collaborations, internationalism also has a political function in the sense that physicists come forward as representatives of their nations. Thus, there is a distinction between internationalism and internationality, the latter implying collaborative efforts devoid of any national reference.[45] One generally finds good examples of internationality in physics. The first physics conferences held in Germany after the war rather involved internationalism, however, and the following will examine the significance of this circumstance.

Max von Laue's inaugural speech at the Göttingen meeting of 1946 already demonstrates the political function of internationalism with its many diplomatic customs. At the very start, von Laue stressed the good

[42] For the presentation of research results in the years before 1945, see the contribution by Gerhard Simonsohn in this volume.

[43] See the circular by Heisenberg, 2 August 1946, DPGA, no. 40034.

[44] The events at the other post-war meetings, particularly the deviating statutes of the Society in Württemberg-Baden, are discussed by Walcher, "Physikalische," F109–F111.

[45] This concept of internationalism and internationality as applied by German historians is discussed in Henning Trüper, "Die Vierteljahresschrift für Sozial- und Wirtschaftsgeschichte und ihr Herausgeber Hermann Aubin im Nationalsozialismus," *Vierteljahresschrift für Sozial- und Wirtschaftsgeschichte*, suppl. issue no. 181 (2005). German physicists are discussed, with an emphasis on their national culture, by Metzler, *Internationale*.

will shown by the military government toward German science. Right afterward, he welcomed the foreign visitors by name: Wolfgang Friedrich Berg, Felix Bloch, Philip Bowden, Antonius Michels, and Nevill Mott.[46] For a *German* physics meeting, the participation of so many foreigners is remarkable and unusual, especially when one considers the obstacles posed by travel and accommodation in 1946. There may well have been another purpose behind this strong foreign presence besides the resumption of international scientific exchange[47] as practiced here in the form of scientific presentations by two of these visitors. This ulterior motive may have been a secret spotting and control of German research. Surveillance of German scientific activities was of central importance to the Allies, and various special units pursued it according to elaborate plans that had been on the drawing board well before Germany's capitulation.[48] But physics conferences presumably were just a subordinate part of this broad surveillance. The international presence at the Göttingen meeting was advantageous to the Germans to the extent that their visitors' interest and the atmosphere of scientific internationalism could be exploited to promote the German research agenda.

The form of internationalism practiced here followed two, not always sharply sundered, goals: professional exchange and political exploitation of international relations for national purposes. This practice certainly agreed with the widespread conception of a boundaryless community of scientists. Von Laue gave one example in his opening speech in 1946. He regarded the DPG "as one part of that Republic of Letters, ardently cherished as an ideal extending beyond all national borders."[49] This view created a plane on which physicists could become acquainted and communicate internationally. In reply to a letter by the Committee for Foreign Correspondence of the Federation of American Scientists, von Laue was able solemnly to aver at the administrative session of 1947 "that the

[46] Opening speech by von Laue, 4 October 1946, MPGA, div. III, rep. 50, von Laue papers, no. 124.

[47] *Neue Physikalische Zeitschrift*, 2 (1946), 178f.

[48] John Gimbel, *Science, Technology, and Reparations. Exploitation and Plunder in Post-War Germany* (Stanford: Stanford University Press, 1990). The surveillance of German research, especially the impact of Control Council Law No. 25, is discussed in Manfred Heinemann, "Überwachung und "Inventur" der deutschen Forschung. Das Kontrollratsgesetz Nr. 25 und die alliierte Forschungskontrolle im Bereich der Kaiser-Wilhelm-/Max-Planck-Gesellschaft (KWG/MPG) 1945–1955," in Lothar Mertens (ed.), *Politischer Systemumbruch als irreversibler Faktor von Modernisierung in der Wissenschaft?* (Berlin: Humblot, 2001), 167–199.

[49] MPGA, div. III, rep. 50, von Laue papers, no. 124.

German Physical Society in the British Zone is dominated by the same mentality of international fraternity and that it imposes on each of its individual members the obligation to bear testimony to this mentality by word and deed at every suitable opportunity."[50]

The British Controller for physical research in Göttingen, Ronald Fraser, likewise made use of this communicative plane. His inaugural speech at the physics meeting in Göttingen in 1947 made reference to a statement by Niels Bohr that physics concerned the treatment of open questions of nature that could only be solved in a milieu of openness. "That is why every true physicist must consciously think supranationally; and my goal as a scientific advisor of the Research Branch of the Control Commission is, accordingly, to constantly increase the possibility of mutual exchange of opinions between German scientists and their colleagues abroad."[51]

Fraser's efforts had an impact. Four fellow physicists came from England to participate even in the less important meeting of the Regional Association of Lower Saxony in April 1947.[52] At the subsequent, larger fall meeting, which again took place in Göttingen, visitors from Amsterdam, Cambridge, Copenhagen, London, and Manchester presented eight scientific talks.[53] This was the culmination of this international scientific exchange, an exceptional phenomenon in Germany that came about thanks to the special support by the Göttingen Research Branch. Fraser's particular engagement on behalf of Göttingen physics was based on the fact that he himself was a physicist and specialist in molecular rays.[54] The collegial relations the Göttingers managed to establish with him ensured his special engagement on their behalf. One letter to von Laue suggests how important Fraser was for Göttingen physics. Erich Regener laments: "We over here have no Dr. Fraser."[55] Wilhelm Hanle from Gießen also

50 Minutes of the administrative session of 6 September 1947, DPGA, no. 40043. The committee's letter is printed in excerpt in *Neue Physikalische Blätter*, 2 (1946), 258f.

51 *Physikalische Blätter*, 2 (1946), 258f.

52 They were Blackett, Burcham, Coulsen, and Rushbroke. See *Physikalische Blätter*, 2 (1946), 129f.

53 See *Physikalische Blätter*, 2 (1946), 317–325; Walcher, "Physikalische," F108f. According to an accommodations roster, the following persons appeared from abroad: Jordan (Copenhagen), Michels (Amsterdam), Pippard (Cambridge), Shoenberg and Sondheimer (Cambridge), Tolansky (London) and Wilson (Manchester). A paper by Békésy (Stockholm) was read by Grützmacher. See DPGA, no. 40044.

54 Fraser authored two books on molecular beams that the Göttingen physicists were also using at the time: Ronald G. J. Fraser, *Molecular Rays* (Cambridge: Cambridge University Press, 1931); Ronald G. J. Fraser, *Molecular Beams* (London: Meuthen & Co., 1937).

55 Quoted from Walcher, "Physikalische," F112.

saw the advantages to the international contacts that Fraser made possible. Thanking Göttingen physicists for the opportunity to take part in the spring meeting in 1947 as one of their invited colleagues from the other zones, he added: "One can only repeatedly congratulate you on the good connections you have with English scientists."[56] Fraser stayed in the Research Branch until the end of 1948, when he became a Paris-based liaison officer of the International Council of Scientific Unions (ICSU) at the United Nations Educational, Scientific and Cultural Organization (UNESCO, founded in 1946).[57]

In 1948, the difficulties German physicists had to contend with in order to travel abroad began to shrink noticeably. They gladly accepted invitations abroad as guest researchers, not just for the sake of more intensive scientific exchange but also to serve as cultural emissaries. In other countries, they became acquainted with different methods of work, contributed their own abilities and competencies, and did their best to leave a "good impression" of German science, so as "to do our bit as well toward strengthening our still so weak ties with kindly disposed foreign countries."[58]

As such visits by Germans abroad became more frequent, the number of foreigners attending the conferences by the DPG in the British Zone fell correspondingly.[59] From this, we may conclude that in the post-war period, too, domestic exchanges among physicists were the primary intention of these physics meetings – notwithstanding all the rhetoric about the international Republic of Letters. A glance at the other western zones also shows that foreign attendance was a peculiarity of Göttingen. As a rule, the Germans were among themselves on these occasions.[60] This

[56] Hanle to Heisenberg, 16 April 1947, Max Planck Institute for Physics (*Max-Planck-Institut für Physik*, hereafter MPIfP), Heisenberg papers. I cordially thank Mark Walker for tips about much relevant correspondence by physicists among the Heisenberg papers.

[57] Heinemann, "Überwachung," 182.

[58] The Göttinger Reinhold Mannkopff wrote this about his research stay in England to Carl Correns on 29 February 1948, MPGA, div. III, rep. 50, von Laue papers, no. 1280.

[59] At the meeting in Clausthal in fall 1948, only one other foreigner presented a talk besides Max Born. There was exactly one foreign speaker at the Hamburg conference of 1949, and none at all at the Bonn conference in fall 1949. See *Physikalische Blätter*, 4 (1948), 391–401, and *Physikalische Blätter*, 5 (1949), 228–236 and 511–524.

[60] There were no contributions by foreigners to the physics conferences in Stuttgart and Heidenheim in 1947, in Frankfurt and Stuttgart in 1948, and in Mainz, Frankfurt, and Munich in 1949. One exception was the talk by Peyrou from Paris at the conference in Heidelberg in 1949. See *Physikalische Blätter*, 3 (1947), 198–205 and 398–407; *Physikalische Blätter*, 4 (1948), 220–223 and 249–259; and *Physikalische Blätter*, 5 (1949), 178–185, 380–383, 422–425, and 426–437.

trend continued after the consolidation into the Union of German Physical Societies. The report in the *Physikalische Blätter* about the physics meeting of 1950 makes no mention of foreign participants, and the corresponding passage in the following year was "foreign guests were present only in low numbers."[61] This just shows that the German conferences did not particularly attract foreigners, and international exchanges took place at other meetings specially conceived for this purpose.

The topic selection at German conferences also eased the restoration of international ties. There was the tendency either not to report about military research at all or to do so in a way that obscured the military context to the point that it looked like basic research. This was possibly a concession to Allied Control Council Law No. 25 from 1946. Its purpose was to secure the disarmament and demilitarization of scientific research. Although research on applied nuclear physics, rocket propulsion, applied aerodynamics and hydrodynamics, and all sorts of applications of electromagnetic, infrared, and acoustic emissions especially suitable for military purposes were forbidden, it was not explicitly prohibited to give a talk on one of these subjects.[62] All the same, Germans seemed to find it appropriate to present their research as pure science.

Examples of such transformations from weapons research to fundamental research include the talks by Erich Bagge on the theory of the mass distribution frequencies of fragments produced in spontaneous fission, by the group led by Erwin Meyer on sound absorption,[63] and by Werner Döring on the structure of an intense shock wave in diatomic gases. All these talks were held at the Göttingen meeting of 1947. The original military contexts of these investigations were, first of all, the uranium project, and second, commissioned research on sound damping for

[61] *Physikalische Blätter*, 6 (1950), 567–569, and *Physikalische Blätter*, 7 (1951), 471f.

[62] This law was reprinted in *Neue Physikalische Blätter*, 2 (1946), 49–52; available at www.verfassungen.de/de/de45-49/kr-gesetz25.htm.

[63] Erich Bagge, Zur Theorie der Massen-Häufigkeitsverteilung der Bruchstücke bei spontaner Kernspaltung. Meyer's talk bore the general title: On recent researches in acoustics (Über neuere Arbeiten in der Akustik). His collaborators spoke about a nonreflective outer surface within a large frequency range in acoustics (Schoch, Über in einem großen Frequenzbereich nicht reflektierende Grenzflächen in der Akustik), the propagation of sound in layered media free from shearing stress (Tamm, Schallausbreitung in geschichteten, schubspannungsfreien Medien), and a dynamic determination of the modulus of elasticity and damping of rubber-like materials in a very large frequency range (Kuhl, Dynamische Bestimmung von Elastizitätsmodul und Dämpfung von gummiartigen Stoffen in einem sehr großen Frequenzbereich). *Physikalische Blätter*, 3 (1947), 322–324.

submarines. The theorist Döring's investigations on the detonation process had been performed for the development of *Hafthohlladung* (shaped hollow explosive charge) at the Reich University in Posen (Poznán).[64] Other speakers chose topics far away from military applications. Examples are talks by Fritz Houtermans on the age of the earth, Kopfermann on isotopic shift effects, von Laue on the theory of superconductivity, Heisenberg on cosmic radiation, Wolfgang Paul on experiments with a betatron, and Richard Becker on a method for the quantization of wave fields.[65]

The Göttingen meetings of 1946 and 1947 made feasible contacts with foreign fellow professionals in the field that would otherwise have been hard to achieve. The travel restrictions imposed on Germans were thus compensated in this way. Genuine interest in the advances in physics attained outside the country went hand in hand with the endeavor to restore German culture, which German science was considered a special part of. It was along these lines that von Laue's activities lay. He saw the first meeting of German physicists after the war, for instance, as a sign of their will "not to let science go under, despite all our privations and problems."[66] Aiming in the same direction, but in somewhat clearer terms, the DPG in the British Zone issued an appeal for support of German science during its conference in Clausthal in 1948:

> We forbear from pointing out the intellectual assets that are lost not only in our own country but around the entire world if German science is sacrificed, which has created so much fundamental knowledge for humanity. We can hardly understand it if, with a rueful shrug of the shoulders, these cultural assets are set behind the immediate necessities of life. What is involved is not just cultural capital but the very existence and future of our people.[67]

The national point of view dominated in such evaluations of scientific developments because the physicist elite tended to have a particularly strong concern for German cultural values. This applied especially to von

64 Werner Döring, Über die Struktur einer intensiven Stoßwelle in zweiatomigen Gasen. According to an interview with Döring, 31 May 2004. See also the overview article: Werner Döring and Hubert Schardin, "Detonationen," in Albert Betz (ed.), *Hydro- and Aerodynamics, Fiat Review of German Science 1939–1946*, vol. 11 (Wiesbaden: Dieterich, 1948), 97–126.

65 All examples are from the Göttingen conferences of 1946 and 1947. *Neue Physikalische Blätter*, 2 (1946), 178f., and *Physikalische Blätter*, 3 (1947), 317–325.

66 Opening speech by von Laue at the meeting in Göttingen, 4 October 1946, MPGA, div. III, rep. 50, von Laue papers, no. 124.

67 Unanimous resolution during the physics meeting in Clausthal, 1948. *Physikalische Blätter*, 4 (1948), 273.

Laue, who maintained this perspective even during his travels abroad. It was during his first trip to the United States after the war that he realized the full scope of the wave of dismissals from 1933. The quite considerable troubles some émigrés had encountered in reestablishing themselves professionally moved von Laue to complain, viewing – through nationalistic-tinted glasses – the loss Germany had suffered by the expulsions: "How ever is one to close these gaps? The most important problem in Germany is, you know, creating a new culture-bearing class. The old one is finished."[68]

Policy toward the Past

To restore Germany's former high international level of physics as an important cultural asset, influential physicists sought to refill major university and research positions with the best scientists. The post-war disruptions in the staffing structure arising from the redrawn national borders and denazification meant that a large number of unemployed physicists were available. In some cases, this was a chance to rectify bad appointments made during the Nazi period. But denazification also led to a confrontation with the past, particularly concerning the political conduct of fellow members of the profession. Private correspondence and individual resolutions by the DPG permit a reconstruction of how this confrontation with the past proceeded.

A major part of the correspondence of the time dealt with personal career prospects, the latest on individual denazification suits, and the existing evidence on personal political involvement. In many cases, a request for an exonerating statement came in conjunction with this information. In summer 1946, Hermann Senftleben, formerly a senior commander of the *Sturmabteilung* (SA) and professor at the University of Münster, complained in a letter to Pohl about his dismissal: "What is to become of me I don't yet know but haven't yet completely lost all courage for now." He added that he was hoping to be able to visit his Göttingen colleagues sometime soon and speak with them.[69] At the end of the same year, Walther Bothe received a report from Helmuth Kulenkampff about the past turbulent months:

> [I]t was right at a time when the much vacillating issue of my appointment as professor once swung strongly toward the positive side. But another swing to the

[68] Von Laue to Hans Wilhelm von Ubisch, 2 January 1949, AIP, MP 126, von Laue.
[69] Senftleben to Pohl, 2 July 1946, DPGA, no. 40028.

other side already followed soon afterwards and the whole thing has been staying completely up in the air for a long while since. The details of this matter are so complicated that I could only tell you about it personally sometime; in some respects they are not uninteresting, at least. You have probably already heard in the meantime that about 2 months ago I landed up here [in Würzburg] in the end.[70]

In such entangled situations, personal encounters with the possibility of confidential conversation were urgently desired. In 1946, despite the mentioned restrictions, around 115 physicists were able to come to the Göttingen meeting and get a better understanding of their colleagues' situations.[71] More than 300 physicists gathered for the fall meeting of

[70] Kulenkampff to Bothe, 17 December 1946, MPGA, div. III, rep. 6, Bothe papers II, 3.

[71] According to the list of participants, the following physicists were present at the meeting in October 1946 (* indicates announced speakers): Arends (Clausthal), von Auwers (Clausthal), Baisch (Leverkusen), Bartels (Göttingen), Bartels* (Hannover), Becker (Göttingen), Betz (Göttingen), Braunsfurth (Hiddesen), Burkhardt* (Kiel), Büttner (Kiel), van Calker (Münster), Cario* (Braunschweig), Dannmeyer (Hamburg), Döring (Göttingen), Eggert (Braunschweig), Eucken (Göttingen), Flügge (Göttingen), Försterling (Cologne), Franz (Münster), Frerichs* (Charlottenburg), Fucks (Aachen), Gerlach (Bonn), Gora (Munich), Gruschke (Hamburg), Grützmacher (Göttingen), Gulbis (Hamburg), Hahn (Göttingen), Hamann (domain of Reinhausen), Hanle (Gießen), Hase (Hannover), Heisenberg* (Göttingen), Hinzpeter (Hannover), Hoffmann (Osnabrück), Jaeckel (Clausthal), Jensen* (Hannover), Justi* (Braunschweig), Kallmann* (Berlin-Dahlem), Kneser (Göttingen), Koehler (Klein Stöckheim), Kollath (Hamburg), Kopfermann* (Göttingen), Kratzer (Münster), Krautz (Braunschweig), Krebs* (Berlin-Charlottenburg), Kremer (Cologne), Kröncke (Clausthal), Kroepelin (Braunschweig), Kuß* (Sack near Alfeld), von Laue* (Göttingen), Lauterjung* (Cologne), Lochte-Holtgreven (Kiel), Maecker* (Kiel), Mannkopff* (Göttingen), Martens (Krefeld), Meissner (Munich), Meixner* (Aachen), Mittelstaedt (Leipzig), Mott* (Bristol), Müller (Hamburg-Groß Flottbek), Nagel (Völkenrode), Pietsch (Clausthal), Pohl (Göttingen), Prandtl (Göttingen), Ramsauer (Berlin), Riezler (Bonn), Rogowski (Aachen), Schaefer (Cologne), Schlechtweg (Essen), Schmieschek (Völkenrode), Schmidt (Völkenrode), Siksna (Hamburg), Steubing (Hildesheim), Stuart* (Hannover), Valentiner (Clausthal), Weizel (Bonn), von Weizsäcker* (Göttingen), Wirtz* (Göttingen), Zahn (Göttingen), Ziegenheim (Clausthal). Bowden came instead of Burcham from Cambridge. From von Laue's opening speech we gather that Bloch was also present. *Neue Physikalische Blätter*, 2 (1946), 178f., and MPGA, div. III, rep. 50, von Laue papers, no. 124.

The following physicists who gave talks or were scheduled to do so are missing from this list: Michels (Eindhoven), Houtermans (Göttingen), König (Göttingen), Meyer (Göttingen), Mollwo (Göttingen), Oetjen (Göttingen), Peetz (Göttingen). See DPGA, no. 40037.

A list signed by those interested in joining the DPG in the British Zone on 5 October 1946 reveals that the following physicists were also in attendance: Krone (Lüneburg), Maue (Wolfenbüttel), Severin (Göttingen), Salow (Bargteheide), Groth (Hamburg), W. F. Berg (Harrow, England), Bagge (Göttingen), Haxel (Göttingen), Jordan (Göttingen), Stenzel (Wolfenbüttel), Rosenhauer (Göttingen), Kehler (Berlin), Rühmkorf (Hellendorf), Schuler (Göttingen), Nagel (Göttingen), Paul (Göttingen), Polley (Göttingen),

1947.[72] Expressions of disappointment when insufficient opportunity arose for such tête-à-têtes indicate the importance attached to them: "I was extremely sorry," Hartmut Kallmann wrote to von Laue, "that I had so little opportunity in Göttingen to speak with you in peace. There would be so many – unfortunately not very pleasant – things to discuss."[73] In summer 1946, Walther Meißner wrote to Pohl from Munich: "Perhaps I could travel to Göttingen sometime soon so that we can talk again properly, which would be a special pleasure for me."[74] Even so, Meißner did not attend the first meeting, probably because of the considerable difficulties attached to traveling in 1946, allowing only a few physicists from outside the British zone to come.

Notably, all those who did come in 1946 were apparently welcome guests. Although the society's board decided on whether to accept new members and according to the newly ratified statutes did not have to give reasons for any rejections, the link with the past was nonetheless seamless: "Former members of the hitherto German Physical Society will be duly accepted as soon as they contact the managing director in writing."[75] This passage reflects the attitude that one could continue on as before, as if nothing had happened. It is surprising to the extent that some physicists had thoroughly discredited themselves by their conduct during National Socialism. Those who had belonged to the "Aryan Physics" group, for instance, were probably not welcome in the society after 1945, particularly considering the internationalism that the DPG was promoting. A closer look at these "Aryan physicists" quickly reveals why the cited passage was unproblematic in this respect.[76] They were outcasts of the physics community even before 1945 and already ostracized by the German physicist elite. The end of National Socialism led to the loss of all their remaining university positions.[77]

Faust (Göttingen), Schaffernicht (Clausthal), Stille (Braunschweig), Hiedemann (Karlsruhe), Gardien (Bremke), Förster (Göttingen), Knauer (Hamburg), Diebner (Hamburg), Walcher (Göttingen), Korsching (Göttingen). See DPGA, no. 40039.

[72] According to the accommodations list, 292 participants came from other parts. See DPGA, no. 40044.

[73] Kallmann to von Laue, 17 November 1947, MPGA, div. III, rep. 50, von Laue papers, no. 991.

[74] Meißner to Pohl, 28 August 1946, DPGA, no. 40028.

[75] Article 5 of the statutes, see *Physikalische Blätter*, 3 (1947), 29.

[76] These "Aryan physicists" are discussed in the contribution by Michael Eckert to this volume.

[77] This applies to August Becker, Alfons Bühl, Hugo Dingler, Ludwig Glaser, Philipp Lenard, Wilhelm Müller, Ferdinand Schmidt, Johannes Stark, Bruno Thüring, Rudolf

After 1945, Aryan physicists had no more influence on German university life, obviating any resolution by the DPG about it. This "group" had little contact with the society anyway. Philipp Lenard and his closer physicist friends had already left it in protest in 1926.[78] In 1933, Johannes Stark failed miserably in his bid for the presidency of the DPG.[79] In 1937, Stark's name disappeared from its membership rolls.[80] The DPG's official stance toward these two discredited physicists after 1945 was mainly to acknowledge their merit as scientists and emphasize their accomplishments in this capacity – they were, after all, Nobel laureates. As renowned scientists, they even continued to benefit from a certain sense of professional solidarity, as a letter by von Laue to Arnold Sommerfeld from summer 1947 shows:

> I hear from Walther Meißner that you are trying to obtain a milder punishment for Johannes Stark. Please tell me whether I can support you in doing so. I have been racking my brains for a few days already about the way one might go about it.
>
> I could, if required, make an attempt during the September meeting of the German Physical Society in the British Zone to move the board to take a step in Stark's favor. I believe the board could be persuaded, particularly since a short while ago it warmly advocated that Philipp Lenard receive an obituary in the "Naturwissenschaften."[81]

Sommerfeld's reply is instructive in more than one respect. He requested that Lenard's obituary indicate "what terrible harm he had done for us abroad ('Aryan' physics) and at home (Sommerfeld succession – Müller)." In the eyes of the physicist elite, one harmful consequence of Lenard's support of National Socialism had been that he had damaged the reputation of physics. In the end, no obituary of Lenard appeared in the *Naturwissenschaften.* But the *Physikalische Blätter* did choose to publish one authored by Brüche that – "bypassing the last few decades" – refrained

Tomaschek, Harald Volkmann, and Ludwig Wesch, who, according to Beyerchen, *Scientists*, all were followers of the "Aryan physics" movement.

78 From then on a sign, hanging on Lenard's door at the Heidelberg institute, warned: "No entry to members of the so-called German Physical Society!" See Armin Hermann, "Die Deutsche Physikalische Gesellschaft 1899–1945," in Mayer-Kuckuk, *Festschrift*, F90.

79 Mayer-Kuckuk, *Festschrift*, F94f.; see also the contribution by Richard H. Beyler to this volume.

80 Ernst Brüche, "Champion of freedom," *Physikalische Blätter*, 5 (1949), 448f., especially 449.

81 Von Laue to Sommerfeld, 16 July 1947, MPGA, div. III, rep. 50, von Laue papers, 1888.

from making any critical comments.[82] The *Blätter* rectified this omission later in a footnote to a report about the physics meeting of 1947 in Göttingen, when von Laue also found a few warm words for Lenard while remembering the dead in his opening speech. This footnote is quoted here in full: "We cannot and do not intend to hide or excuse the mistakes of Lenard, the pseudo-politician; but as a physicist, he ranked among the great."[83] This formulation exhibits the common disjunction of a person's capacity as a physicist from his or her actions in other areas. Instead of just acknowledging scientific accomplishments, Lenard was honored as a physicist. The contemporary perception among the elite was that anyone who had performed research of international fame was automatically a prominent personality. It is also indicative that although Lenard's faults were not supposed to be left unmentioned, they were nevertheless not mentioned. Instead of criticizing his conception of Aryan physics as a pseudo-scientific aberration, he was called a "pseudo-politician."

In Stark's case, the mixing of research accomplishments and personality led to another consequence. At Stark's tribunal, Sommerfeld pleaded mitigating circumstances in Stark's favor "due to the scientific importance of Stark."[84] Thus, the yardstick to be used in determining punishment should depend not just on illegal actions but on the person's scientific rank as well. The punishment was reduced even without the DPG intervening as von Laue envisioned.[85]

But there was another matter that the DPG had to contend with concerning its policy toward the past during the mentioned meeting at Göttingen. Erich Schumann's appearance at the meeting motivated Richard Becker to raise the following motion at the administrative meeting:

> Unsettling reports have recently been made that supposed physicists, who had participated in the war in the worst manner by damaging the professional livelihoods of their colleagues using the power of the party, are now seeking to reconnect and with their proven abilities are trying to regain influence on the current governments. I move that the society resolve to act against such attempts. We aren't

[82] Ernst Brüche, "Philipp Lenard †," *Physikalische Blätter*, 3 (1947), 161.

[83] Brüche, "Lenard," 317.

[84] Sommerfeld to von Laue, 24 July 1947, MPGA, div. III, rep. 50, von Laue papers, no. 2394.

[85] Andreas Kleinert, "Das Spruchkammerverfahren gegen Johannes Stark," *Sudhoffs Archiv*, 67 (1983), 13–24.

delighted with the denazification law but consider it our duty to assure cleanliness among our circle of colleagues.[86]

There are a few noteworthy things about the way this motion was formulated. One of the "worst" actions during the period of National Socialism, in Becker's view, was harming a colleague's professional career. Although it was supposedly clear to those present that the criticism was primarily aimed at the non-DPG-member Erich Schumann, Becker refrained from mentioning names.[87] His reference was to "supposed physicists." By this means, the criticized person was banished from his professional community and for that reason alone no longer deserved the support of his colleagues. Becker perceived as a threat the possibility that he might be successful anyway because – with his "proven ability" – he wanted to save his career with political backing. The strategy Becker was accusing him of dodged the self-governance of academia. So the DPG had to intervene.

In his argumentation, Becker also used the term *denazification*, as if it involved an act against National Socialists. This is misleading because Becker himself signed an exonerating "whitewash certificate" for at least one physicist known among his colleagues, including to Becker himself, to be a National Socialist, and Becker worked ardently to keep the physicist on his university's staff. This physicist, Becker averred, was exceptional for his upright and honest character, presumably quite the opposite of Schumann; besides, he had not sought any professional advantage from his party membership.[88] An Allied report from August 1944 reads: "Informant says that Schumann is perhaps not a Nazi according to the party line, but he is a German nationalist, a pan-Germanist and entirely devoted to the Germany Army."[89] So Becker's petition was not in order to keep former National Socialists away from university positions but, as he put it, "to assure cleanliness among our circle of colleagues."

86 Minutes of the administrative meeting of 6 September 1947, DPGA, no. 40043.

87 The fact that Schumann was the target of Becker's criticism is gathered from von Laue's letter to James Franck of 23 September 1947, RLUC. I thank Klaus Hentschel for pointing out this letter. Board member Fucks referred to the resolution in a letter to von Laue of 27 September 1947 as a "*lex* Schumann put forward by Mr. Becker and accepted by the assembly, which constitutes a precedence case for the exercise of political influence by the society against the person of a nonmember." MPGA, div. III, rep. 50, von Laue papers, no. 2394.

88 Becker's efforts on behalf of his colleague Karl-Heinz Hellwege, who had been dismissed from the Göttingen faculty, are discussed in Rammer, *Nazifizierung*.

89 Office of Strategic Services, Research and Analysis Branch, Biographical Report, 28 August 1944, National Archives, Maryland, RG 77, entry 22, box 167.

The criticism was directed against uncollegial conduct, which counted as the "worst" of actions. This point was most warmly supported at the administrative session. Walther Gerlach agreed: "We must regard it as our duty to secure purity within our ranks, just like Mr. von Laue said."[90]

But because Becker's endeavor was drawn in relation to denazification, a few critical voices were raised as well. They warned against political measures by the DPG that went too far. The private lecturer from Bonn, Rudolf Jaeckel, phrased it this way: "It is to the society's great merit that it kept itself free from all National Socialist influences; however, there is, on the other hand, also the danger of doing too much. I propose that in view of his attitude in the past Mr. von Laue be engaged as a trustee on this question."[91] What would it have meant to assign trusteeship to von Laue? He would have been able to take steps on his own that in his view were necessary, but his actions would have been beholden to the interests of the physics profession. This attempt to bind him to majority opinion on denazification questions reveals a hint of distrust. It is understandable, considering as touchy a topic as this was. Several of the 142 physicists present were themselves politically implicated.

Instead of accepting the role of custodian, von Laue preferred to share the responsibility with the board. The following procedure was agreed:

> The administrative assembly commissions the board, jointly with the Physical Society of Württemberg and Baden, to prevent the possibility that such physicists attain official posts whose conduct during the time of the Nazi regime is known to have been in gross contradiction in scientific or human respects to the standard views of propriety and morals held among scientists.[92]

This resolution avoids the term *denazification* and instead appear the words "propriety and morals," as generally understood by scientists. This formulation begs the question: What was so specific about the moral understanding of scientists? Or put another way: When does a scientist's conduct count as decent? A debate surrounding this issue inspired a newspaper article shortly afterward criticizing the practical policy of physicists toward the past.

90 Minutes of the administrative meeting of 6 September 1947, DPGA, no. 40043.

91 Minutes of the administrative meeting of 6 September 1947, DPGA, no. 40043.

92 Minutes of the administrative meeting of 6 September 1947, DPGA, no. 40043. See also the publication of this resolution in *Physikalische Blätter*, 3 (1947), 281.

PUBLIC CRITICISM

In the November 1947 issue of the *Deutsche Rundschau*, an article appeared by a doctoral student of physics, Ursula Maria Martius. In it, she recorded her negative impressions of the Göttingen physics meeting.

In September of this year I took part in the *annual meeting of the Physical Society in Göttingen*, which was both stimulating and upsetting for me. In the first place, though, it was upsetting because of the constant encounters with the past. People who still appear to me in my nightmares were sitting there alive and unchanged in the front rows. Unchanged, if you don't consider the simple blue suit, instead of the uniform and the missing party badge, a "change." One doesn't appear at the colloquium anymore in an SA uniform (*Stuart*, in Königsberg, then Dresden, now Hannover); one doesn't get addressed as "General" anymore (*E. Schumann*, formerly in Berlin, now striving for an institute in Hamburg); and in the new editions of books, passages like "We are not willing to regard linking science to military power as an abuse, after military power has proven its compelling constructive energy in the creation of a new Europe" (*P. Jordan*, formerly Berlin, today professor in Hamburg; "Physics and the Secret of Organic Life," Braunschweig 1941) and other similar things will be deleted. It will no longer be pleasant to recall having been on the Aryanization panel at the Viennese university and polytechnic (*Schober*, now in Hamburg), or that at one's own institute a doctoral thesis had been written for Miss *Keitel* in 1936 (*E. Schumann*) or that one had denied trainee certificates to those who had not been admitted to the examination for political reasons (*H. Kneser*, formerly in Berlin, now Göttingen).

Otherwise, practical sessions and lectures continue to be held, "objective science" continues to be pursued. And that's where the tolerance of those around one should end![93]

The style of this critique differs significantly from Becker's. Martius specifically mentions five physicists by name and points out their offenses. These were not Becker's "supposed physicists" but ones who could take their seats in the first row again. Established colleagues were the target, not the

[93] Ursula Maria Martius, "Videant consules...," *Deutsche Rundschau*, 70/11 (1947), 99–102; reprinted in the German version of this book: Dieter Hoffmann and Mark Walker (eds.), *Physiker zwischen Autonomie und Anpassung – Die DPG im Dritten Reich* (Weinheim: Wiley-VCH, 2007), 636–640. The title alludes to a Latin saying from ancient Rome. It signifies the procedure that the Roman senate used in times of emergency in exhorting each of the two consuls to ensure that the commonwealth, the state, not suffer harm. Excerpts of this article also appeared on 6 December 1947 in *Berlin am Mittag* under the heading: "The new seeds of discord, professorial reactionaries at western universities" (Die neue Drachensaat. Professorale Reaktion in den westlichen Universitäten). See Pascual Jordan, *Physik und das Geheimnis des organischen Lebens* (Braunschweig: Vieweg, 1941).

odd National Socialist. Letting young students be educated by scientists with such political records Martius perceived as dangerous.

The resolution passed by the administrative assembly is also mentioned, but Martius regretted that at the meeting itself, no form of distancing had been taken. At the end of her article, she again urged those in responsible positions to take action.

> The German intelligentsia has resolutely rejected the idea of collective guilt for the misdeeds of National Socialism. One of the strongest arguments was that at the time nothing could be done. Today, however, something can be done and some of the existing conditions occasionally raise the suspicion that there is a lack of will. If something isn't done very quickly, then the day will come when this suspicion will become a certainty and the accusation that large numbers of Germans made the cause of National Socialism their own, despite knowledge of the facts, will no longer draw its support from the years after 1933 but from the period after 1945.[94]

Martius left open how exactly to understand what the "cause of National Socialism" was. The validity of this serious accusation will be explored at the end of this chapter. Before discussing and analyzing the DPG's reaction to Martius's article, let us first examine the circumstances under which this critical voice was raised.

Personal Motivations and Professional Preconditions for Critique

First of all, it must be pointed out that this article was absolutely exceptional. Not a single comparable case is known to me of a physicist making such a public appeal. Of course, the physicists did know about past offenses committed by some of their colleagues, who were still being held in esteem. Keeping silent, however, was due to the effect of a collegial code of conduct. Tact took precedence over political or personal differences. It was clearly taboo to criticize a colleague's misdeeds publicly. Political criticism was supposed to be kept out of the academic world. "We should all defend ourselves against the invasion of usances from our political life into academic life," Hans Bartels wrote to Heisenberg, for example.[95] Martius was not affected by this taboo insofar as she, being

[94] During the post-war period, National Socialism did in fact still have considerable attraction for Germans. A poll in the American zone in 1946 yielded 40% who thought National Socialism had been a good thing poorly carried out. In 1948, 55.5% were of this opinion. See Michael Balfour and John Mair, *Vier-Mächte-Kontrolle in Deutschland 1945–1946* (Düsseldorf: Droste, 1959), 58.

[95] Bartels to Heisenberg, 8 November 1950, MPIfP, Heisenberg papers.

a student of physics, was not yet a genuine colleague. She did not yet fall under the full force of this code of conduct. As will be shown later, other biographical circumstances motivated her, in addition, to lodge this complaint. The responses she got from her article were guided by an intention to teach her a lesson about more appropriate behavior; that is, as scientists conceived propriety and morals.

Another prerequisite for writing this article was some insight into the physics community. The information that Martius presented was only accessible to those on the inside. Within the field, though, the published knowledge about the behavior of the criticized physicists was already largely known. The experimental physicist Martius aptly described the situation in retrospect:

> This was not a problem or an open question: After all, there had been years during which these men reigned, with photos and documents. *One knew from experience* and, in fact, it were only the big fish that were under discussion after the Göttingen meeting. Their background was known experimentally i.e. from lived evidence. This experiential evidence was as clear as the evidence of the opposite, for instance of von Laue's impeccable integrity.[96]

In her article, Martius only mentioned the "big fish," primarily those she knew from her milieu in Berlin.[97] Despite her infraction against the code of conduct, she managed to obtain her doctorate in Berlin a year after the appearance of her article.[98] This success was owed to favorable local academic circumstances at the time. Her advisor was the formerly persecuted "non-Aryan" theoretical physicist Hartmut Kallmann, who had been appointed by the Americans as professor at the Technical University of Berlin in 1946.[99] He shared the reservations expressed in Martius's

96 Franklin to Rammer, 17–27 November 2002 (original English and emphasis).

97 Jordan became von Laue's successor at the University of Berlin in 1944. Kneser was extraordinary professor at the University of Berlin from 1940 to 1945. Until 1945, Schumann followed his entire academic career in Berlin; beginning as private lecturer in 1929, he rose to full professor and institute director in 1933. Stuart served as a temporary replacement on the chair for theoretical physics at the University of Berlin prior to his appointment in 1939 as full professor at Dresden.

98 Her thesis at the Technical University of Berlin treated the excitation of luminous substances by gamma rays and X-rays of various wavelengths: Ursula Maria Martius, "*Die Anregung von Leuchtstoffen mit Gammastrahlen und Röntgenstrahlen verschiedener Wellenlänge*" (Berlin: Technische Universität Berlin Ph.D., 1948).

99 Hartmut Kallmann had been group leader at the KWI of Physical Chemistry and Electrochemistry from 1920 to 1933 as well as private lecturer at the University of Berlin since 1927. Afterward, he was employed as a staff scientist at I.G. Farben until 1945. He then occupied the chair for theoretical physics at the Technical University of Berlin and was director of the institute until 1948, when he spent a year as a research fellow

article.[100] Their divergent view on the policies of National Socialism arose in part from the fact that the racist measures affected them as "non-Aryans" very differently from their "Aryan" colleagues. Martius, for instance, had been expelled from the university in 1942. Her parents were interned in a concentration camp.[101] Exclusion from the "people's community" (*Volksgemeinschaft*) changed their perception of the conditions. But the period after the liberation in 1945 "was no party" either. Conditions were as hard as before for the victims of discrimination, as a letter by Martius to Otto Hahn from 1946 demonstrates:

> In the summer my mother became very seriously ill from gastric hemorrhages, she was in bed for a long while and we could do nothing to help her recuperate. As generally, care for the racially persecuted exists here only in the papers, insofar as nonmembers of the Jewish religious community are involved. My mother weighed less than 70 pounds – there has been no sign of official aid until now, besides a few questionnaires.[102]

The developments in post-war Berlin only strengthened her distrust in the democratic virtues of Germans. "No matter what these people reconstruct, it always turns into barracks – barracks I'm not all that keen on living in."[103] That was why, very early on, at least since the beginning of 1946, she became intensely preoccupied with the thought of leaving the country.[104] Initial plans to go to France fell through, but in 1949 she succeeded in emigrating to Canada where, not without some difficulty, she eventually obtained a full professorship and became one of Canada's

at the U.S. Army Signal Corps Laboratories in Belmar, N.J. From 1949 to 1968, he was professor at the Physics Department and director of the Radiation and Solid State Laboratories at New York University.

100 Franklin to Rammer, 17–27 November 2002. On 23 April 1946, Ursula Martius wrote to Otto Hahn: "I get on well with Dr. Kallmann, as we have much in common even beyond the professional.... But I can't quite share his optimism and I see things as more difficult than he does by a factor of at least two powers of ten." On 5 August 1946 she wrote: "Even Kallmann's optimism has subsided noticeably in the last few months." MPGA, div. III, rep. 14A, Hahn papers, no. 2726.

101 Franklin to Rammer, 17–27 November 2002.

102 Martius to Hahn, 23 April 1946, MPGA, div. III, rep. 14A, Hahn papers, no. 2726. On 16 December 1946, she wrote: "Absolutely nothing is being done here for the 'victims of fascism,' so the feeling of emotional bitterness is added to the material worries and the great uncertainty defining the general situation. I am still trying all sorts of things to be able to leave Germany."

103 Martius to Hahn, 23 April 1946, MPGA, div. III, rep. 14A, Hahn papers, no. 2726.

104 The gender aspect probably influenced her decision. Being female, the career prospects in the realm of physics in Germany were humble at her time. However, such considerations are not made clear in the sources.

most highly esteemed scientists. Hartmut Kallmann also left Germany in 1948, and his career continued at New York University.

The DPG's Reaction

Negative experiences coaxing Martius to leave her native country included reactions by physicists to her article. Its specific finger-pointing at Pascual Jordan, Hans Otto Kneser, Herbert Schober, Erich Schumann, and Herbert Stuart made von Laue feel compelled to respond as chairman of the DPG in the British Zone. Not wanting to be an autonomous custodian, he had pronounced the entire governing board responsible for such questions and now asked around among its members whether a response by the DPG ought to appear and what such a response might look like. At least six written replies representing various standpoints were received.[105] The one point they all share in common is a lack of full agreement with Martius's critique. It was rather judged an unpleasant and inappropriate attack on universities. Wilhelm Fucks and Josef Meixner came to the conclusion that it would be best to refrain from stating any position at all on it.[106] The other four cautiously favored a riposte. One reason that the tenured professor at the Mining Academy of Clausthal, Otto Auwers, offered was that the article was based on a genuine worry and that Martius's effort to prevent new misfortune appeared to be "worth kindly recognition." But the purpose of a riposte ought to be "neutralization of the article's public impact." However, the core problem of the article lay

> so far beyond the scope of our society that it is difficult to become the judge or advocate of all. If, however, one were to limit oneself to those specifically named cases of relevance to us, then one stoops to a level that I would strictly reject for the society. But dismissing concrete examples with general platitudes would not be serving the purpose well either. So no matter what side one wants to look at the thing: a public answer, which would have to be in basic agreement, yet a rejection of concrete individual cases, is hard to formulate in a way that serves both the cause and ourselves.[107]

105 Written replies exist from the deputy Schaefer as well as from Auwers, Meixner, Unsöld, Schlechtweg, and Fucks. The Göttingen board members Mannkopff and Heisenberg presumably responded orally. There is no evidence of any response by Jensen, Justi, Gerlach, and Kratzer. Von Laue's circular letter of 26 November 1947 is unfortunately also missing from the files. See DPGA, no. 40048.

106 Fucks to von Laue, 12 December 1947; Meixner to von Laue, 14 December 1947, DPGA, no. 40048.

107 Auwers to von Laue, 9 December 1947, DPGA, no. 40048.

The issue addressed here is the delicate task of finding the proper wording, even though the statement sought actually seemed quite straightforward: a rejection of the concrete individual cases. It is a linguistic quest concerning the treatment of implicated fellows of the profession, hence skirting the problem of dealing with them per se, remaining at the meta-level of talking about it. This problem of formulation will be discussed more fully at the end of the current section.

The option of staving off the criticism without going into the individual cases was the suggestion of Heinz Schlechtweg from Aachen. The DPG should refer to its apolitical stance in pointing out in its statement "that such aspersions are settled of themselves in that the society has never yet represented a political tendency and that neither now nor before did any partisan connection exist."[108] By deferring to the DPG's apolitical tradition, Schlechtweg drew the questionable conclusion that it was not the society's responsibility to distance itself from politically strongly implicated members. His conception may well have had self-preservation as one of its underlying motivations. While employed as a guest researcher at Ludwig Prandtl's institution in Göttingen in 1933, Schlechtweg launched an attack on his purported "Jewish" colleagues at the institute without first discussing his differing physical interpretation of their science with them.

The justification he made to Prandtl for his actions was the following: I "as a German always face any German openly and honestly, just as toward you, highly esteemed Professor; truth is the one and only thing that ever matters to me and the honest wish to serve the fatherland to the best of my abilities."[109] Considering that in the Nazi idiom, "Jews" did not count as "Germans," it may be presumed that Schlechtweg did not face them with quite the same openness and honesty. The price he had to pay for this incident, apparently motivated by anti-Semitism, came as early as 1936 when Prandtl successfully prevented his habilitation, the qualification to teach at the University of Göttingen.[110] This prehistory explains why during the period of occupation, Schlechtweg would argue

[108] Schlechtweg to von Laue, 30 November 1947, DPGA, no. 40048. In another letter, he also underscored the view that political discussions should be avoided as much as possible from taking place within the DPG: Schlechtweg to von Laue, 20 September 1947, MPGA, div. III, rep. 50, von Laue papers, no. 2394.

[109] Schlechtweg to Prandtl, 8 April 1934, MPGA, div. III, rep. 61, Prandtl papers, no. 1454.

[110] See Rammer, *Nazifizierung*.

for an "apolitical" attitude and against any distancing from those with politically problematic records.

Two of the board members wanted to differentiate somewhat and brought up the individuals named in their arguments. Strangely enough, no one mentioned the case of Herbert Schober, an extraordinary professor at the Viennese Polytechnic until 1945.[111] The decision on Erich Schumann was most easily reached. The Kiel *Ordinarius* Albrecht Unsöld argued that: "The Physical Society never had anything to do with him and does not need to attribute any value to him." Thereby, he basically merely confirmed the administrative assembly's resolution that had been implicitly directed against Schumann. Unsöld's formulation reveals, though, that he was not interested in what Schumann had actually perpetrated, only in what he was worth to the profession. He provides a typical example of the way criticism was dealt with. Unsöld screened out Schumann's actions, placing only his scientific importance on the scales, just as he did with Stuart and Jordan, albeit in the latter cases the scales tilted on the side of exoneration:

> Both are without a doubt first-class physicists, whose scientific work we all value highly, but both have occasionally let themselves be led in years gone by to say silly things, for which they are now naturally being held to account again. If these things are addressed, it seems to me that the most reasonable stance would be to acknowledge the fact, all right, but to try to also make clear to the people that trivialities are basically what are involved that are insignificant against the scientific importance of the gentlemen concerned.[112]

The typical element of this argument is the dilution of political complicity by scientific merit. In other words, a milder political yardstick should be applied to prominent scientists than that for normal people. This was the precise opposite of Martius's argument that university teachers were "just as much of an agent as any member of parliament" and therefore bore

[111] Herbert (August Walter) Schober was private lecturer at the Viennese polytechnic (TH) from 1933 to 1940, as well as at the University of Vienna since 1937. From 1940 to 1945, he was extraordinary professor at the polytechnic. After a break caused by denazification he returned to academia, teaching at the University of Hamburg from 1953 to 1956 as untenured extraordinary professor and directing the laboratory for radiation physics at the tuberculosis research institute Borstel. From 1957, he was full professor of medical optics at the University of Munich. Schober was a productive physicist, publishing jointly with Edith Evers, E. Fleischer, Hans Jensen, Heribert Jung, Constantin Klett, M. Konasch, D. Lübbers, O. Marchesani, Manfred Monjé, Rudolf Ritschl, Marianne Roggenhausen, U. Schley, Hugo Watzlawek, H. Wenzig, Karl Wittmann, and A. Flesch. He evidently had few contacts among the German physicist elite.

[112] Unsöld to von Laue, 2 December 1947, DPGA, no. 40048.

greater responsibility; so their ideological attitudes ought to be examined that much more closely.[113] Unsöld acted as if those involved had only been saying "silly things" for which they were "now naturally being held to account." This minimizing description does not pinpoint which "facts" Unsöld wanted to have acknowledged and which acts he judged were "trivialities." What diffuse knowledge existed among physicists about the relevant "facts" is now no longer reconstructable. As the foregoing examples demonstrate, the discussion about the behavior of fellow physicists remained superficial. A fence was erected around colleagues deemed worth shielding on the basis of this parsimonious exchange of information. So let us leave aside for now the situation – perhaps those "facts" Unsöld mentioned – reconstructable from sources available today and describe how the DPG came to its decision without further information about those under scrutiny.

The deputy chairman, Clemens Schaefer, used another common pattern of argumentation. This last example taken from the responses by the board members makes an appeal for the human qualities of the physicists concerned; it indicates the concrete importance of the conception scientists commonly held of propriety and morals. Schaefer thought it necessary to give his opinion about the individuals named:

> I would not hesitate to surrender Schumann and Stuart; and the standpoint by the article's author that people like Jordan should be confined to scientific work but kept away from lectureships is also understandable. On the contrary, though, I consider Kneser, whom I know from his youth, an absolutely unobjectionable person.[114]

This defense of Hans Kneser is reminiscent of the emphasis on personal decency often found in whitewash certificates. The concepts there range from "untainted character" to "impeccable personality." The unusual aspect of this opinion was that Schaefer was willing to give up all the named persons, excepting Kneser. As there was evidently no generally accepted form for discussing the political misbehavior of colleagues, the reasons Schaefer had for side-lining three of the physicists remained unsaid. He did not speak strongly in favor of a public answer by the DPG and expressed doubts about the difficult task of drawing distinctions in it.

113 Martius, "Videant," 100f.

114 Schaefer to von Laue, 8 December 1947, DPGA, no. 40048.

This relatively candid statement by a good friend of von Laue's probably particularly affected his deliberations.[115] The opinion that Schumann, Stuart, and Jordan should not get professorships was not able to gain a majority vote, though. Von Laue drafted an official statement that attempted to unite the points of view quoted above:

Like Miss Martius, the society also takes the stance that other means of intervention are necessary in all cases where the denazification process has obviously failed its purpose. Sufficient proof of this is provided by the resolution Miss Martius mentioned, that the administrative assembly commission the board to intervene in such cases. I can add that implementation of this resolution followed in two cases, although it is not academic custom to broadcast such unpleasant matters abroad.

In three cases to which Miss Martius alluded, however, the board regrets that she mentioned names, because they involve people whose political pasts are not incriminating enough to prevent a German university from continuing to avail itself of their proven abilities.[116]

Von Laue presented intervention by the DPG as a corrective for overly mild decisions in denazification cases – in this instance with reference to Schober and Schumann. This was neither adept nor accurate. The board did not attempt to remove National Socialists from universities, because the question whether someone was an avid Nazi was never discussed. The only argument justifying the procedure against Schumann was that he had hurt other colleagues. This argument has nothing to do with his political position. Collegial ties among physicists prevented any grappling with their political attitudes. Any discussion about the political misconduct of others would have been interpreted as an uncollegial act. It would have been perceived at that time as equivalent to a denunciation and would have been rhetorically castigated as reverting back to "Nazi methods." It was not just unacceptable but a generally held taboo. It was a component of the collegial code of conduct applicable to all, even though it was nowhere explicitly spelled out.[117]

[115] Von Laue alluded to his friendship with Schaefer in a letter congratulating Schaefer on his 70th birthday. It mentions how close they had become in Zehlendorf, sharing confidences in hard times. Von Laue to Schaefer, 15 March 1948, MPGA, div. III, rep. 50, von Laue papers, no. 1717.

[116] Von Laue to *Deutsche Rundschau*, 17 December 1947, DPGA, no. 40048.

[117] A written code of conduct with explicit reference to the principle of collegiality is found today, for example in the professional standards of practice for medical doctors, psychologists, therapists, pharmacists, and veterinarians. All these rules include a section on *collegial conduct*, which defines in greater or lesser detail what behavior is desirable and what is not acceptable. Very generally, they call for mutual respect among fellow

Collegiality meant, for instance, that physicists' actions should neither damage the reputation of the "guild" as a whole nor the livelihood of other professionals. During the period of denazification, it was therefore doubly uncollegial to discuss the political pasts of fellows in public. The professional code of conduct created a collegial obligation even between politically estranged scientists. One could say that it acted as a binding force even between "Nazis" and "non-Nazis," but this dichotomy is misleading. A few persons were relatively clearly categorizable under one or the other group. But it was completely impossible to draw a sharp line between these groups. A wide gray area in between was responsible for the surprising stability of the collegial community beyond 1945.

There was one significant exception to the prohibition on discussing political misconduct. It concerned the special case where political involvement led to uncollegial behavior, particularly denunciation. Such a political act constituted a breach of the code of conduct – which was in effect also during the Nazi period – and could consequently lead to sanctions. Clemens Schaefer somewhat candidly pointed out at a board meeting of the DPG in the British Zone in 1948 that it was unpleasant for him personally that Wilhelm Westphal's name appeared on the title page of the *Physikalische Blätter* as one of its advisors.[118] "Mr. Westphal denounced me personally during the first month of the war in 1939. I can recount the matter."

The concrete circumstance that motivated Westphal to report on Schaefer at the Ministry of Culture, as Schaefer continued to explain, is less interesting than Schaefer's statement that he could relate the incident. To be more precise, the notable thing is the fact that the statement "I can recount the matter" was included in the minutes. It means that the statement was not insignificant. How are we to understand it?

professionals. Specifically, for example, no public criticism about a colleague's methods or professional knowledge and expertise should be made, particularly in front of patients.

Matters are different for expert opinions. What is sought are statements about the acts of a colleague to the best of one's knowledge and conviction as an expert. Criticism may also be put forward in a factual manner at professional forums. Any professional infraction should first be privately pointed out to the colleague. The aim of these rules is to preserve the public prestige of the profession and to prevent any acts that could harm this prestige. General standards of collegiality safeguard individual practitioners and the reputation of the profession as a whole.

[118] In 1948, the names of advisors of the *Physikalische Blätter* began to appear on the front page: H. Ebert, J. Eggert, E. Fues, H. Görtler, P. Jordan, K. Philipp, W. Lietzmann, C. Ramsauer, H. Schimank, and W. Westphal.

Above all, what meaning should be attached to the word "can"? Certainly not "capable," for example, in that Schaefer could still remember the events of 1939 and was therefore in a position to tell about them. The very fact that this statement was recorded in the minutes indicates that the intended meaning was "justified" or "permitted."

Besides recording Westphal's denunciation, the minutes also let its readers know that Schaefer was making Westphal's misconduct known in a permissible way. It could be mentioned because it had been a gross breach of the collegial code of conduct. Even in this instance, other political activities by Westphal stayed unsaid. They remained unmentionable outside of confidential conversations. Because Schaefer had gotten off lightly at the time, the sanction he demanded was modest: "I have no intention of doing anything nasty against Westphal, it's just that the name on the title page bothers me."[119]

Westphal was experiencing difficulties in his denazification case – Kallmann and Friedrich Möglich were opposing his employment at the Technical University in Berlin[120] – and approached von Laue with the request that he issue him a certificate of good character. The statement has unfortunately not been preserved among von Laue's papers, but Westphal's reaction to it, "Your letter to me I naturally cannot very well present," shows that his past misdeeds had not been forgiven. It is, in fact, the sole refusal or useless whitewash certificate I could find among von Laue's papers.[121]

119 Minutes of the board meeting of 8 September 1948 in Clausthal-Zellerfeld, DPGA, no. 40051. Westphal continued to be listed as advisor on the title page of the *Physikalische Blätter*, nonetheless.

120 Westphal to von Laue, 19 March 1948, MPGA, div. III, rep. 50, von Laue papers, no. 2125. On Möglich, see Dieter Hoffmann and Mark Walker, "Der Physiker Friedrich Möglich (1902–1957) – Ein Antifaschist?" in Dieter Hoffmann and Kristie Macrakis (eds.), *Naturwissenschaft und Technik in der DDR* (Berlin: Akademie-Verlag, 1997), 361–382.

121 See the correspondence von Laue–Westphal. The quote originates from: Westphal to von Laue, 16 May 1948, MPGA, div. III, rep. 50, von Laue papers, no. 2125. Another passage of that letter reads: "Your letter has naturally shaken me somewhat. My conscience tells me that I have truly not deserved being repeatedly subjected these last few years to misunderstandings, misappreciations and worse. That I must now experience this from you, after 44 years of acquaintance, is particularly painful to me.... How precarious the situation I was in during the Hitler period was, because of my wife, nobody knows about: that I could only just barely escape being assigned to hard labor, having to pick up our food coupons with the Jews for many months long – truly not a sign of goodwill by the NSDAP [Nazi party]!"

The bad feelings of 1948 soon gave way again to their former friendly terms. Von Laue again "counted among his friends" and he was able to write after his death: "Laue

Von Laue's statement about Martius's article was part of an endeavor to stabilize the physics community. Jordan, Kneser, and Stuart got some shielding without being specifically named. The link von Laue's text drew with the purpose of denazification threatened this stability, however. It was possibly because of this that the statement was never published. This reason was nowhere put in writing in any of the sources I have examined; nonetheless, I will argue its plausibility here from the context.

In my opinion, the leading figures of the DPG regarded Martius as a troublemaker because, on one hand, she had access to relevant information and, on the other hand, did not feel bound by the code of conduct. Clemens Schaefer had already suggested to von Laue whether "it might not perhaps be more effective to contact the author in private?"[122] A private avenue already existed with Otto Hahn. When he informed her that the DPG would reply to her article, she welcomed it enthusiastically. But rather than complying with the silence physicists generally maintained about the political dealings of their colleagues of "proven abilities," she indicated to Hahn her intention to continue the debate and to add more precision to her points of criticism. "So it will be possible to nail down the society's attitude in principle on past and future cases. Moreover, it would also offer me the opportunity to define the scope and consequences of the problem more precisely."[123]

This reference to potential future cases seems to have made Hahn and von Laue nervous about new unpleasant problems involving other physicists vulnerable to criticism. It would have been a difficult debate to conduct, owing to the strictures of their code of conduct. Hahn had seen Martius's article before it was published, had discussed it with von Laue, and in the end had advised her not to let it appear. The only valid reason Martius could see for withdrawing the article would have been that it had become superfluous, because the consequences she deemed necessary had already been drawn.

> I had been hoping for that since Göttingen and that is why I waited so long after the meeting with writing. I would have preferred best to bring up a couple of things then and there, but I always told myself: "Ursula, you are 26 years old,

was a person of incorruptible love of justice and absolute purity of feeling, thought and deed. He applied a very strict scale to a person's character – for all his understanding of little human failings." Wilhelm Westphal, "Der Mensch Max v. Laue," *Physikalische Blätter*, 16 (1960), 549–551, quote on 549.

122 Schaefer to von Laue, 8 December 1947, DPGA, no. 40048.

123 Martius to Hahn, 27 November 1947, MPGA, div. III, rep. 14A, Hahn papers, no. 2726.

others are present here whose responsibility and prerogative it is to settle such things." Only when nobody said anything and then, in the months that followed, nobody wrote anything, did I start.[124]

Hahn answered Martius that he had spoken with Kneser and asked him about her accusations. As Kneser was not able to recall the case, Hahn supposed that "perhaps a general prohibition existed from issuing trainee certificates to anyone who was no longer permitted to continue studying." If Kneser "believed he had to do his duty . . . ," one should not "resent him for it anymore today." Besides, Kneser had formerly collaborated with Hahn's friend Eduard Grüneisen in Marburg. "Grüneisen was anything but a National Socialist. But he had, as far as I can remember, a very high opinion of Dr. Kneser, so this already argues strongly in favor of a decent mentality."[125] This somewhat construed-sounding, but surely earnestly meant attempt at saving Kneser's reputation Martius set against a detailed account of what had actually taken place, relying on a good reconstruction from entries in her diary. Kneser had denied her a trainee certificate even though, based on information provided by the rector and the Ministry of Culture, being of "mixed race" she still had a right to receive the certificate.

Von Laue read Martius's letter to Hahn, and it was presumably the reason why he decided not to publish his reply in the end. He certainly did not want to offer Martius the opportunity to give more particulars about her criticism and thus jeopardize Kneser's career. At that time, Kneser's denazification proceedings had not yet been decided; many of his colleagues had put in a good word for him. At the fall meeting of 1948, Schaefer urged at the board meeting: "We must intervene on Kneser's behalf in any case. We cannot keep Schumann."[126] Hans Bartels wanted to place Kneser on a list of candidates for a position at the Hannover Polytechnic and thought that because of his "personal prestige," von Laue would be in a position to clear away the "impediment" of the

[124] Martius to Hahn, 27 November 1947, MPGA, div. III, rep. 14A, Hahn papers, no. 2726.

[125] Hahn to Martius, 12 November 1947, MPGA, div. III, rep. 14A, Hahn papers, no. 2726. An obituary on Grüneisen by E. Huster (*Physikalische Blätter*, 5 [1949], 378f.) states: "It is thanks to Grüneisen, for whom probity in every regard was a necessity of life, that political influences were kept away from the research at the institute even during the past epoch, and 'racially undesirables' were working in harmony with members of Nazi organizations there until shortly before the war."

[126] Minutes of the board meeting of 8 September 1948 in Clausthal-Zellerfeld, DPGA, no. 40051.

pending denazification.[127] Von Laue replied that at the end of 1947, he had already gone to see the British authorities with Hahn, Heisenberg, and Adolf Windaus and ask for milder treatment of former party members "who through their scientific merit are valuable for universities. The case of Professor Kneser was among the individual cases mentioned then."[128]

Von Laue also had his own reasons for supporting Kneser. He was energetically trying to preserve the Imperial Physical-Technical Institute (*Physikalisch-Technische Reichsanstalt*, PTR), to which he planned to recruit Kneser. In early 1948, he certified that Kneser had never spread any political propaganda. He was additionally "a man of very distinguished mentality, who is always extremely tactfully reserved in all matters."[129] If Martius had had an opportunity to publish her experience with Kneser, then the extremely tactful reserve von Laue was attesting to would not have been so credible anymore. Thus, however, Kneser was able eventually to return to the university landscape. First he directed a laboratory at the federalized bureau of standards, the Federal Physical-Technical Institute (*Physikalisch Technische Bundesanstalt*). He was also allowed to give lectures again at the Hannover Polytechnic, serving at the same time as a substitute at the University of Göttingen. In 1950, he was appointed to an extraordinary professorship at the University of Tübingen and 2 years later received a tenured position at the Technical University of Stuttgart.[130]

Since there was not going to be any official response by the DPG, von Laue wrote a private letter to Martius.[131] He first went into the Schumann case. As soon as he started discussing the other cases, though, the style changed abruptly. Without mentioning any more names, von Laue reverted to preaching. The guiding principle he placed at the head of this

[127] Bartels to von Laue, 4 February 1948, MPGA, div. III, rep. 50, von Laue papers, no. 210.

[128] Von Laue to Bartels, 10 February 1948, MPGA, div. III, rep. 50, von Laue papers, no. 210. See also his certification for Kneser from 10 February 1948, in which the request to the military government is mentioned, MPGA, div. III, rep. 50, von Laue papers, no. 1065.

[129] Certification, MPGA, div. III, rep. 50, von Laue papers, no. 1065.

[130] On Kneser's career, see: H. Oberst, "Hans Otto Kneser 60 Jahre," *Physikalische Blätter*, 17 (1961), 328f.; Günther Laukien, "Hans Otto Kneser 80 Jahre," *Physikalische Blätter*, 37 (1981), 274f.; W. Eisenmenger, "Hans Kneser zum Gedenken," *Physikalische Blätter*, 41 (1985), 320.

[131] Von Laue to Martius, 26 December 1947, MPGA, div. III, rep. 50, von Laue papers, no. 2395. He had written the official response 1 week before, on 17 December. It is possible that he saw Martius's letter to Hahn only after that date. At that time, letters from Berlin to Göttingen were en route for roughly 2 weeks.

section was the following: "In any organized state the principles affecting the judiciary must be strictly adhered to, first and foremost." The judiciary was an unassailable authority for von Laue. After a verdict had been reached in accordance with the law, a case should not be publicly rolled out again. Criticism was only permissible if it concerned the underlying legal process, therefore the denazification procedure, not the individual cases. "*The defendants have a right to be left in peace following official settlement of their cases. And articles like the one you have written infringe on this right.* This consideration is, incidentally, also the reason why we do not ring the bells about our moves in the Schumann case but deal with it quietly." This principle was so important to von Laue because it was only by its compliance that one could "arrive at a really *orderly* state." He emphasized this point again at the end of the letter: "In the establishment of order it is less a matter of personal issues than the reintroduction of *principles* of law. I do not, by the way, mistake the good intention that motivated you to write your article, by any means."[132]

Interpreting this letter is difficult because von Laue's position is intrinsically inconsistent. One gets the feeling that von Laue wanted to shift the responsibility onto the judiciary. But his point is not that straightforward. On one hand, one was not supposed to reopen decided cases because this would injure the rights of those concerned. On the other hand, his intervention in the Schumann case had the precise purpose of preventing the consequences of a legal verdict that was deemed wrong. Von Laue and Martius were presumably of one mind in the belief that in certain cases such preventive measures were necessary. He nonetheless criticized her article on a matter of principle. Not a single word was wasted on the problem of how to prevent National Socialists from obtaining university positions. He merely divulged the steps he had taken against Schumann, whom he referred to as a charlatan. When Schumann was attempting to open an institute at Kiel, von Laue arranged for "sufficient enlightenment of the Kiel Ministry of Culture." His justification for this interference was the concern that physicists in the western zones had "that such strongly implicated persons not lodge themselves in the scientific establishment again."[133] No concrete incriminating facts were mentioned in this case either.

132 Von Laue to Martius, 26 December 1947, MPGA, div. III, rep. 50, von Laue papers, no. 2395 (original emphasis).

133 Von Laue to Martius, 26 December 1947, MPGA, div. III, rep. 50, von Laue papers, no. 2395.

Why Schumann was referred to as a "supposed physicist" by Becker and as a "charlatan" by von Laue may have had something to do with his professional background. He completed his studies in mathematics, physics, musicology, psychology, and medicine at the University of Berlin in 1922 with a thesis on systematic musicology. Entering the civil service as a physicist of the Reich Ministry of Defense in 1926, he became head of the Central Science Office for Army Physics in 1929 and was promoted to ministerial advisor in 1932. In 1928, he qualified for academic teaching with a habilitation thesis on tone color (*Über Klangfarben*) in the field of systematic musicology, with Max Planck being the third member on his jury. After teaching as private lecturer in systematic musicology for 3 years, he was granted permission to teach physics generally in 1931, at the suggestion of the entire permanent faculty of physics.

When attempts to confer an honorary professorship on Schumann in the Faculty of Philosophy fell flat in early 1933, he was appointed shortly afterward to a personal chair. At the beginning of 1934, Schumann became director of the newly established Physics Institute II at the *Friedrich-Wilhelms-Universität* in Berlin. In September 1933, his teaching assignment in physics and systematic musicology was expanded to include the physics of military defense. According to a report by a former student, Werner Luck, his "character make-up" (*Persönlichkeitsstruktur*) included "accumulation of power and an ambition for prestige."[134] Under National Socialism, he assumed important functions as a science organizer, particularly for the furtherance of defense research.[135] He was department head of the Science Office at the Reich Ministry for Science, Education and Culture (*Reichsministerium für Wissenschaft, Erziehung und Volksbildung*, REM), member of the presiding board

[134] Werner Luck, "Erich Schumann und die Studentenkompanie des Heereswaffenamtes – Ein Zeitzeugenbericht," *Dresdener Beiträge zur Geschichte der Technikwissenschaften*, 27 (2001), 27–45.

[135] For his own view of the connection between the Armed Forces and research and its necessary guidance, see Erich Schumann, "Wehrmacht und Forschung," in Richard Donnevert (ed.), *Wehrmacht und Partei*, 2nd ed. (Leipzig: Barth, 1939), 133–151. Richly annotated excerpts of it appear in English translation in Klaus Hentschel (ed.), *Physics and National Socialism: An Anthology of Primary Sources* (Science Networks. Historical Studies, vol. 18) (Basel–Boston–Berlin: Birkhäuser, 1996), doc. 75, 207–220. For Schumann's participation in the reconstruction of the Faculty of Defense Technology at the Berlin Polytechnic, see: Hans Ebert and Hermann J. Rupieper, "Technische Wissenschaft und nationalsozialistische Rüstungspolitik. Die Wehrtechnische Fakultät der TH Berlin 1933–1945," in Reinhard Rürup (ed.), *Wissenschaft und Gesellschaft. Beiträge zur Geschichte der Technischen Universität Berlin 1879–1979*, vol. 1 (Berlin: Springer, 1979), 469–491.

of the Reich Research Council (*Reichsforschungsrat*, RFR), as well as Reich plenipotentiary for the physics of explosives, head of the research departments at the Army's Ordnance Office and High Command, ministerial director of the Science Department of the Reich Ministry of War, and head of the Science Department of the Supreme Command of the Armed Forces (*Oberkommando der Wehrmacht*, OKW). In 1938, he was promoted to general and in 1940 also organized the Army Ordnance Office's company of students.[136]

Whether this impressive array of titles as a research policy maker could adequately explain Schumann's banishment from the community must remain an open question for lack of instructive sources. One argument on his side was his opposition to Stark; Schumann could not be counted among the members of the "Aryan physics group."[137] The DPG's board apparently did not make any further inquiries into Martius's accusation that he had arranged to have a doctoral thesis written for a daughter of his superior at the OKW, the later Field Marshal Keitel.[138] As all of Schumann's other machinations, this incident, too, was shrouded in silence – only when fellow scientists had suffered was the shroud lifted a little.[139] Concrete or diffuse awareness of his deeds seems to have been general knowledge, though. According to Hahn, "all" physicists noticed with "embarrassed astonishment" Schumann's appearance at the Göttingen meeting.[140]

But Schumann was not left to himself at Göttingen. Martius, who had taken part in all the communal meals during the conference, reported that she never once saw Schumann sitting alone. "But I did see the greetings by those embarrassed and astonished colleagues and also how they came to sit down at the general's table and even drew up a few more chairs."[141]

136 On Schumann's career, see: Luck, "Erich Schumann"; Hentschel, *Physics*, appendix F, XLV f. and *Kürschner's Gelehrtenkalender*, as well as Rainer Karlsch, *Hitlers Bombe* (Munich: DVA, 2005).

137 Werner Luck, "Erich Schumann," 32.

138 *Generalfeldmarschall* Wilhelm Keitel was at the top of the OKW, the highest command and administrative authority of the *Wehrmacht*, until the end of the war; the High Commands of the Army, the Air Force, and the Navy were all subordinate to it.

139 Appolonia Keitel submitted to Schumann the dissertation: *Subjektive und objektive Vokalanalysen*. From an analysis of datings and redatings of the oral examination, Werner Luck also discovered a violation of the procedural rules in the conferral of her doctoral degree.

140 Hahn to Martius, 12 November 1947, MPGA, div. III, rep. 14A, Hahn papers, no. 2726.

141 Martius to Hahn, 27 November 1947, MPGA, div. III, rep. 14A, Hahn papers, no. 2726. As a contrast to the lack of estrangement from Schumann, I mention the

There was many a hearty handshake for him on the first evening. The main reason why Hahn and von Laue did not notice these things themselves was that "the 'Göttingen celebrities' rarely joined the convivial meals, preferring instead to form their own circles" – at least according to the critical view in the *Physikalische Blätter*.[142] Schumann's lack of connections among the "Göttingen celebrities" was a crucial condition for his banishment. Another of their colleagues to feel this post-war frostiness was Wolfgang Finkelnburg. After Pohl had denied him a university position, he glumly generalized in August 1947: only "when one has been ordained in Göttingen" does one have a chance to find a job in Germany.[143]

The thesis argued thus far can now be narrowed down further, that during the post-war period collegial ties were a defining quantity deciding between integration or ostracism. Having a number of influential contacts was not enough. As direct a link as possible to the prominent figures in the field was what was needed. Right after the war, the majority of this prominent elite had assembled in Göttingen. It was there that important decisions were made about the futures of German physicists. What the verdict was in the cases of Schumann and Kneser was clear enough. How Herbert Stuart was handled reveals a more ambivalent attitude.

VON LAUE'S ATTITUDE TOWARD THE POLITICS OF THE PAST

As coeditor of the journal *Zeitschrift für Physik*, von Laue was also occasionally confronted with the decision on whether or not to accept papers submitted by Stuart. The approach he chose was political caution. The majority of the articles in the first post-war issues of the journal appearing in 1948 concerned research that had been completed while the war was still in progress. In September 1944, Stuart had submitted a paper about a model experiment on vaporization, condensation, saturation pressure, and the critical point, which had been recorded on film for teaching purposes. As von Laue informed Stuart in August 1946 in his capacity as

one experienced by the "non-Aryan" Richard Gans at the physics meeting in 1936 in Bad Salzbrunn: "Gans sat completely alone at his table!" Quoted from Edgar Swinne, *Richard Gans. Hochschullehrer in Deutschland und Argentinien* (Berlin: ERS-Verlag, 1992), 87.

142 *Physikalische Blätter*, 3 (1947), 288.

143 Finkelnburg to Sommerfeld, 8 August 1947, Archives of the Deutsches Museum, Munich (Deutsches Museum Archiv, DMA), papers 89 (Sommerfeld), no. 008. I thank Klaus Hentschel for the reference to this letter. Pohl's actions against Finkelnburg are discussed in Rammer, *Nazifizierung*.

editor, it unfortunately could not be included in the first issues because, "as far as we know, no permission to resume employment in the profession has been granted to you yet. We must consider it a priority that our first issues not be the subject of any kind of objections. We hope, however, that these difficulties will eventually be lifted."[144] The hope von Laue alluded to presumably was that soon all capable physicists would be able to take up their research and publishing activities again without having to worry about politics anymore. "My opinion on the future of physics is that we are going to build it up again here in Germany, all right, if politics leaves us alone," he wrote in October 1947 to his Dutch fellow physicist Hendrik Kramers.[145]

In 1946, Stuart was employed at the Hannover Polytechnic, but during the physics meeting in Göttingen in 1947, a physicist from Amsterdam complained that during the war Stuart had accused him of sabotage. This physicist Antonius Michels had barely been able to escape arrest by the secret police as a refugee. He was supposedly able to provide documentary proof of these allegations from a file he had found in Paris in summer 1945. The military government had thereupon informed the rector of the Hannover Polytechnic that Stuart was no longer permitted to teach there.[146] Stuart petitioned the DPG to have a disciplinary committee deliberate on the accusation against him.[147] As the society's statutes made no provision for such a jury, his petition was rejected. Two months later, Martius's article appeared. How did von Laue react to his colleague now that he had become the target of multiple complaints and lost his job as a consequence?

Von Laue decided to write him a letter, using not the formal address "Esteemed Colleague,"[148] as formerly, but the more familiar "Dear

[144] Von Laue to Stuart, 30 August 1946, MPGA, div. III, rep. 50, von Laue papers, no. 1951, regarding the paper: Herbert A. Stuart, Verdampfung, Kondensation, Sättigungsdruck und kritischer Punkt im Modellversuch und für den Unterricht gefilmt. From this information historians of science may conclude that delayed publication of submitted manuscripts potentially signal political reservations (in the eyes of the editors) about their physicist authors.

[145] Von Laue to Kramers, 23 October 1947, MPGA, div. III, rep. 50, von Laue papers, no. 1114.

[146] Stuart was forced to relinquish his post as assistant professor as of 1 December 1947. See Bartels letter to von Laue, 4 February 1948, MPGA, div. III, rep. 50, von Laue papers, no. 210.

[147] Stuart to von Laue, 9 September 1947, MPGA, div. III, rep. 50, von Laue papers, no. 2394.

[148] Von Laue to Stuart, 30 August 1946 and 11 June 1947, MPGA, div. III, rep. 50, von Laue papers, no. 1951.

Colleague."[149] So von Laue wanted to maintain friendly relations – not studied reserve – with his colleague Stuart. Von Laue informed him that he had discussed his situation with Pohl, who had given the advice that Stuart submit an application to the Bayer dye works, where problems in physics also figured in their research. A student of Pohl's was already working there. Stuart's application was successful. This was a tried and true strategy from the Nazi period that Pohl was applying; namely, shifting physicists into industry who for political reasons were momentarily not retainable at universities. This, then, is one structural continuity extending past 1945.

Von Laue's sense of collegial obligation toward Stuart was, I would think, rooted in Stuart's scientific origins and professional merits. Stuart had taken his doctorate under James Franck at Göttingen in 1925 with a thesis on the resonance fluorescence of vaporized mercury and had subsequently become assistant to Otto Stern and Richard Gans. Another important fact in the favorable assessment may also have been that Stuart had received his training almost entirely from Jewish professors. In 1934 his first book appeared, on molecular structure.[150] It remained for decades the most important handbook for physicists and chemists working in this field.[151] The single political inference in this clearly research-oriented study is in the preface. "Non-Aryans" as well as "Aryans" are named in the acknowledgments: "I likewise cordially thank Professor Gans and Dr. Volkmann, who have read through the manuscript and suggested many improvements. I thank the professors Mssrs. Born, Franck, Hund and Miss Sponer as well as, in particular, Dr. Teller for many a valuable piece of advice."

Apparent opportunistic anti-Semitism was left aside in favor of the conventional form of acknowledgment in use in science. Even so, Stuart's attitude toward those shunned under National Socialism gave some cause for criticism, as a letter by Gans to Gerlach of 4 October 1934 demonstrates: "Then Stuart informed me that I was not allowed to give examinations anymore. The rector had instructed him to suggest other examiners and he wanted to propose himself, Steinke and Kretschmann.

149 Von Laue to Stuart, 16 January 1948, MPGA, div. III, rep. 50, von Laue papers, no. 1951.

150 Herbert A. Stuart, *Molekülstruktur. Bestimmung von Molekülstrukturen mit physikalischen Methoden*. It was vol. 14 of the series *Struktur und Eigenschaften der Materie in Einzeldarstellungen* edited by Born, Franck, and Mark.

151 This evaluation is taken from E.W. Fischer, "Herbert Arthur Stuart 1899–1974," *Physikalische Blätter*, 30 (1974), 510ff., especially 510.

Characters like that, who aren't equal to the temptations of the current day, do exist."[152]

In 1935, Stuart received an extraordinary professorship at the University of Königsberg; a year later he substituted for the vacant chair in theoretical physics at the University of Berlin that Erwin Schrödinger had given up in 1933.[153] It was during this certainly important period of his career that he came into closer contact with the elite among German physicists in Berlin. He was among the skit actors with Sommerfeld, Peter Debye, Ernst Ruska, Heisenberg, and Gerlach in a humorous performance in honor of Planck's 80th birthday.[154] In 1939, Stuart accepted a tenured professorship at the Dresden Polytechnic. In 1942, Springer published a brief physics textbook by him. His first book had proved him to be a serious researcher; his second demonstrated his skill as a teacher with a broad overview of physics. This textbook was reprinted many times, and its revised edition by Gerhard Klages is still popular.[155]

In the post-war period, Stuart wrote a contribution for the *Fiat Review of German Science* series on alternative methods of researching electron shells and molecular forms.[156] Von Laue's ambivalence toward him was based, on one hand, on his respect for Stuart's scientific talent, on the other, on the admission that Stuart's "behavior in Berlin gave all sorts of cause for suspicion." When Stuart was dismissed in Hannover, von Laue saw "no possibility at the moment... of keeping him in a university position," either.[157] Nevertheless, at the end of 1947, von Laue still held the view that Stuart's proven abilities warranted his being allowed to teach at a German university. The military government got in the way of this, however.

[152] Quoted from Swinne, *Gans*, 88.

[153] See Hentschel, *Physics*, 182 and appendix F, XLVIII.

[154] Dieter Hoffmann, Hole Rößler, and Gerald Reuther, "'Lachkabinett' und 'großes Fest' der Physiker. Walter Grotrians 'physikalischer Einakter' zu Max Plancks 80. Geburtstag," *Berichte zur Wissenschaftsgeschichte*, 33 (2010), 30–53. See also the photos from this event in the German version of this book: Hoffmann and Walker, *Physiker*, 514f.

[155] See the reviews to the 2nd and 3rd editions of: Herbert A. Stuart, *Kurzes Lehrbuch der Physik* (Berlin: Springer, 1949) by Friedrich Asselmeyer in *Zeitschrift für angewandte Physik*, 1 (1949), 579, and by W. Braunbeck, *Zeitschrift für Naturforschung*, 5a (1950), 177.

[156] Herbert Stuart, "Erforschung der Elektronenhüllen und der Molekülgestalt mit anderen Methoden," in Hans Kopfermann (ed.), *Physics of the Electron Shells, Fiat Review of German Science 1939–1946*, vol. 12 (Wiesbaden: Dieterich, 1948), 69–91.

[157] Von Laue to Bartels, 10 February 1948, MPGA, div. III, rep. 50, von Laue papers, no. 210.

In April 1951, Article 131 of the Basic Law acknowledged the right to suitable employment of civil servants who had been dismissed in the denazification process.[158] With this legal backing, Stuart tried to revive his academic career in 1951; the University of Mainz wanted to appoint him to an extraordinary professorship for experimental physics. Considerable resistance by certain groups of physicists made itself felt, though. The Göttinger Hans Kopfermann was among these dissenting voices. Stuart wanted to know from Kopfermann what people had against him. When Kopfermann had no qualms about airing his opinion, Stuart demanded evidence. This Kopfermann succeeded in obtaining from the émigré Samuel Goudsmit.[159]

He procured a copy of Stuart's letter to Georg Stetter from March 1939 in which Stuart announced his intention to arrange for the election of more party members on a panel concerned with modifying the DPG's statutes. "It is actually a disgrace," it reads, "that we have to deal with such trifles and these bourgeois, while our Führer is making history and pointing out to us one great mission after the next.... It is good and necessary that the Nazis at universities keep in close touch." It also mentions arrangements he made with Wilhelm Schütz and Wilhelm Orthmann to incorporate the DPG into the National Socialist League of German Engineers (*Nationalsozialistischer Bund Deutscher Technik*, NSBDT), which would have subordinated the society's activities to centralized state control.[160] Kopfermann's request to Goudsmit for a copy

[158] Article 131 of the Federal Republic's constitution fully reinstated the eligibility of all public employees on 8 May 1945 who had left the civil service for reasons other than those based on the laws governing their employment or compensation. See Frei, *Vergangenheitspolitik*.

[159] Kopfermann to Goudsmit, 7 May 1951; Goudsmit to Kopfermann, 16 May 1951, AIP, Goudsmit papers, box 12, folder 120.

[160] Stuart to Stetter, 17 March 1939, AIP, Goudsmit papers, box 28, folder 53 (transcribed in Hoffmann and Walker, *Physiker*, 566f.). In 1938, after the Jewish staff members had been driven out of the University of Vienna, Stetter took over the directorship of the 2nd Institute of Physics. He had been an illegal National Socialist in Austria even before 1938. See Robert Rosner and Brigitte Strohmaier (eds.), *Marietta Blau–Sterne der Zertrümmerung. Biographie einer Wegbereiterin der modernen Teilchenphysik* (Vienna: Böhlau, 2003), especially 50f. On their plans to incorporate the DPG into the NSBDT, see the contribution by Dieter Hoffmann to this volume as well as his publications: Hoffmann, *Zwischen Autonomie*, 6–14; Dieter Hoffmann, "Carl Ramsauer, die Deutsche Physikalische Gesellschaft und die Selbstmobilisierung der Physikerschaft im 'Dritten Reich,'" in Helmut Maier (ed.), *Rüstungsforschung im Nationalsozialismus: Organisation, Mobilisierung und Entgrenzung der Technikwissenschaften* (Göttingen: Wallstein, 2002), 273–304, especially 276. Karl-Heinz

of this letter was soon followed by von Laue's in the same matter.[161] Together they prevented Stuart's appointment – but only initially so. In 1955, Stuart finally succeeded in rising to a full professorship again at the University of Mainz after being engaged there as guest professor in 1952.

What Stuart had specifically perpetrated only became the subject of debate in the early 1950s, however, not in 1947, when he was under fire from two sides. Goudsmit had been in possession of the letter since 1945; so consulting him would have been possible then, as many Americans actually had done before inviting their German colleagues over to the United States.[162] But there were Germans in the field who could also have been interviewed about Stuart's conduct under National Socialism. At that time, the DPG's board members were evidently not interested in any of the details. They clung to an assumption of innocence or played down the criticism. It was only during their first visits to the United States in 1948 and 1950 that the "prominent Göttingers" learned that Stuart had tried "to bring [the DPG] into Nazi hands" – as Kopfermann loosely phrased it[163] – when Goudsmit showed them the mentioned letter. Goudsmit sent them other incriminating evidence besides that exposed the network of National Socialist physicists collaborating with Stuart politically. A report on armaments research that Stuart had conducted was also enclosed.

But it appears as if prior to 1949, even von Laue thought discussions about such relationships were kicking up too much dust at a time when denazification still posed an – albeit rapidly diminishing – threat to physicists with troubling records. Goudsmit's material on Stuart suggested a particular anti-Semitic attitude among Nazi physicists that, although exhibiting a form of animosity toward Jews, did not exclude a high regard for scientific achievements by Jewish members of the profession. This could not be attacked with the same arguments as the particularly prominent form of anti-Semitism represented by Lenard, Stark, and

Ludwig, *Technik und Ingenieure im Dritten Reich* (Düsseldorf: Athenäum-Droste, 1979), especially chap. 5, discusses the NSBDT's function.

161 Von Laue to Goudsmit, 2 June 1951, AIP, Goudsmit papers, box 28, folder 53.

162 See the numerous inquiries among Goudsmit's papers, AIP, Goudsmit papers, box 14, folder 142. By chance, von Laue happened to write Goudsmit a letter regarding an entirely different matter on 17 December 1947, the day on which he formulated the DPG's official statement acquitting Stuart, Jordan, and Kneser; see AIP, Goudsmit papers, box 14, folder 142.

163 Kopfermann to Goudsmit, 7 May 1951, AIP, Goudsmit papers, box 12, folder 120.

Wilhelm Müller,[164] which boiled down to a fabrication of so-called Jewish physics.

The report about Stuart's research on the viscosity and compressibility of various fluids at low temperatures, a part of the V-1 rocket development program, could have stimulated a discussion about the wartime activities of German physicists that might well have led to different results than the ones von Laue presented in his article bearing that title.[165] He saw it as "the particular curse of such a time" that in order to "preserve throughout the war years the foundation," research had to be oriented toward military purposes. A few scientists had succeeded nevertheless in avoiding "being drawn with his work into the maelstrom." The possibility that German physicists had conducted weaponry research on their own initiative and with enthusiasm remained unexplored. The intention of von Laue's account had been to defuse suspicions from abroad that a link could be drawn between German physicists, Himmler, and Auschwitz.

As the above examples have just clearly shown, von Laue's actions were very influential in the restoration of internationalism by the DPG and the formulation and implementation of the society's policy on the past. Von Laue occupied a pioneering place in the reestablishment of international relations. His election to this role at home and abroad was owed to his personality and to an integrity attested to by all. German and Austrian émigrés, who were able to observe those colleagues who had "stayed over there" from a critical distance, played a decisive part in this acknowledgment. To them, von Laue was "the only German physicist who had behaved thoroughly decently and, from the point of view of the people here [in the United States], unobjectionably, whereas all the other people of rank are counted to a greater or lesser degree among the compromisers and are rejected."[166] In considering which German physicists deserved support, even the Dutch émigré Goudsmit arrived at the conclusion: "Laue certainly tops the list."[167]

[164] On Müller see: Freddy Litten, *Mechanik und Antisemitismus. Wilhelm Müller (1880–1968)* (Munich: Institut für Geschichte der Naturwissenschaften, 2000).

[165] Max von Laue, "The Wartime Activities of German Scientists," *Bulletin of the Atomic Scientists*, 4/4 (1948), 103. This article originally appeared in German: Max von Laue, "Die Kriegstätigkeit der deutschen Physiker," *Physikalische Blätter*, 3 (1947), 424–425 (reprinted in Hoffmann and Walker, *Physiker*, 640–643).

[166] Finkelnburg to Heisenberg, 6 February 1948, MPIfP, Heisenberg papers.

[167] Goudsmit to Ladenburg, 14 October 1946, AIP, Samuel Goudsmit papers, box 14, folder 138.

Von Laue applied his political capital toward helping German physics regain its former reputation. To this end, he made use of a policy on the past that did not permit probing criticism of the behavior of German physicists. A final example will underscore this. To refute the charge by Roger Murray from Britain that scientists were partly to blame for the Nazis' crimes,[168] von Laue used the image of the apolitical scientist. He countered in 1949 that only those people were responsible who had taken political decisions, if, indeed, the responsibility did not lie with the "nation" as a whole. Von Laue drew no distinctions between individual scientists: between scientists who had only been conducting "apolitical" science and, as he thought, therefore surely did not bear any specific responsibility, versus scientists who, like Jordan, for example, condoned devoting their science to military applications and saw in it a "compelling constructive energy in the creation of a new Europe." This passage was surely known to von Laue, but even physicists like Jordan were relegated among the large group of scientists who had been engaged in an ethically neutral – a "value-free" – inquiry of natural phenomena.[169] Research and politics were entirely separate spheres.[170] Accordingly, a scientist remained apolitical even when he occasionally got publicly involved in political activities; in that case, he was just acting outside of his profession.

With this argument, von Laue rehabilitated not just individuals but the entire guild of professional physicists. This was a relativization of political implication no different from many a whitewash certificate issued by others as well. Membership in the *Schutz-Staffel* (SS), for example, did not bear very heavily on von Laue's scale. This comes through in his assessment of a former assistant of his, Max Kohler, in March 1948: "Kohler had belonged to the S.S. and therefore unfortunately doesn't

[168] Roger C. Murray, "Social Action," *Göttinger Universitäts-Zeitung*, no. 4 (2 September 1949), 13.

[169] Von Laue did not use the term "value-free." He circumscribed the point by asserting that all sorts of purposes, peaceful just as warlike ones, can be satisfied on the basis of scientific findings. Scientific research had nothing to do with the issue of war and peace. That was why it was not permissible to evaluate pure scientific research by its conduciveness to peace. Max von Laue, "Public relations?" *Göttinger Universitäts-Zeitung*, no. 18 (23 September 1949), 11f.

[170] This opinion of von Laue is expressed in his frequently cited letter to Einstein from 14 May 1933: "But why did you have to come forward *politically* as well! I don't intend to reproach you at all for your statements, far from it. I just think that scholars should be reserved. Political contests demand different methods and different interests than scientific research." (Published in Herneck, *Laue*, 59f.; this translation from: Hentschel, *Mental*, 138.)

yet fall under consideration for a university post.... He isn't particularly seriously implicated at all."[171] Fritz Bopp, for whom von Laue had likewise issued a whitewash certificate, got to the core of von Laue's service to physics in Germany in his congratulatory letter on the occasion of the latter's 70th birthday:

> May you... be granted a long period of productive activity yet, in your very own science and in your solicitude for the whole of physics! I think, this is the most appropriate birthday wish for a man of action, even if there should be a ring of anxious egoism to it, just as once upon a time in a decennial tribute to Planck: 'We still need you!'[172]

Von Laue's personal engagement on behalf of German physics during the post-war period did not always take on a tone one would expect of a man who in 1933 had consistently fought against Nazi influences in science.[173] This certainly lay in the interest of his German colleagues, as can be gathered from a remark Ernst Brüche made to von Laue in 1949: "Today, too, you are one of the few who can be 'somewhat bold in certain regards' and physicists are placing their hopes on it."[174] Notwithstanding the negative tone regarding issues like inadequate support of German science, which was also directed against the Allies, von Laue executed an important function in the restoration of internationalism in physics.

Erwin Fues's wish for his 70th birthday was "that by the strength of your personality you help retie the threads linking German physics with

[171] Von Laue to Westphal, 6 March 1948, MPGA, div. III, rep. 50, von Laue papers no. 2125. Kohler had been assistant to von Laue at the University of Berlin from 1933 to 1943; thereafter extraordinary professor at the University of Greifswald. On Kohler's curriculum vitae, see the obituary: Hubert Goenner and R. Klein, "Nachruf auf Max Kohler," *Physikalische Blätter*, 38 (1982), 298f.

[172] Bopp to von Laue on the occasion of von Laue's 70th birthday, 9 October 1949, MPGA, div. III, rep. 50, von Laue papers, no. 311.

[173] Besides the above-mentioned papers: von Laue, "Public relations?" and von Laue, "The wartime activities," see also his article: "Zum Jahresbeginn 1947," *Physikalische Blätter*, 3 (1947), 1.

[174] Brüche to von Laue, 13 May 1949, MPGA, div. III, rep. 50, von Laue papers, no. 370. Brüche's words related to von Laue's address at the physics meeting in Hamburg in 1949, a "reconstructed wording" of which was published in: *Physikalische Blätter*, 5 (1949), 228f. On that occasion, von Laue warned against highly unwelcome political consequences occurring if foreign colleagues were to lose interest in physics in Germany, because German cultural institutions, such as universities, were not receiving the necessary funding and so could not operate successfully.

the world again."[175] In this von Laue was extremely successful, as the award of his honorary doctorate in Chicago in 1948, for instance, proves. The diploma describes him as a "physicist and resolute champion of freedom."[176] This was in acknowledgment of von Laue's unbending resistance to Stark's attempts to draw physics in Germany under his control during the first years of the dictatorship. This and his courageous voice on behalf of persons suffering from persecution under National Socialism earned him the exceptionally good reputation he enjoyed in the post-war period inside and outside the country. His conception of scientific freedom was that an elite of scientists steer the development of their disciplines and that politically partisan influences have no business interfering with this autonomy.[177] That was why from 1933 on, he repelled whatever he understood as National Socialist influencing, and after 1945, he rebuffed any interference from denazification that conflicted with the goals of the scientific elite. It explains why he was held in such high regard in Germany among both active opponents and profiteers of National Socialism. Take, for example, one of the clearest opponents of the regime among German physicists: Paul Rosbaud. During the dictatorship, he worked for the British as an active spy, resettling in England at the end of 1945.[178] His congratulatory wish on von Laue's 70th birthday was not only for von Laue's sake "but also for the sake of us all and particularly for my own, that your good spirit long remains safely with us, this mind of character, to whom I owe more strength, courage and faith in past years than to anyone else."[179]

CONCLUSION

The DPG in the British Zone resolved in 1947 "to assure cleanliness among our circle of colleagues." This political act regarding the past led to exclusion of physicists who had politically damaged the reputations of their colleagues under National Socialism. The DPG acted on its own

175 Fues to von Laue, 7 October 1949, MPGA, div. III, rep. 50, von Laue papers, no. 658.
176 Brüche, "Champion," 448 (original English).
177 See the contribution by Richard H. Beyler to this volume.
178 Arnold Kramish, *Der Greif. Paul Rosbaud – der Mann, der Hitlers Atompläne scheitern ließ* (Munich: Kindler, 1986).
179 Rosbaud to von Laue, 5 October 1949, MPGA, div. III, rep. 50, von Laue papers, no. 1666.

initiative in this matter, without any invitation to do so from outside. "Cleanliness" did not, however, mean banishing those who had supported National Socialism out of conviction, in word and deed, from the academic world. During the period of occupation, the issue of political misconduct by colleagues was avoided. This contribution proposes as an explanation that discussion of the political conduct of physicists was not possible because an established code of conduct prohibited it. Such statements about a fellow colleague would have amounted to denunciation. Even when the deeds by colleagues began to be broached more seriously, as denazification was nearing its close, only deeds that had caused damage to a fellow professional or a scientific institution remained a basis for expressible and negotiable criticism.

As chairman of the DPG in the British Zone between 1946 and 1949, Max von Laue was instrumental in shaping and implementing the policy on the past sketched above. Besides the sidelining of individual physicists deemed uncollegial, one consequence of this policy was the return of professionally highly qualified National Socialists to the university landscape. The justification offered was that excellent scientific achievement outweighed political misconduct. The DPG's leadership staved off any criticism of this policy that demanded a consistent removal of former Nazis from universities.

Ursula Martius's interpretation that this post-war policy took up the "cause of National Socialism" was erroneous to the extent that this "politics of the past" (*Vergangenheitspolitik*) was not really about National Socialism. It was rather a call for the promotion of physics in Germany or, more generally, the preservation of German science. It was part of a larger program for the reconstruction of German culture. The fact that some use was found along the way for politically implicated individuals in staffing decisions happened in the legitimate belief that they would primarily serve the promotion of science and not become active again in a National Socialist vein. Their continued employment was, however, a sign to those who had been politically or "racially" persecuted by the dictatorship that the lessons of the past had not been learned. Martius left Germany in disappointment while the country was still under occupation also because her criticism had driven her into the position of an outsider. This decision to emigrate out of resignation about the political developments in post-war Germany was not unique. Hartmut Kallmann and other physicist outcasts from the people's community during the dictatorship must have had similar motivations. Future historical research will have to explore this phenomenon for a more reliable assessment.[180]

180 One unquestionable member of this group of post-war émigrés was the physicist Kurt Hohenemser. He moved away to the United States in 1947 after 2 years of futile efforts to regain his former position at the University of Göttingen; see Rammer, *Nazifizierung*. In Richard Gans's case, who emigrated to Argentina in 1947, personal motives were added to a lack of integration in the German academic world after 1945; see Swinne, *Gans*. One example from medicine is the post-war rector at Hamburg. Rudolf Degwitz, a victim of Nazism, resignedly emigrated to the United States in 1948 because he could not prevent the return of politically seriously incriminated colleagues to the university. See Tobias Freimüller, "Mediziner: Operation Volkskörper," in Frei, *Karrieren* 13–69, especially 32f.

Appendix

DPG Members Who Left the Society as Victims of "Racial" or Political Discrimination

Name	Last City of Residence before Leaving DPG	Birth and Death	Discipline	Year Left DPG	Emigration (Year, Land Fate)	Source
Courant, Richard	Göttingen	1888–1972	Mathematics	1933	1933 England, 1934 USA	HB
Drucker, Karl	Leipzig	1876–1959	Physical Chemistry	1933	1933 Sweden	Not; P7a
Einstein, Albert	Berlin	1879–1955	Physics	1933	1933 USA	HB
Gemant, Andreas (Andrew)	Berlin-Charlottenburg	1895–1983	Electronics	1933	1934 England and USA	Not, SPSL, P7b
Hertz, Paul	Göttingen	1881–1940	Theoretical Physics	1933	1934 Geneva, Switzerland, 1936 Prague, Czechoslovakia, 1938 USA	HB
Herzog, Reginald Oliver	Berlin-Dahlem	1878–1935	Chemistry	1933	1934 Turkey; Suicide while vacationing in Zurich	HB
Houtermans, Fritz	Berlin-Charlottenburg	1903–1966	Physics	1933	1933 England, 1935 USSR	HB, SPSL
Jakob, Max	Berlin-Charlottenburg	1879–1955	Electronics	1933	1936 USA	HB
Konstantinowsky, Kurt	Bratislava	1892–	Physics	1933	1938 England (cable factory)	P6[1]
Kornfeld, Gertrud	Berlin	1891–1955	Physical Chemistry	1933	1933 England, 1935 Vienna, Austria, 1937 USA	HB, NDB
Landé, Alfred	Columbus	1888–1976	Theoretical Physics	1933	Migrant to USA	NDB, Not
Mendelssohn, Kurt	Neubabelsberg near Berlin	1906–1980	Physics	1933	1933 England	HB, SPSL

(continued)

Name	Last City of Residence before Leaving DPG	Birth and Death	Discipline	Year Left DPG	Emigration (Year, Land Fate)	Source
Paneth, Friedrich	Königsberg	1887–1958	Chemistry	1933	1933 England	HB, NDB
Reissner, Hans J	Berlin-Charlottenburg	1874–1967	Mechanics	1933	1938 USA	HB
Riesenfeld, Ernst Hermann	Berlin	1877–1957	Physical Chemistry	1933	Date unknown, Stockholm, Sweden	Not, P7a
Rosenthal, Adolf H	Frankfurt a.M.	1906–1962	Astronomy	1933	1934 Utrecht, Netherlands	Not, SPSL
Sack, Heinrich (Henri Samuel)	Leipzig	1903–1972	Theoretical Physics	1933	1933 Belgium, 1940 USA	HB, SPSL
Weissenberg, Karl	Berlin-Dahlem	1893–1976	Physics	1933	1933 Paris, France, 1934 England	SPSL, Not, P7a
Wolfsohn, Günther	Berlin-Charlottenburg	1901–1948	Physics	1933	1933 Netherlands and Palestine	Not, SPSL
Kolben, Emil	Prague	1862–1943	Factory Owner	1934	Deported to Theresienstadt	[2]
Kuhn, Heinrich	Oxford	1904–1994	Physics	1934	1933 England	HB, SPSL
Lion, Kurt	Darmstadt	1904–1980	Medical Physics	1934	1935 Turkey, 1937 Switzerland, 1941 USA	HB, SPSL
Placzek, George	No information	1905–1955	Theoretical Physics	1934	1933 USSR, 1934 Palestine, 1937 USA	NDB
Alexander, Ernst	Berlin	1902–	Physics	1935	1933 Palestine	HB, SPSL
Beck, Guido	Prague	1903–1988	Theoretical Physics	1935	1934 USA, 1935 USSR	HB; SPSL
Born, Max	Cambridge	1882–1970	Theoretical Physics	1935	1933 England	HB, SPSL
Bredig, Georg	Karlsruhe	1868–1944	Physical Chemistry	1935	1939 Netherlands, 1940 USA	HB
Elsasser, Walter	Paris	1904–1991	Theoretical Physics	1935	1933 France, 1936 USA	HB, SPSL
Epstein, Paul	Pasadena	1883–1966	Theoretical Physics	1935	Migrant to USA	P7b
Fajans, Kasimir	Munich	1887–1975	Physical Chemistry	1935	1936 USA	HB
Frankenthal, Max	Jerusalem	Not avail.	Mathematics	1935	Migrant to Jerusalem, Palestine	
Hellmann, Hans	No information	1903–1938	Physics	1935	1934 USSR, murdered there	Not[3]
Herzberger, Maximilian Jakob	Jena	1899–1982	Physics	1935	1934 Netherlands, 1935 England/ USA	HB
Jentzsch, Felix	Jena	1882–1946	Optics	1935	1935 retired; remained in Germany	Not
Klemperer, Otto Ernst	Cambridge	1899–1987	Physics	1935	1933 England	HB, SPSL

Name	Last City of Residence before Leaving DPG	Birth and Death	Discipline	Year Left DPG	Emigration (Year, Land Fate)	Source
Kottler, Friedrich	Vienna	1886–1965	Theoretical Physics	1935	1939 USA	SPSL, P7a
Liebreich–(Landolt), Erik	Berlin	1884–1946	Metallurgy	1935	Remained in Germany	NDB
Marx, Erich	Leipzig	1874–1956	Physics	1935	1933 fired, private radio–physics institute, 1941 USA	SPSL, HB
Minkowski, Rudolph	Hamburg	1895–1976	Astrophysics	1935	1935 USA	HB, SPSL
Polanyi, Michael	Manchester	1891–1976	Physical Chemistry	1935	1933 England	NDB, HB
Reichenbach, Hans	Istanbul	1891–1953	Theoretical Physics, Philosophy	1935	1933 Turkey, 38 USA	HB
Reinheimer, Hans	Brighton	Not avail.	Physics	1935	1934(?) England	
Reis, Alfred	Vienna	1882–1951	Physics	1935	1933 France, 1941 USA	Not, P7a
Rosenberg, Hans	Kiel	1879–1940	Astronomy	1935	1934 USA, 1938 Istanbul, Turkey	HB, NDB
Scharf, Karl	No Information	1903–	Physics	1935		Not
Schocken, Klaus	Bad Nauheim	1905–	Physics	1935	1935 USA	Not, SPSL
Sitte, Kurt	Prague	1910–	Physics	1935	Survived Buchenwald (political prisoner)	SPSL, P7a
Stobbe, Martin	Bristol	1903–1945	Theoretical Physics	1935	1933 England, USA, returned to Germany	Not
Szilard, Leo	London	1898–1964	Theoretical Physics	1935	1933 England, 1939 USA	HB, SPSL
Wohl, Kurt	Berlin	1896–1962	Physical Chemistry	1935	1939 England, 1942 USA	HB
Archenhold, F Simon	Berlin-Treptow	1861–1939	Astronomy	1936	Died in Germany	HB[4]
Estermann, Immanuel	Pittsburgh	1900–1973	Physics	1936	1933 England and USA, 1964 Haifa, Israel	HB
Fröhlich, Herbert	Leningrad	1905–1991	Theoretical Physics	1936	1933 USSR, 1935 England	HB, SPSL
Traube, Isidor	"England"	1860–1943	Physical Chemistry	1936	1934 England	Not, P7a
Blüh (Bluh), Otto	Prague	1902–	Physical Chemistry	1937	1939 England	SPSL, P7b
Brück, Hermann	Castel Gandolfo	1905–2000	Astronomy	1937	1936 Italy, 1937 England	HB
Dessauer, Friedrich	Istanbul	1881–1963	Medical Physics	1937	1934 Turkey, 1937 Switzerland	HB, NDB, Not, P7a

(continued)

Name	Last City of Residence before Leaving DPG	Birth and Death	Discipline	Year Left DPG	Emigration (Year, Land Fate)	Source
Freudenberg, Karl	Köln	1892–1966	Medical Statistics	1937	1935 fired, 1939 Netherlands, 1947 remigration to Germany	HB
Heitler, Walter	Bristol	1904–1981	Theoretical Physics	1937	1933 England, 1941 Ireland	SPSL, HB
Ludloff, Hanfried (John Frederick)	Leipzig	1899–	Theoretical Physics	1937	1936 Vienna, Austria, 1939 USA	HB
Peierls, Rudolf	Manchester	1907–1996	Theoretical Physics	1937	1933 England	HB, SPSL
Segré, Emilio	Palermo	1905–1989	Theoretical Physics	1937	1938 USA	SPSL, P7b
Abel, Emil	Vienna	1875–1958	Physical Chemistry	1938	1938 England	HB
Alterthum, Hans	Berlin	1890–1955	Industry	1938	1939 England, 1940 Argentina	P7a
Berg, Otto	Berlin	1874–1939	Physics	1938	England	HB
Berliner, Arnold	Berlin	1862–1942	Publisher	1938	1942 Suicide	NDB
Blau, Marietta	Vienna	1894–1970	Physics	1938	1938 Norway, 1939 Mexico	P7a
Boas, Hans Adolf	Berlin	1869–	Factory Owner	1938	Remained in Germany	SPSL
Byk, Alfred	Berlin	1878–1942	Theoretical Physics	1938	Deported and murdered	P7a,
Cohn, Emil	Heidelberg	1854–1944	Theoretical Physics	1938	1939 Switzerland	HB
Deutsch, Walter	Frankfurt a.M.	1885–1947	Industry	1938	1939 England	HB[5]
Ehrenhaft, Felix	Vienna	1879–1952	Theoretical Physics	1938	1938 England, 39 USA	HB, SPSL
Frank, Philipp	Prague	1884–1966	Theoretical Physics	1938	1938 USA	HB, SPSL
Frankenburg(er), Walter G.	Ludwigshafen	1893–1959	Physical Chemistry	1938	1938 USA	P7a
Gans, Richard	Berlin	1880–1954	Theoretical Physics	1938	Survived in Germany	NDB
Graetz, Leo	Munich	1856–1941	Physics	1938	Died in Germany	NDB
Haas, Arthur, Erich	Notre Dame	1884–1941	Theoretical Physics	1938	1935 USA	HB
Hirsch, Rudolf von Baron or Freiherr	Planegg bei Munich	1875–1975	Physics	1938	Survived Theresienstadt	
Hopf, Ludwig	Aachen	1884–1939	Mathematics	1938	1939 England/Ireland	HB
Jaffé, Georg(e) Cecil	Freiburg	1880–1965	Theoretical Physics	1938	1939 USA	HB, SPSL
Kallmann, Hartmut	Berlin-Charlottenburg	1896–1978	Physical Chemistry	1938	Survived in Germany	SPSL, Not, P7b
Kárman, Theodor von	Pasadena	1881–1963	Aerodynamics	1938	Emigrant to USA	NDB
Kaufmann, Walter	Freiburg	1871–1947	Theoretical Physics	1938	Survived in Germany	

Name	Last City of Residence before Leaving DPG	Birth and Death	Discipline	Year Left DPG	Emigration (Year, Land Fate)	Source
Kohn, Hedwig	Breslau	1887–1964	Physics	1938	1940 USA	HB, SPSL
Koref, Fritz	Berlin-Charlottenburg	1884–	Industry	1938	1938 France	P7a
Korn, Arthur	Berlin	1870–1945	Electronics	1938	1939 USA	HB
Kürti, Gustav	Vienna	1903–1978	Physics	1938	1938 England, 1939 USA, Visiting Professor Rostock 1968, died in Vienna, Austria	HB, SPSL
Lehmann, Erich	Berlin-Charlottenburg	1878–	Chemistry	1938	1933 fired; Fate unknown	Not, P7a,
Lessheim, Hans	No information	1900–	Physics	1938	Since 1932 in India	SPSL, Not
London, Fritz	Paris	1900–1954	Theoretical Physics	1938	1933 England, 1936 France, 1939 USA	HB, SPSL
Meitner, Lise	Berlin–Dahlem	1878–1968	Physics	1938	1938 Sweden	HB, SPSL
Meyer, Stefan	Vienna	1872–1949	Physics	1938	Survived in Austria	SPSL, NDB
Nordheim, Lothar	Durham	1899–1988	Theoretical Physics	1938	1933 France, 1934 Netherlands, 1935 USA	HB, SPSL
Pauli, Wolfgang	Zürich	1900–1958	Theoretical Physics	1938	1940 USA	HB[7]
Pel(t)zer, Heinrich	Vienna	1903–	Physics	1938	1938 England	SPSL[8]
Pelz, Stefan	Wattens, Österreich	1908–1973	Physics	1938	1938 England	[9]
Pirani, Marcello	Wembley	1880–1968	Technical Physics	1938	1936 England	HB
Pollitzer, Franz	Großhesselohe near Munich	1885–1942	Industry	1938	1939 France, murdered in Auschwitz	P7a
Przibram, Karl	Vienna	1878–1973	Physics	1938	1939 Belgium; survived in the Belgium underground	HB, SPSL
Reiche, Fritz	Berlin	1883–1969	Theoretical Physics	1938	1941 USA	HB, SPSL
Reichenheim, Otto	Berlin	1882–1950	Physics	1938	1939(?) England Industry	SPSL, P7a
Rona, Elisabeth	Vienna	1890–1981	Physics	1938	1941 USA	HB[10] SPSL
Rosenthal, Arthur	Heidelberg	1887–1959	Mathematics	1938	1939 Netherlands, 40 USA	HB
Rosenthal–Schneider	Berlin–Charlottenburg	1891–1990	Philosophy	1938	1939(?) Australia	
Sachs, George	Düren	1896–1960	Industry	1938	1936 USA	HB
Salinger, Hans	Philadelphia	1891–1965	Electronics	1938	1938 USA	SPSL, Not, P7a

(continued)

Name	Last City of Residence before Leaving DPG	Birth and Death	Discipline	Year Left DPG	Emigration (Year, Land Fate)	Source
Simon, Franz	Oxford	1893–1956	Physics	1938	1933 England	HB, SPSL
Urbach, Franz	Vienna	1902–1969	Physics	1938	1939 USA	HB, SPSL
Weigert, Fritz	Markkleeberg	1876–1947	Physical Chemistry	1938	1935 England	HB
Wigner, Eugene	Madison	1902–1995	Theoretical Physics	1938	1933 USA	Not, P7a

Identification through the *Biographische Handbuch der deutschsprachigen Emigration nach 1933* is denoted with (HB). If there is no entry there, then the *Neue Deutsche Biographie* (NDB) or the published list from the *Notgemeinschaft List of Displaced Scholars* from 1936 with a supplement from 1937 is used, here denoted as (Not). In a few cases, where there was no entry in the HB, NDB or Not, but a file was kept by the emigrant supporting organization *Society for the Protection of Science and Learning* (Bodleian Library, Oxford), then (SPSL) appears as a source. Not and SPSL were expanded as much as possible through reference to the last volume of the *Poggendorff, Biographisch–Literarisches Handwörterbuch der Exakten Naturwissenschaften* (P), in which the relevant person was included. The individuals who left the DPG 1938 are also in the internet list compiled by Klaus Hentschel.[11]

1 Reference from Wolfgang Reiter based on archival studies in Vienna.

2 Österreichische Akademie der Wissenschaften (ed.), *Österreichisches biographisches Lexikon 1815–1950*, Volume 4 (Vienna: Verlag der Österreichischen Akademie der Wissenschaften, 1969), 76.

3 Short biography: http://www.tc.Chemistry.uni–siegen.de/hellmann/hellbiod.html.

4 Entry of his son Günther Hermann.

5 Entry of his son, Ernst Robert.

6 UFA, NL Kaufmann, C139.

7 Entry of his sister, Hertha Paul.

8 Biographical section in Johannes Feichtinger, "Die Viennaer Hochpolymerforschung in England und Amerika," http://gewi.kfunigraz.ac.at/~johannes/HPF.htm.

9 Reference from Wolfgang Reiter based on archival studies in Vienna. Also see his article: "Die Vertreibung der jüdischen Intelligenz: Verdopplung eines Verlustes 1938/45," *Internationale Mathematische Nachrichten*, No. 187 (2001), pp. 1–20.

10 In the *Handbuch* the emigration is mistakenly dated as 1935; in contrast see Dorothee Mussgnug, *Die vertriebenen Heidelberger Dozenten* (Heidelberg: Winter, 1988), 155.

11 http://www.uni–stuttgart.de/hi/gnt/hentschel/Dpg38–39.htm, accessed 20 December 2010.

Bibliography

Finn Aaserud, *Redirecting Science* (Cambridge: Cambridge University Press, 1990).

Helmuth Albrecht, "'Max Planck: Mein Besuch bei Adolf Hitler'–Anmerkungen zum Wert einer historischen Quelle," in Helmuth Albrecht (ed.), *Naturwissenschaft und Technik in der Geschichte: 25 Jahre Lehrstuhl für Geschichte der Naturwissenschaft und Technik am historischen Institut der Universität Stuttgart* (Stuttgart: Verlag für Geschichte der Naturwissenschaften und der Technik, 1993), 41–63.

Helmuth Albrecht (ed.), *Naturwissenschaft und Technik in der Geschichte: 25 Jahre Lehrstuhl für Geschichte der Naturwissenschaft und Technik am historischen Institut der Universität Stuttgart* (Stuttgart: Verlag für Geschichte der Naturwissenschaften und der Technik, 1993).

Helmuth Albrecht and Armin Hermann, "Die Kaiser-Wilhelm-Gesellschaft im Dritten Reich (1933–1945)," in Rudolf Vierhaus and Bernhard vom Brocke (eds.), *Forschung im Spannungsfeld von Politik und Gesellschaft* (Stuttgart: DVA, 1990), 356–406.

Ulrich Albrecht, Andreas Heinemann-Grüder, and Arend Wellmann, *Die Spezialisten. Deutsche Naturwissenschaftler und Techniker in der Sovietunion nach 1945* (Berlin: Dietz, 1992).

William Sheridan Allen, *The Nazi Seizure of Power: The Experience of a Single German Town 1922–1945*, rev. ed. (New York: Franklin Watts, 1984; orig. 1966).

Hannah Arendt, *Besuch in Deutschland* (Berlin: Rotbuch-Verlag, 1993).

Philippe Ariès, "L'histoire des mentalités," in Roger Chartier and Jacques Revel (eds.), *La nouvelle histoire* (Paris: Retz, 1978), 402–422.

Mitchell G. Ash, "Wissenschaft und Politik als Ressourcen für einander," in Rüdiger vom Bruch and Brigitte Kaderas (eds.), *Wissenschaften und Wissenschaftspolitik: Bestandaufnahmen zu Formationen, Brüchen, und Kontinuitäten im Deutschland des 20. Jahrhunderts* (Stuttgart: Franz Steiner Verlag, 2002), 32–51.

Mitchell G. Ash and Alfons Söllner (eds.), *Forced Migration and Scientific Change* (Washington, DC: German Historical Institute, 1996).

Alan Beyerchen, *Scientists under Hitler: Politics and the Physics Community in the Third Reich* (New Haven: Yale University Press, 1977).

Michael Balfour and John Mair, *Four-Power Control in Germany and Austria 1945–1946* (Oxford: Oxford University Press, 1956).

Michael Balfour and John Mair, *Vier-Mächte-Kontrolle in Deutschland 1945–1946* (Düsseldorf: Droste, 1959).

Zygmunt Baumann, *Modernity and the Holocaust* (Ithaca: Cornell University Press, 1989).

Karen Bayer, Frank Sparing, and Wolfgang Woelk (eds.), *Universitäten und Hochschulen im Nationalsozialismus und in der frühen Nachkriegszeit* (Stuttgart: Steiner, 2004).

Ulrich Benz, *Arnold Sommerfeld. Lehrer und Forscher an der Schwelle zum Atomzeitalter 1868–1951* (Stuttgart: Wissenschaftliche Verlagsgesellschaft, 1975).

Wolfgang Benz (ed.), *Deutschland unter alliierter Besatzung 1945–1949. Ein Handbuch* (Berlin: Akademie Verlag, 1999).

Wolfgang Benz, "Emil J. Gumbel: Die Karriere eines deutschen Pazifisten," in Ulrich Walberer (ed.), *10. Mai 1933. Bücherverbrennung in Deutschland und die Folgen* (Frankfurt am Main: Fischer Taschenbuch Verlag, 1983), 160–198.

Richard Bessell (ed.), *Life in the Third Reich* (New York: Oxford University Press, 1987).

Alan D. Beyerchen, *Scientists under Hitler: Politics and the Physics Community in the Third Reich* (New Haven: Yale University Press, 1977).

Richard H. Beyler, "Maintaining Discipline in the Kaiser Wilhelm Society during the National Socialist Regime," *Minerva*, 44/3 (2006), 251–266.

Richard H. Beyler, "'Reine' Wissenschaft und personelle 'Säuberungen.' Die Kaiser-Wilhelm/Max-Planck-Gesellschaft 1933 und 1945," in Carola Sachse (ed.), *Ergebnisse. Vorabdrucke aus dem Forschungsprogramm "Geschichte der Kaiser-Wilhelm-Gesellschaft im Nationalsozialismus*, No. 16 (Berlin: Forschungsprogramm, 2004).

Richard H. Beyler, "Targeting the Organism: The Scientific and Cultural Context of Pascual Jordan's Quantum Biology, 1932–1947," *Isis*, 87 (1996), 248–273.

Richard H. Beyler and Morris F. Low, "Science Policy in Post-War West Germany and Japan between Ideology and Economics," in Mark Walker (ed.), *Science and Ideology: A Comparative History* (London: Routledge, 2003), 97–123.

Richard H. Beyler, Alexei Kojevnikov, and Jessica Wang, "Purges in Comparative Perspective: Rules for Exclusion and Inclusion in the Scientific Community under Political Pressure," in Carola Sachse and Mark Walker (eds.), *Politics and Science in Wartime*, Volume 20 of *Osiris*, (Chicago: University of Chicago Press, 2005), 23–48.

Mario Biagioli, *Galileo, Courtier: The Practice of Science in the Culture of Absolutism* (Chicago: University of Chicago Press, 1993).

H. Birret, K. Helbig, W. Kertz, and U. Schmucker (eds.), *Zur Geschichte der Geophysik* (Berlin: Springer, 1974).

Heinz Boberach (ed.), *Meldungen aus dem Reich. Die Geheimen Lageberichte des Sicherheitsdienstes der SS* (Herrsching: Pawlak, 1984).

Joseph Borkin, *Die unheilige Allianz der I.G. Farben. Eine Interessengemeinschaft im Dritten Reich*, 3rd ed. (Frankfurt am Main: Campus Verlag, 1990).

Max Born, *Mein Leben. Die Erinnerungen des Nobelpreistägers* (Munich: Nymphenburger Verlag, 1975).

Tom Bower, *The Paperclip Conspiracy: The Battle for the Spoils and Secrets of Nazi Germany* (London: Michael Joseph, 1987).

Martin Broszat et al., *Bayern in der NS-Zeit*, 6 vols. (Munich: R. Oldenbourg, 1977–1983)

Martin Broszat, *The Hitler State: The Foundation and Development of the Internal Structure of the Third Reich*, trans. John W. Hiden (London: Longman, 1981).

Martin Broszat, "Ein Landkreis in der Fränkischen Schweiz: Der Bezirk Ebermannstadt 1929–1945" in Martin Broszat, Elke Fröhlich, and Falk Wiesemann (eds.), *Bayern in der NS-Zeit I: Soziale Lage und politisches Verhalten der Bevölkerung im Spiegel vertraulicher Berichte* (Munich: R. Oldenbourg, 1977), 21–192.

Martin Broszat, "Resistenz und Widerstand: Eine Zwischenbilanz des Forschungsprojekts," in Martin Broszat, Elke Fröhlich, and Anton Grossmann (eds.), *Bayern in der NS-Zeit IV. Herrschaft und Gesellschaft im Konflikt, Teil C* (Munich: R. Oldenbourg, 1981), 691–709.

Elke Brychta, Anna-Maria Reinhold, and Arno Mersmann (eds.), *Mutig, streitbar, reformerisch. Die Landés – Sechs Biografien 1859–1977* (Essen: Klartext-Verlag, 2004).

Bundesarchiv Koblenz (ed.), *Gedenkbuch: Opfer der Verfolgung der Juden unter der nationalsozialistischen Gewaltherrschaft in Deutschland 1933–45*, Vol. 1 (Frankfurt am Main: Weisbecker, 1995).

Cathryn Carson, "New Models for Science in Politics: Heisenberg in West Germany," *Historical Studies in the Physical and Biological Sciences*, 31 (1999), 115–171.

Cathryn Carson, "Science Advising and Science Policy in Postwar West Germany: The Example of the Deutscher Forschungsrat," *Minerva*, 40/2 (2002), 147–179.

David Cassidy, *Beyond Uncertainty: Heisenberg, Quantum Physics, and the Bomb* (New York: Bellevue Literary Press, 2009).

David Cassidy, "Controlling German Science I: U.S. and Allied Forces in Germany, 1945–1947," *Historical Studies in the Physical Sciences*, 24 (1994), 197–235.

David Cassidy, "Controlling German Science II: Bizonal Occupation and the Struggle over West German Science Policy, 1946–1949," *Historical Studies in the Physical Sciences*, 26 (1996), 197–237.

David Cassidy, *Uncertainty: The Life and Science of Werner Heisenberg* (New York: W. H. Freeman, 1992).

Geoffrey Cocks and Konrad H. Jarausch (eds.), *The German Professions 1800–1950* (New York: Oxford University Press, 1990).

John Connelly and Michael Grüttner (eds.), *Zwischen Autonomie und Anpassung. Universitäten in den Diktaturen des 20. Jahrhunderts* (Paderborn: Schöningh, 2002).

John Cornwell, *Hitler's Scientists: Science, War and the Devil's Pact* (London: Viking, 2003).

Elizabeth Crawford, "German Scientists and Hitler's Vendetta against the Nobel Prizes," *Historical Studies in the Physical Sciences*, 31 (2000), 37–53.

Per F. Dahl, *Superconductivity. Its Historical Roots and Development* (New York: American Institute of Physics, 1992).

Hans Joachim Dahms, "Die Universität Göttingen 1918 bis 1989: Vom "Goldenen Zeitalter" der zwanziger Jahre bis zur "Verwaltung des Mangels" in der Gegenwart," in Rudolf von Thadden and Günter J. Trittel (eds.), *Göttingen. Geschichte einer Universitätsstadt*, Vol. 3 (Göttingen: Vandenhoeck & Ruprecht, 1999), 395–456.

Ute Deichmann, *Biologists under Hitler* (Cambridge, MA: Harvard University Press, 1996).

Ute Deichmann, "Dem Vaterlande – solange es dies wünscht. Fritz Habers Rücktritt 1933, Tod 1934 und die Fritz Haber-Gedächtnisfeier 1935," *Chemie in unserer Zeit*, 30 (1996), 141–149.

Ute Deichmann, *Flüchten, Mitmachen, Vergessen. Chemiker und Biochemiker in der NS-Zeit* (Weinheim: Wiley-VCH, 2001).

Ute Deichmann, "Kriegsbezogene biologische, biochemische und chemische Forschung an den Kaiser Wilhelm-Instituten für Züchtungsforschung, für Physikalische Chemie und Elektrochemie und für Medizinische Forschung," in Doris Kaufmann (ed.), *Geschichte der Kaiser-Wilhelm-Gesellschaft im Nationalsozialismus. Bestandsaufnahme und Perspektiven der Forschung*, Vol. 1 (Göttingen: Wallstein, 2000), 231–257.

Peter Dinzelbacher, "Zu Theorie und Praxis der Mentalitätsgeschichte," in Peter Dinzelbacher (ed.), *Europäische Mentalitätsgeschichte* (Stuttgart: Kröner, 1993), XV–XXXVII.

Matthias Dörries (ed.), *Michael Frayn's "Copenhagen" in Debate. Historical Essays and Documents on the 1941 Meeting Between Niels Bohr and Werner Heisenberg* (Berkeley: Office for History of Science and Technology, 2005).

Wolfgang Drechsler and Helmut Rechenberg, "Herbert Jehle (5.3.1907–14.1.1983)," *Physikalische Blätter*, 39 (1983), 71.

Ernst Dreisigacker and Helmut Rechenberg, "50 Jahre Physikalische Blätter," in *Physikalische Blätter*, 50 (1994), 21–23.

Ernst Dreisigacker and Helmut Rechenberg, "Karl Scheel, Ernst Brüche und die Publikationsorgane," in Theo Mayer-Kuckuk (ed.), *150 Jahre Deutsche Physikalische Gesellschaft*, special issue 51 (1995), F135–F142.

Hans Ebert and Hermann Rupieper, "Technische Wissenschaft und nationalsozialistische Rüstungspolitik: Die Wehrtechnische Fakultät der TH Berlin 1933–1945," in Reinhard Rürup (ed.), *Wissenschaft und Gesellschaft. Beiträge zur Geschichte der Technischen Universität Berlin 1879–1979*, Vol. 1 (Berlin: Springer, 1979), 469–491.

Michael Eckert, *Die Atomphysiker. Eine Geschichte der theoretischen Physik am Beispiel der Sommerfeldschule* (Braunschweig: Vieweg, 1993).

Michael Eckert, "Primacy doomed to failure: Heisenberg's role as scientific advisor for nuclear policy in the Federal Republic of Germany," *Historical Studies in the Physical Sciences*, 21 (1990), 29–58.

Michael Eckert, "Theoretical Physics at War: Sommerfeld Students in Germany and as Emigrants," in Paul Forman and José M. Sánchez-Ron (eds.), *National Military Establishments and the Advancement of Science and Technology: Studies in 20th Century History* (Dordrecht: Kluwer Academic Publishers, 1996), 69–86.

Michael Eckert, "Theoretische Physiker in Kriegsprojekten. Zur Problematik einer internationalen vergleichenden Analyse," in Doris Kaufmann (ed.), *Geschichte der Kaiser-Wilhelm-Gesellschaft im Nationalsozialismus. Bestandsaufnahme und Perspektiven der Forschung*, Vol. 1 (Göttingen: Wallstein, 2000), 296–308.

Michael Eckert and Karl Märker (eds.), *Arnold Sommerfeld. Wissenschaftliche Briefwechsel, Vol. 2: 1919–1951* (Berlin: Verlag für Geschichte der Naturwissenschaften und der Technik, 2004).

Michael Eckert, Willibald Pricha, Helmut Schubert, and Gisela Torkar (eds.), *Geheimrat Sommerfeld – Theoretischer Physiker: Eine Dokumentation aus seinem Nachlaß* (Munich: Deutsches Museum, 1984).

Moritz Epple, "Rechnen, Messen, Führen. Kriegsforschung am Kaiser-Wilhelm-Institut für Strömungsforschung 1937–1945," in Helmut Maier (ed.), *Rüstungsforschung im Nationalsozialismus. Organisation, Mobilisierung und Entgrenzung der Technikwissenschaften* (Göttingen: Wallstein, 2002), 305–356.

Moritz Epple, Andreas Karachalios, and Volker R. Remmert, "Aerodynamics and Mathematics in National Socialist Germany and Fascist Italy: A Comparison of Research Institutes," in Carola Sachse and Mark Walker (eds.), *Politics and Science in Wartime*, Volume 20 of *Osiris*, (Chicago: U. of Chicago Press, 2005), 131–158.

Paul Erker, *Deutsche Unternehmer zwischen Kriegswirtschaft und Wiederaufbau: Studien zur Erfahrungsbildung von Industrie-Eliten* (Munich: Oldenbourg, 1999).

Ruth Federspiel, "Mobilisierung der Rüstungsforschung? Werner Osenberg und das Planungsamt im Reichsforschungsrat 1943–1945," in Helmut Maier (ed.), *Rüstungsforschung im Nationalsozialismus. Organisation, Mobilisierung und Entgrenzung der Technikwissenschaften* (Göttingen: Wallstein, 2002), 72–105.

Klaus Fischer, "Die Emigration von Wissenschaftlern nach 1933. Möglichkeiten und Grenzen einer Bilanzierung," *Vierteljahrshefte für Zeitgeschichte*, 39 (1991), 535–549.

E.W. Fischer, "Herbert Arthur Stuart 1899–1974," *Physikalische Blätter*, 30 (1974), 510–511.

Wolfram Fischer (ed.), *Die Preußische Akademie der Wissenschaften zu Berlin 1914–1945* (Berlin: Akademie-Verlag, 2000).

Sören Flachowsky, *Von der Notgemeinschaft zum Reichsforschungsrat. Wissenschaftspolitik im Kontext von Autarkie, Aufrüstung und Krieg* (Stuttgart: Franz Steiner Verlag, 2008).

Donald Fleming and Bernhard Bailyn (eds.), *The Intellectual Migration: Europe and America 1930–1960* (Cambridge, MA: Belknap Press, 1969).

Andreas Flitner (ed.), *Deutsches Geistesleben und Nationalsozialismus: Eine Vortragsreihe der Universität Tübingen* (Tübingen: Rainer Wunderlich, 1965).

Paul Forman, "Die Naturforscherversammlung in Nauheim im September 1920," in Dieter Hoffmann and Mark Walker (eds.), *Physiker zwischen Autonomie und Anpassung – Die DPG im Dritten Reich* (Weinheim: Wiley-VCH, 2007), 29–58.

Paul Forman, "The Financial Support and Political Alignment of Physicists in Weimar Germany," *Minerva*, 12 (1974), 39–66.

Paul Forman and José M. Sánchez-Ron (eds.), *National Military Establishments and the Advancement of Science and Technology: Studies in 20th Century History* (Dordrecht: Kluwer Academic Publishers, 1996).

Norbert Frei, *Adenauer's Germany and the Nazi Past. The Politics of Amnesty and Integration* (New York: Columbia University Press, 2002).

Norbert Frei (ed.), *Karrieren im Zwielicht. Hitlers Eliten nach 1945* (Frankfurt am Main: Campus Verlag, 2001).

Norbert Frei, *Vergangenheitspolitik. Die Anfänge der Bundesrepublik und die NS-Vergangenheit* (München: DTV, 1999).

Freie Universität Zentralinstitut für sozialwissenschaftliche Forschung (ed.), *Gedenkbuch Berlins der jüdischen Opfer des Nationalsozialismus* (Berlin: Edition Hentrich, 1995).

Tobias Freimüller, "Mediziner: Operation Volkskörper," in Norbert Frei (ed.), *Karrieren im Zwielicht. Hitlers Eliten nach 1945* (Frankfurt am Main: Campus Verlag, 2001), 13–69.

Saul Friedländer, *Das Dritte Reich und die Juden* (Munich: Beck, 1998).

Elke Fröhlich (ed.), *Die Tagebücher von Joseph Goebbels*, part I, vol. 4, 1 January 1940–8 July 1941 (Munich: Saur, 1987).

Elke Fröhlich, "Stimmung und Verhalten der Bevölkerung unter den Bedingungen des Krieges," in Martin Broszat, Elke Fröhlich, and Falk Wiesemann (eds.), *Bayern in der NS-Zeit I: Soziale Lage und politisches Verhalten der Bevölkerung im Spiegel vertraulicher Berichte* (Munich: R. Oldenbourg, 1977), 571–688.

Helmuth Gericke, *50 Jahre GAMM*, supplementary issue of *Ingenieurarchiv*, 41 (1972).

Helmuth Gericke, *Aus der Chronik der Deutschen Mathematiker-Vereinigung* (Stuttgart: Teubner, 1980).

Thomas F. Gieryn, "Boundary-Work and the Demarcation of Science from Non-Science: Strains and Interests in Professional Ideologies of Scientists," *American Sociological Review*, 48 (1983), 781–795.

John Gimbel, *Science, Technology, and Reparations. Exploitation and Plunder in Postwar Germany* (Stanford: Stanford University Press, 1990).

Johann Wolfgang Goethe, *Gesammelte Werke* (Berlin: Tillgner, 1981).

C. J. Gorter, "Superconductivity until 1940 in Leiden and as seen from there," *Reviews of Modern Physics*, 36 (1964), 3–7.

Samuel A. Goudsmit, *Alsos* (New York: H. Schuman, 1947; reprinted Woodbury, NY: AIP Press, 1996).

Bernd Grün, "Das Rektorat des Mathematikers Wilhelm Süss in den Jahren 1940–1945 und seine Wiederwahl 1958/59," *Freiburger Universitätsblätter*, 145 (1999), 171–191.

Michael Grüttner, "Schlussüberlegungen: Universität und Diktatur," in John Connelly and Michael Grüttner (eds.), *Zwischen Autonomie und Anpassung.*

Universitäten in den Diktaturen des 20. Jahrhunderts (Paderborn: Schöningh, 2002), 266–276.

Michael Grüttner, *Studenten im Dritten Reich* (Paderborn: Schöningh, 1995).

Michael Grüttner, "Wissenschaftspolitik im Nationalsozialismus," in Doris Kaufmann (ed.), *Geschichte der Kaiser-Wilhelm-Gesellschaft im Nationalsozialismus. Bestandsaufnahme und Perspektiven der Forschung*, Vol. 1 (Göttingen: Wallstein, 2000), 557–585.

Siegfried Grundmann, *The Einstein Dossiers* (Berlin: Springer, 2005).

Klaus Habetha (ed.), *Wissenschaft zwischen technischer und gesellschaftlicher Herausforderung. Die Rheinisch-Westfälische Technische Hochschule Aachen 1970–1995* (Aachen: Einhard, 1995), 208–215.

Notker Hammerstein, *Die Deutsche Forschungsgemeinschaft in der Weimarer Republik und im Dritten Reich* (Munich: Beck, 1999).

Notker Hammerstein, "Die Geschichte der Deutschen Forschungsgemeinschaft," in Doris Kaufmann (ed.), *Geschichte der Kaiser-Wilhelm-Gesellschaft im Nationalsozialismus, Volume I* (Göttingen: Wallstein, 2000), 600–609.

Kai Handel, "Die Arbeitsgemeinschaft Rotterdam und die Entwicklung von Halbleiterdetektoren. Hochfrequenzforschung in der militärischen Krise 1943–1945," in Helmut Maier (ed.), *Rüstungsforschung im Nationalsozialismus. Organisation, Mobilisierung und Entgrenzung der Technikwissenschaften* (Göttingen: Wallstein, 2002), 250–270.

Helmut M. Hanko, "Kommunalpolitik in der 'Hauptstadt der Bewegung' 1933–1935: Zwischen 'revolutionärer' Umgestaltung und Verwaltungskontinuität," in Martin Broszat, Elke Fröhlich, and Anton Grossmann (eds.), *Bayern in der NS-Zeit III. Herrschaft und Gesellschaft im Konflikt, Teil B* (Munich: R. Oldenbourg, 1981), 329–441.

Jonathan Harwood, "'Mandarine' oder 'Außenseiter'? Selbstverständnis deutscher Naturwissenschaftler (1900–1933)," in Jürgen Schriewer et al. (eds.), *Sozialer Raum und akademische Kulturen* (Frankfurt am Main: Lang, 1993), 183–212.

Jonathan Harwood, *Styles of Scientific Thought: The German Genetics Community 1900–1933* (Chicago: University of Chicago Press, 1993).

Jonathan Harwood, "The Rise of the Party-Political Professor? Changing Self-Understandings among German Academics, 1890–1933," in Doris Kaufmann (ed.), *Geschichte der Kaiser-Wilhelm-Gesellschaft im Nationalsozialismus, Volume I* (Göttingen: Wallstein, 2000), 21–45.

John L. Heilbron, "The Earliest Missionaries of the Copenhagen Spirit," in Edna Ullmann-Margalit (ed.), *Science in Reflection* (Dordrecht: Kluwer Academic, 1988), 201–233.

John L. Heilbron, *The Dilemmas of an Upright Man. Max Planck as Spokesman for German Science* (Berkeley: University of California Press, 1986).

Susanne Heim, Carola Sachse, and Mark Walker (eds.), *The Kaiser Wilhelm Society under National Socialism* (Cambridge: Cambridge University Press, 2009).

Manfred Heinemann, "Der Wiederaufbau der Kaiser-Wilhelm-Gesellschaft und die Neugründung der Max-Planck-Gesellschaft (1945–1949)," in Rudolf

Vierhaus and Bernhard vom Brocke (eds.), *Forschung im Spannungsfeld von Politik und Gesellschaft* (Stuttgart: DVA, 1990), 407–470.

Manfred Heinemann, "Überwachung und "Inventur" der deutschen Forschung. Das Kontrollratsgesetz Nr. 25 und die alliierte Forschungskontrolle im Bereich der Kaiser-Wilhelm-/Max-Planck-Gesellschaft (KWG/MPG) 1945–1955," in Lothar Mertens (ed.), *Politischer Systemumbruch als irreversibler Faktor von Modernisierung in der Wissenschaft?* (Berlin: Humblot, 2001), 167–199.

Rudolf Heinrich and Hans-Reinhard Bachmann, *Walther Gerlach: Physiker – Lehrer – Organisator* (Munich: Deutsches Museum, 1989).

Klaus Hentschel, "Finally, some historical polyphony!" in Matthias Dörries (ed.), *Michael Frayn's "Copenhagen" in Debate. Historical Essays and Documents on the 1941 Meeting Between Niels Bohr and Werner Heisenberg* (Berkeley: Office for History of Science and Technology, 2005), 31–37.

Klaus Hentschel, *Physics and National Socialism: An Anthology of Primary Sources* (Basel: Birkhäuser, 1996).

Klaus Hentschel, *The Mental Aftermath: The Mentality of German Physicists 1945–1949* (Oxford: Oxford University Press, 2007).

Klaus Hentschel, "What History of Science Can Learn from Michael Frayn's 'Copenhagen,'" *Interdisciplinary Science Reviews*, 27 (2002), 211–216.

Klaus Hentschel, *Zur Mentalität deutscher Physiker in der frühen Nachkriegszeit* (Heidelberg: Spektrum Verlag, 2005).

Klaus Hentschel and Gerhard Rammer, "Kein Neuanfang: Physiker an der Universität Göttingen 1945–1955," *Zeitschrift für Geschichtswissenschaft*, 48 (2000), 718–741.

Klaus Hentschel and Gerhard Rammer, "Nachkriegsphysik an der Leine: Eine Göttinger Vogelperspektive," in Dieter Hoffmann (ed.), *Physik im Nachkriegs-Deutschland* (Frankfurt: Harri Deutsch, 2003), 27–56.

Klaus Hentschel and Gerhard Rammer, "Physicists at the University of Göttingen, 1945–1955," *Physics in Perspective*, 3/1 (2001), 189–209.

Klaus Hentschel and Monika Renneberg, "Eine akademische Karriere. Der Astronom Otto Heckmann im Dritten Reich," *Vierteljahrshefte für Zeitgeschichte*, 43 (1995), 581–610.

Armin Hermann, "Die Deutsche Physikalische Gesellschaft 1899–1945," in Theo Mayer-Kuckuk (ed.), *Festschrift 150 Jahre Deutsche Physikalische Gesellschaft*, special issue of the *Physikalische Blätter*, 51 (1995), F61–F105.

Armin Hermann, "Science under Foreign Rule; Policy of the Allies in Germany 1945–1949," in Fritz Krafft and Christoph J. Scriba (eds.), *XVIIIth Int. Congress of History of Science Hamburg–Munich. Final Report* (Stuttgart: Steiner, 1993), 75–86.

Armin Hermann, *Werner Heisenberg 1901–1976* (Bonn: Inter Nationes, 1976).

Friedrich Herneck, *Max von Laue* (Leipzig: Teubner, 1979).

Sébastien Hertz, *Emil Julius Gumbel (1891–1966) et la statistique des extrêmes*, unpublished dissertation (Lyon: University of Lyon, 1997).

Guntolf Herzberg and Klaus Meier: *Karrieremuster* (Berlin: Aufbau-Taschenbuch-Verlag, 1992).

Thomas Hochkirchen, "Wahrscheinlichkeitsrechnungen im Spannungsfeld von Maß- und Häufigkeitstheorie – Leben und Werk des "Deutschen" Mathematikers Erhard Tornier (1894–1982)," *NTM. Internationale Zeitschrift für Geschichte und Ethik der Naturwissenschaften, Technik und Medizin, new series*, 6 (1998), 22–41.

Dieter Hoffmann, "Between Autonomy and Accommodation: The German Physical Society during the Third Reich," *Physics in Perspective*, 7/3 (2005), 293–329.

Dieter Hoffmann, "Carl Ramsauer, die Deutsche Physikalische Gesellschaft und die Selbstmobilisierung der Physikerschaft im 'Dritten Reich,'" in Helmut Maier (ed.), *Rüstungsforschung im Nationalsozialismus. Organisation, Mobilisierung und Entgrenzung der Technikwissenschaften* (Göttingen: Wallstein, 2002), 273–304.

Dieter Hoffmann, "Das Verhältnis der Akademie zu Republik und Diktatur. Max Planck als Sekretär," in Wolfram Fischer (ed.), *Die Preußische Akademie der Wissenschaften zu Berlin 1914–1945* (Berlin: Akademie-Verlag, 2000), 53–85.

Dieter Hoffmann, "Die Physikerdenkschriften von 1934/36 und zur Situation im faschistischen Deutschland, Wissenschaft und Staat," *itw-kolloquien*, No. 68 (1989), 185–211.

Dieter Hoffmann, "Einsteins politische Akte," *Physik in unserer Zeit*, 35, No. 2 (2004), 64–69.

Dieter Hoffmann, "Kriegsschicksale: Hans Euler," *Physikalische Blätter*, 45, No. 9 (1989), 382–383.

Dieter Hoffmann, *Max Planck: Die Entstehung der modernen Physik* (Munich: Beck, 2008).

Dieter Hoffmann (ed.), *Operation Epsilon. Die Farm-Hall-Protokolle oder Die Angst der Allierten vor der deutschen Atombombe* (Berlin: Rowohlt, 1993).

Dieter Hoffmann, "Peter Debye (1884–1966): Ein Dossier," *Max Planck Institut für Wissenschaftsgeschichte, Preprint 314* (Berlin: Max Planck Institute for History of Science, 2006).

Dieter Hoffmann (ed.), *Physik im Nachkriegs-Deutschland* (Frankfurt: Harri Deutsch, 2003).

Dieter Hoffmann and Kristie Macrakis (eds.), *Naturwissenschaft und Technik in der DDR* (Berlin: Akademie-Verlag, 1997).

Dieter Hoffmann and Ulrich Schmidt-Rohr (eds.), *Wolfgang Gentner (1906–1980). Festschrift zum 100. Geburtstag* (Heidelberg: Springer, 2006).

Dieter Hoffmann and Thomas Stange, "East-German Physics and Physicists in the Light of the 'Physikalische Blätter,'" in Dieter Hoffmann (ed.), *The Emergence of Modern Physics* (Pavia: Universita degli Studi di Pavia, 1997), 521–529.

Dieter Hoffmann and Rüdiger Stutz, "Grenzgänger der Wissenschaft: Abraham Esau als Industriephysiker, Universitätsrektor und Forschungsmanager," in Uwe Hossfeld, Jürgen John, Oliver Lemuth, and Rüdiger Stutz (eds.), *Kämpferische Wissenschaft: Studien zur Universität Jena im Dritten Reich* (Cologne: Böhlau, 2003), 136–179.

Dieter Hoffmann and Mark Walker, "Der gute Nazi: Pascual Jordan und das Dritte Reich," in *Pascual Jordan (1902–1980), Mainzer Symposium zum 100.*

Geburtstag. Preprint 329 (Berlin: Max Planck Institute for History of Science, 2007), 83–112.

Dieter Hoffmann and Mark Walker, "Der Physiker Friedrich Möglich (1902–1957) – Ein Antifaschist?" in Dieter Hoffmann and Kristie Macrakis (eds.), *Naturwissenschaft und Technik in der DDR* (Berlin: Akademie-Verlag, 1997), 361–382.

Dieter Hoffmann and Mark Walker (eds.), *"Fremde" Wissenschaftler im Dritten Reich. Die Debye-Affäre im Kontext* (Göttingen: Wallstein, 2011).

Dieter Hoffmann and Mark Walker (eds.), *Physiker zwischen Autonomie und Anpassung – Die DPG im Dritten Reich* (Weinheim: Wiley-VCH, 2007).

Dieter Hoffmann and Mark Walker, "The German Physical Society under National Socialism," *Physics Today*, December (2004), 52–58.

Dieter Hoffmann, Fabio Bevilacqua, Roger H. Stuewer (eds.), *The Emergence of Modern Physics* (Pavia: Universita degli Studi die Pavia, 1996).

Dieter Hoffmann, Hole Rößler, and Gerald Reuther, "'Lachkabinett' und 'großes Fest' der Physiker. Walter Grotrians 'physikalischer Einakter' zu Max Plancks 80. Geburtstag," *Berichte zur Wissenschaftsgeschichte*, 33/1 (2010), 30–53.

Uwe Hossfeld, Jürgen John, Oliver Lemuth, and Rüdiger Stutz (eds.), *Kämpferische Wissenschaft: Studien zur Universität Jena im Dritten Reich* (Cologne: Böhlau, 2003).

Christian Jansen, *Emil Julius Gumbel: Portrait eines Zivilisten* (Heidelberg: Wunderhorn, 1991).

Konard H. Jarausch, *The Unfree Professions: German Lawyers, Teachers, and Engineers, 1900–1950* (New York: Oxford University Press, 1990).

Georg Joos, "Robert Pohl 70 Jahre," *Zeitschrift für angewandte Physik*, 6 (1954), 339.

Ulrich Kalkmann, *Die Technische Hochschule Aachen im Dritten Reich (1933–1945)* (Aachen: Aachener Studien zu Technik und Gesellschaft, 2003).

Horst Kant, "Peter Debye als Direktor des Kaiser-Wilhelm-Instituts für Physik in Berlin," in Dieter Hoffmann and Mark Walker (eds.), *"Fremde" Wissenschaftler im Dritten Reich. Die Debye-Affäre im Kontext* (Göttingen: Wallstein, 2011), 76–109.

Horst Kant, "Peter Debye und die Deutsche Physikalische Gesellschaft," in Dieter Hoffmann, Fabio Bevilacqua, and Roger H. Stuewer (eds.), *The Emergence of Modern Physics* (Pavia: Universita degli Studi die Pavia, 1996), 505–520.

Rainer Karlsch, *Hitlers Bombe* (Munich: DVA, 2005).

Doris Kaufmann (ed.), *Geschichte der Kaiser-Wilhelm-Gesellschaft im Nationalsozialismus*, 2 volumes (Göttingen: Wallstein, 2000).

Henry Kellermann, *Cultural Relations as an Instrument of U.S. Foreign Policy: The Educational Exchange Program between the United States and Germany, 1945–1954* (Washington, DC: U.S. Government Printing Office, 1978).

Ulrich Kern, *Forschung und Präzisionsmessung. Die Physikalisch-Technische Reichsanstalt zwischen 1918 und 1948* (Weinheim: VCH, 1994).

Walter Kertz (ed.), *Hochschule und Nationalsozialismus* (Braunschweig: TU Braunschweig, 1994).

Ian Kershaw, *Popular Opinion and Political Dissent in the Third Reich: Bavaria 1933–1945* (Oxford: Clarendon Press, 1983).

Serge Klarsfeld, *Le Memorial de la Déportation des Juifs de France* (Paris: Klarsfeld, 1978).

Serge Klarsfeld, *Memorial to the Jews Deported from France 1942–44* (New York: B. Klarsfeld Foundation, 1983).

Andreas Kleinert, "Das Spruchkammerverfahren gegen Johannes Stark," *Sudhoffs Archiv*, 67 (1983), 13–24.

Andreas Kleinert, "Der Briefwechsel zwischen Philipp Lenard (1862–1947) und Johannes Stark (1874–1957)," *Jahrbuch 2000 der Deutschen Akademie der Naturforscher Leopoldina*, 3rd series, 46 (2001), 243–261.

Christian Kleint, Helmut Rechenberg, and Gerald Wiemers, *Werner Heisenberg 1901–1976, Beiträge, Berichte, Briefe. Festschrift zu seinem 100. Geburtstag* (Leipzig: Verlag der Sächsische Akademie der Wissenschaften, 2005).

Victor Klemperer, *Ich will Zeugnis ablegen. Tagebücher*, Vol. 1 (Berlin: Aufbau-Verlag, 1997).

Victor Klemperer, *I Will Bear Witness* (New York: Random House, 1998).

Victor Klemperer, *The Language of the Third Reich: LTI–Lingua Tertii Imperii. A Philologist's Notebook* (New York: Continuum, 2006).

Martin Kneser and Norbert Schappacher, "Fachverband – Institut – Staat," in Gerd Fischer, Friedrich Hirzebruch, Winfried Scharlau, and Willi Törnig (eds.), *Ein Jahrhundert Mathematik 1890–1990* (Braunschweig: Vieweg, 1990), 1–82.

Michael Knoche, "Scientific Journals under National Socialism," *Libraries & Culture*, 26 (1991), 415–426.

Jürgen Kocha (ed.), *Die Königlich Preußische Akademie der Wissenschaften zu Berlin im Kaiserreich* (Berlin: Akademie-Verlag, 1999).

Helmut König, Wolfgang Kuhlmann, and Klaus Schwabe (eds.), *Vertuschte Vergangenheit. Der Fall Schwerte und die NS-Vergangenheit der deutschen Hochschulen* (Munich: Beck, 1997).

Heinz Kolben, "Dr. h.c. Ing. Emil Kolben zum Gedächtnis," *Bohemia*, 26 (1985), 111–121.

Rudy Koshar, *Social Life, Local Politics, and Nazism: Marburg, 1880–1935* (Chapel Hill: University of North Carolina Press, 1986).

Fritz Krafft, *Im Schatten der Sensation* (Weinheim: VCH, 1981).

Arnold Kramish, *Der Greif. Paul Rosbaud – der Mann, der Hitlers Atompläne scheitern ließ* (Munich: Kindler, 1986).

Bernhard R. Kroener, Rolf-Dieter Müller, and Hand Umbreit, *Das Deutsche Reich und der Zweite Weltkrieg* (Stuttgart: DVA, 1999).

Conrad F. Latour and Thilo Vogelsang, *Okkupation und Wiederaufbau. Die Tätigkeit der Militärregierung in der amerikanischen Besatzungszone Deutschlands 1944–1947* (Stuttgart: DVA, 1973).

Jacques Le Goff, "Les mentalités. Une histoire ambigüe," in Jacques Le Goff and Pierre Nora (eds.), *Faire de l'histoire* (Paris: Gallimard, 1974), 76–94.

Jacques Le Goff and Pierre Nora (eds.), *Faire de l'histoire* (Paris: Gallimard, 1974).

Jost Lemmerich, *Aufrecht im Sturm der Zeit: Der Physiker James Franck 1882–1964*, (Stuttgart: Verlag für Geschichte der Naturwissenschaften und der Technik, 2007).

Jost Lemmerich, "Ein Angriff von Johannes Stark auf Werner Heisenberg über das Reichsministerium für Wissenschaft, Erziehung und Volksbildung (REM)," in Christian Kleint, Helmut Rechenberg, and Gerald Wiemers, *Werner Heisenberg 1901–1976, Beiträge, Berichte, Briefe. Festschrift zu seinem 100. Geburtstag* (Leipzig: Verlag der Sächsische Akademie der Wissenschaften, 2005), 213–221.

Jost Lemmerich (ed.), *Lise Meitner – Max von Laue. Briefwechsel 1938–1948* (Berlin: ERS Verlag, 1998).

Jost Lemmerich (ed.), *Max Born, James Franck. Physiker in ihrer Zeit. Der Luxus des Gewissens*, exhibition catalog (Wiesbaden: Reichert, 1982).

Helmut Lindner, "'Deutsche' und 'gegentypische' Mathematik. Zur Begründung einer 'arteigenen Mathematik' im Dritten Reich," in Herbert Mehrtens and Steffen Richter (eds.), *Naturwissenschaft, Technik und NS-Ideologie. Beiträge zur Wissenschaftsgeschichte des Dritten Reiches* (Frankfurt am Main: Suhrkamp, 1980), 88–115.

Konrad Lindner, *Carl Friedrich von Weizsäckers Wanderung ins Atomzeitalter. Ein dialogisches Selbstporträt* (Paderborn: Mentis, 2002).

Freddy Litten, *Astronomie in Bayern 1914–1945* (Stuttgart: Steiner, 1992).

Freddy Litten, *Mechanik und Antisemitismus. Wilhelm Müller (1880–1968)* (München: Institut für Geschichte der Naturwissenschaften, 2000).

Ludwig Lochner (ed.), *Goebbels Tagebücher 1942/43*, (Zurich: Atlantis Verlag, 1948).

Wilfried Loth and Bernd-A. Rusinek (eds.), *Verwandlungspolitik. NS-Eliten in der westdeutschen Nachkriegsgesellschaft* (Frankfurt am Main: Campus Verlag, 1998).

Werner Luck, "Erich Schumann und die Studentenkompanie des Heereswaffenamtes – Ein Zeitzeugenbericht," *Dresdener Beiträge zur Geschichte der Technikwissenschaften*, 27 (2001), 27–45.

Karl-Heinz Ludwig, *Technik und Ingenieure im Dritten Reich* (Düsseldorf: Droste, 1974).

Kristie Macrakis, *Surviving the Swastika: Scientific Research in Nazi Germany* (Cambridge, MA: Harvard University Press, 1993).

Kurt Magnus, *Raketensklaven. Deutsche Forscher hinter rotem Stacheldraht* (Stuttgart: DVA, 1993).

Helmut Maier, "Einleitung," in Helmut Maier (ed.), *Rüstungsforschung im Nationalsozialismus. Organisation, Mobilisierung und Entgrenzung der Technikwissenschaften* (Göttingen: Wallstein, 2002), 7–29.

Helmut Maier (ed.), *Rüstungsforschung im Nationalsozialismus. Organisation, Mobilisierung und Entgrenzung der Technikwissenschaften* (Göttingen: Wallstein, 2002).

Heinz Maier-Leibnitz, "Die große Zeit in Göttingen. Robert W. Pohl, ein Patriarch der Physik, wird neunzig," *Frankfurter Allgemeine Zeitung*, 10 August (1974).

Peter Mattes, "Die Charakterologen. Westdeutsche Psychologie nach 1945," in Walter H. Pehle and Peter Sillem (eds.), *Wissenschaft im geteilten Deutschland. Restauration oder Neubeginn nach 1945?* (Frankfurt am Main: Fischer Taschenbuch Verlag, 1992), 125–135.

Max-Planck-Gesellschaft zur Förderung der Wissenschaften (ed.), *Max Planck. Vorträge und Ausstellung zum 50. Todestag* (Munich: Max-Planck-Gesellschaft, 1997).

Theo Mayer-Kuckuk (ed.), *Festschrift 150 Jahre Deutsche Physikalische Gesellschaft*, special issue of *Physikalische Blätter*, 51 (1995).

Charles E. McClelland, *The German Experience of Professionalization: Modern Learned Professions and Their Organizations from the Early Nineteenth Century to the Hitler Era* (Cambridge: Cambridge University Press, 1991).

Herbert Mehrtens, "Angewandte Mathematik und Anwendungen der Mathematik im nationalsozialistischen Deutschland," *Geschichte und Gesellschaft*, 12 (1986), 317–347.

Herbert Mehrtens, "Das 'Dritte Reich' in der Naturwissenschaftsgeschichte: Literaturbericht und Problemskizze," in Herbert Mehrtens and Steffen Richter (eds.), *Naturwissenschaft, Technik, und NS-Ideologie: Beiträge zur Wissenschaftsgeschichte des Dritten Reich*, (Frankfurt am Main: Suhrkamp, 1980), 15–87.

Herbert Mehrtens, "Die 'Gleichschaltung' der mathematischen Gesellschaften im nationalsozialistischen Deutschland," *Jahrbuch Überblicke Mathematik* (1985), 83–103.

Herbert Mehrtens, "Irresponsible Purity: The Political and Moral Structure of Mathematical Sciences in the National Socialist State," in Monika Renneberg and Mark Walker (eds.), *Science, Technology, and National Socialism* (Cambridge: Cambridge University Press, 1993), 324–338, 411–413.

Herbert Mehrtens, "Kollaborationsverhältnisse: Natur- und Technikwissenschaften im NS-Staat und ihre Historie," in Christoph Meinel and Peter Voswinckel (eds.), *Medizin, Naturwissenschaft, Technik und Nationalsozialismus: Kontinuitäten und Diskontinuitäten*, (Stuttgart: Verlag für Geschichte der Naturwissenschaft und der Technik, 1994), 13–31.

Herbert Mehrtens, "Ludwig Bieberbach and 'Deutsche Mathematik,'" in Esther R. Phillips (ed.). *Studies in the History of Mathematics* (Washington, DC, Mathematical Association of America, 1987), 195–241.

Herbert Mehrtens, "Mathematics and war: Germany 1900–1945," in Paul Forman and José M. Sánchez-Ron (eds.), *National Military Establishments and the Advancement of Science and Technology: Studies in Twentieth Century History* (Dordrecht: Kluwer, 1996), 87–134.

Herbert Mehrtens and Steffen Richter (eds.), *Naturwissenschaft, Technik, und NS-Ideologie: Beiträge zur Wissenschaftsgeschichte des Dritten Reich*, (Frankfurt am Main: Suhrkamp, 1980).

Christoph Meinel and Peter Voswinckel (eds.), *Medizin, Naturwissenschaft, Technik und Nationalsozialismus: Kontinuitäten und Diskontinuitäten*, (Stuttgart: Verlag für Geschichte der Naturwissenschaft und der Technik, 1994).

Kurt Mendelssohn, *Walther Nernst und seine Zeit. Aufstieg und Niedergang der deutschen Naturwissenschaften* (Weinheim: Physik-Verlag, 1973).

Gabriele Metzler, *Internationale Wissenschaft und nationale Kultur. Deutsche Physiker in der internationalen Community 1900–1960* (Göttingen: Vandenhoeck & Ruprecht, 2000).

Swantje Middeldorff, "Ernst Brüche und die Geschichte der Physikalischen Blätter 1944–1974," (Diplom-Thesis: University of Hamburg, 1993).

Alexander Mitscherlich and Margarete Mitscherlich, *Die Unfähigkeit zu trauern. Grundlagen kollektiven Verhaltens* (Frankfurt: Piper, 1967).

Horst Nelkowski, "Die Physikalische Gesellschaft zu Berlin in den Jahren nach dem Zweiten Weltkrieg," in Theo Mayer-Kuckuk (ed.), Festschrift 150 Jahre Deutsche Physikalische Gesellschaft, special issue of *Physikalische Blätter*, 51 (1995), F-143–F-156.

Alfred Neubauer, *Bittere Nobelpreise* (Norderstedt: Books on Demand, 2005).

Lutz Niethammer, *Entnazifizierung in Bayern. Säuberung und Rehabilitierung unter amerikanischer Besatzung* (Frankfurt am Main: Fischer, 1972).

Otto Gerhard Oexle, "Hahn, Heisenberg und die anderen. Anmerkungen zu 'Kopenhagen,' 'Farm Hall' und 'Göttingen,'" *Ergebnisse. Vorabdrucke aus dem Forschungsprogramm "Geschichte der Kaiser-Wilhelm-Gesellschaft im Nationalsozialismus,"* No. 9 (Berlin: Forschungsprogramm, 2003).

Otto Gerhard Oexle, "Wie in Göttingen die Max-Planck-Gesellschaft entstand," *Max-Planck-Gesellschaft* Jahrbuch, (1994), 43–60.

Karin Orth and Willi Oberkrome (eds.), *Die Deutsche Forschungsgemeinschaft 1920–1970* (Stuttgart: Franz Steiner Verlag, 2010).

Walter H. Pehle and Peter Sillem (eds.), *Wissenschaft im geteilten Deutschland. Restauration oder Neubeginn nach 1945?* (Frankfurt am Main: Fischer Taschenbuch Verlag, 1992).

Lilli Peltzer, *Die Demontage deutscher naturwissenschaftlicher Intelligenz nach dem 2. Weltkrieg – Die Physikalisch-Technische Reichsanstalt 1945/1948* (Berlin: ERS-Verlag, 1995).

Detlev J. K. Peukert, *Inside Nazi Germany*, trans. Richard Deveson (New Haven: Yale University Press, 1987).

Gerhard Rammer, *Die Nazifizierung und Entnazifizierung der Physik an der Universität Göttingen*, unpublished dissertation (Göttingen: University of Göttingen, 2004).

Gerhard Rammer, "Göttinger Physiker nach 1945. Über die Wirkung kollegialer Netze," *Göttinger Jahrbuch*, 51 (2003), 83–104.

Cornelia Rauh-Kühne, "Die Entnazifizierung und die deutsche Gesellschaft," *Archiv für Sozialgeschichte*, 35 (1995), 35–70.

Helmut Rechenberg, "Vor fünfzig Jahren," *Physikalische Blätter*, 44 (1988), 418.

Ralf Reichardt, "Für eine Konzeptualisierung der Mentalitätstheorie," *Ethnologica Europea*, 11 (1978/80), 234–241.

Constance Reid, *Courant in Göttingen and New York: The Story of an Improbable Mathematician* (New York: Springer-Verlag, 1976).

Nathan Reingold, "Refugee Mathematicians in the United States of America, 1933–1941: Reception and Reaction," *Annals of Science*, 38 (1981), 313–338.

Wolfgang Reiter, "Die Vertreibung der jüdischen Intelligenz: Verdopplung eines Verlustes – 1938/1945," *Internationale Mathematische Nachrichten*, 187 (2001), 1–20.

Wolfgang Reiter, "Stefan Meyer: Pioneer in Radioactivity," *Physics in Perspective*, 3 (2001), 106–127.

Volker R. Remmert, *Findbuch des Bestandes C 89 – Nachlaß Wilhelm Süss: 1913–1961* (Freiburg: Uni-Archiv, 2000).

Volker R. Remmert, *Findbuch des Bestandes E 4 – Deutsche Mathematiker-Vereinigung (1889–1987)* (Freiburg: Uni-Archiv, 1999).

Volker R. Remmert, "In the Service of the Reich: Aspects of Copernicus and Galileo in Nazi Germany's Historiographical and Political Discourse," *Science in Context*, 14 (2001), 333–359.

Volker Remmert, "Mathematical Publishing in the Third Reich: Springer-Verlag and the Deutsche Mathematiker-Vereinigung," *The Mathematical Intelligencer*, 22/3 (2000), 22–30.

Volker R. Remmert, "Mathematicians at War. Power Struggles in Nazi Germany's Mathematical Community: Gustav Doetsch and Wilhelm Süss," *Revue d'histoire des mathématiques*, 5 (1999), 7–59.

Volker R. Remmert, "Vom Umgang mit der Macht. Das Freiburger Mathematische Institut im 'Dritten Reich,'" *Zeitschrift für Sozialgeschichte des 20. und 21. Jahrhunderts*, 14 (1999), 56–85.

Volker R. Remmert, "Wilhelm Süss," in Bernd Ottnad and Fred Ludwig Sepainter (eds.), *Baden-Württembergische Biographien*, vol. III (Stuttgart: Kohlhammer, 2002), 418–421.

Volker R. Remmert, "Zwischen Universitäts- und Fachpolitik: Wilhelm Süss, Rektor der Albert-Ludwigs-Universität Freiburg (1940–1945) und Vorsitzender der Deutschen Mathematiker-Vereinigung (1937–1945)," in Karen Bayer, Frank Sparing, and Wolfgang Woelk (eds.), *Universitäten und Hochschulen im Nationalsozialismus und in der frühen Nachkriegszeit* (Stuttgart: Steiner, 2004), 147–165.

Jürgen Renn, Giuseppe Castagnetti, and Peter Damerow, "Albert Einstein. Alte und neue Kontexte in Berlin," in Jürgen Kocha (ed.), *Die Königlich Preußische Akademie der Wissenschaften zu Berlin im Kaiserreich* (Berlin: Akademie-Verlag, 1999), 333–354.

Monika Renneberg, "Die Physik und die physikalischen Institute der Hamburger Universität im 'Dritten Reich,'" in Eckhart Krause, Ludwig Huber, and Holger Fischer (eds.), *Hochschulalltag im Dritten Reich: Die Hamburger Universität 1933–1945*, Vol. 3 (Hamburg/Berlin: Reimer, 1991), 1097–1118.

Monika Renneberg and Mark Walker (eds.), *Science, Technology, and National Socialism* (Cambridge: Cambridge University Press, 1993).

Monika Renneberg and Mark Walker, "Scientists, Engineers, and National Socialism," in Monika Renneberg and Mark Walker (eds.), *Science, Technology, and National Socialism* (Cambridge: Cambridge University Press, 1993), 1–17, 339–346.

Steffen Richter, "Die 'Deutsche Physik,'" in Herbert Mehrtens and Steffen Richter (eds.), *Naturwissenschaft, Technik, und NS-Ideologie: Beiträge zur Wissenschaftsgeschichte des Dritten Reich*, (Frankfurt am Main: Suhrkamp, 1980), 116–139.

Fritz Ringer, *Die Gelehrten. Der Niedergang der deutschen Mandarine 1890–1930* (Stuttgart: Klett-Cotta, 1983).

Alan J. Rocke, *The Quiet Revolution. Hermann Kolbe and the Science of Organic Chemistry* (Berkeley: University of California Press, 1993).

Robert Rosner and Brigitte Strohmaier (eds.), *Marietta Blau–Sterne der Zertrümmerung. Biographie einer Wegbereiterin der modernen Teilchenphysik* (Vienna: Böhlau, 2003).

David Rowe and Robert Schulman (eds.), *Einstein on Politics: His Private Thoughts and Public Stands on Nationalism, Zionism, War, Peace, and the Bomb* (Princeton: Princeton University Press, 2007).

Walter Ruske, *100 Jahre Deutsche Chemische Gesellschaft* (Weinheim: Verlag Chemie, 1967).

Carola Sachse, "'*Persilscheinkultur.*' Zum Umgang mit der NS-Vergangenheit in der Kaiser-Wilhelm/ Max-Planck-Gesellschaft," in Bernd Weisbrod (ed.), *Akademische Vergangenheitspolitik* (Göttingen: Wallstein, 2002), 217–245.

Carola Sachse and Mark Walker (eds.), *Politics and Science in Wartime*, Volume 20 of *Osiris*, (Chicago: University of Chicago Press, 2005).

Heinz Sarkowski, *Der Springer Verlag. Stationen seiner Geschichte*, part I: *1842–1945* (Heidelberg: Springer, 1992).

Oliver Schael, "Die Grenzen der akademischen Vergangenheitspolitik: Der Verband der nicht-amtierenden (amtsverdrängten) Hochschullehrer und die Göttinger Universität," in Bernd Weisbrod (ed.), *Akademische Vergangenheitspolitik. Beiträge zur Wissenschaftskultur der Nachkriegszeit* (Göttingen: Wallstein, 2002), 53–72.

Norbert Schappacher and Erhard Scholz, "Oswald Teichmüller – Leben und Werk," *Jahresbericht der Deutschen Mathematiker-Vereinigung*, 94 (1992), 1–39.

Otto Scherzer, "Physik im Totalitären Staat," in Andreas Flitner (ed.), *Deutsches Geistesleben und Nationalsozialismus: Eine Vortragsreihe der Universität Tübingen* (Tübingen: Rainer Wunderlich, 1965), 47–57.

Kay Schiller, "Review Article. The Presence of the Nazi Past in the Early Decades of the Bonn Republic," *Journal of Contemporary History*, 39 (2004), 286–294.

Rudolf Schottlaender (ed.), *Verfolgte Berliner Wissenschaft* (Berlin: Edition Hentrich, 1988).

Jürgen Schriewer et al. (eds.), *Sozialer Raum und akademische Kulturen* (Frankfurt am Main: Lang, 1993).

Sanford Segal, *Mathematicians under the Nazis* (Princeton: Princeton University Press, 2003).

Reinhard Siegmund-Schultze, "The Effects of Nazi Rule on the International Participation of German Mathematicians: An Overview and Two Case Studies," in Karen Hunger Parshall and Adrian C. Rice (eds.), *Mathematics Unbound: The Evolution of an International Mathematical Research Community, 1800–1945* (Providence: American Mathematical Society, 2002), 335–351.

Reinhard Siegmund-Schultze, *Mathematicians Fleeing Nazi Germany: Individual Fates and Global Impact* (Princeton: Princeton University Press, 2009).

Reinhard Siegmund-Schultze, "*Mathematische Berichterstattung*; Faschistische Pläne zur 'Neuordnung' der europäischen Wissenschaft. Das Beispiel Mathematik," *NTM. Internationale Zeitschrift für Geschichte und Ethik der Naturwissenschaften, Technik und Medizin*, 23 (1986), 1–17.

Reinhard Siegmund-Schultze, *Mathematische Berichterstattung in Hitlerdeutschland. Der Niedergang des 'Jahrbuchs über die Fortschritte der Mathematik'* (Göttingen: Vandenhoek & Ruprecht, 1993).

Skúli Sigurdsson, "Physics, Life, and Contingency: Born, Schrödinger and Weyl in Exile," in Mitchell Ash and Alfons Söllner (eds.), *Forced Migration and Scientific Change* (Washington: German Historical Institute, 1996), 48–70.

Ruth Sime, *Lise Meitner: A Life in Physics* (Berkeley: University of California, 1996).

Gerhard Simonsohn, "Physiker in Deutschland," *Physikalische Blätter*, 48 (1992), 23–28.

John Stachel, "Lanczos's Early Contributions to Relativity and His Relationship with Einstein," in J. David Brown et al. (eds.), *Proceedings of the Cornelius Lanczos International Centenary Conference* (Philadelphia: SIAM, 1993), 201–221.

Thomas Stamm, *Zwischen Staat und Selbstverwaltung. Die deutsche Forschung im Wiederaufbau 1945–1965* (Cologne: Verlag Wissenschaft und Politik, 1981).

Johannes Stark, *Erinnerungen eines deutschen Naturforschers* (ed.) Andreas Kleinert (Mannheim: Bionomica-Verlag, 1987).

Johannes Stark, "Zu den Kämpfen in der Physik während der Hitler-Zeit," *Physikalische Blätter*, 3 (1947), 271–272

Max Steenbeck, *Impulse und Wirkungen* (Berlin: Verlag der Nation, 1978).

Walter Stöcker, *Der Nobelpreisträger Johannes Stark (1874–1957). Eine politische Biographie* (Tübingen: MVK, 2001).

Dietrich Stolzenberg, *Fritz Haber. Chemiker, Nobelpreisträger, Deutscher, Jude* (Weinheim: VCH, 1994).

Herbert Strauss (ed.), *Emigration. Deutsche Wissenschaftler nach 1933. Entlassung und Vertreibung* (Berlin: Technische Universität, 1987).

Edgar Swinne, *Richard Gans. Hochschullehrer in Deutschland und Argentinien* (Berlin: ERS-Verlag, 1992).

Anikó Szabó, *Vertreibung, Rückkehr, Wiedergutmachung. Göttinger Hochschullehrer im Schatten des Nationalsozialismus* (Göttingen: Wallstein, 2000).

Margit Szöllösi-Janze, *Fritz Haber 1868–1934* (Munich: Beck, 1998).

Levi Tansjö, "Die Wiederherstellung von freundschaftlichen Beziehungen zwischen Gelehrten nach dem 1. Weltkrieg. Bestrebungen von Svante Arrhenius und Ernst Cohen," in Gerhard Pohl (ed.), *Naturwissenschaften und Politik. Proceedings of papers presented at the University of Innsbruck, April 1996* (Vienna: Gesellschaft Österreicher Chemiker, 1997), 72–80.

James F. Tent (ed.), *Academic Proconsul. Harvard Sociologist Edward Y. Hartshorne and the Reopening of German Universities 1945–46* (Trier: Wissenschaftlicher Verlag Trier, 1998).

James F. Tent, *Mission on the Rhine. Reeducation and Denazification in American-Occupied Germany* (Chicago: University of Chicago Press, 1982).

Jean-Michel Thiriet, "Methoden der Mentalitätsforschung in der französischen Sozialgeschichte," *Ethnologica Europea*, 11 (1978/80), 208–225.

Helmuth Trischler, "'Big Science' or 'Small Science'? Die Luftfahrtforschung im Nationalsozialismus," in Doris Kaufmann (ed.), *Geschichte der Kaiser-Wilhelm-Gesellschaft im Nationalsozialismus. Bestandsaufnahme und Perspektiven der Forschung*, Vol. 1 (Göttingen: Wallstein, 2000), 328–362.

Helmuth Trischler, *Luft- und Raumfahrtforschung in Deutschland 1900–1970* (Frankfurt Main: Campus-Verlag, 1992).

Henning Trüper, "Die Vierteljahresschrift für Sozial- und Wirtschaftsgeschichte und ihr Herausgeber Hermann Aubin im Nationalsozialismus," *Vierteljahresschrift für Sozial- und Wirtschaftsgeschichte*, supplementary issue no. 181 (2005).

Elisabeth Vaupel and Stefan L. Wolff (eds.), *Das Deutsche Museum in der Zeit des Nationalsozialismus* (Göttingen: Wallstein 2010).

Rudolf Vierhaus and Bernhard vom Brocke (eds.), *Forschung im Spannungsfeld von Politik und Gesellschaft* (Stuttgart: DVA, 1990).

Johanna Vogel-Prandtl, *Ludwig Prandtl. Ein Lebensbild. Erinnerungen. Dokumente.* No. 107 (Göttingen: Mitteilungen aus dem Max-Planck-Institute für Strömungsforschung, no. 107, 1993), 210–214.

Rüdiger vom Bruch and Brigitte Kaderas (eds.), *Wissenschaften und Wissenschaftspolitik. Wissenschaften und Wissenschaftspolitik: Bestandsaufnahmen zu Formationen, Brüchen und Kontinuitäten im Deutschland des 20. Jahrhunderts* (Stuttgart: Steiner, 2002).

Max von Laue, "Bemerkung zu der vorstehenden Veröffentlichung von J. Stark," *Physikalische Blätter*, 3 (1947), 272–273.

Karl von Meyenn (ed.), *Wolfgang Pauli. Wissenschaftlicher Briefwechsel*, Vol. 2 (Berlin: Springer, 1985).

Rudolf von Thadden and Günter J. Trittel (eds.), *Göttingen. Geschichte einer Universitätsstadt*, Vol. 3 (Göttingen: Vandenhoeck & Ruprecht, 1999).

Ulrich Walberer (ed.), *10. Mai 1933. Bücherverbrennung in Deutschland und die Folgen* (Frankfurt am Main: Fischer Taschenbuch Verlag, 1983).

Wilhelm Walcher, "Fünfzigste Physikertagung," *Physikalische Blätter*, 42 (1986), 218.

Wilhelm Walcher, "Physikalische Gesellschaften im Umbruch (1945–1963)," in Theo Mayer-Kuckuk (ed.), *150 Jahre Deutsche Physikalische Gesellschaft*, special issue 51 (1995), F107–F133.

Joseph Walk, *Das Sonderrecht für die Juden im NS-Staat* (Heidelberg: Müller, 1981).

Mark Walker, *German National Socialism and the Quest for Nuclear Power, 1939–1949* (Cambridge: Cambridge University Press, 1989).

Mark Walker, "Introduction: Science and Ideology," in Mark Walker (ed.), *Science and Ideology: A Comparative History* (London: Routledge, 2003), 1–16.

Mark Walker, *Nazi Science: Myth, Truth, and the German Atomic Bomb* (New York: Perseus Publishing, 1995).

Mark Walker, "Otto Hahn: Responsibility and Repression," *Physics in Perspective*, 8/2 (2006), 116–163.

Mark Walker (ed.), *Science and Ideology: A Comparative History* (London: Routledge, 2003).

Mark Walker, "The Nazification and Denazification of Physics," in Walter Kertz (ed.), *Hochschule und Nationalsozialismus* (Braunschweig: TU Braunschweig, 1994), 79–89.

Mark Walker, "Von Kopenhagen bis Göttingen und zurück. Verdeckte Vergangenheitspolitik in den Naturwissenschaften," in Bernd Weisbrod (ed.), *Akademische Vergangenheitspolitik. Beiträge zur Wissenschaftskultur der Nachkriegszeit* (Göttingen: Wallstein, 2002), 247–259.

Jessica Wang, "Merton's Shadow: Perspectives on Science and Democracy," *Historical Studies in the Physical and Biological Sciences*, 30 (1999), 279–306.

Valentin Wehefritz, *Gefangener zweier Welten* (Dortmund: Univ.-Bibliothek, 1999).

Valentin Wehefritz, *Verwehte Spuren – Prof. Dr. phil. Fritz Reiche 1883–1969: ein deutsches Schicksal im 20. Jahrhundert* (Dortmund: Univ.-Bibliothek, 2002).

Charles Weiner, "A New Site for the Seminar. The Refugees and American Physics in the Thirties," in Donald Fleming and Bernhard Bailyn (eds.), *The Intellectual Migration: Europe and America 1930–1960* (Cambridge, MA: Belknap Press, 1969), 190–234.

Max Weinreich, *Hitler's Professors* (New York: YIVO, 1946; reprinted New Haven: Yale University Press, 1999).

Bernd Weisbrod (ed.), *Akademische Vergangenheitspolitik. Beiträge zur Wissenschaftskultur der Nachkriegszeit* (Göttingen: Wallstein, 2002).

Burghard Weiss, "Forschung zwischen Industrie und Militär. Carl Ramsauer und die Rüstungsforschung am Forschungsinstitut der AEG," *Physik Journal*, 4 (2005), 53–57.

Burghard Weiss, "Rüstungsforschung am Forschungsinstitut der Allgemeinen Elektricitätsgesellschaft bis 1945," in Helmut Maier (ed.), *Rüstungsforschung im Nationalsozialismus. Organisation, Mobilisierung und Entgrenzung der Technikwissenschaften* (Göttingen: Wallstein, 2002), 109–141.

Ulrich Wengenroth (ed.), *Technische Universität München. Annäherungen an ihre Geschichte* (Munich: Technische Universität, 1993).

Ulrich Wengenroth, "Zwischen Aufruhr und Diktatur. Die Technische Hochschule 1918–1945," in Ulrich Wengenroth (ed.), *Technische Universität München. Annäherungen an ihre Geschichte* (Munich: Technische Universität, 1993), 215–260.

Gerhard Wilke, "Village Life in Nazi Germany," in Richard Bessell (ed.), *Life in the Third Reich* (New York: Oxford University Press, 1987), 17–24.

Brenda Winnewisser, "Hedwig Kohn – eine Physikerin des zwanzigsten Jahrhunderts," *Physik Journal*, 2 (2003), 51–55.

Norton Wise, "Pascual Jordan: Quantum Mechanics, Psychology, National Socialism," in Monika Renneberg and Mark Walker (eds.), *Science, Technology and National Socialism* (Cambridge: Cambridge University Press, 1994), 224–254, 391–396.

Stefan L. Wolff, "Das Vorgehen von Debye bei dem Ausschluss der jüdischen Mitglieder aus der DPG," in Dieter Hoffmann and Mark Walker (eds.), *"Fremde" Wissenschaftler im Dritten Reich. Die Debye-Affäre im Kontext* (Göttingen: Wallstein, 2011), 106–130.

Stefan L. Wolff, "The Establishment of a network of Reactionary Physicists in the Weimar Republic," in Cathryn Carson, Alexei Kojevnikov, and Helmuth Trischler (eds.), *Weimar Culture and Quantum Mechanics* (London: Imperial College Press, 2011), 293–318.

Stefan L. Wolff, "Frederick Lindemanns Rolle bei der Emigration der aus Deutschland vertriebenen Physiker," *Yearbook of the Research Center for German and Austrian Exile Studies*, 2 (2000), 25–58.

Stefan L. Wolff, "Hartmut Kallmann (1896–1978) – ein während des Nationalsozialismus verhinderter Emigrant verlässt Deutschland nach dem Krieg," in Dieter Hoffmann and Mark Walker (eds.), *"Fremde" Wissenschaftler im Dritten Reich. Die Debye-Affäre im Kontext* (Göttingen: Wallstein, 2011), 310–334.

Stefan L. Wolff, "Jonathan Zenneck als Vorstand im Deutschen Museum," in Elisabeth Vaupel and Stefan L. Wolff (eds.), *Das Deutsche Museum in der Zeit des Nationalsozialismus* (Göttingen: Wallstein 2010), 78–126.

Stefan L. Wolff, "Vertreibung und Emigration in der Physik," *Physik in unserer Zeit*, 24 (1993), 267–273.

Gereon Wolters, "Opportunismus als Naturanlage. Hugo Dingler und das 'Dritte Reich,'" in Peter Janich (ed.), *Entwicklungen der methodischen Philosophie* (Frankfurt am Main: Suhrkamp, 1992), 257–327.

Katharina Zeitz, *Max von Laue (1879–1960). Seine Bedeutung für den Wiederaufbau der deutschen Wissenschaft nach dem Zweiten Weltkrieg* (Stuttgart: Steiner, 2006).

Zdenek Zofka, "Dorfeliten und NSDAP: Fallbeispiele der Gleichschaltung aus dem Bezirk Günzberg" in Martin Broszat, Elke Fröhlich, and Anton Grossmann (eds.), *Bayern in der NS-Zeit IV. Herrschaft und Gesellschaft im Konflikt, Teil C* (Munich: R. Oldenbourg, 1981), 383–433.

Index